T0329315

THE NEUROBIOLOGY OF BRAIN AND BEHAVIORAL DEVELOPMENT

THE NEUROBIOLOGY OF BRAIN AND BEHAVIORAL DEVELOPMENT

Edited by

ROBBIN GIBB AND BRYAN KOLB

University of Lethbridge, Lethbridge, AB, Canada

Academic Press is an imprint of Elsevier
125 London Wall, London EC2Y 5AS, United Kingdom
525 B Street, Suite 1800, San Diego, CA 92101-4495, United States
50 Hampshire Street, 5th Floor, Cambridge, MA 02139, United States
The Boulevard, Langford Lane, Kidlington, Oxford OX5 1GB, United Kingdom

Notices
Knowledge and best practice in this field are constantly changing. As new research and experience broaden our understanding, changes in research methods, professional practices, or medical treatment may become necessary.

Practitioners and researchers must always rely on their own experience and knowledge in evaluating and using any information, methods, compounds, or experiments described herein. In using such information or methods they should be mindful of their own safety and the safety of others, including parties for whom they have a professional responsibility.

To the fullest extent of the law, neither the Publisher nor the authors, contributors, or editors, assume any liability for any injury and/or damage to persons or property as a matter of products liability, negligence or otherwise, or from any use or operation of any methods, products, instructions, or ideas contained in the material herein.

British Library Cataloguing-in-Publication Data
A catalogue record for this book is available from the British Library

Library of Congress Cataloging-in-Publication Data
A catalog record for this book is available from the Library of Congress

ISBN 978-0-12-804036-2

For Information on all Academic Press publications
visit our website at https://www.elsevier.com/books-and-journals

Publisher: Nikki Levy
Acquisition Editor: Natalie Farra
Editorial Project Manager: Kathy Padilla
Production Project Manager: Stalin Viswanathan
Cover Designer: Christian Bilbow

Typeset by MPS Limited, Chennai, India

CONTENTS

LIST OF CONTRIBUTORS

Brandon Almy
University of Minnesota, Minneapolis, MN, United States

Vicki Anderson
Murdoch Childrens Research Institute, Parkville, VIC, Australia; University of Melbourne, Melbourne, VIC, Australia

Lindsay C. Bowman
University of California, Davis, CA, United States; Harvard Medical School, Boston, MA, United States

Maryam Faiz
University of Toronto, Toronto, ON, Canada

Robbin Gibb
University of Lethbridge, Lethbridge, AB, Canada

Claudia L.R. Gonzalez
University of Lethbridge, Lethbridge, AB, Canada

Mardee Greenham
Murdoch Childrens Research Institute, Parkville, VIC, Australia; University of Melbourne, Melbourne, VIC, Australia

Celeste Halliwell
University of Lethbridge, Lethbridge, AB, Canada

Allonna Harker
University of Lethbridge, Lethbridge, AB, Canada

Takao K. Hensch
Harvard University, Cambridge, MA, United States; Harvard Medical School, Boston, MA, United States

Brett T. Himmler
University of Minnesota, Minneapolis, MN, United States

Stephanie M. Himmler
University of Lethbridge, Lethbridge, AB, Canada

Sumaiya A. Islam
University of British Columbia, Vancouver, BC, Canada

Serena Jenkins
University of Lethbridge, Lethbridge, AB, Canada

Michael S. Kobor
University of British Columbia, Vancouver, BC, Canada

Bryan Kolb
University of Lethbridge, Lethbridge, AB, Canada

Anna Kovalchuk
University of Lethbridge, Lethbridge, AB, Canada

Terri L. Lewis
McMaster University, Hamilton, ON, Canada; The Hospital for Sick Children, Toronto, ON, Canada

Alexandre A. Lussier
University of British Columbia, Vancouver, BC, Canada

Daphne Maurer
McMaster University, Hamilton, ON, Canada; The Hospital for Sick Children, Toronto, ON, Canada

Cindi M. Morshead
University of Toronto, Toronto, ON, Canada

Charles A. Nelson, III
University of California, Davis, CA, United States; Harvard Medical School, Boston, MA, United States; Harvard University, Cambridge, MA, United States; Canadian Institutes for Advanced Research, Toronto, ON, Canada

Sergio M. Pellis
University of Lethbridge, Lethbridge, AB, Canada

Vivien C. Pellis
University of Lethbridge, Lethbridge, AB, Canada

Sammy Perone
Washington State University, Pullman, WA, United States

Lara J. Pierce
University of California, Davis, CA, United States; Harvard Medical School, Boston, MA, United States

Nicholas P. Ryan
Murdoch Childrens Research Institute, Parkville, VIC, Australia; University of Melbourne, Melbourne, VIC, Australia

Lori-Ann R. Sacrey
University of Alberta, Edmonton, AB, Canada

Rachel Stark
University of Lethbridge, Lethbridge, AB, Canada

Janet F. Werker
Canadian Institutes for Advanced Research, Toronto, ON, Canada; University of British Columbia, Vancouver, BC, Canada

Philip D. Zelazo
University of Minnesota, Minneapolis, MN, United States

PREFACE

In the past century the scientific understanding of brain development has grown exponentially. The now classic studies of Santiago Ramon y Cajal focused on descriptions of the emergence of anatomical structures in the nervous system and suggested that development was an unfolding of an innate sequence of events. Later works by many anatomists such as Joe Altman and Shirley Bayer as well as Pasko Rakic continued the anatomical tradition using a range of neuroanatomical techniques to expand Ramon y Cajal's anatomical understanding to the mammalian brain, and especially the neocortex. Behavioral studies by Harry Harlow and others began to show that the development was not just an unfolding of a genetic blueprint but that experiences, including sensory experiences and caregiver–infant relations, could profoundly change the course of functional development. Work on restriction of sensory input during development by Austin Riesen and his colleagues showed that sensory deprivation had profound effects of perceptual development. The physiological consequences of restricted experience were dramatically shown by Wiesel and Hubel in the 1960s in their studies of the effects of monocular deprivation on the development of visual acuity. Such studies emphasized the role of critical periods that had been described earlier by ethologists such as Konrad Lorenz. Our understanding of the role of molecules in controlling brain development can be traced back to the pioneering work of investigators such as Rita Levi-Montalcini on the chick embryo and the discovery of nerve growth factor in the 1950s and Roger Sperry's parallel work on the development of nerves by chemical codes that were under genetic control. Pioneering studies on the effects of early brain injury by Margaret Kennard in the 1940s led to the idea that the effects of early brain injury during development were different than the effects seen after later injuries—a view that was followed up and clarified in the 1970s by investigators such as Patricia Goldman. More recently, the emergence of noninvasive imaging including magnetic resonance imaging, electroencephalography (event-related potentials), and forms of optical imaging has provided a revolution in our thinking about how the cerebral hemispheres change over the first 2–3 decades of life. Taken together, these developmental pioneers have left us with a rich broth of information and ideas that have set the stage for a new rapidly changing field of brain and behavioral development. But for newcomers to the field the sheer volume of information can be overwhelming. Where to start and what to learn?

This book emerged from discussions with Elsevier about writing or editing a volume on brain development. Although we toyed with the idea of writing an extensive monograph that could tell a pedagogical story, the challenge to develop the breadth to do justice to the field appeared overwhelming. We thus settled on editing a volume that would include authors whose expertise went well beyond our own. Of course, the challenge for editors is to convince busy colleagues that time spent on writing reviews for books justifies the time commitment in a time of intense competition for research dollars. For those who agreed with us (the authors in this volume), we are immensely grateful. We recognize that time is precious for active bench researchers.

Our goal in editing this volume was twofold. First, we wanted to initiate senior undergraduate or graduate students in neuroscience or psychology to the issues and questions regarding how the brain develops and adapts to the environment that it finds itself in. Although earlier volumes by others have focused either on mechanisms driving the anatomical, neurochemical, and physiological organization of the developing brain or on the behavioral correlates of brain development, our goal was somewhat different. Our experience with our students in behavioral neuroscience over the past 40 years has shown us that the interplay of these brain–behavior interrelationships during development provides a good entry point to expand to more specialized topics. Our hope is that if students can get a sense of the breadth of the questions and research related to brain development, they will learn to ask, and hopefully try to answer, the big questions about brain and behavioral development.

Second, by covering a broad range of topics in both psychology and neuroscience, we hoped to stimulate discussions and thinking by readers as they investigated topics well beyond their comfort zones. We apologize to readers who were hoping to find a volume restricted to their favorite developmental topic, be it molecules or minds. We hope that we might be able to broaden reader's view of the complexity and excitement of the broader field of brain and behavioral development.

We have divided the book into four parts. The first gives overviews about the historical context of brain and behavioral development. The second provides broad discussions of powerful molecular concepts (stem cells and epigenetics) and an example of the application of molecular methods to the fundamental issues of critical periods. The third part focuses on topics in behavioral development, while the final part examines general factors that influence brain development.

We have both been in the field for a long time and must point to a few who have strongly influenced our views. These include especially our long-time friends and colleagues Paul Cornwell (Penn State) and Bill Greenough (Illinois), both of whom died too young. We must also thank Fraser Mustard who had the foresight over 25 years ago to invent the Canadian Institute for Advanced Research and especially its program in Child Brain Development. This program has been an example of the power of multidisciplinary discussion in getting to the big questions in brain and behavioral development. Fraser spent the last decade of his life dedicated to changing public policy related to early child development around the world. As we continue to learn more about brain and behavioral development, his message becomes more and more important.

Robbin Gibb and Bryan Kolb

PART I

General Perspectives in Brain Development

CHAPTER 1

Brain Development

Robbin Gibb and Anna Kovalchuk
University of Lethbridge, Lethbridge, AB, Canada

ABBREVIATIONS

bRGC basal radial glia cell
CR Cajal Retzius
CNS central nervous system
Cl^- chloride
E embryonic
GABA γ-aminobutyric acid
ICI inhibitory cortical interneuron
IPC intermediate precursor cell
Na^+ sodium
NEC neuroepithelial cell
OPC oligodendrocyte progenitor cell
RGC radial glial cell
RMS rostral migratory stream
SNP short neural precursor
SVZ subventricular zone
VZ ventricular zone

1.1 INTRODUCTION

Human brain development is a protracted process that begins soon after conception and continues at least into the third decade of life. The brain is crafted by the lifelong interplay of generative (cell birth and synapse formation) and degenerative processes (cell death and synaptic pruning), which are modulated to varying degrees by an individual's experiences. The process of brain development is programed by information encoded on DNA, a double-helical molecule that provides a genetic blueprint for construction. Genes, key units of inheritance, are said to be expressed—turned on or off, giving rise to gene products—proteins that influence the function of the brain and the cells that comprise it. Genes are expressed in an organized and intricately controlled manner, governing the proper and structured step-wise development and maturation of cells and brain areas. The control of gene expression is regulated through epigenetic (above-genetic) phenomena—methylation of DNA, modifications of histone proteins and chromatin remodeling, and noncoding RNA-mediated effects. (For a detailed discussion of epigenetics see Chapter 7: Epigenetics and Genetics of

The Neurobiology of Brain and Behavioral Development
DOI: http://dx.doi.org/10.1016/B978-0-12-804036-2.00001-7

Brain Development.) Epigenetic changes are flexible and reversible, and are influenced by physical, chemical, biological and social environmental factors, and life experiences. Epigenetic changes are also heritable and therefore might carry an imprint of exposures and experiences of previous generations. Although the process of brain development is fundamentally the same for each individual, environmental exposures can alter the way in which the developmental program manifests. This environmental modulation of genetic expression is called epigenetic programing and its effects can be seen in the preconception period (e.g., Mychasiuk, Harker, Ilnytskyy, & Gibb, 2013), the prenatal period (e.g., Mychasiuk et al., 2012), and through the lifespan of the individual (e.g., Harker et al., 2015). Mounting evidence demonstrates that the experiences of our forbearers can and often do result in epigenetic changes in gene expression ultimately changing the expression of proteins. Because proteins are the building blocks of the brain, different proteins build different brains by modifying cell number, cell connectivity, brain size, and ultimately, behavior. Epigenetic programing thus provides an adaptive means for an organism to prepare its brain for the unique environmental challenges that it will face without changing its genetic blueprint. Our experiences and the experiences of our predecessors are able to turn certain genes "on" or "off" as per specific mechanisms of epigenetic regulation, thus regulating brain development and function (Fig. 1.1).

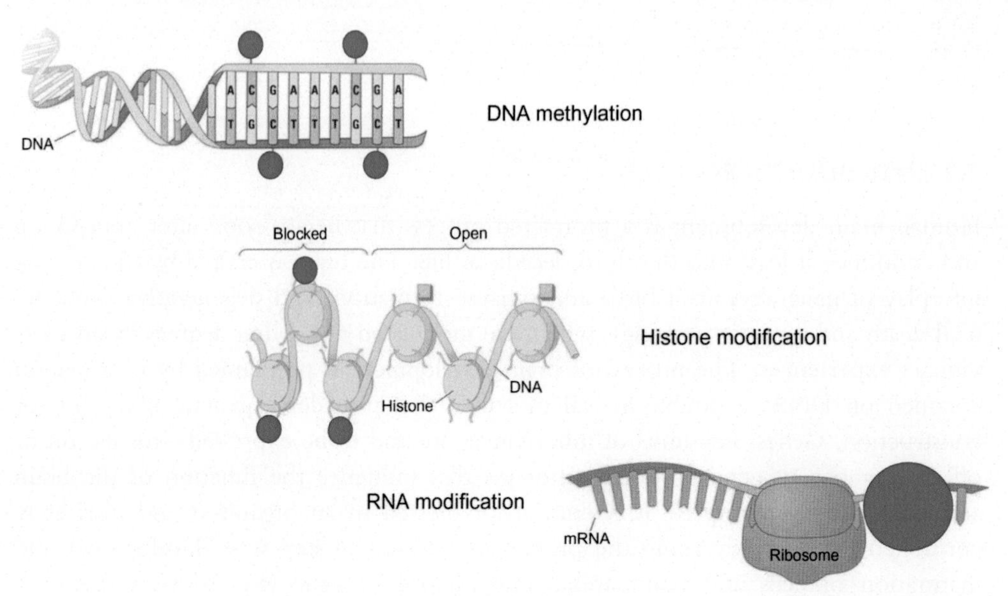

Figure 1.1 Epigenetic mechanisms. The blue circles represent methyl groups for the DNA methylation and histone modification figures. In the case of RNA modification, the blue circle represents a non-coding RNA bound to the mRNA (After Kolb & Whishaw, 2014).

1.2 CELLS OF THE BRAIN

The brain is composed of two main classes of cells: neurons and glia. Neurons are electrically active cells that form connections with other neurons (synapses) and communicate in a systematic fashion to produce behavior. Neurons have a cell body, dendrites and an axon. The dendrites receive incoming information from other connected cells. The cell body is the control center of the cell and it integrates the information it receives. An electrochemical (neurotransmitter) response is then transmitted along the axon to cells connected further down the line. Neurons can be classified as one of two fundamental types: excitatory projection neurons and inhibitory interneurons. Excitatory cells typically bear dendritic spines, protrusions on the dendrite, which form at sites of synaptic contact. They can be of a pyramidal or stellate (star-shaped) form. Inhibitory interneurons do not have spines and usually appear in a stellate form Fig. 1.2.

Glial cells are cells that support neural function. They are smaller than neurons and lack both dendrites and axons. The name glia derives from the Greek word meaning glue. It was originally proposed that these cells were responsible for holding the brain together, yet no evidence exists for any glue-like or binding function. Despite our long-held view that glial cells are "helper cells" for neurons, current evidence points to a much larger role for these cells in modulation and maintenance of neural function. In addition, glia are involved in processes of reuptake of neurotransmitters, providing a structural scaffolding for neural migration, and supporting recovery after brain injury.

1.2.1 Neural cells

All cells in the mature brain arise from the precursor neuroepitheal cells that are expressed during development. Once in their end stage of development neural cells fall into one of two principal cell classes.

Excititory projection neurons are spiny and typically communicate with other cells by use of the neurotransmitter glutamate. Inhibitory interneurons are smooth and use γ-aminobutyric acid (GABA) as their neurotransmitter (Fig. 1.2).

1.2.2 Glial cells

Mature glial cells can be classified as either macroglia or microglia. This distinction of glial subtypes arises not only on the basis of their size but also by their site of origin. Macroglia are derived from the neuroectodermal layer of the embryo whereas microglia originate in the mesodermal layer.

1.2.2.1 Macroglia

There are three main types of macroglia: astrocytes, oligodendrocytes, and ependymal cells.

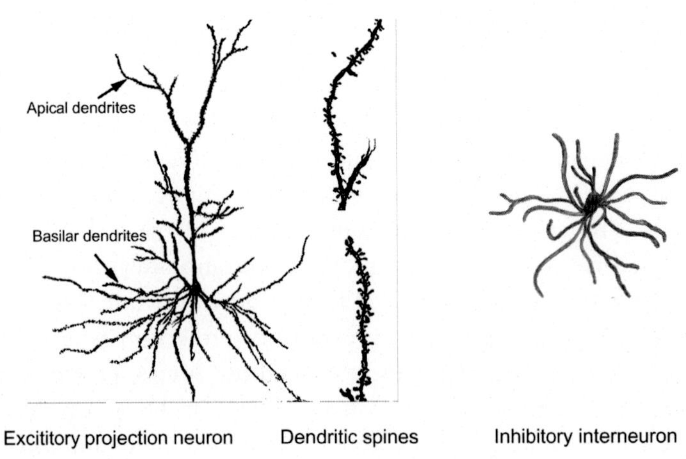

Figure 1.2 Neural cell types.

Astrocytes are the most plentiful form of macroglia. Their roles include regulation of neuron production, maintainence of neural networks, and modulation of neural activity and communication (Jernigan & Stiles, 2017). Oligodendrocytes are cells that produce myelin in the central nervous system (CNS). Myelin is a fatty substance that insulates axonal fibers, thereby enhancing the speed of neural transmission.

Ependymal cells are found in the ventricular walls of the brain and in conjunction with local capillary beds comprise the choroid plexus. The choroid plexus produces the cerebrospinal fluid that fills the ventricular system providing a cushion that serves to protect the brain.

1.2.2.2 Microglia

Microglia have been traditionally considered the sole provider of immune support to an immunodeficient Brain. They are known to monitor for signs of infection, clear debris, and support the inflammation and Repair response to brain injury and disease. New research has expanded the role of microglia to include cell proliferation, synaptic pruning, and sex-specific changes in brain development (Shafer & Stevens, 2015: see Chapter 14: Hormones and Development for a more detailed discussion).

1.3 PHASES OF BRAIN DEVELOPMENT

Genetically preset, the process of brain construction occurs in seven well-defined phases that extend over a prolonged period of brain development (Kolb & Whishaw, 2014). Some phases are more or less confined to restricted periods whereas others are in play for extended periods of time (Table 1.1).

Table 1.1 Seven phases of brain development

Developmental phase	Process
1. Cell birth	Genesis of neurons and glia
2. Cell migration	Movement of cells to their functional position
3. Cell differentiation	Precursors cells transform to specified cell type
4. Cell maturation	Growth of dendrites and axons
5. Synaptogenesis	Formation of cell to cell communication sites – synapses
6. Cell death and synaptic pruning	Programed cell death and dismantling of unused circuitry
7. Myelination	Formation of myelin sheath to enhance speed of neurotransmission

Each phase will be considered in depth in the developmental period(s) during which it predominately occurs and unless otherwise specified should be considered typical of *human* development.

1.4 CONSTRUCTING THE HUMAN BRAIN

1.4.1 Construction of the brain: embryonic development (conception to week 8)

By embryonic day 13 (E17), the process of replacing the single layer blastula with a trilayer structure known as the gastrula begins. By the end of gastrulation (E20) the endodermal, mesodermal, and ectodermal layers of the trilamina have formed. The ectodermal layer gives rise to both the skin and the CNS. Importantly, the *neuroepithelial cells* (NECs), which eventually produce cells that make the neurons and glia are found within this layer. The neural plate forms on E21 and by E22 the neural groove becomes apparent. The neural groove fuses starting on E23 to form the neural tube (neurulation) with the central section closing first. The rostral portion of the neural tube is inhabited by the earliest migrating cells and becomes the brain while the caudal portion receives later migrating cells and forms the spinal cord (Stiles & Jernigan, 2010). The rostral and caudal regions of the neural tube are the last to close.

Neural tube defects are birth defects that affect the brain or spinal cord. Failure of the rostral section of the neural tube to close on E23–E26 results in a condition known as anencephaly where the brain and skull fail to fully develop. Infants born with anencephaly usually lack sensory processing abilities and are unable to feel pain. They generally die soon after birth. Spina bifida is the result of incomplete closure of a portion of the caudal end of the neural tube on about E28. The symptoms of spina bifida range in severity but it is most commonly characterized by paralysis of the legs. Both anencephaly and spina bifida often result from a lack of the B vitamin, folate, in the diet.

The cylindrical cavity in the neural tube eventually becomes the ventricular system and the NECs inhabit this region, also known as the ventricular zone (VZ) (**Phase 1—cell birth**). The NECs are aligned in the VZ in an apico-basilar fashion. At the ventricular (apical) surface the NECs are bound together by both *tight* and *adherens junctions* which provide important adhesive contact between neighboring cells. Adherins junctions mediate the maturation and maintenance of the contact while tight junctions regulate the transport of ions and other molecules between the cells (Hartsock & Nelson, 2008). At the pial (basal) surface the NECs are bound by *integrins*. Integrins act to "velcro" the cell to the extracellular matrix (ECM) and activate intracellular signaling pathways to familiarize the cell with the ECM on which it is bound.

During neurulation the neural stem cell population undergoes rapid expansion by symmetric cell division of the NECs. The NECs born at this time are influenced by the expression of the signaling molecules Emx2 and Pax6 to produce progenitors destined for specified brain regions. These two transcription factors are expressed in opposing gradients from the anterior to posterior regions of the proliferative zone. Emx2 is highly expressed in the posterior and medial areas of the proliferative zone whereas Pax2 is more abundantly expressed in the anterior and lateral areas. The differential expression of transcription factor proteins that emerges at this stage of development produces a primitive map that provides a basic blueprint for brain organization.

In order to supply the demand for neurons to populate the brain, on E25 the NECs divide in a symmetrical fashion. Symmetrical cell division produces two NECs with each division and this process continues until E42. By E28 the rudimentary structures of the forebrain (prosencephalon), midbrain (mesencephalon), and hindbrain (rhombencephalon) are visible in the human embryo (Stiles & Jernigan, 2010). Just prior to neurogenesis, NECs lose their tight junctions, start to express glial genes and begin the process of transformation into *radial glial cells* (RGCs). Recent work on neural progenitors has led to an understanding that diverse populations of progenitors exist. Three new subclasses: the *intermediate progenitor cell* (IPC), the *short neural precursor* (SNP) with an apical attachment at the VZ but no basal attachment at the pial surface, and *basal RGCs* (bRGCs) with basal attachments at the pial surface but no apical attachment in the VZ, have been described (Franco & Müller, 2013). RGCs divide asymmetrically to self-renew and provide a non-RGC daughter cell (E42—E108: Fig. 1.3). It turns out that some but not all of the daughter cells are postmitotic neurons but most of these daughter cells are IPCs. IPCs primarily undergo symmetrical division to produce pairs of neurons but some undergo symmetrical divisions to produce more IPCs. IPCs lose apical contact with the VZ and move to a more basal position. Basally dividing IPCs undergo symmetrical divisions to self-renew but most supply pairs of neurons. The IPCs occupying a basal position in the VZ ultimately form the proliferative region known as the sub-VZ. SNPs are similar in function to the IPCs in that they produce neurons but they differ in that they maintain an apical ventricular attachment and remain in the VZ. The bRGCs that occupy the outermost

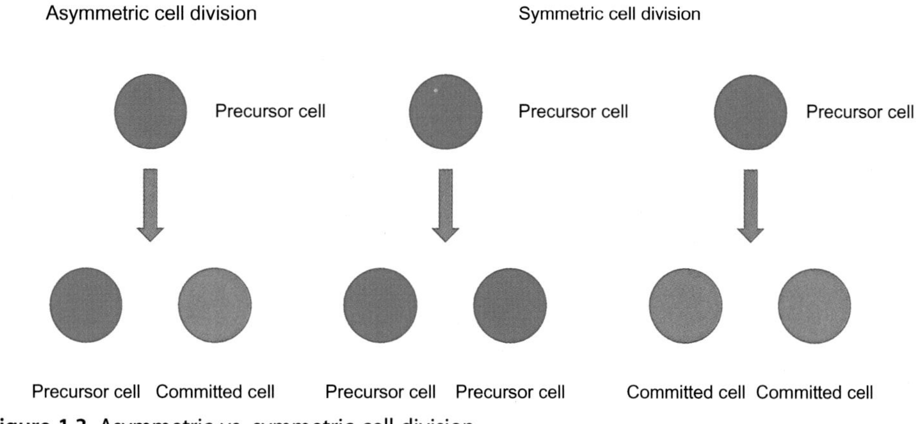

Figure 1.3 Asymmetric vs. symmetric cell division.

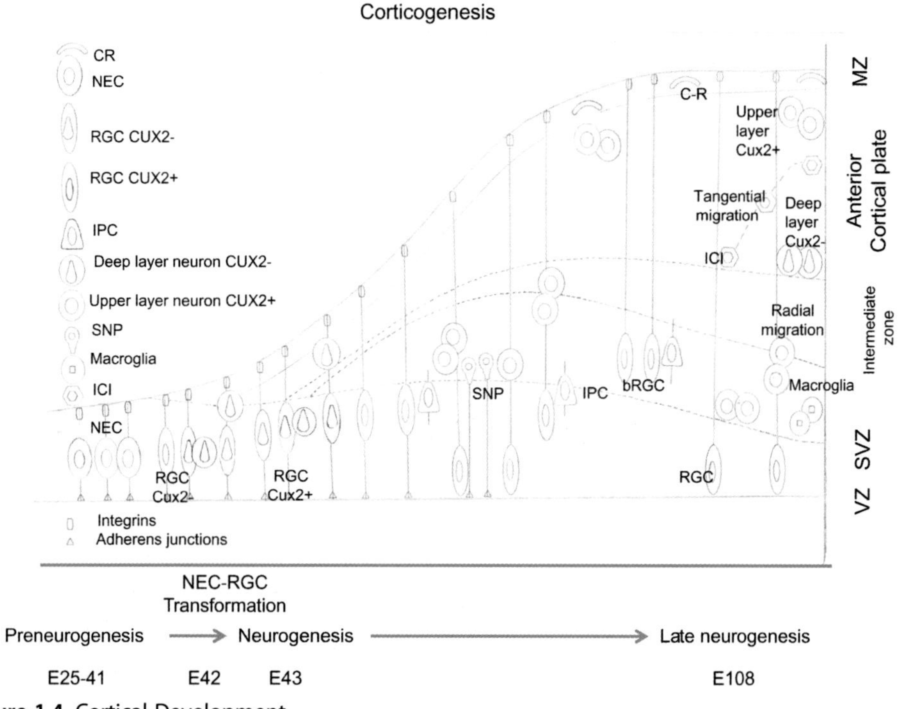

Figure 1.4 Cortical Development.

region of the SVZ self-renew and generate both IPCs and neurons. At the end of neurogenesis (E108; Clancy, Darlington, & Finlay, 2001) both RGCs and bRGCs transform into progenitors that produce astrocytes. The extensive amplification of bRGCs is thought to account for the increased number of neocortical neurons in the upper layers of the primate brain. (For more details on precursor cells see Chapter 5: Stem Cells to Function) (Fig. 1.4.)

1.4.2 Construction of the brain: fetal development (week 9 to birth)

During this phase of brain development the previously smooth brain acquires the folding pattern with gyri (bumps) and sulci (grooves) typically seen in the mature brain. After cells are born in the VZ or SVZ, they migrate away in a radial fashion along the RGCs and cells that form the deepest layers of the cerebral cortex leave first (**Phase 2—cell migration**). Most radially migrating cells are glutamatergic (use glutamate as their principal neurotransmitter) and comprise approximately 80% of cortical neurons.

As cells continue to form, the layers gradually build with the newest arrivals seated atop those most recently placed in a so-called "inside out" fashion. A notable exception to the inside out rule is the first wave of cells that leave the proliferative zone. These cells form a primordial structure known as the preplate (Bayer & Altman, 1991). Once formed the preplate is split in two by the next wave of arriving cells producing a transient subplate and marginal zone (layer 1). The new region that these cells form is known as the cortical plate and the first cells to populate the cortical plate are the cells destined to become layer 6 cells. The marginal zone contains a population of cells called Cajal Retzius (CR) cells. These cells provide important signaling information for newly arriving neurons to position them in their correct layer in the cortex. The CR cells produce a substance called *reelin* that signals migrating neurons to stop their journey and take a position in the developing cerebral cortex. CR cells appear early in the brain development process and are highly expressed in the immature brain. As an individual matures, the levels of CR cells drop but they are retained, albeit in reduced numbers, in the mature brain. Animals with reelin deficiencies fail to develop laminar structure in the cerebral cortex. In humans, abnormalities in the reelin signaling cascade are implicated in a variety of neurodevelopmental disorders including schizophrenia, autism, major depression, and bipolar disorder (Folsom & Fatemi, 2013). RGCs that are anchored at the VZ and the pial surface form the guideposts for the migrating cells. Newly born neurons climb along the radial glia fibers to reach their destination. In the year 2000 it was discovered that some RGCs are actually neural stem cells and produce the progeny that climb up them. This finding caused a huge uproar in the neuroscience community as it was previously believed that neural progenitors gave birth to neurons and glial progenitors produced only glia. RGCs are a heterogeneous population and some give birth to glial cells, some to neural cells, and yet others give rise to a mix of both (Malatesta & Golz, 2013). Interestingly, most current neuroscience textbooks currently label RGCs as neural stem cells even though only a sub-population of RGCs have this role in the developing brain (**Phase 3—cell differentiation**). IPCs produce neurons whereas RGCs give rise to both projection neurons and astrocytes (Franco & Müller, 2013). As successive layers of the cortical mantle are formed the progenitors become more limited in the cell types that they are able to construct. This process is known as fate restriction and it is thought to account for how progenitors are able to manufacture the variety of cells required to build the brain. In a recent paper, Franco et al. (2012) describe work done in a murine model that demonstrates that birthdate alone does not account for the ultimate position of neurons in particular cortical layers. Rather, they have identified RGCs sub-lineages that are responsible for producing either the lower (V and VI) or upper (II, III, and IV) cortical lamina. IPCs and some RGCs appear to bear the same transcription factor (Cux2 and are thus Cux2 +) and produced neurons destined for the upper cortical layers. RGCs that lack expression of Cux2 (Cux2 −) generate neurons destined for lower cortical layers. Results of this work suggest that birthdate may not be the cause for layer specification of the cerebral cortex. It appears that IPCs and RGCs that produce upper layer cortical neurons just happen to do so at a later time than the RGCs that generate the lower layer neurons. This work suggests that cell fate specification drives birth order and not the reverse as was previously believed. Cux2 − RGCs stop symmetrical divisions and start to produce neurons before Cux2 + cells, allowing expansion of the Cux2 + progenitor pool. The authors of this work note that upper cortical layers are expanded in primates and are necessary for sophisticated associative connectivity. They propose defects in the generation of these cells could result in conditions such as autism and schizophrenia (Fig. 1.4).

Inhibitory cortical interneurons (ICIs), which comprise 20% of cortical neurons, arise from different origins. Three ganglionic eminences, the medial, lateral, and caudal are confirmed sources of ICIs whereas the rostral migratory stream (RMS), the septal region and the anterior region of the cortical plate may be additional generative sites (Wonders & Anderson, 2006; Al-Jaberi et al., 2013) ICIs have multiple phenotypes based on their morphology, neurochemical constituents, and physiology. A uniting feature of ICIs is their use of the inhibitory GABA molecule as their neurotransmitter. Postmitotic ICIs migrate tangentially to reach their destination by relying on a number of molecular guides or tropic factors. The guidance molecules arise from local regions of the cortex and help the interneurons find their way to their final destination (Stiles & Jernigan, 2010). In addition to the key role that ICIs play in modulating cortical output, they are also involved in developmental processes including regulation of neuron proliferation, migration, and plasticity of cortical circuitry (Wonders & Anderson, 2006). A fascinating feature of the developing cortex is the depolarizing (i.e., excitatory) effects of GABA in the immature brain. Although this phenomenon has been recently called into question (Bregestovski & Bernard, 2012), overwhelming evidence seems to support the finding that in immature ICI's, GABA has an excitatory action by depolarizing membranes and sometimes even causing sodium (Na^+) mediated action potentials (Ben-Ari et al., 2012). Excessive accumulation of the chloride ion (Cl^-) in immature neurons may mediate the excitatory action of GABA and indeed the shift in GABA polarity is concurrent with a progressive reduction of intracellular Cl^-. It appears that this inverted excitatory action of GABA during development may have an organizational role in cell birth, migration (especially the tangentially migrating ICIs), synaptogenesis, and network formation. Tyzio et al. (2006) describe the influence of oxytocin during delivery as the cellular switch that changes the polarity of GABA function to inhibition. Failure to assume the fully mature GABA system is associated with the appearance of epileptic seizures (Aronica, Iyer, Zurolo, & Gorter, 2011).

Once neurons reach their final destination they extend dendrites and an axon (**Phase 4—cell maturation**) in an attempt to make connections with other cells and become an integral part of a communication network. Dendrites gather information from other neurons and the axon provides a means of sending information to neurons further down the communication line. Many dendrites extend from a neuron to receive input from cells in the information network but a single axon conveys information processed by the cell. In order to make appropriate contacts the axon has a growth cone on its leading edge. The growth cone is guided by sampling locally produced tropic molecules that ultimately assist the axon in finding its intended target. Once that target is identified, a connection called a synapse is formed (**Phase 5—synaptogenesis**). The synapse provides the means for cell-to-cell communication. In the context of the synapse, the axon is considered the presynaptic terminal and the dendrite, the postsynaptic terminal.

Interestingly, the formation of the brain also depends on degenerative processes that commence in the prenatal period of life. Programmed cell death or apoptosis (**Phase 6—cell death and synaptic pruning**) is instigated to reduce the number of cells in the brain that have failed to make useful connections or have connections that are underutilized (Chan, Lorke, Tiu, & Yew, 2002). During this process a cascade of death promoting genes are expressed that initiate a suicide program in the cell. Cells that have established stable and vibrant connections are protected from apoptosis by neurotrophic factors supplied by other cells in the circuit.

1.4.3 Construction of the brain: postnatal development

The process of neurogenesis (**Phase 1—cell birth**) persists through the postnatal period in a restricted manner. Cells continue to form in the SVZ which are destined to populate the olfactory bulb and they then migrate **(Phase 2—cell migration**) via the RMS to their target site.

In humans newly generated cells only migrate to the olfactory bulb vis-a-vis the RMS in the perinatal period, whereas rodents and nonhuman primates show new cells inhabiting the olfactory bulb through the lifespan. In humans it appears that following the perinatal period, the migration target for neural cells arising from the SVZ is the nearby striatum. This area demonstrates significant neural proliferation even in adults

(Ernst et al., 2014). Another site of significant postnatal neurogenesis is the dentate granule layer in the hippocampus. Hippocampus neurogenesis occurs throughout the lifespan and appears to contribute in a very significant way to hippocampal plasticity.

Proliferation and migration of cells destined to become glial cells begins in the prenatal period and continues for an extended time in the postnatal period. Glial progenitors multiply in the SVZ and then migrate out to the white matter eventually reaching the cortex, striatum, and hippocampus. The mode of emigration for glial progenitors is dependent on their ultimate destination. Cells destined for dorsal areas migrate in a radial fashion to the overlying corpus callosum where they remain or continue on along RGCs to their target in cerebral cortex. Cells destined for lateral cortex follow the white matter laterally and then exit to follow RGCs to their terminal location. Some glial progenitors track along the corpus callosum to reach the opposite hemisphere and others follow blood vessels to their target (Cayre, Canoll, & Goldman, 2009). It appears that glial progenitors, particularly *oligodendrocyte progenitor cells* (OPCs) are retained in the cerebral cortex for an indefinite period of time. Even the adult brain has the ability to produce astroglia and oligodendroglia and both can arise from OPCs.

Synaptogenesis (**Phase 5—synaptogenesis**) that began in the prenatal period continues in the postnatal period and throughout the lifespan of an individual. Dendrites on a single neuron can make tens of thousands of connections with other neurons and it appears that the number of connections varies with the cortical area. At the site of contact, the neuron usually elaborates a dendritic spine to enhance the effectiveness of the synaptic site. Fewer dendritic spines are observed in posterior regions of the brain (visual areas) than in anterior regions of the brain (frontal lobe) and in the most anterior areas (prefrontal cortex), it is estimated that there are 23 times more synapses than in primary visual cortex (Elston, 2003). Most dendritic spines are sites of excitatory neural transmission.

Apoptosis (**Phase 6—cell death and synaptic pruning**) continues to play a major role in brain development in the postnatal period. "Use it or lose it" becomes the rationale for brain architecture at this juncture; and it does not just apply to dormant cells, but also to the exuberant production of synapses. Underused synapses are removed in a process called synaptic pruning. Recently, the role of microglia in both apoptosis and synaptic pruning has been described (Paolicelli et al., 2011; Wake et al., 2013). Microglia are not derived through the same precursor cells as neurons and astroctyes but rather originate from mesodermal myeloid progenitors that also produce macrophages and blood cells. Microglia immigrate to the CNS early in embryogenesis where they proliferate. During development microglia typically display an amoeboid appearance in contrast to the ramified phenotype that characterizes microglia in the adult CNS. Estimates are that approximately 50%–70% of the cells born into the brain are removed through apoptosis (the number varying with the location within the CNS; Stiles & Jernigan, 2010; Rabinowcicz et al., 1996) and approximately

40%–50% of synaptic connections are lost through synaptic pruning (Low & Cheng, 2006). Cell loss and synaptic pruning provides a means for tuning brain connections by ramping up those that are important and eliminating those that are not. This provides a mechanism for experiences to modify brain development.

Although some myelination **(Phase 7—myelination)** occurs in the prenatal period, it ramps up after birth and continues well into the third decade of life. The process of myelination frequently heralds the maturation of cortical areas. Primary motor and sensory areas of the brain myelinate first and association areas myelinate last. For example, the first cortical area to myelinate is motor cortex and myelination of the dorsolateral prefrontal cortex occurs last (Flechsig, 1901). OPCs differentiate upon reaching their final destination. Processes are extended and begin to form myelin that wraps around axons and acts as insulation to speed up the transmission of electrical signals. For myelination to occur, an ample supply of an energy substrate needs to be readily available. In the absence of a readily available energy source, myelination is impaired and OPC's risk enacting the apoptotic pathway. Although glucose is often that substrate under certain conditions of intense neural activity, lactate may provide another resource for the energy required. In cases of anoxia during myelination resulting in cerebral palsy or periventricular leukomalacia, lactate may be an important factor in minimizing the devastating outcomes by supporting integrity of immature oligodendrocytes (Rinholm et al., 2011). It is clear that the role of oligodendroglia extends far beyond that of just insulating axonal fibers. In a recent paper (Morrison et al., 2013) the role of oligodendroglia in supporting axonal health and function was examined. It appears that axons are metabolically costly to maintain and once myelinated, have limited access to the extracellular fluid to fulfill their energy requirements. Oligodendroglia may provide an important link to energy supplies for neurons. Another recent study compared the time course of myelination in humans and chimpanzees, our closest living relatives (Miller et al., 2012). The authors in this study discovered that while chimpanzees completed the process of cortical myelination by the time of sexual maturity, in humans this process extended well beyond late adolescence. They conclude that extended timing of myelination in the human brain may be in part responsible for the greater degree of cognitive development and cortical plasticity observed in humans. They also propose that this extended period of maturation makes the human brain more vulnerable to psychiatric disorders that have typical onset in adolescence or early adulthood like schizophrenia.

1.5 BRAIN SYSTEMS CONSTRUCTION AND EMERGING BEHAVIOR

With the advent of powerful, noninvasive imaging techniques it has become feasible to study the development of neural circuitry in the fetal and postnatal brain that supports emergent behavior.

Although a variety of tools are available that enable the study of functional brain connectivity (electroencephalography, magnetoencephalography, functional near-infrared spectroscopy) one of the most used methodologies is resting state functional magnetic resonance imaging. Resting state functional magnetic resonance imaging (rs_fMRI) assays the fluctuations in blood oxygenation levels that occur spontaneously. These signal changes can be detected at very low frequencies (0.01 to 0.1 Hz) with millimeter resolution (Keunen et al., 2017). Interhemispheric connections have been studied as early as 24 weeks postconception and have been shown to strengthen with increasing gestational age. Similarly, thalamocortical and intrahemispheric connections that span long distances become more robust as development proceeds (Thomason et al., 2013).

Rs_fMRI has been used to study brain regions that support primary sensory and motor functions and higher order cognitive functions in healthy adults. In these studies, resting state networks supporting primary function are more localized whereas networks supporting complex cognitive abilities are distributed across the neocortical mantle and encompass multiple brain regions.

Studies of newborn infants using rs_fMRI reveal similar findings. The primary networks more extensively developed and localized compared to those that support higher cognitive abilities. The higher order cognitive networks observed in newborns were fragmented and incomplete. At birth the default mode network (DMN) showed limited functional connectivity. Although the network was immature it showed involvement of the medial prefrontal and posterior cingulate cortices. A disjointed precursor of the DMN was recently described at 35 weeks postconception (Thomason et al., 2015). In keeping with the primary to higher order sequence of maturation in functional networks, immature forms of sensorimotor, visual, and auditory networks have been detected in fetus's as young as 30 weeks postconception (Thomason et al., 2015).

The maturational trajectories of resting state networks were recently examined in the perinatal period at 3 month intervals. This study revealed that at birth, a number of networks supporting primary functions sensorimotor, visual processing and auditory/language networks mimicked the topology observed in adults and changed very little in the ensuing year. The dorsal attention network and DMN were not mature at birth but developed into their mature forms by the end of the first postnatal year. Higher order networks that support executive function were the last to develop and did not show a mature topology by 1 year post birth (Gao, Alcauter, Smith, Gilmore, & Lin, 2014). It is important to note that the functional connectivity patterns of primary and associative networks in the brain, correlate with myelination patterns and synaptogenesis within the network. Developmental milestones observed in visual and sensorimotor function in the first year of life correlate with the completion of the brain networks that support these functions. Alternately, higher order cognition such as executive control and social understanding continue to develop well beyond adolescence and their associated brain networks are immature until later in adulthood (Casey, Giedd, & Thomas, 2000).

1.6 THE GENETIC BLUEPRINT FOR BRAIN CONSTRUCTION

Given the complexity of the brain, it is not surprising that there is a plethora of genes that govern its development, and recent advances in genomic technologies have allowed us to identify and analyze known and novel genes and their effects on the brain. Comparisons between the human genome and the genomes of other species have begun to shed light on the role genetic changes play in the evolution of the human brain. Furthermore, the human brain has evolved dramatically over the past 2–3 million years, during which time we have acquired novel and often exclusive cell types, circuits, and signaling pathways, which are either rare or absent in other animals (Bae, Jayaraman, & Walsh, 2015).

Nonetheless, it has become increasingly apparent that no distinct human-specific genetic change can explain the human brain's evolution into a highly functional and robust "bio computer" that coordinates our bodies and societies. Thus, through the identification of critical neural genes shared by species and through the study of their functions in the brains of model organisms, we can gain novel insights into our own development and behavioral outcomes (Bae et al., 2015; Geschwind & Rakic, 2013; Molnar et al., 2014; Paabo, 2014).

Furthermore, 99% of our genome does not code for proteins; the 3 billion base pairs (bp) of the human genome contain only about 21,000 protein-coding genes. The rest encode for an array of RNA genes (such as short and long noncoding RNA molecules), regulatory elements, and transposable elements. These assist in the regulation of gene expression and genome stability (Bae et al., 2015).

Although protein-coding genes only constitute approximately 1% of the human genome, they may produce different proteins through processes of alternative splicing that yield multiple functional protein-coding RNAs (mRNAs) and proteins from the same gene. These proteins often play diverse cellular roles. Alternative splicing occurs in the synapse during the production and formation of complex synapse-specificity and circuit-assembly molecules—protocadherin (Pcdh) proteins from protocadherin genes (Zipursky & Sanes, 2010).

It is estimated that through alternative splicing, about 60 Pcdh loci can produce to nearly 350,000 possible protocadherin proteins. Alternative splicing has been reported to occur in neurexins, a family of synaptic receptors important in neurogenesis (Bae et al., 2015).

Along with alternative splicing, the majority of human protein-coding genes use alternative promoters, regulatory regions of DNA that initiate and control the transcription of particular genes. Promoters serve for binding of general and gene-specific activating or inhibitory transcription factors, thereby allowing for diverse and dynamic cell- and tissue-type specific gene expression patterns (Davuluri, Suzuki, Sugano, Plass, & Huang, 2008). Furthermore, mutations in promoter regions may lead to aberrant gene expression and disease. One of the key neurogenesis genes, the human brain-derived neurotrophic factor (BDNF) gene has nine promoters that coordinate tissue and brain-region specific gene expression (Pruunsild, Kazantseva, Aid, Palm, & Timmusk, 2007). However, genetic changes alone cannot explain the complexity, plasticity, and high adaptive capacity of the human brain, as well as its susceptibility to environmental effects and how it is shaped by human experience (Bae et al., 2015).

1.7 EPIGENETIC EDITS TO THE BLUEPRINT FOR BRAIN CONSTRUCTION

A step-wise and highly coordinated brain development program is genetically predetermined and executed via timely and precisely orchestrated gene activity. While each cell of an organism carries the same amount and sequence of DNA, cell and tissues vary greatly in their structure and function. For example, liver cells are very different from skin, muscle, or other cells. These structural and functional differences result from differential gene expression in various cell types. Gene expression programs underlie and determine organismal development, growth, functioning, and aging, as well as environmental interactions.

It is estimated that if uncoiled, human cellular DNA molecules would be about 5 cm on average. Yet, this DNA is carefully packed in cellular nuclei in a highly compact and organized manner and is tightly packaged into a DNA-protein complex known as chromatin with the help of small proteins known as histones. Initially, each 200 bp of DNA wrap twice around an octamer (a complex of eight molecules) of histones, two each of H2A, H2B, H3, and H4. Histones are small basic proteins that have a high affinity to DNA based on their positive charges and DNA's negative charge. The small histone H1 is positioned outside the octamer and helps stabilize the structure; these are termed "beads on the string." Histone beads on the DNA string further fold to form a solenoid-like structure, which folds even further to create radial loops. This occurs with the help of nonhistone scaffolding proteins. In the end, each chromosome constitutes a small unit. Chromatin is very flexible, and the tightness of the interaction between histones and DNA may change, allowing for the formation of loose, genetically active chromatin (euchromatin) and tightly packaged, genetically inactive heterochromatin (Kovalchuk & Kovalchuk, 2012).

Epigenetic mechanisms set and maintain meiotically and mitotically heritable and stable patterns of gene expression and regulation (for more detail see Chapter 10:

Language and Cognition). These occur without changing the DNA sequence. Epigenetic regulation controls gene expression, chromatin structure, and genome functioning through processes that include DNA methylation, histone modifications, chromatin remodeling, and noncoding RNAs (Jaenisch & Bird, 2003; Sandoval & Esteller, 2012) (see Chapter 7: Epigenetics and Genetics of Brain Development for a detailed discussion of the Genetics and Epigenetics of Brain Development).

1.7.1 Altering brain construction with environmental cues

Gene expression programs are a foundation for brain development, and because epigenetic changes control gene expression, the latter govern brain development and integrates environmental cues.

Initial evidence indicating that epigenetic mechanisms are crucial for brain development was provided by Michael Meaney and Moshe Szyf's seminal rodent model-based experiments, which established the precise dependence of offspring phenotype on variations in maternal care (Kovalchuk & Kovalchuk, 2012; Nugent & McCarthy, 2011). This was the first experimental evidence showing that the effects of maternal care on developing progeny are mediated by epigenetic mechanisms, particularly DNA methylation (Meaney & Szyf, 2005; Weaver et al., 2001; Weaver, Szyf, & Meaney, 2002). Meaney and Szyf analyzed variations in maternal licking/grooming (LG) and evaluated their effects on offspring. In rodents, LG is a form of tactile stimulation that activates critical endocrine and metabolic responses and therefore regulates somatic growth and development (Kappeler & Meaney, 2010). Meaney and Szyf's data showed that variations in maternal care influence behavioral and hypothalamic–pituitary–adrenal responses to stress in the adult offspring through epigenetic mechanisms. Their experimental results yielded clear evidence that progeny of high LG mothers had decreased methylation of glucocorticoid receptors (GRs) in the hippocampus, leading to increased GR expression and associated with a decreased stress response (Meaney, Szyf, & Seckl, 2007; Weaver et al., 2004).

In contrast, the progeny of low LG mothers showed increased GR methylation, decreased GR expression, and, as a result, an increased stress response. Furthermore, maternal care in early life produced a longstanding effect on the progeny, but this epigenetically-mediated effect was reversed using crossfostering experiments (Caldji, Diorio, & Meaney, 2003; Francis, Diorio, Liu, & Meaney, 1999). The progeny of low LG mothers reared by high LG mothers demonstrated very similar stress responses to those of the high LG mothers' biological offspring and *vice versa*; Furthermore, the effects of maternal care were transmitted to the next generation (Francis et al., 1999). Altered patterns of gene expression, including differences in the expression of genes involved in stress responses [GR in the hippocampus, central benzodiazepine receptor (CBZ) in the amygdala, and corticotropin-releasing factor (CRF) in the hypothalamus] were subject to nongenomic transmission from one generation to the next through behavior (Francis et al., 1999). As proven by elegant handling experiments, the careful postnatal handling of offspring by the researchers altered stress responses in pups and increased mothers' LG behavior (Francis et al., 1999). Consequently, expression levels of GR, CBZ, and CRF in the brains of the handled offspring of low LG mothers were comparable to those in the offspring of handled or nonhandled high LG mothers (Francis et al., 1999; Kovalchuk & Kovalchuk, 2012). All of the aforementioned effects are based on the epigenetic reprograming of gene expression as a function of early care (Francis et al., 1999).

Many initial studies analyzed the epigenetic reprograming of gene expression as a function of early-life maternal care, focusing on just a few genes. For example, it has also been shown in rats that variations in maternal care in the first week of life are associated with alterations in DNA methylation and H3K9 acetylation of the promoter region of GR *NR3C1* gene, and consequently with the expression of the $GR1_7$ splice variant of the *NR3C1* gene in the hippocampus of adult offspring (Weaver et al., 2004). However, altered levels of numerous genes are required to achieve large-scale and concerted behavioral outcomes (Weaver, Meaney, & Szyf, 2006).

Therefore, a follow-up landmark study by the Meaney and Szyf research groups addressed large-scale epigenetic changes in offspring and provided evidence for the global epigenome changes that occur in the brain as a function of the quality of maternal care (McGowan et al., 2011). The group performed in-depth analyses of hippocampal samples from the adult offspring of female rats with different LG behavior in the first week of their offspring' life (i.e., high vs low LG adult offspring). They analyzed DNA methylation, H3K9 acetylation, and gene expression in a 7-million base pair region of chromosome 18, which carries the *NR3C1* gene. In this study, variations in maternal care were associated with notable epigenetic changes spanning more than 100 kilobase pairs on the entire studied chromosome. The adult offspring of high (compared to low) LG mothers demonstrated major epigenetic changes in promoters, exons, and gene ends, as well as elevated expression of many genes within the examined region. This suggested an extensive epigenomic and gene expression response by the progeny to maternal care variations, which affected numerous genes, including Pcdh gene family implicated in synaptogenesis. This gene family exhibited the highest differential epigenetic response to maternal care variations (McGowan et al., 2011).

These seminal experiments laid the foundation for understanding the effects of early life experiences on the developing brain. For the first time, they proved that the quality of maternal care in rodents has a widespread impact on the progeny's phenotype, which persists into adulthood. Meaney and Szyf developed an excellent model for studying how gene expression and epigenetic mechanisms underlie and govern the way early life experiences impact the brain later in life. They also provided significant and novel insight into the role of epigenetic mechanisms, and especially DNA methylation, on the regulation of gene expression, the governing of neurological processes, and behavior (Meaney & Szyf, 2005). As a result in the past years increasing evidence has accumulated demonstrating that the quality of parental care significantly influences mental health, including the risk of psychopathology, as well as stress responses, emotional function, learning and memory, and neuroplasticity.

During early development, the brain is responsive to experiences and environmental changes. Also, more than any other organ, the brain is placed under heavy social and environmental influences. External experiences (e.g., stress, nutrition, abusive environment, and toxins) can trigger signals between neurons that, reacting to an external stimulus or environment change, respond through the alterations in epigenetic marks, activation of gene expression and production of proteins (Fagiolini, Jensen, & Champagne, 2009).

These proteins, known as gene regulatory proteins, travel to the nucleus within the neural cell where they can draw in or prevent other proteins from interacting with regulatory regions of the genes, including those that are capable of modifying epigenetic markers crucial for generating certain cellular responses. Experiences considered positive, such as enriched learning opportunities, can cause epigenetic changes, alter neural chemistry and impact gene expression. Similarly, negative experiences, such as malnutrition or exposure to environmental toxins, cause gene expression changes. Together, positive and negative experiences have an effect on the brain—often the opposite—and, through experience-induced epigenetic changes, can control which genes are turned on or off. This is called experience-mediated epigenetic modification (Fagiolini et al., 2009).

Besides maternal care as Meaney and Szyf describe, prenatal maternal stress also causes significant, persistent epigenetic changes in the developing brain. Progeny of gestationally-stressed females exhibited decreased levels of promoter methylation of the corticotrophin-releasing-factor (CRF) gene and increased levels of methylation of the GR exon 17 promoters in hypothalamic tissues of adult progeny (Oberlander et al., 2008). Maternal depression as well as the adverse nutritional conditions experienced during fetal development have also been shown to affect growth and metabolism and brain development. Levels of key DNA methyltransferase DNMT1 were shown to decrease with restricted dietary protein, and the altered prenatal regulation of DNA methylation has been seen in brain tissue in association with DNMT1 expression levels (Fagiolini et al., 2009).

Exposure to environmental chemicals, such as Bisphenol A (a well-known endocrine disrupter that profoundly affects DNA methylation), may affect parenting patterns as well as offspring brain and behavior (Kundakovic et al., 2015). These chemicals also negatively impact material behaviors, disrupt neurodevelopment and lead to long-term behavioral effects in animals and humans (Fagiolini et al., 2009).

More recently it has become clear that perinatal mother–infant interactions are not limited to GR regulation. Indeed, rats exposed to abusive maternal care, such as dragging about and general rough handling, demonstrated an increase in the methylation of the promoter of the BDNF gene and a decrease in the levels of BDNF in the prefrontal cortex. Decreased levels of BDNF, a key protein involved in the growth and differentiation of new neurons and synapses and the survival of existing neurons, is likely to have long-term negative effects on brain development (Roth, Lubin, Funk, & Sweatt, 2009). The offspring's exposures manifested epigenetic effects that, while emerging in infancy, lasted into adulthood—and were even carried to the next generations (Champagne, 2008). (For a more information on social influence on brain development, see Chapter 16: Socioeconomic Status.)

Epigenetic changes also underlie experience-dependent plasticity and help regulate synaptic transmission. Exposure to high maternal care levels and an enriched juvenile environment was shown to improve offspring's aptitude for learning and memory through the modulation of gene expression and histone modifications. An elegant study by Monteggia and colleagues showed that alterations in DNA methylation regulate spontaneous synaptic transmission in hippocampal neurons, as evidenced by treatments with compounds that inhibited DNA methyltransferases.

While epigenetic mechanisms have been implicated to be involved in mediating high levels of neuronal plasticity in the early stages of development, epigenetics can be seen as contributing to decreased plasticity later in development—especially during the critical windows of postnatal brain development (so-called "critical periods"). For example, patterns of histone modification–regulated activation and inhibition of numerous molecular pathways, including those involved in myelin maturation, characteristically define the period for ocular dominance.

Furthermore, a wide array of recent studies suggest that epigenetic changes are important in establishment and maintenance of sex differences in the brain, including sex-specific developmental trajectories and sex-specific environmental stress responses (McCarthy & Nugent, 2015).

1.7.2 Environmental cues: lessons learned from animals models

The vast majority of studies investigate the effects of experiences and environmental exposures on developing brain using animal models. These provide solid evidence that exposures to positive and negative factors reshape the brain and affect development trajectories via the modulation of epigenetic mechanisms and associated alterations of gene expression. In humans, mounting evidence suggests that negative experiences and especially early life adversity lead to an increased predisposition to behavioral and psychiatric disorders, including depression, anxiety, and schizophrenia. While epigenetic mechanisms are proposed as plausible candidates, primarily based on the outcomes of rodent studies, the precise mechanism remains elusive. Several seminal studies by Meaney and Turecki groups that used brain samples from suicide victims have shown importance of epigenetic deregulation in human brain in association with childhood abuse and suicide (Labonte et al., 2012, 2013; McGowan et al., 2008, 2009; Suderman et al., 2012). Indeed, studies in humans are limited due to the scarcity of target brain tissues.

Ethical issues prevent human brain analysis; thus, *it remains a question whether we will ever be able to fully discern the mechanisms and underpinnings of the exposure-associated human brain and behavior outcomes.*

With the advent of novel technologies, new evidence suggests that epigenetic changes in the brain may in turn be reflected by epigenetic biomarkers in peripheral tissues, such as blood, saliva, or buccal cells that are easily accessible and provide opportunities for noninvasive or minimally invasive collection procedures.

Indeed, environmental exposures that impact brain development (e.g., prenatal stress, famine and malnutrition, environmental toxicants) were reported to cause epigenetic changes in human peripheral tissues (Kundakovic et al., 2015). DNA methylation and small RNA changes were reported in the peripheral blood of psychiatric patients (D'Addario et al., 2012; Fuchikami et al., 2011; Ikegame et al., 2013). Yet, the question is if epigenetic markers in blood are truly representative of those in the brain and if they can be used to determine epigenetic changes occurring in the brain and their future relationship to behavioral outcomes. The question is particularly important for prenatal diagnostics and the analysis of epigenetic changes induced by adverse prenatal exposures. To address this question, Kundakovic and colleagues used in-utero bisphenol A (BPA) exposure as a model environmental adverse effect. BPA is a well-documented potent endocrine disruptor and ubiquitous environmental toxicant that exerts negative effects on the neurodevelopment and behavior of animals and humans. It acts through epigenetic modulation, and when administered to pregnant animals, it was shown to cause profound epigenetic deregulation in the prefrontal cortex and hypothalamus of the offspring. Researchers explored long-term epigenetic effects of BPA, analyzed and compared molecular changes in the hippocampus and blood, and conducted an in-depth behavioral assessment of offspring prenatally exposed to BPA. Prenatal BPA exposure caused profound and persistent DNA methylation changes in the regulatory region of the Bdnf gene both in the hippocampus and blood of experimental animals. Most importantly, changes observed in the mouse model were considered, with BDNF changes noted in the cord blood of humans exposed to high maternal BPA levels during gestation (Kundakovic et al., 2015). Therefore, a comparison of brain and blood tissues proved that blood can be used as surrogate markers to analyze brain changes caused by prenatal exposure, and high congruence with human cord blood results confirms the validity of rodent prenatal toxicology and epigenetic studies. (For a more detailed discussion of animal models, see Chapter 4: The Role of Animal Models in Studying Brain Development.)

BPA exposure caused epigenetic changes in the BDNF promoter and altered levels of BDNF in the brain and peripheral blood. Interestingly, BDNF levels in the blood were recently suggested to constitute good markers of depression (Molendijk et al., 2014) and schizophrenia (Green, Matheson, Shepherd, Weickert, & Carr, 2011). Moreover, BDNF DNA methylation changes were reported in the peripheral blood of patients with depression (Fuchikami et al., 2011), bipolar disorder (D'Addario et al., 2012), schizophrenia (Ikegame et al., 2013), and eating disorders (Thaler et al., 2014). High BDNF DNA methylation was strongly associated with a history of child maltreatment (Perroud et al., 2013; Thaler et al., 2014). Interestingly, early studies of the offspring of high LG mothers showed elevated BDNF expression, increased cholinergic innervation of the hippocampus, and enhanced spatial learning and memory. Thus, the epigenetic regulation of the expression of neurotrophic factors may underlie some of the molecular regulation of brain development and brain-environment interactions (Kundakovic et al., 2015).

1.8 SUMMARY

Brain development has a protracted time course in humans beginning in the third week post conception and extending to the third decade of life. The brain develops in 7 well-defined phases that overlap and are repeated over the course of development. Neurodevelopmental disorders arise when processes engaged during construction of the brain go awry. Experiences of an individual play a fundamental role in producing a brain uniquely constructed to complement the environmental niche that individual occupies. Overall, experiences that change the epigenome early in life can have a powerful effect on an individual's health and fitness (both physical and mental) for their lifetime! Even small environmental variations in utero can result in recognizable differences in cognitive functionality and in the structure of the postnatal brain. Physical, chemical, biological, and social environment changes that occur postnatally also profoundly affect developing brains. Such changes are epigenetically regulated. While epigenetics modifies the brain's development to increase adaptability and resilience in preparation for future environmental challenges, the resulting developmental trajectory might not be ideal as it is impossible to predict future circumstances to precise levels. For example, adulthood might bring circumstances for which an epigenetic profile isn't optimized, thereby rendering it ineffective in the face of new conditions. If the individual has developed in an ideal environment, they might be unprepared for an introduction into stressed or challenged conditions. On the other hand, development characterized by stress and adversity can also lead to adult individuals who are easily stressed and hypervigilant, even in relatively ideal conditions (e.g., prepared by epigenetics for potential starvation in an adult environment where food is plentiful and nutrients are ample) (Nugent & McCarthy, 2011). Luckily, epigenetic modifications are not permanent in the brain, meaning that individuals are still able to adapt to their environments as adults. There is also potential to change previous epigenetic modifications and, thus, reshape the plastic brain.

REFERENCES

Al-Jaberi, N., Lindsay, S., Sarma, S., Bayatti, N., & Clowry, G. J. (2013). The early fetal development of human neocortical GABAergic interneurons. *Cerebral Cortex, 25*(3), 631–645. Available from http://dx.doi.org/10.1093/cercor/bht254.

Aronica, E., Iyer, A., Zurolo, E., & Gorter, J. A. (2011). Ontogenetic modifications of neuronal excitability during brain maturation: developmental changes of neurotransmitter receptors. *Epilepsia, 52* (Suppl 8), 3–5.

Bae, B. I., Jayaraman, D., & Walsh, C. A. (2015). Genetic changes shaping the human brain. *Developmental Cell, 32*, 423–434.

Bayer, S. A., & Altman, J. (1991). *Neocortical development*. New York: Raven Press.

Ben-Ari, Y., Woodin, M. A., Sernagor, E., Cancedda, L., Vinay, L., Rivera, C., ... Cherubini, E. (2012). Refuting the challenges of the developmental shift of polarity of GABA actions: GABA more exciting than ever!. *Frontiers in Cellular Neuroscience, 6*, 35. Available from http://dx.doi.org/10.3389/fncel.2012.00035.

Bregestovski, P., & Bernard, C. (2012). Excitatory GABA: How a correct observation may turn out to be an experimental artifact. *Frontiers in Pharmacology, 3*, 65.

Caldji, C., Diorio, J., & Meaney, M. J. (2003). Variations in maternal care alter GABA(A) receptor subunit expression in brain regions associated with fear. *Neuropsychopharmacology, 28*, 1950–1959.

Casey, B. J., Giedd, J. N., & Thomas, K. N. (2000). Structural and functional brain development and its relation to cognitive development. *Biological Psychology, 54*, 241–257.

Cayre, M., Canoll, P., & Goldman, J. E. (2009). Cell migration in the normal and pathological postnatal mammalian brain. *Progress in Neurobiology, 88*, 41–63.

Champagne, F. A. (2008). Epigenetic mechanisms and the transgenerational effects of maternal care. *Frontiers in Neuroendocrinology, 29*, 386–397.

Chan, W. Y., Lorke, D. E., Tiu, S. C., & Yew, D. T. (2002). Proliferation and apoptosis in the developing human neocortex. *Anatomical Record, 267*, 261–276.

Clancy, B., Darlington, R. B., & Finlay, B. L. (2001). Translating developmental time across mammalian species. *Neuroscience, 105*, 7–17.

D'Addario, C., Dell'osso, B., Palazzo, M. C., Benatti, B., Lietti, L., & Altamura, A. C. (2012). Selective DNA methylation of BDNF promoter in bipolar disorder: differences among patients with BDI and BDII. *Neuropsychopharmacology, 37*, 1647–1655.

Davuluri, R. V., Suzuki, Y., Sugano, S., Plass, C., & Huang, T. H. (2008). The functional consequences of alternative promoter use in mammalian genomes. *Trends in Genetics, 24*, 167–177.

Elston, G. N. (2003). Cortex, cognition and the cell: new insights into the pyramidal neuron and prefrontal function. *Cerebral Cortex, 13*(11), 1124–1138.

Ernst, A., Alkass, K., Bernard, S., Salehpour, M., Perl, S., & Frisen, J. (2014). Neurogenesis in the adult human striatum. *Cell, 156*, 1072–1083.

Fagiolini, M., Jensen, C. L., & Champagne, F. A. (2009). Epigenetic influences on brain development and plasticity. *Current Opinion in Neurobiology, 19*, 207–212.

Flechsig, P. (1901). Developmental (myelogenetic) localisation of the cerebral cortex in the human subject. *Lancet, 158*, 1027–1030.

Folsom, T. D., & Fatemi, S. H. (2013). The involvement of Reelin in neurodevelopmental disorders. *Neuropharmacology, 68*, 122–135.

Francis, D., Diorio, J., Liu, D., & Meaney, M. J. (1999). Nongenomic transmission across generations of maternal behavior and stress responses in the rat. *Science*, *286*, 1155–1158.

Franco, S. J., Gil-Sanz, C., Martinez-Garay, I., Espinosa, A., Harkins-Perry, S. R., & Muller, U. (2012). Fate-restricted neural progenitors in the mammalian cerebral cortex. *Science*, *337*, 746–749.

Franco, S. J., & Müller, U. (2013). Shaping our minds: stem and progenitor cell diversity in the mammalian neocortex. *Neuron*, 77, 19–34.

Fuchikami, M., Morinobu, S., Segawa, M., Okamoto, Y., Yamawaki, S., & Terao, T. (2011). DNA methylation profiles of the brain-derived neurotrophic factor (BDNF) gene as a potent diagnostic biomarker in major depression. *PLoS ONE*, *6*, e23881.

Gao, W., Alcauter, S., Smith, J. K., Gilmore, J. H., & Lin, W. (2014). Development of human brain cortical network architecture during infancy. *Brain Structure and Function*, 1–14. Available from http://dx.doi.org/10.1007/s00429-014-0710-3.

Geschwind, D. H., & Rakic, P. (2013). Cortical evolution: judge the brain by its cover. *Neuron*, *80*, 633–647.

Green, M. J., Matheson, S. L., Shepherd, A., Weickert, C. S., & Carr, V. J. (2011). Brain-derived neurotrophic factor levels in schizophrenia: a systematic review with meta-analysis. *Molecular Psychiatry*, *16*, 960–972.

Harker, A., Raza, S., Williamson, K., Kolb, B., & Gibb, R. (2015). Preconception paternal stress in rats alters dendritic morphology and connectivity in the brain of developing male and female offspring. *Neuroscience*, *303*, 200–210 . PMID: 26149350.

Hartsock, A., & Nelson, W. J. (2008). Adherents and tight junctions: Structure, function and connections to the actin cytoskeleton. *Biochimica et Biophysica Acta*, *1778*, 660–669.

Ikegame, T., Bundo, M., Sunaga, F., Asai, T., Nishimura, F., & Iwamoto, K. (2013). DNA methylation analysis of BDNF gene promoters in peripheral blood cells of schizophrenia patients. *Neuroscience Research*, 77, 208–214.

Jaenisch, R., & Bird, A. (2003). Epigenetic regulation of gene expression: how the genome integrates intrinsic and environmental signals. *Nature Genetics*, *33* (Suppl), 245–254.

Jernigan, T. L., & Stiles, J. (2017). Construction of the human forebrain. *WIREs Cognitive Science*, e1409. Available from http://dx.doi.org/10.1002/wcs.1409.

Kappeler, L., & Meaney, M. J. (2010). Epigenetics and parental effects. *Bioessays*, *32*, 818–827.

Keunen, K., Counsell, S. J., & Benders, M. J. N. L. (2017). The emergence of functional architecture during early brain development. *NeuroImage*. Available from http://dx.doi.org/10.1016/j.neuroimage.2017.01.047.

Kolb, B., & Whishaw, I. Q. (2014). *An introduction to brain and behavior.* New York, NY: Worth Publishers.

Kovalchuk, I., & Kovalchuk, O. (2012). *Epigenetics in health and disease.* Upper Saddle River, N.J.: FT Press.

Kundakovic, M., Gudsnuk, K., Herbstman, J. B., Tang, D., Perera, F. P., & Champagne, F. A. (2015). DNA methylation of BDNF as a biomarker of early-life adversity. *Proceedings of the National Academy of Sciences of the United States of America*, *112*, 6807–6813.

Labonte, B., Suderman, M., Maussion, G., Lopez, J. P., Navarro-Sanchez, L., & Turecki, G. (2013). Genome-wide methylation changes in the brains of suicide completers. *American Journal of Psychiatry*, *170*, 511–520.

Labonte, B., Yerko, V., Gross, J., Mechawar, N., Meaney, M. J., & Turecki, G. (2012). Differential glucocorticoid receptor exon 1(B), 1(C), and 1(H) expression and methylation in suicide completers with a history of childhood abuse. *Biological Psychiatry*, *72*, 41–48.

Low, L. K., & Cheng, H. J. (2006). Axon pruning: an essential step underlying the developmental plasticity of neuronal connections. *Philosophical Transactions of the Royal Society B: Biological Sciences*, *361* (1473), 1531–1544.

Mccarthy, M. M., & Nugent, B. M. (2015). At the frontier of epigenetics of brain sex differences. *Frontiers in Behavioral Neuroscience*, *9*, 221.

Malatesta, P., & Gotz, M. (2013). Radial glia-from boring cables to stem cell stars. *Development*, *140*, 483–486. Available from http://dx.doi.org/10.1242/dev.085852.

McGowan, P. O., Sasaki, A., D'alessio, A. C., Dymov, S., Labonte, B., & Meaney, M. J. (2009). Epigenetic regulation of the glucocorticoid receptor in human brain associates with childhood abuse. *Nature Neuroscience, 12*, 342–348.

McGowan, P. O., Sasaki, A., Huang, T. C., Unterberger, A., Suderman, M., & Szyf, M. (2008). Promoter-wide hypermethylation of the ribosomal RNA gene promoter in the suicide brain. *PLoS ONE, 3*, e2085.

McGowan, P. O., Suderman, M., Sasaki, A., Huang, T. C., Hallett, M., & Szyf, M. (2011). Broad epigenetic signature of maternal care in the brain of adult rats. *PLoS ONE, 6*, e14739.

Meaney, M. J., & Szyf, M. (2005). Environmental programming of stress responses through DNA methylation: life at the interface between a dynamic environment and a fixed genome. *Dialogues in Clinical Neuroscience*, 7, 103–123.

Meaney, M. J., Szyf, M., & Seckl, J. R. (2007). Epigenetic mechanisms of perinatal programming of hypothalamic–pituitary–adrenal function and health. *Trends in Molecular Medicine, 13*, 269–277.

Miller, D. J., Duka, T., Stimpson, C. D., Schapiro, S. J., Baze, W. B., & Sherwood, C. C. (2012). Prolonged myelination in human neocortical evolution. *Proceedings of the National Academy of Sciences of the United States of America, 109*, 16480–16485.

Molendijk, M. L., Spinhoven, P., Polak, M., Bus, B. A., Penninx, B. W., & Elzinga, B. M. (2014). Serum BDNF concentrations as peripheral manifestations of depression: evidence from a systematic review and meta-analyses on 179 associations (*N* = 9484). *Molecular Psychiatry, 19*, 791–800.

Molnar, Z., Kaas, J. H., de Carlos, J. A., Hevner, R. F., Lein, E., & Nemec, P. (2014). Evolution and development of the mammalian cerebral cortex. *Brain, Behavior and Evolution, 83*, 126–139.

Mychasiuk, R., Harker, A., Ilnytskyy, S., & Gibb, R. (2013). Paternal stress prior to conception alters DNA methylation and behaviour of developing rat offspring. *Neuroscience, 241*, 100–105.

Morrison, B. M., Lee, Y., & Rothstein, J. D. (2013). Oligodendroglia: metabolic supporters of axons. *Trends Cell Biol, 23*(12), 644–651. http://dx.doi.org/10.1016/j.tcb.2013.07.007S0962-8924(13)00120-7.

Mychasiuk, R., Zahir, S., Schmold, N., Ilnytskyy, S., Kovalchuk, O., & Gibb, R. (2012). Parental enrichment and offspring development: modifications to brain, behavior and the epigenome. *Behavioural Brain Research, 228*, 294–298.

Nugent, B. M., & Mccarthy, M. M. (2011). Epigenetic underpinnings of developmental sex differences in the brain. *Neuroendocrinology, 93*, 150–158.

Oberlander, T. F., Weinberg, J., Papsdorf, M., Grunau, R., Misri, S., & Devlin, A. M. (2008). Prenatal exposure to maternal depression, neonatal methylation of human glucocorticoid receptor gene (NR3C1) and infant cortisol stress responses. *Epigenetics, 3*, 97–106.

Paabo, S. (2014). The human condition-a molecular approach. *Cell, 157*, 216–226.

Paolicelli, R. C., Bolasco, G., Pagani, F., Maggi, L., Scianni, M., & Gross, C. T. (2011). Synaptic pruning by microglia is necessary for normal brain development. *Science, 333*, 1456–1458.

Perroud, N., Salzmann, A., Prada, P., Nicastro, R., Hoeppli, M. E., & Malafosse, A. (2013). Response to psychotherapy in borderline personality disorder and methylation status of the BDNF gene. *Translational Psychiatry, 3*, e207.

Pruunsild, P., Kazantseva, A., Aid, T., Palm, K., & Timmusk, T. (2007). Dissecting the human BDNF locus: bidirectional transcription, complex splicing, and multiple promoters. *Genomics, 90*, 397–406.

Rinholm, J. E., Hamilton, N. B., Kessaris, N., Richardson, W. D., Bergersen, L. H., & Attwell, D. (2011). Regulation of oligodendrocyte development and myelination by glucose and lactate. *Journal of Neuroscience, 31*, 538–548.

Rabinowicz, T., de Courten-Myers, G. M., Petetot, J. M., Xi, G., & de los Reyes, E. (1996). Human cortex development: estimates of neuronal numbers indicate major loss late during gestation. *Journal of Neuropathology & Experimental Neurology, 55*(3), 320–328.

Roth, T. L., Lubin, F. D., Funk, A. J., & Sweatt, J. D. (2009). Lasting epigenetic influence of early-life adversity on the BDNF gene. *Biological Psychiatry, 65*, 760–769.

Sandoval, J., & Esteller, M. (2012). Cancer epigenomics: Beyond genomics. *Current Opinion in Genetics & Development, 22*, 50–55.

Shafer, D. P., & Stevens, B. (2015). Microglia function in central nervous system plasticity. *Cold Spring Harbor Perspectives in Biology*, 7, a020545.

Stiles, J., & Jernigan, T. L. (2010). The basics of brain development. *Neuropsychology Review*, *20*, 327–348.

Suderman, M., McGowan, P. O., Sasaki, A., Huang, T. C., Hallett, M. T., & Szyf, M. (2012). Conserved epigenetic sensitivity to early life experience in the rat and human hippocampus. *Proceedings of the National Academy of Sciences of the United States of America*, *109* (Suppl 2), 17266–17272.

Thaler, L., Gauvin, L., Joober, R., Groleau, P., de Guzman, R., & Steiger, H. (2014). Methylation of BDNF in women with bulimic eating syndromes: associations with childhood abuse and borderline personality disorder. *Progress in Neuro-Psychopharmacology and Biological Psychiatry*, *54*, 43–49.

Thomason, M. E., Dassanayke, M. T., Shen, S., Katkuri, Y., Hassan, S. S., & Romero, R. (2013). Cross-hemispheric functional connectivity in the human fetal brain. *Science Translational Medicine*, *24*. http://dx.doi.org/10.1126/scitranslmed.3004978.

Thomason, M. E., Lozon, L. E., Vila, A. M., Ye, Y., Nye, M. J., & Romero, R. (2015). Age-related increases on long-range connectivity in fetal functional neural connectivity networks in utero. *Developmental Cognitive Neuroscience*, *11*, 96–104.

Tyzio, R., Cossart, R., Khalilov, I., Minlebaev, M., Hubner, C. A., & Khazipov, R. (2006). Maternal oxytocin triggers a transient inhibitory switch in GABA signaling in the fetal brain during delivery. *Science*, *314*, 1788–1792.

Wake, H., Moorhouse, A. J., Miyamoto, A., & Nabekura, J. (2013). Microglia: actively surveying and shaping neuronal circuit structure and function. *Trends in Neurosciences*, *36*(4), 209–217. http://dx.doi.org/10.1016/j.tins.2012.11.007S0166-2236(12)00204-4.

Weaver, I. C., Cervoni, N., Champagne, F. A., D'alessio, A. C., Sharma, S., & Meaney, M. J. (2004). Epigenetic programming by maternal behavior. *Nature Neuroscience*, 7, 847–854.

Weaver, I. C., La Plante, P., Weaver, S., Parent, A., Sharma, S., & Meaney, M. J. (2001). Early environmental regulation of hippocampal glucocorticoid receptor gene expression: characterization of intracellular mediators and potential genomic target sites. *Molecular and Cellular Endocrinology*, *185*, 205–218.

Weaver, I. C., Meaney, M. J., & Szyf, M. (2006). Maternal care effects on the hippocampal transcriptome and anxiety-mediated behaviors in the offspring that are reversible in adulthood. *Proceedings of the National Academy of Sciences of the United States of America*, *103*, 3480–3485.

Weaver, I. C., Szyf, M., & Meaney, M. J. (2002). From maternal care to gene expression: DNA methylation and the maternal programming of stress responses. *Endocrine Research*, *28*, 699.

Wonders, C. P., & Anderson, S. A. (2006). The origin and specification of cortical interneurons. *Nature Reviews Neuroscience*, 7, 687–696.

Zipursky, S. L., & Sanes, J. R. (2010). Chemoaffinity revisited: dscams, protocadherins, and neural circuit assembly. *Cell*, *143*, 343–353.

FURTHER READING

Ben-Ari, Y., Woodin, M. A., Sernagor, E., Cancedda, L., Vinay, L., & Cherubini, E. (2012). Refuting the challenges of the developmental shift of polarity of GABA actions: GABA more exciting than ever!. *Frontiers in Cellular Neuroscience*, *6*, 35.

Gray, E. G. (1959). Axo-somatic and axo-dendritic synapses of the cerebral cortex: An electron microscope study. *Journal of Anatomy*, *93*, 420–433.

Greenough, W. T., Black, J. E., & Wallace, C. S. (1987). Experience and brain development. *Child Development*, *58*, 539–559.

Shatz, C. J. (1992). The developing brain. *Scientific American*, *267*, 60–67.

CHAPTER 2

Perspectives on Behavioral Development

Serena Jenkins
University of Lethbridge, Lethbridge, AB, Canada

2.1 HISTORICAL PERSPECTIVES ON BEHAVIORAL DEVELOPMENT

Perspectives on the nature of behavioral development have an interesting history that has arisen from two very different traditions: ethology and psychology. While both fields were interested in understanding how complex behaviors arise in animals, historically there was an artificial dichotomy that slowly melted away as contemporary perspectives of gene–environment interactions began to take form. This chapter begins with a historical consideration of the ethological and psychological positions before reviewing the emergence of a new perspective that emphasizes the role of experience in shaping gene expression throughout development.

2.1.1 The innate-learned dichotomy

The field of Ethology is the scientific study of animal behavior, and is most often attributed to the works of Konrad Lorenz (1903–89). Lorenz is known for popularizing the still prevalent distinction between behaviors that are "innate" and those that are "learned". To Lorenz, innate, or instinctual behaviors are those that are present at birth, and are invariably programed by the animal's genetic composition. For a behavior to be innate, it must satisfy several criteria, and satisfaction of the criteria was both necessary and sufficient for the behavior to be deemed "genetically programed" by Lorenz. Innate behaviors must: (1) be ubiquitous in all typical individuals of a particular species; (2) be heritable; (3) be sometimes displayed randomly in seemingly inappropriate situations; and (4) be resistant to modification by the environment (Richards, 1974). Lorenz developed a model of motivation to explain the occurrence of innate behaviors. "Excitation substance" collects over time into "reservoirs" that correspond to specific innate behaviors. Under appropriate conditions, an "innate releasing mechanism" (IRM) will trigger the opening of the reservoir, resulting in the innate behavior to be displayed (Kalikow, 1976; Richards, 1974). Given Lorenz's background in ornithology (Brigandt, 2005), many of the observations that were cornerstone to the development of his theory were based on avian behavior. One such

The Neurobiology of Brain and Behavioral Development
DOI: http://dx.doi.org/10.1016/B978-0-12-804036-2.00002-9

example is egg rolling in the gray goose (summarized in Lehrman, 1953). When a goose notices an egg outside of the nest, it will extend its head toward the egg, rest the underside of its bill on the egg, and use a series of movements of neck to roll the egg back into the nest. Lorenz and his collaborator Niko Tinbergen concluded that this act involves two stages. The first stage is instinctive and relies on the perception of particular physical properties of the egg, namely the smooth hard surface. The appearance of the egg is the IRM that triggers the egg rolling behavior. The second stage is the actual sequence of neck movements that move the egg into the nest; these movements rely on sensory feedback from the egg on the underside of the bill so that the goose can adjust its behavior accordingly. Lorenz's motivational model could explain both spontaneous behavior and habituation of response. In cases where the animal does not experience the IRM for an extended period of time, the reservoir over flows with excitation substance and the behavior is displayed under inappropriate circumstances. When the IRM is present repeatedly, the reservoir is unable to refill so the behavior is no longer displayed, even when the IRM is present (Richards, 1974).

Learned behaviors are all other behaviors that the animal acquires throughout its lifetime via experience, trial-and-error learning, observation, or imitation. Lorenz's dichotomy was strict in the sense that there was no middle ground; all behaviors fit into one of two categories, innate or learned. In cases where it appeared a behavior had elements that were both innate and learned, Lorenz argued that this was because simple elementary behaviors are linked together in chains, with some links being innate and others being learned. Lorenz called this linkage instinct-learning intercalation. However, the "gaps" in which learned elements could be inserted were innately decided by the animal's evolutionary history, which constrained possible modifications to a behavior sequence (Richards, 1974).

The classical experiment used by Lorenz to determine if a behavior was innate or learned was the deprivation experiment. In brief, an animal would be raised without the opportunity to practice or observe a behavior, or without being exposed to the IRM that would normally trigger the behavior. Once the deprived animal was of the age that it would normally be able to display the behavior in full, it was given opportunity to do so. If the behavior was performed normally, Lorenz reasoned that this behavior must be innate. If the behavior was not performed, it was a learned behavior that depended on prior experience or practice to develop. Because Lorenz believed in a strict dichotomy between these two categories of behavior, in cases where the animal would perform the behavior in an altered or impoverished manner, Lorenz attributed this to poor experimental design (Richards, 1974).

Lorenz's views of instinctual behaviors were considered radical by some and were received with criticism by several others in the field of animal behavior. One critic was Daniel Lehrman, who published a thorough evaluation of Lorenz's theory in 1953 (Lehrman, 1953). Lehrman, among other critics of the innate-learned dichotomy, saw no value in preconceived categories that attempted to group together vastly different behaviors based on weak, often theoretical, evidence that some require learning and others do not. Lehrman also criticized Lorenz's use of the term "maturation," arguing that such a label offers no explanation to the origins of behavior and actually discourages further study of behavioral development. To Lorenz, innate behaviors become more competent and effective with age through the process of maturation, not development due to learning (Brigandt, 2005). Lehrman conceded that certain systems that are involved in the performance of a behavior sequence may be maturing, such as the muscular or vestibular systems, resulting in increased competency in performing a behavior, but this does not mean the behaviors themselves mature (Lehrman, 1953).

Lehrman (1953) also took issue with Lorenz's deprivation experiments that he used to justify the innateness of a behavior. He argued that simply isolating an individual from conspecifics or necessary space is not sufficient to void the animal of all factors that may influence its development. Lorenz's studies could only provide insight into what influences are not important in the emergence of a behavior, and did not provide any information regarding what factors are involved. One of the sources of this criticism came from experiments conducted by Kuo and reported on in a series of papers published in the 1930s called "Ontogeny of embryonic behavior in Aves." In the fourth installment of this series, Kuo (1932) summarizes the embryonic age that various bodily movements and physiological activities emerge. He then explicitly explains why one behavior that is commonly thought to be "innate" and "unlearned," pecking, is in fact the result of embryonic behavior. Pecking involves three components: vertical movement of the head, opening and closing of the bill, and swallowing. Starting around embryonic day six, the bill begins to open and close, as well as thrust. This opening of the bill inevitably leads to the chick taking in amniotic fluid, which is then swallowed as is evident from fluid found in the crop and the gizzard. The vertical movements of the head are practiced due to the head being bent downwards and resting on the breast of the chick, where it is passively moved by the heartbeat. Kuo's studies are important in demonstrating that behavior is subjected to influences that are not necessarily directly related to the behavior, but are still important in shaping its development. Deprivation experiments may be able to isolate an individual from certain experiences, but not necessarily all the experiences that shape the behavior, rendering them ineffective at determining the innateness of a given behavior. We may ask "from what is the animal isolated?", but not "is the animal isolated?" (Lehrman, 1953). A solution to this issue is provided by Hogan (1994) and discussed in a later section.

Lorenz defended his theory of instinctual behaviors, and implemented some changes to appease his opposition. Although he maintained that the terms innate and learned were still viable, he admitted that a behavior is not innate in and of itself, but that the range of possible behavior is determined genetically, and therefore innate. He continued to believe that only certain aspects of a behavior sequence are modifiable by learning, whereas other components inherently have such a narrow range that any modification due to learning is negligible (Richards, 1974). He now admitted that even innately determined behaviors are influenced by numerous external factors, but that it would be "pathological" to attempt to study all such factors. Lorenz contended that it is the ethologist's responsibility to identify *adaptive* modifications to behavior; in this sense, "innate" is intended to convey phylogenetic adaptiveness, and "learned" individual adaptiveness (Richards, 1974).

Lorenz's critics clearly had a large impact on his theory of instinct. Overall, the criticisms forced Lorenz to reevaluate the dichotomizing of innate and learned behaviors, and to adjust his theory to allow for a middle ground. Although he continued to advocate for the categories of innate and learned, he conceded that all behaviors are influenced by innumerable external factors.

2.1.2 Tinbergen's ethology

Niko Tinbergen (1907–88) is most well-known for his seminal paper "On Aims and Methods of Ethology," published in 1963 (Tinbergen, 1963). It was with this paper that Tinbergen formally defined the field of ethology as the scientific study of animal behavior, and gave Konrad Lorenz credit for its conception. Tinbergen was a long-time collaborator with Lorenz and used this paper to honor the contributions Lorenz had made to the field. Unfortunately, intellectual disagreements regarding how to modify the theory of instinct forced a wedge between the two over time (Burkhardt, 2014).

"On Aims and Methods of Ethology" had three main goals. First, Tinbergen felt that if ethology was to become a true science, thorough descriptions of all behaviors being studied must precede explanations. He felt that failure to describe was a shortcoming of disciplines such as Psychology, which attempted to explain complex phenomenon without first clearly stating what such phenomenon entail. Tinbergen was a proponent of naturalistic observation and detailed note taking in order to get a complete picture of a behavior in its natural setting. The paper's second aim was to introduce the four main questions in the field of Ethology. Tinbergen was heavily influenced by several of his predecessors in that he borrowed three of the four questions: causal analysis, function or survival value, and evolutionary history. However, Tinbergen was the first to suggest that the ontogeny of a behavior should also be analyzed, which is of particular interest to the present chapter. The third goal of "...Aims and Methods..." was to emphasize the importance of integrating all four of the questions regarding the study of behavior. Too often, researchers isolate one of the questions and attempt to find a meaningful answer without regard for the others. Tinbergen argued that understanding how a behavior develops in an individual requires also understanding how it evolved in that species, what survival value it offers the species, and its causal mechanisms.

Over the years, interest in each of the four questions central to Ethology has varied, despite Tinbergen's insistence that all four questions be studied together. Tinbergen himself was guilty of having disproportionate investments in certain questions throughout the course of his career. Unfortunately for us and the present topic, the question of development was never his main focus (Burkhardt, 2014). However, we do have Tinbergen to thank for first suggesting that ontogeny be studied when studying animal behavior.

2.1.3 Conditioned learning theories

Although Lorenz acknowledged the existence of behaviors that are learned during the development of an animal, his main focus was on those behaviors that were innate. Meanwhile, others were investigating the mechanisms of learning. Within Psychology, much learning that is observed in humans and nonhuman animals is categorized as Classical Conditioning or Operant Conditioning. Classical conditioning, pioneered by Ivan Pavlov (1849–1936), consists of forming an association between two stimuli that are commonly presented together. One stimulus in the pair in the unconditioned stimulus (US); the US elicits an unconditioned response (UR), which is commonly reflexive or physiological. The second stimulus is the conditioned stimulus (CS). With repeated pairings of the US and the CS, the CS becomes able to elicit the same response as the US, which is now called the conditioned response (CR). Pavlov discovered this phenomenon in his famous study on salivation in dogs. A dog naturally salivates (UR) upon smelling food (US). If a bell (CS) is paired with the presentation of food repeatedly, the bell becomes sufficient to elicit salivation (CR). Classical conditioning is often used to explain phobias, which likely has its roots in the experiments conducted by John B. Watson (1878–1958) on "Little Albert." Watson set out to test experimentally if a fearful response could be conditioned in an infant (Watson & Rayner, 1920). When Albert was 9 months old, he was exposed to a white rat, a rabbit, a dog, and several other objects; he did not show fear toward any of the objects. Watson then placed Albert is a room and exposed him to a known fear-inducing stimulus, a sudden loud sound; Albert reacted violently to the sound being presented behind him. The conditioning experiments began when Albert was 11 months old. A white rat was placed in front of the infant, and as before he showed no fear toward the animal. The experimenters then made the loud noise behind to infant. After two pairings of the rat and the sound, the rat alone was sufficient to induce a fear reaction. Moreover, other stimuli that shared common features with the white rat also elicited the fear response, such as a rabbit, cotton wool, and a Santa Claus mask.

Operant conditioning was first described by B.F. Skinner (1904–90) in his 1938 book "The Behavior of Organisms" (Skinner, 1938/1991). Operant conditioning increases the frequency of a behavior through reinforcement; an animal receives a reward upon completion of a behavior and increases how often it performs the behavior in order to increase frequency of the reward. Unlike classical conditioning, operant conditioning is reversible due to extinction; if the animal performs the behavior without being rewarded by the reinforcer, frequency of the behavior will decrease (Skinner, 1938/1991). Skinner famously demonstrated operant conditioning in a series of experiments using an apparatus now called the Skinner Box, but initially called the operant conditioning chamber. The chamber was equipped with a bar (when used with rats) or a disc (when used with pigeons) that when pressed, released a small amount of food. A naïve animal would be placed in the box and allowed to explore. Eventually, the rat or pigeon would accidentally press the bar or disc, respectively, releasing the food. The pressing behavior would increase due to the reinforcement. Skinner (1948) also demonstrated that false operant conditioning can occur. A pigeon placed in a Skinner box that was not equipped with a disc to press received food on a regular interval. Over time, it associated the arbitrary behavior it happened to be performing immediately prior to the food release with the food, and therefore increased this behavior. This is false conditioning in the sense that the pigeon would have received the reward regardless of performing the behavior, or any other behavior. Skinner (1948) likened this to superstition; the pigeon believed the food was contingent on the behavior, when in reality the behavior had no bearing on the reward.

The innate-learned dichotomy is important when considering the development of behavior, as is the goal of the present chapter. It highlights the history of the intellectual endeavor to understand how complex behaviors arise in animals and humans and provides an appropriate transition into the discussion of more contemporary theories on behavioral development, which were undoubtedly influenced by these earlier ways of thinking. Lorenz and the early learning theorists continue to have a large impact on how both the scientific community and the general public view the development of behavior.

2.2 CONTEMPORARY PERSPECTIVES ON BEHAVIORAL DEVELOPMENT

2.2.1 Gene–environment interaction

Along with objections to the sharp contrast between behaviors that are phylogenetically or ontogenetically determined came the notion that behavioral development is determined by the interaction of genes and the environment. The "nature-nurture" debate has more or less been discarded, although there remain remnants of such a way of thinking in how we talk about development to this day (for review of this topic, see Johnston, 1987). It is now widely accepted that the development of behavior is the product of input from the environment and genetic influences. Fig. 2.1 depicts a model of behavioral development, adapted from Johnston and Edwards (2002).

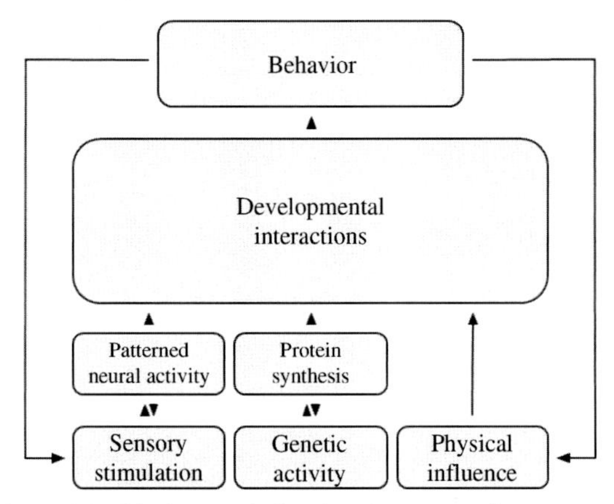

Figure 2.1 A simplified model of behavioral development. Development is influenced by sensory stimulation, genetic activity, and all other physical influences. Behavior has direct effects on both sensory stimulation and physical influences, shown by arrows. The effect of behavior on genetic activity is indirect and not shown. There exists reciprocal relationships between sensory stimulation and its immediate effect, patterned neural activity, and between genetic activity and protein synthesis. *Refer to Johnston, T.D., & Edwards, L. (2002). Genes, interactions, and the development of behavior. Psychological Review, 109(1): 26–34. DOI:10.1037//0033-295X.109.1.26 for a more complete version of the model.*

Evidence to suggest that genes and the environment interact is not limited to recent research. For example, Cooper and Zubek (1958) conducted a study in which they manipulated both the genotype of the rats they studied and the environment to which the rats were exposed. Rats that were either "bright" or "dull" were raised in environments that were either "enriched" or "restricted." Following 40 days in their respective environment, rats were tested in a maze and their number of errors was recorded. When considering genotype alone, the bright rats made significantly fewer errors than the dull rats, which is the anticipated outcome. However, the interaction between genotype and environment became apparent when statistical interactions were analyzed. Dull rats raised in enriched environments performed equally well as bright rats raised in normal environments; these rats showed significant improvement in their performance. Bright rats raised in enriched environments did not perform significantly better than their counterparts raised in normal conditions. Bright rats raised in restricted environments performed equally poorly as dull rats raised in normal environments; these rats made significantly more errors than bright rats from normal conditions. Dull rats raised in restricted environments did not experience any further deficit, and performed the same as dull rats in normal conditions. Fig. 2.2 summarizes the findings of this study. The results of this experiment nicely demonstrate that the influence of genotype on behavior is moderated by environmental conditions and that there are ceiling effects for behavior.

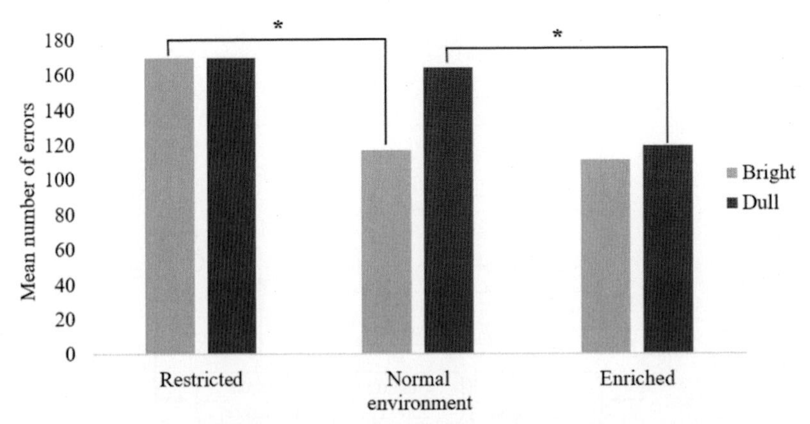

Figure 2.2 Mean number of maze errors committed by bright and dull rats that were raised in either restricted, normal, or enriched environments for 40 days prior to testing. Bright rats raised in restricted environments performed significantly worse than bright rats in normal environments. Dull rats raised in enriched environments performed significantly better than dull rats in normal environments. However, bright rats in enriched environments and dull rats in restricted environments did not perform differently than their counterparts in normal environments, demonstrating a ceiling effect. *Data from Cooper, R.M., & Zubek, J.P. (1958). Effects of enriched and restricted early environments on the learning ability of bright and dull rats. Canadian Journal of Psychology, 12(3): 159–164. DOI:10.1037/h0083747.*

The interaction between genes and the environment is exemplified by epigenetics, which examines reversible changes in gene expression that are not the result of modification to the actual DNA sequence (i.e., mutations). Epigenetic modifications function to maintain differential gene expression in different tissues, allowing each tissue type to express the appropriate phenotype based on its function (Kobor & Weinberg, 2011; see also Chapter 7: Epigenetics and Genetics of Brain Development), and play a major role in cell differentiation during development (Keverne, Pfaff, & Tabansky, 2015). Extraneous environmental factors can also influence gene expression through DNA methylation, in which methyl groups attach to cytosine residues that are abundant in gene regulatory regions and typically suppress gene expression by preventing transcription factors from binding (Roth, 2013), or histone modification, which alters the accessibility of DNA by changing the "tightness" of the chromatin through acetylation or methylation, among other modifications. Changes in DNA methylation can occur throughout the lifespan of an animal, beginning prenatally and continuing to senescence, and can therefore alter behavior throughout the lifespan as well (e.g., Roth, 2013). Furthermore, recent evidence from preconception studies indicates that parental experiences prior to conception can also influence gene expression and behavior in the offspring (e.g., Huang et al., 2010; Mychasiuk et al., 2012; Mychasiuk, Harker, Ilnytskyy, & Gibb, 2013).

Preconception experiences. The preconception period is the time immediately preceding conception and has recently gained interest as an influential period in the life of a developing animal. Empirical studies using animals and opportunistic studies with humans elucidate the importance of this developmental stage. Offspring may be affected by positive and aversive experiences that the mother or father was subjected to during the preconception period.

Offspring of male rats that experience enrichment immediately prior to conception exhibit decreased levels of DNA methylation in both the frontal cortex and the hippocampus (Mychasiuk et al., 2012), suggesting an increase in gene expression. Conversely, the offspring of male rats that endured a negative experience prior to conception, namely elevated platform stress, showed increased DNA methylation in the hippocampus, whereas methylation patterns in the frontal cortex were sex-dependent. Male offspring of preconception stressed fathers were less anxious compared to control offspring when tested in an open field, and both male and female offspring showed delays in sensorimotor development (Mychasiuk et al., 2013). Fetal offspring of female rats exposed to chronic unpredictable preconception stress exhibit decreased expression of a serotonin receptor (5-HT1A) in the hippocampus and of serotonin transporter in both the hippocampus and hypothalamus when compared to control offspring, suggesting alterations to the serotonergic system (Huang et al., 2010). Adult offspring of preconception stressed dams show decreased expression of brain-derived neurotrophic factor and *N*-methyl-D-aspartate receptor in the hippocampus, accompanied by impaired spatial memory compared to control offspring (Huang et al., 2010). Alcohol abuse by fathers prior to conception also has negative impacts on offspring behavior, with studies demonstrating hyperactivity, learning and memory deficits, and hyper-responsiveness to stress; epigenetics is a likely mechanism behind these behavioral aberrations (Abel, 2004).

Prenatal experiences. Prenatal experiences have substantial effects on later gene expression in offspring. Offspring rats of dams who experienced prenatal enrichment had reduced DNA methylation in both the hippocampus and frontal cortex. These offspring showed increased activity in the open field and a delay in sensorimotor development, which the authors speculated may be due to the slowing of maturation in order to enhance brain plasticity in adulthood (Mychasiuk et al., 2012). Prenatal stress has opposing effects on DNA methylation depending on the severity of the stressor. When stress is mild (i.e., 10-min elevated platform exposure twice daily for 27 days), DNA methylation increases in both the hippocampus and frontal cortex. Conversely, severe stress (i.e., 30-min elevated platform exposure twice daily for 27 days) results in decreased methylation in the same two regions (Mychasiuk, Ilnytskyy, Kovalchuk, Kolb, & Gibb, 2011). Offspring of mild prenatal stressed mothers are hypoactive, whereas male offspring of severely stressed mothers are hyperactive relative to controls and female offspring. Sensorimotor development is impaired due to both mild and severe prenatal stress, although the effects are different. Mild prenatal stress results in a consistent impairment in sensorimotor ability compared to controls that persists throughout early development. Severe prenatal stress seems to have no impact on baseline sensorimotor ability, but significantly stunts development (Mychasiuk et al., 2012). A review by Roth (2013) summarizes other consequences of prenatal stress including altered expression of genes involved in synaptic plasticity, neurotransmitter reception, and protection from damage induced by stress hormones.

Prenatal alcohol use is accompanied by a variety of behavioral problems in offspring, that when considered with physical and physiological abnormalities constitute fetal alcohol spectrum disorder (FASD; O'Leary, 2004). Epigenetic modifications may be responsible for some of the behavioral deficits associated with FASD; three epigenetic mechanisms (DNA methylation, histone modification, and regulation by microRNAs) are disrupted by prenatal ethanol (Ungerer, Knezovich, & Ramsay, 2013). In utero alcohol exposure results in global DNA hypomethylation and altered histone methylation, likely because folate deficiency is a consequence of alcohol abuse and folate is a major methyl group donor (Ungerer et al., 2013). In mice given 4% v/v ethanol prenatally during the late first and second trimesters, altered DNA methylation in the hippocampus is correlated with delayed hippocampal development (Chen, Ozturk, & Zhou, 2013), which may help explain learning impairments and memory deficits observed in FASD (O'Leary, 2004). MicroRNA expression is generally reduced due to ethanol exposure, resulting in an imbalance in apoptosis and cell proliferation that may lead to aberrant development (Ungerer et al., 2013).

Postnatal experiences. Along with the prenatal period, the perinatal and postnatal periods are especially sensitive to epigenetic reprograming by external experiences. Maternal care during the postnatal period influences gene expression and subsequent behavior in developing offspring (see Roth, 2013 for a review of rodent and human studies). Gene expression continues to be regulated via epigenetic mechanisms throughout the lifespan, although the most pronounced effects occur during times of rapid growth and differentiation (Kobor & Weinberg, 2011; Roth, 2013). Roth (2013) summarizes rodent studies that indicate the role of DNA methylation within the hippocampus during fear conditioning, and within the hippocampus, hypothalamus, and nucleus accumbens in response to various stressors.

As alcohol consumption causes epigenetic disruption during most life stages, it is not surprising that it continues to affect gene expression following parturition. In rats, perinatal ethanol exposure models third trimester fetal alcohol exposure in humans; ethanol exposure during this time alters gene expression in the hippocampus, resulting in spatial memory deficits that persist into adulthood (Zink et al., 2011), and in the cerebellum, providing a possible explanation for FASD-related motor difficulties (Guo et al., 2011). Furthermore, alcohol-induced liver disease in adults is characterized by global changes in DNA methylation, and these epigenetic alterations may sensitize the liver to later stressors (Kobor & Weinberg, 2011).

Many disorders have associated perturbations in epigenetic regulation. Psychiatric, neurodevelopmental, neurodegenerative, and neurological disorders, as well as cancer are characterized by changes in gene expression (Abel & Zukin, 2008; Keverne et al.,

2015; Roth, 2013; Urdinguio, Sanchez-Mut, & Esteller, 2009). Fortunately, epigenetic mechanisms may also be utilized in the management and partial reversal of many conditions, notably neurodevelopmental, neurodegenerative, and psychiatric disorders (Abel & Zukin, 2008).

2.2.2 Gottlieb's probabilistic epigenesis

Gilbert Gottlieb is one proponent of the interactionist perspective on development, and has termed his specific theory probabilistic epigenesis. Probabilistic epigenesis proposes that all levels of a developing system, from genes to the whole organism in its environment, interact reciprocally to determine the trajectory of development. Bidirectional relationships exist between the four levels of a developing organism: genetic activity, neural activity, behavior, and the physical, social, and cultural environment (Gottlieb, 2007). Gottlieb also postulates that it is the interaction between components of the developing system and not the components themselves that cause development. This coaction between at least two components can be horizontal (i.e., two components at the same level, such as cell—cell) or vertical (i.e., two components across different levels, such as organism—environment). By defining coaction as the cause of development, it becomes impossible to separate the effects of either genes or environment on development, which necessitates the interactionist approach.

The interactions between varying levels of the developing system do not necessarily produce intuitive outcomes. Experience can have indirect effects that produce "nonlinear and nonobvious" changes in development (Gottlieb, 1991). This finding further emphasizes the inadequacy of Lorenz's deprivation experiments; it is not always possible to predict which experiences are necessary or sufficient to allow a behavior to develop normally. Gottlieb (1991) provides several examples of this phenomenon, including one from his own work on mallard duckling responsiveness to maternal calls. He found that ducklings must hear the species-typical embryonic calls from themselves or their siblings to differentially respond to their species-typical maternal call. A hypothetical deprivation experiment would simply deprive a hatchling from hearing the appropriate maternal call for a period of time, then introduce the hatchling to two maternal calls, including the species-typical call. The hatchling would respond preferentially to its own species due to having heard its own embryonic calls. However, the behavior would inappropriately be categorized as "innate" using Lorenz's criteria.

2.2.3 Hogan's behavior mechanism

Jerry Hogan postulated that behavior is the result of behavior mechanisms within the central nervous system. A behavior mechanism corresponds with a particular neural network within the CNS = Central Nervous System; HPA = Hypothalamic-Pituitary-Adrenal; AMPA = α-amino-3-hydroxy-5-methyl-4-isoxazolepropionic acid that when activated results in the specified behavior. There are three categories of behavior mechanisms that produce different behaviors when activated: (1) motor mechanisms that produce a specified action; (2) central mechanisms that produce a specified internal state; and (3) perceptual mechanisms that produce a perception. Behavior mechanisms can be combined into higher order behavior systems, such as reproductive behavior or feeding (Hogan, 1994). A hypothetical behavior system is depicted in Fig. 2.3. Information from the external environment is detected by perceptual mechanisms, such as visual and olfactory stimuli. Output from the perceptual mechanisms is integrated by central mechanisms, perhaps resulting in the recognition of a food item which produces the internal state of hunger. Perceptual mechanisms can also directly influence motor mechanisms, which together with the output of central mechanisms produce an action toward the external environment which constitutes the feeding behavior system (adapted from Hogan, 1994).

Behavior mechanisms, and therefore the behavior systems they comprise, develop over time. Hogan (1994) stated that when studying development, it is first important to describe the changes in the mechanisms and the connections among them, and secondly to investigate the cause of these changes. In the examination of cause, we

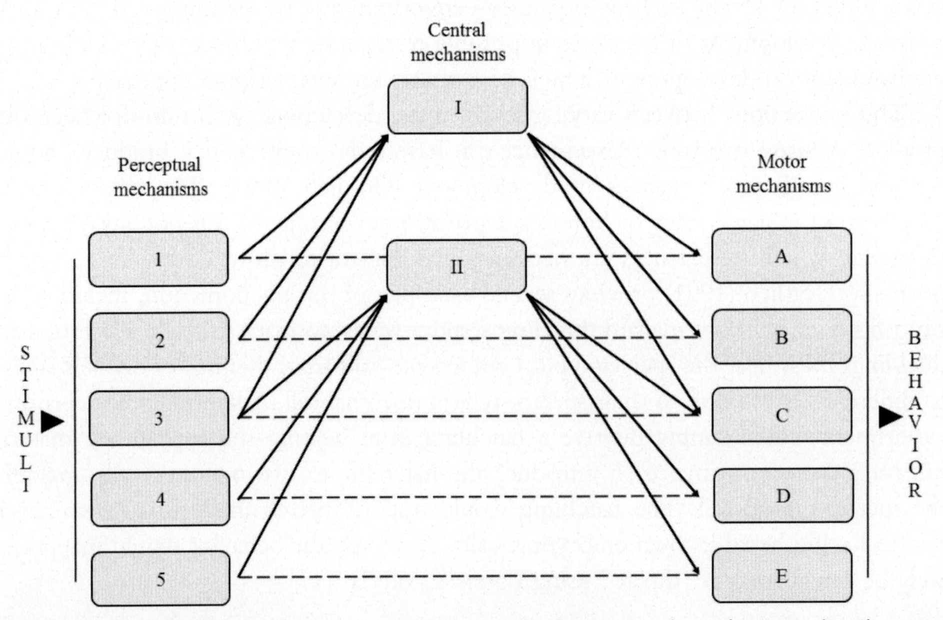

Figure 2.3 A hypothetical behavior system comprised of perceptual mechanisms (1–5), central mechanisms (I and II) and motor mechanisms (A–E). Stimuli from the external environment are processed by perceptual mechanisms, integrated into an internal state by central mechanisms, and an appropriate behavioral response is produced by motor mechanisms. Perceptual mechanisms can also directly influence motor mechanisms, depicted by dashed lines. *Adapted from Hogan, J.A. (1994). Structure and development of behavior systems. Psychonomic Bulletin and Review, 1(4): 439–450. DOI:10.3758/BF03210948.*

unavoidably return to the innate-learned distinction. Hogan circumvented this issue by employing the concept *prefunctional*. A prefunctional behavior mechanism is one that does not require *functional* experience to develop normally, emphasizing that only certain experiences are functional for a given behavior mechanism and that other, nonfunctional or indirect experiences may still be necessary. The various behavior mechanisms develop at different rates and independently of each other, and are later integrated into a coordinated behavior system.

To demonstrate the different behavior mechanisms and the concept of prefunctional, Hogan has described several behavior systems, one of which is dustbathing in fowl (Hogan, 1994). Dustbathing involves a series of movements beginning with pecking and scratching a sandy substrate, and ending with the bird rolling in the substrate and rubbing sand into its feathers. The sand removes excess lipids from the feathers, such that this behavior is important in maintaining the bird's plumage. Both the motor mechanisms and the central mechanisms that control dustbathing develop prefunctionally. Hatchlings can perform all motor elements regardless of the substrate within the first 10 or 12 days of life. During this time, hatchlings will spontaneously produce all the motor sequences involved, and later integrate the movements into the complete behavior. Also, the motivation to dustbathe is intact as soon as hatchlings are able to perform the behavior; as in adults, dustbathing in hatchlings is controlled by the length of time that has elapsed since the previous bout. However, functional experience is necessary for the development of perceptual mechanisms and the connections between perceptual and central mechanisms, which is manifested by the chick's preference to dustbathe in the substrate on which it was raised.

In addition to his theory of behavior mechanisms and systems, Hogan (2015) offered some modifications to the approach of studying behavior recommended by Tinbergen (1963) that are worth mentioning. Borrowing from Aristotle's four causes (material, efficient, formal, and final), Hogan expanded the study of behavior to include all of Tinbergen's questions, as well as his own formulations. The *material cause* of behavior depends on the definition of behavior being used, which for Hogan is "the expression of the activity of the nervous system" (Hogan, 2015); therefore, the material is the neurons and the tissues they act upon. The *efficient cause* can be considered the cause of behavior, and Hogan distinguishes between three categories of causation: (1) motivation causes have immediate effects on behavior and correspond to Tinbergen's mechanism question; (2) development causes have prolonged, irreversible effects on behavior throughout the lifespan and correspond to Tinbergen's ontogenetic question; and (3) phylogenetic causes affect change in behavior across generations, such as through genetic modifications, and correspond to Tinbergen's phylogenetic question. The *formal cause* refers to the structure of behavior, which to Hogan is comprised of behavior mechanisms and systems. Finally, the *final cause* refers to the consequences of behavior and corresponds to Tinbergen's adaptive value.

This section was intended to introduce more contemporary views of behavioral development. Modern conceptualizations of behavioral development rely on the interaction of all levels of the developing system. The developing animal does not reside in a vacuum, isolated from the innumerable external experiences that influence its behavior. Rather, development occurs within the rich context of the environment surrounding the animal and it is unavoidably shaped by countless experiences. The study of behavioral development is undoubtedly complicated by this complexity. However, only by embracing the complexity of the developing system will we be able to gain a full understanding of how behavior develops, what hinders this development, and what can be done to enhance development.

2.3 THE IMPORTANCE OF ATTACHMENT

As was evidenced by the preceding section, the development of behavior depends on numerous external factors that make up the environment of the animal. One experience that has been studied extensively in both humans and animals is attachment, which is the close bond that forms between two individuals, commonly studied between mother and infant. More specifically, Mary Ainsworth, a prominent researcher in human attachment, describes attachment as a discriminatory, affectionate relationship with a person or object (or animal) in which the subject evokes a response from the object that serves over time to strengthen the relationship (Ainsworth, 1964).

2.3.1 Attachment in humans

John Bowlby (1907–90) postulated that attachment behavior is a composite of initially independent instinctual drives that elicit a response from the primary caregiver and over time serve to form a bond between caregiver and infant. He lists five behaviors that are important in forming this bond: sucking, crying, smiling, clinging, and following, and placed special emphasis on the latter two (Bowlby, 1958). Bowlby also advocated that attachment was an active behavior that served the adaptive function of protection for the infant, as opposed to passive dependency of the infant upon the caregiver (Ainsworth & Bowlby, 1991). As such, Bowlby approached the study of attachment from an evolutionary perspective, drawing on research from several ethologists already discussed in this chapter.

Mary Ainsworth (1913–99) is perhaps most well-known for the Strange-Situation, the experimental paradigm she used to understand the attachment relationships between human mothers and their infants (Ainsworth, 1979). In the Strange-Situation, mother and infant are brought into a viewing room filled with many stimulating objects. After a short time, the mother leaves the room, leaving the infant alone (but still under supervision by the experimenters). The infant is allowed to explore the room for a period of time, after which the mother returns. The interactions of interest during the Strange-Situation are how the infant reacts when the mother first leaves and how he or she reacts upon the mother's return (Ainsworth, 1979). Using this paradigm, Ainsworth identified three primary attachment styles between mother and infant: secure, ambivalent, and avoidant. Secure infants use their mother as a base for exploration; they actively explore the objects in the room when the mother is present, show distress and reduced exploration when she is absent, and are comforted by her arrival. Ambivalent infants are anxious even when mother is present, are highly distressed by her absence, and simultaneously seek proximity to and resist the affections of the mother upon her return. Avoidant infants appear unaffected by the absence of the mother and usually completely ignore her when she returns (Ainsworth, 1979).

Ainsworth suggested that attachment relations are determined by the responsiveness and sensitivity of the mother (Ainsworth, 1979). Secure babies have mothers who respond immediately and appropriately to them, allowing them to form the expectation that their mother will respond to them in this way in later situations, resulting in a secure relationship between infant and mother. Insecurely attached infants have mothers who either respond inappropriately (i.e., who do not correctly interpret the needs of the infants), respond long after the infant requests help, or do not respond at all. Therefore, these infants regard their mother as a source of uncertainty and show anxiety toward her (Ainsworth, 1979).

Early attachment style is a predictor of later behavioral outcomes. Children that were securely attached as infants have superior social skills than their insecure counterparts. They are more cooperative, interested in play, and have a more positive affect. Their problem-solving ability is more advanced and they show enhanced language development. They are also more confident and show indicators of higher self-esteem. Insecurely attached infants grow into children who are generally more aggressive and noncompliant in cases of avoidant attachment, or irritable, incompetent, and less persistent in cases of ambivalent attachment (Ainsworth, 1979).

The situation that occurred in Romanian orphanages is useful to demonstrate the paramount importance of early attachments. Over-populated orphanages provided extremely impoverished environments for young children, who were under-stimulated and barely handled by caretakers. Children spent the vast majority of their time in cribs staring in silence at blank walls. Children were responsible for feeding themselves as soon as they were able, in some cases as young as one and a half years old, using bottles that were held or attached to the crib (Fisher, Ames, Chisholm, & Savoie, 1997).

Fisher and colleagues (1997) analyzed behavioral problems in a sample of Romanian orphans adopted into Canadian families. All orphans had spent a minimum of eight months in the orphanage and their age at adoption was strongly correlated with the length of their institutionalization ($r = 0.97$). The Romanian orphans were compared to two groups: (1) a matched sample of Canadian-born, nonadopted children and (2) a matched sample of adopted Romanian children that would have been institutionalized if they had not been adopted prior to four months of age. Parents of Romanian orphans reported the highest incidence of eating problems (e.g., over-eating), medical problems (e.g., Hepatitis B, intestinal parasites), sleep problems (e.g., lying silently upon waking up), stereotyped behaviors (e.g., rocking), and interpersonal problems (e.g., problems with sibling and/or peer interactions) compared to at least one of the comparison groups. Romanian orphans were not different from Romanian comparison children with respect to interpersonal problems. The two comparison groups had similar incidences across all behavioral problems, suggesting that it was the orphanages, and not the general hardships facing Romania, that caused the increased incidence of problems in the orphans. Fortunately, for all areas of behavioral problems, parents of Romanian orphans did report improvements compared to when their child first arrived in Canada.

Rutter and the English and Romanian Adoptees (ERA) Study Team (1998) found severe physical and cognitive impairments among children adopted from Romanian orphanages. Romanian orphans adopted into UK families had significantly lower weight, height, and head circumference compared to UK-born adoptees upon entry into the UK Romanian orphans also performed significantly worse on measures of cognitive development (Denver Quotient), with over half of the children performing within the retarded range. Fortunately, developmental catch-up was observed for these children. By the age of 4 years, Romanian orphans who had been in institution less than 6 months did not differ from UK-born adoptees or noninstitutionalized Romanian adoptees with respect to cognitive functioning, demonstrating remarkable improvements. Romanian orphans who had been in institution longer than 6 months also improved drastically, although they remained at a lower level of functioning compared to UK-born adoptees. The strongest predictor of later functioning was age at adoption, indicating that the magnitude of the deficits was related to the time spent in the deprived environment.

Romanian orphans also tend to exhibit attachment disorders with higher frequency. O'Conner, Rutter, and the ERA Study Team (2000) compared Romanian orphans who had been in institution for 0–6 months, 6–24 months, or 24–48 months with UK-born adopted children. The incidence of severe attachment disorder increased with the duration of institutionalization. When assessed at 6-years old, approximately one third of Romanian orphans who had been institutionalized for 24–48 months displayed severe attachment disorder, compared to roughly 20% of those who were in institutions for 6–24 months. Romanian orphans who spent only 0–6 months in institutions did not differ from UK-born adopted children. Severe attachment disorders include behaviors such as undifferentiating between adults, no apprehension toward strangers, lack of consultation with parent in potentially dangerous situations, and lack of comfort seeking from parents when distressed.

Infants who are unable to form early attachments, primarily with the mother, or who experience unstable and unpredictable attachments display a range of impairments later in life. Physical contact between mother and infant is believed to be one of the key factors in forming a secure early attachment. In the late 1970s, Dr. Edgar Rey Sanabria, a pediatrician in Columbia, recommended "Kangaroo Mother Care" (KMC) as an alternative for over-crowded hospitals (e.g., Charpak et al., 2005; Tessier et al., 1998). KMC consists of placing the naked, unwashed newborn directly between the breasts of the mother immediately following birth, allowing the infant to smell the mother and feel her heartbeat and warmth. Initially suggested as a method to decrease mortality in low birth-weight infants, KMC is now being recommended to replace typical infant handling in Western hospitals, in which the infant is taken away from the mother immediately following birth to clean and weigh (Vetulani, 2013). Given our current understanding of the importance of skin-to-skin contact in forming early attachments, the movement toward KMC is encouraging.

2.3.2 Attachment in animals

In nonhuman animals, the simplest form of attachment is imprinting. Lorenz first used the term to refer to the following behavior exhibited by hatchlings and determined that there are critical periods in which imprinting can occur (Hess, 1959). Eckhard Hess (1916–86) has since completed a body of research on imprinting, primarily in birds (see Hess, 1959 for a review). Hess helped to solidify the concept of a critical period for imprinting, demonstrating experimentally that chicks imprint most strongly between 13 and 16 hours post hatching. The onset of the critical period is determined by locomotor competency and the offset by the development of fear, since fearful chicks, like other young animals, will avoid rather than follow. Chicks also demonstrate color and shape preferences; hatchlings imprinted more strongly to blue spheres compared to objects of other colors or shapes that more closely modeled an adult conspecific. A further finding from Hess's experiments is that the strength of imprinting obeys a law of effort in that it is a function of the effort expended by the chick to follow the object upon which it imprinted. Imprinting is stronger when chicks must follow for longer distances, overcome hurdles, or follow up an inclined plane.

Harry Harlow (1905–81) has helped to elucidate the factors important in forming attachment bonds in primates. His studies with infant macaque monkeys have built on theoretical work by researchers such as Bowlby and have discredited some earlier theories of attachment, such that infants bond to mothers for the sole purpose of fulfilling physiological needs such as hunger. In his classic experiment, Harlow (Harlow & Zimmerman, 1958) separated newborn rhesus macaques from their mother and raised them on surrogate mothers. The monkeys were given the choice of two surrogate mothers, one that was constructed of wire mesh, and the other constructed of a wood block covered in sponge rubber and terry cloth. Half of the infant monkeys could receive food from the wire mother and the other half from the cloth mother. Regardless of which mother provided sustenance, the infant monkeys showed a strong preference for the cloth mother, suggesting that comfort takes precedence over nourishment when determining attachment bonds.

The importance of forming early attachments cannot be overstated in the development of later behavior. Harlow completed a series of experiments that highlight the importance of mother-infant interaction as well as peer—peer interactions in the normal development of behavior. In the absence of the mother, infant monkeys will cling to each other and will stay in this position if left uninterrupted. If monkeys are raised with no maternal figure but have access to peers, the development of prosocial interactions is greatly delayed and clinging is exaggerated, although it does lessen over time. However, this peer-rearing situation does not appear to have any negative consequences on sexual behavior; sexual activities begin sooner than in mother-raised infants and maternal behavior appears normal (Harlow & Harlow, 1966).

Infant macaques raised in social isolation, either partial or complete, do not experience the buffering effects of peer-rearing. Partial social isolation refers to infants that are housed in individual cages adjacent to other isolated monkeys. There is no contact, but the monkeys can see and hear one another. Complete social isolation occurs when an individual monkey is caged and does not hear or see any other monkeys or humans during the isolation. After a relatively short period of social isolation, the monkeys experience "emotional shock" (Harlow & Harlow, 1966), characterized by self-clutching, but can be effectively resocialized. However, if the isolation is prolonged, rehabilitation is unlikely. Following 6 months of total social isolation from birth, infants are highly stunted in their ability to form social bonds with other monkeys. Within the 6 to 7 months of social rehabilitation that these monkeys experienced, they were only able to marginally improve their interactions with other previously isolated monkeys, and showed no improvement interacting with normal monkeys (Harlow & Harlow, 1966). Infant monkeys that experienced total social isolation for 12 months were later unable to engage in even the simplest play with peers and became targets for abuse by normal monkeys (Harlow & Harlow, 1966).

Complete social isolation for 6 to 12 months also has long-term consequences. During preadolescence/adolescence, previously isolated monkeys were pair housed with either a normal adult, a normal peer, or a normal juvenile. In all cases, the isolated monkeys demonstrated fear of their cage-mate. Interestingly, the 6-month isolated monkeys also displayed atypical aggression toward the juveniles and the adults, the latter resulting in brutal retaliation. The 12-month isolated monkeys appeared too afraid of their cage-mate for any form of outward aggression (Harlow & Harlow, 1966).

The legacy of Harlow's studies continues under the direction of one of his last students, Stephen Suomi (e.g., Dettmer & Suomi, 2014; Suomi, 2002). Taken together, the Suomi studies reveal that monkeys with some sort of maternal deprivation exhibit higher than normal anxiety and behavioral inhibition than normally reared monkeys. They are also at high risk for developing high levels of alcohol consumption and showing high levels of aggression toward conspecifics. These behavioral abnormalities are associated with differential gene methylation patterns in early adulthood in the dorsolateral prefrontal cortex as well as chronic abnormalities in the HPA axis activity (Dettmer et al., 2017; Provencal et al., 2012).

More recent studies have continued to demonstrate the detrimental effects of early social isolation. Rat pups that are removed from their mother and siblings for 6 hours/day from postnatal days 7 to 11 (P7—P11) demonstrate decreased long-term potentiation (LTP), a mechanism involved in learning (Miyazaki et al., 2012). LTP depends on the transport of AMPA glutamate receptors into the synaptic membrane. Chronic early stress interrupts the delivery of AMPA receptors, likely mediated by increased levels of corticosterone. The disruption of AMPA receptor synaptic delivery in somatosensory cortex, namely the barrel cortex which serves as the sensory map of the vibrissae, interferes with circuitry formation and indirectly affects how the pup perceives and interacts with its environment, potentially resulting in abnormal behavioral development (Miyazaki et al., 2012). These effects were long-term and continued into the juvenile period (P27).

In a review of rodent literature on maternal separation, Vetulani (2013) summarizes the long-term effects of repeated acute (15-min/day) or prolonged (3-h/day) maternal separation. Acute maternal separation in rodents appears to have a "hardening" effect that diminishes responsiveness to stress in adulthood. However, prolonged maternal separation has the opposite effect. Pups become hyper-responsive to stress and display anxiety- and depressive-like behaviors as adults, as well as increased ethanol intake. When dams are returned to their litters following prolonged separation, they are inattentive and their pups appear distressed.

Early attachment between caregiver and infant and other forms of early socialization are critical in the normal development of behavior in both humans and animals. It is apparent that developing animals require experience with others of their species to become competent adults. Secure attachment between caregiver and infant provides the infant with a secure base from which to explore, providing the infant with opportunities for growth and learning. Infants who are able to manipulate their environment become more competent adults than those not given this opportunity due to deprived rearing, or those who are too fearful of their environment. Although attachment is just one factor that influences later behavioral development, the magnitude of its impact rendered it a valuable topic to include in this chapter.

2.4 SUMMARY

This chapter began with a historical overview of the study of behavioral development. The innate-learned dichotomy and Lorenz's theory of instinct postulated that all animal behavior can be categorized as either arising due to genetic programing or learning. Instinctually innate behaviors require no prior experience in order to be executed, whereas learned behaviors develop due to experience during the lifespan of the animal. Over time, modifications were made to these theories to allow for innate behaviors to be shaped by external factors as well. Learning theorists focused on the mechanisms through which behavior is shaped by the environment. Two major categories of learning within Psychology were classical and operant conditioning. More recent advances in the study of behavioral development have focused on the interaction between genes and the environment, and epigenetics research has elucidated much about how various experiences during develop shape behavior by altering gene expression. It is now widely accepted that behavioral development occurs due to events ranging from the cellular level to the external environment of the animal. The final section of this chapter looked at the importance of attachment in the development of behavior, using human and animal studies to demonstrate the consequences of not forming attachment relationships on later behavior.

REFERENCES

Abel, E. L. (2004). Paternal contribution to fetal alcohol syndrome. *Addiction Biology, 9*, 127–133. Available from http://dx.doi.org/10.1080/13556210410001716980.

Abel, T., & Zukin, R. S. (2008). Epigenetic targets of HDAC inhibition in neurodegenerative and psychiatric disorders. *Current Opinion in Pharmacology, 8*, 57–64. Available from http://dx.doi.org/10.1016/j.coph.2007.12.002.

Ainsworth, M. D. (1964). Patterns of attachment behavior shown by the infant in interaction with his mother. *Merrill-Palmer Quarterly of Behavior and Development, 10*(1), 51–58.

Ainsworth, M. D. S. (1979). Infant–mother attachment. *American Psychologist, 34*(10), 932–937.

Ainsworth, M. D. S., & Bowlby, J. (1991). An ethological approach to personality development. *American Psychologist, 46*(4), 333–341. Available from http://dx.doi.org/10.1037/0003-066X.46.4.333.

Bowlby, J. (1958). The nature of the child's tie to his mother. *International Journal of Psycho-Analysis, 39*, 350–373.

Brigandt, I. (2005). The instinct concept of the early Konrad Lorenz. *Journal of the History of Biology, 38*, 571–608. Available from http://dx.doi.org/10.1007/s10739-005-6544-3.

Burkhardt, R. W. (2014). Tribute to Tinbergen: Putting Niko Tinbergen's 'four questions' in historical context. *Ethology, 120*, 215–223. Available from http://dx.doi.org/10.1111/eth.12200.

Charpak, N., Ruiz, J. G., Zupan, J., Cattaneo, A., Figueroa, Z., ... Worku, B. (2005). Kangaroo mother care: 25 years after. *Acta Paediatrica, 94*, 514–522. Available from http://dx.doi.org/10.1080/08035250515527381.

Chen, Y., Ozturk, N. C., & Zhou, F. C. (2013). DNA methylation program in developing hippocampus and its alteration by alcohol. *PLoS ONE, 8*(3), e60503. Available from http://dx.doi.org/10.1371/journal.pone.0060503.

Cooper, R. M., & Zubek, J. P. (1958). Effects of enriched and restricted early environments on the learning ability of bright and dull rats. *Canadian Journal of Psychology, 12*(3), 159–164. Available from http://dx.doi.org/10.1037/h0083747.

Dettmer, A. M., & Suomi, S. J. (2014). Nonhuman primate models of neuropsychiatric disorders: Influences of early rearing, genetics, and epigenetics. *ILAR Journal, 55*(2), 361–370. Available from http://dx.doi.org/10.1093/ilar/ilu025.

Dettmer, A. M., Wooddell, L. J., Rosenberg, K. L., Kaburu, S. S., Novak, M. A., ... Suomi, S. J. (2017). Associations between early life experience, chronic HPA axis activity, and adult social rank in rhesus monkeys. *Social Neuroscience, 12*(1), 92–101. Available from http://dx.doi.org/10.1080/17470919.2016.1176952.

Fisher, L., Ames, E. W., Chisholm, K., & Savoie, L. (1997). Problems reported by parents of Romanian orphans adopted to British Columbia. *International Journal of Behavioral Development, 20*(1), 67–82. Available from http://dx.doi.org/10.1080/016502597385441.

Gottlieb, G. (1991). Experiential canalization of behavioral development: Theory. *Developmental Psychology, 27*(1), 4–13. Available from http://dx.doi.org/10.1037/0012-1649.27.1.4.

Gottlieb, G. (2007). Probabilistic epigenesis. *Developmental Science, 10*(1), 1–11. Available from http://dx.doi.org/10.1111/j.1467-7687.2007.00556.x.

Guo, W., Crossey, E. L., Zhang, L., Zucca, S., George, O. L., & Zhao, X. (2011). Alcohol exposure deceases CREB binding protein expression and histone acetylation in the developing cerebellum. *PLoS ONE, 6*(5), e19351. Available from http://dx.doi.org/10.1371/journal.pone.0019351.

Harlow, H. F., & Harlow, M. (1966). Learning to love. *American Scientist, 54*(3), 244–272.

Harlow, H. F., & Zimmerman, R. R. (1958). The development of affectional responses in infant monkeys. *Proceedings from the American Philosophical Society, 102*(5), 501–509.

Hess, E. H. (1959). Imprinting. *Science, 130*(3368), 133–141. Available from http://dx.doi.org/10.1126/science.130.3377.733.

Hogan, J. A. (1994). Structure and development of behavior systems. *Psychonomic Bulletin and Review, 1* (4), 439–450. Available from http://dx.doi.org/10.3758/BF03210948.

Hogan, J. A. (2015). A framework for the study of behavior. *Behavioural Processes, 117*, 105–113. Available from http://dx.doi.org/10.1016/j.beproc.2014.05.003.

Huang, Y., Shi, X., Xu, H., Yang, H., Chen, T., & Chen, X. (2010). Chronic unpredictable stress before pregnancy reduce the expression of brain-derived neurotrophic factor and *N*-methyl-D-aspartate receptor in hippocampus of offspring rats associated with impairment of memory. *Neurochemical Research, 35*, 1038–1049. Available from http://dx.doi.org/10.1007/s11064-010-0152-0.

Johnston, T. D. (1987). The persistence of dichotomies in the study of behavioral development. *Developmental Review, 7*, 149–172. Available from: http://dx.doi.org/10.1016.0273-2297(87)90011-6.

Johnston, T. D., & Edwards, L. (2002). Genes, interactions, and the development of behavior. *Psychological Review, 109*(1), 26–34. Available from http://dx.doi.org/10.1037//0033-295X.109.1.26.

Kalikow, T. J. (1976). Konrad Lorenz's ethological theory, 1939–1943: 'Explanations' of human thinking, feeling, and behaviour. *Philosophy of the Social Sciences, 6*(1), 15–34. Available from http://dx.doi.org/10.1177/004839317600600102.

Keverne, E. B., Pfaff, D. W., & Tabansky, I. (2015). Epigenetic changes in the developing brain: Effects on behavior. *PNAS, 112*(22), 6789–6795. Available from http://dx.doi.org/10.1073/pnas.1501482112.

Kobor, M. S., & Weinberg, J. (2011). Focus on: Epigenetics and fetal alcohol spectrum disorders. *Alcohol Research and Health, 34*(1), 29–37.

Kuo, Z. Y. (1932). Ontogeny of embryonic behavior in Aves IV: The influence of embryonic movements upon the behavior after hatching. *Journal of Comparative Psychology, 14*(1), 109–122. Available from http://dx.doi.org/10.1037/h0071451.

Lehrman, D. S. (1953). A critique of Konrad Lorenz's theory of instinctive behavior. *The Quarterly Review of Biology, 28*(4), 337–363. Available from http://dx.doi.org/10.1086/399858.

Miyazaki, T., Takase, K., Nakajima, W., Tada, H., Ohya, D., ... Takahashi, T. (2012). Disrupted cortical function underlies behavior dysfunction due to social isolation. *The Journal of Clinical Investigation, 122*(7), 2690–2701. Available from http://dx.doi.org/10.1172/JCI63060.

Mychasiuk, R., Harker, A., Ilnytskyy, S., & Gibb, R. (2013). Paternal stress prior to conception alters DNA methylation and behaviour of developing rat offspring. *Neuroscience, 241*, 100–105. Available from http://dx.doi.org/10.1016/j.neuroscience.2013.03.025.

Mychasiuk, R., Ilnytskyy, S., Kovalchuk, O., Kolb, B., & Gibb, R. (2011). Intensity matters: Brain, behaviour and the epigenome of prenatally stressed rats. *Neuroscience, 180*, 105–110. Available from http://dx.doi.org/10.1016/j.neuroscience.2011.02.026.

Mychasiuk, R., Zahir, S., Schmold, N., Ilnytskyy, S., Kovalchuk, O., & Gibb, R. (2012). Parental enrichment and offspring development: Modifications to brain, behaviour, and the epigenome. *Behavioural Brain Research*, *228*, 294–298. Available from http://dx.doi.org/10.1016/j.bbr.2011.11.036.

O'Connor, T. G., Rutter, M., & The English and Romanian Adoptees Study Team (2000). Attachment disorder behavior following early severe deprivation: Extension and longitudinal follow-up. *Journal of the American Academy of Child and Adolescent Psychiatry*, *39*(6), 703–712. Available from http://dx.doi.org/10.1097/00004583-200006000-00008.

O'Leary, C. M. (2004). Fetal alcohol syndrome: Diagnosis, epidemiology, and developmental outcomes. *Journal of Pediatrics and Child Health*, *40*, 2–7. Available from http://dx.doi.org/10.1111/j.1440-1754.2004.00280.x.

Provencal, N., Suderman, M. J., Cuillemin, C., Massart, R., Ruggiero, A., ... Szyf, M. (2012). The signature of maternal rearing in the methylome in rhesus macaque prefrontal cortex and T cells. *Journal of Neuroscience*, *32*, 15626–15642.

Richards, R. J. (1974). The innate and the learned: The evolution of Konrad Lorenz's theory of instinct. *Philosophy of the Social Sciences*, *4*, 111–133. Available from http://dx.doi.org/10.1177/004839317400400102.

Roth, T. L. (2013). Epigenetic mechanisms in the development of behavior: Advances, challenges, and future promises of a new field. *Development and Psychopathology*, *25*, 1279–1291. Available from http://dx.doi.org/10.1017/S0954579413000618.

Rutter, M., & The English and Romanian Adoptees Study Team (1998). Developmental catch-up, and deficit, following adoption after severe global early privation. *Journal of Child Psychology and Psychiatry*, *39*(4), 465–476. Available from http://dx.doi.org/10.1111/1469-7610.00343.

Skinner, B. F. (1948). Superstition in the pigeon. *Journal of Experimental Psychology*, *38*, 168–172. Available from http://dx.doi.org/10.1037//0096-3445.121.3.273.

Skinner, B. F. (1991). *The Behavior of Organisms*. Cambridge: B. F: Skinner Foundation, (Original work published 1938).

Suomi, S. J. (2002). How gene–environment interactions can shape the development of socioemotional regulation in rhesus monkeys (pp. 5–26). In B. S. Zuckerman, A. F. Lieberman, & N. A. Fox (Eds.), *Emotional Regulation and Developmental Health: Infancy and Early Childhood*. New York: Johnson & Johnson Pediatric Institute.

Tessier, R., Cristo, M., Velez, S., Girón, M., Ruiz-Paláez, J. G., & Charpak, N. (1998). Kangaroo mother care and the bonding hypothesis. *Pediatrics*, *102*, e17. Available from http://dx.doi.org/10.1542/peds.102.2.e17.

Tinbergen, N. (1963). On aims and methods of ethology. *Zeitschrift für Tierpsychologie*, *20*, 410–433.

Ungerer, M., Knezovich, J., & Ramsay, M. (2013). In utero alcohol exposure, epigenetic changes, and their consequences. *Alcohol Research: Current Reviews*, *35*(1), 37–46.

Urdinguio, R. G., Sanchez-Mut, J. V., & Esteller, M. (2009). Epigenetic mechanisms in neurological diseases: Genes, syndromes, and therapies. *The Lancet Neurology*, *8*(11), 1056–1072. Available from http://dx.doi.org/10.1016/S1474-4422(09)70262-5.

Vetulani, J. (2013). Early maternal separation: A rodent model of depression and a prevailing human condition. *Pharmacological Reports*, *65*(6), 1451–1461. Available from http://dx.doi.org/10.1016/S1734-1140(13)71505-6.

Watson, J. B., & Rayner, R. (1920). Conditioned emotional reactions. *Journal of Experimental Psychology*, *3*(1), 1–14. Available from http://dx.doi.org/10.1037/h0069608.

Zink, M., Ferbert, T., Frank, S. T., Seufert, P., Gebicke-Haerter, P. J., & Spanagel, R. (2011). Perinatal exposure to alcohol disturbs spatial learning and glutamate transmission-related gene expression in the adult hippocampus. *European Journal of Neuroscience*, *34*, 457–468. Available from http://dx.doi.org/10.1111/j.1460-9568.2011.07776.x.

FURTHER READING

Huang, Y., Xu, H., Li, H., Yang, H., Chen, Y., & Shi, X. (2012). Pre-gestational stress reduces the ratio of 5-HIAA to 5-HT and the expression of 5-HT1A receptor and serotonin transporter in the brain of foetal rat. *BMC Neuroscience*, *13*, 22.

CHAPTER 3

Overview of Factors Influencing Brain Development

Bryan Kolb
University of Lethbridge, Lethbridge, AB, Canada

3.1 INTRODUCTION

We saw in Chapter 1, Overview of Brain Development that brain development is a prolonged process beginning in utero and continuing in humans until the end of the third decade. Brain development is guided not only by a basic genetic blueprint but also is shaped by a wide range of experiences, ranging from sensory stimuli to social relationships to stress, throughout the lifetime (Table 3.1). Brain development in humans can be measured directly by neuroimaging (structural MRIs, functional MRIs), electrophysiology, and behavior. Behavioral measures include neuropsychological measures of cognitive functioning as well as measures of motor, perceptual, and social functioning. The behavioral measures present a challenge because tests need to be minimally culturally biased and age-appropriate to allow generalizations in a global context. One difficulty is that behavioral tests are difficult to administer to young children yet we know that experience sets children on trajectories that can be seen by 2 years of age (e.g., Hart & Risley, 1995), meaning that there is a need to intervene very early to set more optimistic trajectories for children at risk.

An important challenge is to identify mechanisms that may underlay modifications in brain development and behavior. Ultimately, the mechanisms will be molecular but there is a significant gap in our understanding of how molecular changes affect behavior, presumably via changes in neural networks and/or neural activity. Nonetheless, the emergence of epigenetics is providing evidence that pre- and postnatal, and even preconception, experiences modify gene expression, both developmentally and later in life.

The relationship between molecular or cellular changes, neural networks, and behavior is by no means clear and is plagued by the problems inherent in inferring causation from correlation. By its very nature, behavioral neuroscience searches for epigenetic and neural correlates of behavior. Some of the molecular and neural changes are most certainly directly associated with behavior, but others are more ambiguous. Consider an example. When a person learns to play tennis there is an

The Neurobiology of Brain and Behavioral Development
DOI: http://dx.doi.org/10.1016/B978-0-12-804036-2.00003-0

Table 3.1 Summary of factors affecting brain development

Factors
Environmental factors
Sensory and motor experience
Language and cognitive experience
Music
Stress
Psychoactive drugs
Parent–child and peer relationships
Diet
Poverty
Brain injury
Internal factors
Microbiome
Immune system

obvious change in the ability to make smooth and accurate movements that become so fast that a novice player would see them as being impossible. But what is the relationship between the molecular, behavioral, and cerebral changes? We can conclude that the tennis training directly caused the behavioral change, but it is less clear how molecular changes altered the brain or how the neural changes relate to the behavior.

The improved behavior may have preceded some of the changes in the brain or perhaps induced molecular changes. Thus, a common criticism of studies trying to link experience, neuronal and molecular changes, and behavior is that "they are only correlates." This is true but it is hardly a reason to dismiss such studies. Ultimately, the proof would be in showing how the neural changes arose, which would presumably involve molecular analysis such as a change in gene transcription. For many studies using humans this would be an extremely difficult challenge and often impractical. It is our view that once we understand the "rules" that govern how different factors lead to changes in brain structure and function, we will be better able to look for molecular changes. A certain level of ambiguity in the degree of causation is perfectly justifiable at this stage of our knowledge. Understanding the precise mechanism whereby the synaptic changes might occur is not necessary to proceed with further studies aimed at improving functional outcomes in children.

The goal of this chapter is to consider the manner in which a wide range of factors influence both brain and behavioral development as well as how the brain responds to other experiences later in life. We begin with a brief discussion of epigenetics and how changes in gene expression can modify brain development in order to provide a general framework for understanding how experience can translate into brain and behavioral changes.

3.2 EPIGENETICS AND BRAIN DEVELOPMENT

The genes expressed within a cell are influenced by factors inside the cell and in the cell's environment. Cells within the body find themselves in different environments (e.g., bone, heart, brain) and this environment will determine which genes are expressed and what kind of tissue it becomes, including what type of nervous system cell. But environmental influence on cells does not end at birth because our environment changes daily throughout our lives, providing opportunity to influence our gene expression (for a more extensive discussion, see Chapter 7, Epigenetics and Genetics of Brain Development).

Epigenetics can be viewed as a second genetic code, the first one being the genome, which is an organism's complete set of DNA. Epigenetics refers to the changes in gene expression that do not involve change to the DNA sequence but rather the processes whereby enzymes read the genes within the cells. Thus, epigenetics describes how a single genome can code for many phenotypes, depending upon the internal and external environments.

Epigenetic mechanisms can influence protein production either by blocking a gene to prevent transcription or by unlocking a gene so that it can be transcribed (see Fig. 3.1). Chromosomes are comprised of DNA wrapped around supporting molecules

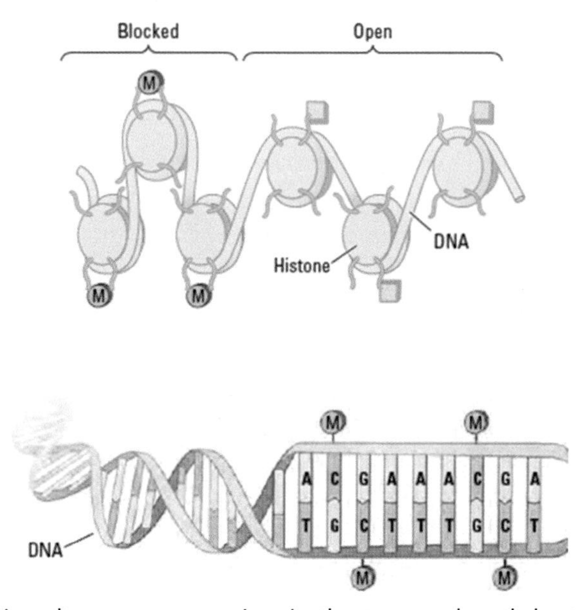

Figure 3.1 Methylation alters gene expression. In the top panel methyl groups (CH_3) or other molecules bind to tails of histones, either blocking them from opening (orange circles) or allowing them to open for transcription (green squares. In the bottom panel, methyl groups bind to CG base pairs to block transcription (after Kolb, Whishaw, & Teskey, 2016).

of a protein called histone, which allows many meters of DNA to be packaged in a small space. For any gene to be transcribed into messenger (mRNA), its DNA must be unspooled from the histones. Thus, one measure of changes in gene expression is histone modification through methylation, phosphorylation, ubiquinylation, or acetylation. Methyl groups (CH_3) or other molecules bind to the histones, either blocking them from opening or allowing them to unravel. Methylation can also occur in CpG islands on the DNA that are usually located near promoter sites for genes (see Fig. 3.1). (CpG refers to cytosine and guanine that appear in a row on a strand of DNA. The "p" refers to the phosphodiester bond between them.) When methylation occurs at CpG sites gene transcription cannot proceed. In mammals, about 60%–90% of CpG sites are methylated. Changes in the amount of methylation in a tissue thus can give an indirect estimate of the amount of gene expression. As a rule of thumb more methylation means fewer genes expressed whereas reduced methylation means more genes are expressed.

Epigenetic mechanisms act to influence protein production in a cell either by blocking a gene, and thus preventing transcription, or by unlocking a gene to allow transcription. This is where experiential and environmental influences come into play to influence brain and behavior. Consider an example. There is a growing literature showing that maternal smoking during pregnancy is associated with adverse effects on neurodevelopment and abnormal cognitive development in the offspring but it has been difficult to identify a direct link until recently. Several studies now have examined the epigenetic impact of a pregnant mother's smoking on her child. For example, Joubert, Feliz, Yousefi, and Bakulski (2016) studied over 6600 mothers and their newborns, comparing which genes were methylated. The methylation pattern in the infants was similar to what is seen in adult smokers, even though the infant was never exposed directly to cigarette smoke. The newborns of smoking mothers had a methylation pattern that differed from newborns of nonsmoking mothers at about 6000 locations, about half of which were associated with a particular gene. Many of these genes are related to nervous system development.

Holz, Boecker, Baumeister, and Hohm (2014) examined the long-term effect of gestational exposure to smoking by measuring fMRI (functional Magnetic Resonance Imaging) activity during the performance of a test of inhibitory control in young adult participants. They found that there was reduced volume and activity in the anterior cingulate gyrus, inferior frontal gyrus (IFG), and the supramarginal gyrus, patterns often associated with attention-deficit hyperactivity disorder (ADHD) symptoms, in a dose-dependent manner. Other studies in laboratory animals have found decreased brain-derived neurotrophic factor mRNA and protein, which is also associated with neurodevelopmental disorders (e.g., Yochum et al., 2014).

But what about smoking grandmothers? Golding, Northtone, Gregory, Miller, and Pembrey (2014) have studied a large group of over 20,000 children in whom there are data regarding smoking in grandmothers and mothers. In one study, they found that if paternal grandmothers had smoked in pregnancy, but their mother had not, the girls were taller and both genders had greater bone and lean mass at 17 years of age. In contrast, if the maternal grandmother had smoked prenatally but the mother had not, the boys were heavier than expected due to lean rather than fat mass, which was reflected in increased strength and fitness. But when both grandmothers had smoked the girls had reduced height and weight when compared to girls whose grandmothers had not smoked.

Taken together, these studies provide a link between gestational experience (exposure to smoking), behavioral and brain changes, and changes in gene expression, thus providing a putative mechanism for factors to alter brain and behavioral development. A similar logic can be applied to the other factors that we will now consider.

3.3 SENSORY EXPERIENCES AFFECTING BRAIN DEVELOPMENT

Animals are born into and develop in varied environments. After birth, animals immediately begin to adapt to the more specific conditions of a particular environment. As development proceeds each individual begins to lose degrees of freedom for behavioral and biological adaptations but become adapted to the limited conditions of a given environmental niche (Riesen, 1975).

It is self-evident that infants and young children are like sponges and soak up the stimulation in their environment to learn to recognize faces, understand language, walk, and so on. Children differ in how much they learn, however, in part because their environments are so different across cultures, socioeconomic (SES) status, diet, etc. It has been clear from at least the 1940s (e.g., Hebb, 1949) that impoverished environments have deleterious effects on brain and behavioral development and the earliest laboratory studies on experience and brain development focused on deprivation. We begin with deprivation studies before moving to other types of experiences. Many of the other experiences are discussed in more detail in the following chapters so in those cases the discussion will be relative brief.

3.3.1 Effects of sensory deprivation on brain development

Many of the early demonstrations of the role of sensory input on brain development came from studies in which afferent input was reduced or blocked early in development. One of the earliest studies was Levi-Montalcini's (1949) demonstration that surgically cutting the afferent fibers to the "acoustic-vestibular centers" of the chicken embryo blocked the development of the neurons. Parallel studies in mammals showed that interfering with inputs to the cerebellar Purkinje cells in early development has similar effects (for a review see Berry, McConnell, & Sievers, 1980). But what is more important is evidence that reduced sensory input has similar effects. For example, there were many studies in the 1950s and 1960s examining the effects of raising chimpanzees, cats, rabbits, rats, and mice in the dark on the organization of the retina, thalamus, and cortex (e.g., Coleman & Riesen, 1968). In general, although the effects appear to be larger in the retina and subcortical regions than in the cortex, there is still evidence that the intact nervous system may fail to develop in a normal fashion as a consequence of disuse or decreased sensory input.

Although laboratory animal studies have shown significant deficits in visual processing, the most spectacular evidence came from the studies of von Sendon (1932) who studied the vision in people with congenital cataracts removed at about 40 years of age. None of his participants were able to achieve anything like normal vision. More recently, Daphne Maurer has done a long-prospective study of babies with cataract removal at different ages (see Chapter 8: Visual Systems). Her results have shown that even with only a few months or years of abnormal vision results in a wide range of chronic deficits in visual processing. Cataract removal in infants is now done as soon as possible to avoid these visual disturbances.

Laboratory animal studies have shown that like the children with cataracts, monkeys, cats, or rats raised in the light but without patterned visual stimulation show poor visual guidance of behavior (for a review see Riesen & Zilbert, 1975). Furthermore, cats raised with goggles with either vertical or horizontal lines are only able to perceive lines of the orientation they were exposed to (e.g., Tieman & Hirssch, 1982).

Although visual deprivation is the most studied of the senses, several laboratory studies have shown that degradation of auditory perception result from auditory deprivation. For example, Tees (1967) used earplugs and a sound attenuation chamber to reduce auditory input during development. When the animals were allowed exposure to sound after 60 days of deprivation they were chronically impaired in learning sound patterns or tonal sequences, although they could respond to a change of pitch of one full octave.

The reduction in auditory input does not have to be extreme to have obvious effects, the best example being language experience. The number of sound segments in the languages of the world is very large but any particular language uses only a subset to phonetically differentiate meaning in speech. At birth, infants are able to discriminate virtually all speech sounds but over the first year they lose this ability because normally they are only exposed to a subset of these sounds (e.g., Werker & Tees, 1984). Thus, in the absence of specific auditory inputs, the ability to perceive the information is degraded (see Chapter 10: Language and Cognition).

Studies of tactile/somatosensory deprivation in the early weeks of development have shown that early experience is critical for normal sensorimotor development. Nissen, Chow, and Semmes (1951) raised a male chimpanzee with cardboard mailing tubes over all four extremities from 1 to 31 mo. Although the chimp could later walk, initially it could not grasp with its digits, climb, self-groom, or sit normally. Even with an extended recovery period the chimp never groomed nor sat as normal chimps would. Thompson and Melzack (1956) reared Scotch terriers in sensory isolation from birth, protected from any painful experiences by padding. One surprising finding was that the dogs had no reaction to painful stimuli and failed to learn to avoid painful stimuli. Early experience with pain appears necessary for the normal development of the nociceptive system. The Scotties had other peculiarities as well including the development of stereotyped behaviors such as tail chasing and they became highly agitated when exposed to novelty (Melzack, 1968). (For additional examples, see Riesen & Zilbert, 1975.)

Harlow and his colleagues and students systematically studied the effects of social deprivation in rhesus monkeys beginning about 1955 (e.g., Harlow & Harlow, 1965; Harlow, 1959). In the first studies baby monkeys were provided with two wire "mother" monkeys, one of which was covered with soft terrycloth but provided no food whereas the other, plain wire, mother provided food from an attached baby bottle. If the role of mothers was to provide food then one might expect the infants to spend more time near the food-providing mother but instead, the baby monkeys spent significantly more time attached to the cloth mother rather than the food-providing mother. Harlow concluded: "These data make it obvious that contact comfort is a variable of overwhelming importance in the development of affectional response, whereas lactation is a variable of negligible importance" (Harlow, 1959).

Further studies showed that the contact played an important role in reducing fear and providing security. Harlow placed the baby monkeys in strange rooms in which they could explore either in the presence of the surrogate cloth mother or in her absence. The monkeys used their surrogate mother as a secure base to explore the room by leaving her briefly and then rushing back to be comforted before making longer forays around the room. In the absence of the surrogate mother the monkeys were distraught and often would crouch and clasp themselves and rock, scream, and cry (Harlow, 1959).

Although the effects of being raised in orphanages is not normally seen as a case of sensory deprivation, the Romanian orphanages under the communist regime featured what one would have to call sensory deprivation. These institutions were grim and the infants and children living there had a barren existence that was better than Harlow's surrogate-reared monkeys—but not much. Children were housed in groups with a ratio of about 1:25 caregiver to infant ratio, meaning that there was little one-on-one with a caregiver and little contact comfort for the infants. The result was a dramatic stunting of physical and cognitive development in these unfortunate children.

When the Communist government fell, children were adopted out to loving families in western countries such as the United Kingdom, United States, Canada, and Australia with the expectation that the effects of the early deprived experiences could be reversed. Unfortunately, this was only true in children adopted out before 12–18 months of age. After this age the children remained severely scarred and over 25 years later they still have significant cognitive and emotional deficits including an IQ drop of about 15 or more points, smaller brains, abnormal brain electrical activity, and a host of serious chronic cognitive and social deficits that do not appear to be easily reversed (e.g., Johnson et al., 2010; Lawler, Hostinar, Mliner, & Gunnar, 2014; Rutter, O'Connor, & The English & Romanian Adoptees ERA Study Team, 2004).

3.3.2 Effects of sensory "enrichment" on brain development

Appropriate sensory experience is essential for adaptive development of brain and behavior. But is it possible to provide enhanced sensory stimulation during development? This is a difficult question, in part because of the problem of identifying an appropriate control group for comparison. There is a long tradition in behavioral neuroscience of placing animals in environments that are much more complex than standard laboratory caging, with one of the early examples being described by Hebb (1947). He brought laboratory rats into his home where they were treated more like family pets than research subjects. When the rats were returned to the laboratory for testing on various maze problems, they outperformed their lab-reared littermates. Hebb's conclusion was that a more stimulating environment enhanced brain function and this enhancement was the basis of improved performance on cognitive tasks. Hebb suggested that the more stimulating environment of his home must have changed circuits in the brain, and there is now a wealth of evidence showing that animals living in complex environments have very different brains from those living in laboratory caging (see Chapter 13: Brain Plasticity and Experience). But one difficulty is that so-called "enrichment" studies are actually studies in which deficient environments (lab caging) have been made more "normal." Riesen (1975) notes that "normal" has never really been adequately identified so our best position is to define environments in terms of their qualitative and quantitative characteristics. Certainly, the "normal" environments of the early *Homo sapiens* are nothing like the environment that most of us find ourselves in today. Thus, our definition of what is normal likely must be couched in terms of a healthy balance of various types of stimulation. Seen in this way, comparisons of the effects of complex, or enriched, housing to laboratory housing in smaller cages and smaller social groups, may still yield significant insights into factors that influence brain and behavioral development.

A landmark paper in this regard by Krech, Rosenzweig, and Bennett (1960) placed young animals in complex environments for 80 days following weaning. When they subsequently examined the rats' brains they found that the cerebral cortex was heavier and its acetylcholinesterase (AChE) activity was increased. Thus experience changed the structure of the brain and its chemistry. In a parallel study they also showed that the higher AChE levels predicted improved problem solving ability. As their experiments continued, they developed an experimental protocol to try to tease out exactly what the key experiences might be. "Control" animals were housed in standard cages in groups of three and exposed to ongoing activity in the colony room but received no special treatment, a condition they labeled a "social condition." The animals receiving "enriched experience" were housed in groups of 10–12 in large cages provided with "toys." The general finding of the literally hundreds of studies on the effects of enriched housing is that animals have significant sensory, motor, and cognitive advantages over their lab-reared littermates, and that these behavioral advantages are correlated with a wide range of neurochemical, morphological, and epigenetic differences (see Chapter 13: Brain Plasticity and Experience).

The magnitude of enrichment effects is impressive. Just 30 days of enriched housing in young rats can increase brain weight by at least 5% and these changes appear to persist for the rest of the life of the animals. And although the original studies were done with rats, there are now studies across a wide range of animal taxa, including insects, fish, birds, and mammals (including humans) showing the power of this type of experience to change the brain.

A legitimate question to ask is which experiences in the enrichment studies produce the widespread changes in brain and behavior? This is impossible to answer because it is likely the interaction of motor, sensory, social, and cognitive events experienced in complex housing. But it is possible to separate out some of the experiences to get a sense of their role. One of the most powerful appears to be tactile stimulation.

3.3.3 Effects of tactile stimulation on brain development

The skin is extremely sensitive to touch even in utero and at birth tactile stimulation provides a powerful stimulation related to bonding with the parents and exploring the environment. Tactile stimulation (or massage therapy), sometimes in combination with kinesthetic stimulation (moving of the limbs) has been investigated as a treatment for preterm infants since the 1970s (see reviews by Field, Diego, & Hernandez-Reif, 2010; Pepino & Mezzaccappa, 2015). The general finding is that tactile stimulation several times a day for about 15 min leads to weight gain and increased bone density in preterm infants. A formal program, the Newborn Individualized Developmental Care and Assessment Program was developed to improve the quality of neurodevelopmental functioning of infants in newborn intensive and special care nurseries and a central component is extensive tactile stimulation. Longitudinal randomized controlled trials have shown enhanced motor, affective, and cognitive development correlated with significant improvements in EEG (electroencephalography) and brain morphology (MRI) over what is typically observed in untreated preterm infants (Als, Duffy, McAnulty, & Butler, 2012).

Laboratory animal studies have confirmed that tactile stimulation in the infant period of "normal rats" significantly alters brain and behavioral development. For example, Guzzetta, Baldini, Bancale, and Baroncelli (2009) showed that tactile stimulation in rat pups accelerated the maturation of visual function, which was associated with an increased level of IGF-1 Insulin-Growth Factor-1 (ILG-1) in the cortex. Tactile stimulation in infancy has also been shown to enhance motor and cognitive behavior in adulthood, which was correlated with changes in cortical dendritic organization and the expression of Fibroblast Growth Factor-2 (FGF-2) (Gibb et al., 2016; Kolb & Gibb, 2010; Muhammad & Kolb, 2011a, 2011b). In addition, the tactile stimulation dramatically ameliorated the effects of perinatal brain injury and in another study it reversed the effects of stress from maternal deprivation (van Oers et al., 1998). Tactile stimulation during gestation in rats has also been shown to have beneficial behavioral and brain morphological effects in adult rats (Muhammad & Kolb, 2011a; 2011b).

We noted earlier (see above) that the absence of tactile stimulation between infants and mothers alters social and affective development in infant monkeys (Harlow & Harlow, 1965) and more recently Suomi and his colleagues (e.g., Suomi, 2002) have shown the powerful influence of touch in infant monkeys in reducing fear and providing security. Early life experiences such as the amount of licking and grooming of infant rats by their mothers has been shown to alter gene expression profiles (Weaver, Cervoni, & Champagne, 2004) and Provencal, Suderman, Guillemin, and Massart (2012) showed that infant–mother interactions influence DNA methylation in the prefrontal cortex of monkeys.

It is clear that stimulation of the skin has a profound effect on brain development and in remediation of negative perinatal experiences, with no reported negative effects. Indeed, massage therapy appears to have beneficial effects at all ages and can reduce pain, anxiety, depression, and enhance immune function (Field, 2014).

3.3.4 Early multilingual experiences

Learning more than one language is typical of the majority of people in the world at some time in their life. Typically, the language proficiency is greater as languages are learned early and simultaneously in development. Studies of people who differ in the timing of second or third language acquisition provide a unique opportunity to examine the influence of age at language acquisition on shaping brain structure and cognition. Two different groups examined the cortical structure in individuals who learned two languages simultaneously from birth versus those who learned languages sequentially or were monolingual. Klein, Mok, Chen, and Watkins (2014) reported that simultaneous acquisition was associated with thinner cortex in the left IFG (part of Broca's area) and thicker cortex in the right IFG. The later the second language was learned, the larger the effect. Curiously, the monolinguals and simultaneous language learners did not differ. Parallel results were found by Kaisser, Eppenberger, Smieskova, and Borgwardt (2015) in their measure of gray matter volume, but in addition they also showed reduced cortical volume in posterior language zones of simultaneous language learners. Growing up in a multilingual environment in early childhood may allow the brain to build more efficient networks for language processing.

One characteristic of being multilingual is that individuals must routinely switch between their languages. Although there is some controversy over exactly what benefits accrue, Bialystock and her colleagues (e.g., Barac & Bialystok, 2012; Wiseheart, Viswanathan, & Bialystok, 2016) suggest that this mental switching has a significant impact on cognitive abilities, and especially in enhancing attentional and executive functions, and in increasing cognitive reserve in aging. In an early study, Peal and Lambert (1962) gave a battery of cognitive tests to monolingual or bilingual children, expecting to show that monolinguals would perform better. Instead, the authors found that the bilingual children were superior on most tests given, especially those requiring symbol manipulation and reorganization, and executive control. But bilingual children generally have weaker language abilities in either language compared to monolinguals (for a review, see Bialystok, Craik, & Luk, 2012). This may result from the joint activation of languages in linguistic tasks, rather than activation of just one of the languages, creating an attention problem in bilinguals. However, in tasks in which participants had to generate words in response to a cue signaling which language to use in bilinguals but not monolinguals, there was activation of the dorsolateral prefrontal cortex in the bilinguals but not the monolinguals (Hernandez, Martinez, & Kohnert, 2000) and later studies have expanded this finding to include the anterior cingulate and posterior parietal cortex, striatum, and right IFG in bilinguals (e.g., Garbin, Sanjuan, Forn, & Bustamanted, 2010). Taken together these regions are essential for general attention and cognitive control and it is suggested that bilingual and monolingual people use somewhat different neural networks to solve both verbal and nonverbal tasks, and that the bilingual networks are more extensive (Bialystok et al., 2012). The finding that bilingualism enhances certain cognitive functions has led to the hypothesis that lifelong bilingualism protects against age-related cognitive decline by increasing cognitive reserve and this appears to be the case (Olsen et al., 2015; Schweizer, Ware, Fischer, Craik, & Bialystok, 2012).

3.3.5 Early musical experiences

Music, like language, is a cornerstone of what defines humans and is found in everyday life in all societies. Trehab, Schellenberg, and Ramenetsky (1999) have argued that like language the capacity for music is innate. Young infants show preferences for musical scales versus random noises and, like adults, young children are sensitive to musical errors, presumably because they are biased for perceiving regularity in rhythms. Although comparing the brains of those with and without language abilities would be very difficult, it is possible to compare the brains of musicians and nonmusicians and infer that musical training must have changed the brain, and especially the young brain.

Adult musicians have increased gray matter density in widespread regions of the cortex including regions involved in auditory processing as well as cognitive processing, including right mid-orbital frontal gyrus, left IFG, left posterior parietal cortex, posterior cingulate cortex, right prefrontal cortex, and cerebellum (Groussard et al., 2014; James et al., 2014; Schlaug, 2015). The extent of increased gray matter density varied with the age of onset and extent of musical training. Conversely, gray matter density was reduced bilaterally in sensorimotor cortex and striatum, possibly reflecting high automation of motor skills.

Musical training usually begins at an early age and presumably plays an important role in shaping brain development. Although the musician nonmusician differences are obviously related to experiences in development, less is known about the effects of early musical exposure or training on the brain than is known about language. Nonetheless, there are some clues. Elbert, Heim, and Rockstroh (2001) correlated the age at which adult musicians began practicing on stringed instruments with the amount of neural activation that they showed in response to tactile stimulation of the left hand, showing a clear effect of age at inception of musical training and (see Kolb & Whishaw, 2015, Figure 23.15). Trainor, Marie, Gerry, Whiskin, and Unrau (2012) examined the effect of random assignment of 6-month-old Western infants to either an active participatory music class or a class in which they experienced music passively while playing. Active music participation resulted in large and/or earlier Event Related Potential (ERP) responses to musical tones and earlier acculturation to Western tonal pitch structure. Musical acculturation is a process whereby a society's music undergoes changes attributable to culture. Changes in the brain related to musical acculturation may be parallel to brain changes related to hearing speech sounds from different languages (see above).

Learning music early in life appears to have broader benefits similar to simultaneous learning of languages. Early musical training appears to enhance a wide range of cognitive skills including reading, vocabulary, mathematics, working memory, and spatial skills (see review by Schlaug, 2015). These effects may appear fairly quickly as 1 year of Suzuki music lessons shows evidence of improved reading in preschool children (Anvari, Trainor, Woodside, & Levy, 2002). One intriguing study showed that the level of engagement in musical training during childhood predicted academic performance in university, even when SES status and parent education were controlled (Schellenberg, 2011). One hypothesis is that like multiple language learning, learning to play music read musical notes in scores may enhance executive functions, leading to a general improvement in cognitive functioning. Finally, like multilingualism, musical training in early to midlife is associated with enhanced cognitive reserve and likely reduced incidence of dementia (e.g., Gooding, Abner, Jicha, Krylscio, & Schmitt, 2014).

3.3.6 Effect of early stress

Stress has significant effects on the brain throughout the life course (e.g., McEwen & Morrison, 2013). Historically, although most studies in the literature have emphasized the effect of stressful experiences on the hippocampus (HPC), there is a growing literature on the effects of adult stress on the prefrontal cortex (PFC) and amygdala, and the interactions between the HPC, medial Prefrontal Cortex (mPFC), Orbital Frontal Cortex (OFC) and amygdala. Functionally, chronic stress has been associated with a multitude of cognitive, social, and physical symptoms that include deficits in emotional regulation, impaired motor function, impaired executive function including short-term memory, diminished self-regulatory behavior, and immunological impairment (e.g., Cohen, Janicki-Deverts, & Miller, 2007; McEwen, 2008; Metz, 2007; Segerstrom & Miller, 2004). It has been proposed that many of the cognitive symptoms, especially those related to cognitive functioning, are directly linked to dysfunctional connectivity between regions of the HPC, mPFC, and OFC (McEwen, 2007).

Over the past 15 years there has been a growing literature on the effects of gestational and early life stress. For example, prenatal stress is now known to be a risk factor in the development of disorders such as schizophrenia, depression, drug addiction, and ADHD (e.g., Anda et al., 2006; van den Bergh & Marcoen, 2004). Studies with laboratory animals have shown that perinatal stress produces a wide range of behavioral changes when measured in adulthood, including impaired learning and memory, high anxiety, altered social behavior, and a preference for alcohol (e.g., review by Weinstock, 2008). These behavioral changes are correlated with synaptic changes in many cerebral structures including the prefrontal cortex, HPC, and amygdala (e.g., Muhammad & Kolb, 2011a; 2011b; Murmu et al., 2006) (for a more extensive discussion, see Chapter 13: Brain Plasticity and Experience on brain plasticity and development).

Although there is little doubt that the resultant offspring of women stressed while pregnant may suffer from significant cognitive sequelae, no studies have been able to relate the stressors to effects on specific brain regions. Recent studies of natural disasters may shed light on this. Perhaps the most extensive prospective study is on the effects of the Quebec, Canada, Ice Storm of 1998. This storm was the largest natural disaster in Canadian history (until the wildfire in Fort MacMurray in 2016) as power was knocked out for up to 6 weeks, affecting approximately 2 million people, during the coldest month of the year. Many women were pregnant at different stages of gestation and had varying levels of hardship ranging from severe to none. There are now several prospective studies on the children of a large cohort of women beginning shortly after the storm that still continuing today. Overall, the results show that the offspring of the women exposed to high levels of objective stress show significant developmental delays in language, cognition (IQ), motor skills, and play (e.g., King, Dancause, Turcotte-Tremblay, Veru, & Laplante, 2012). This is correlated with widespread effects on DNA methylation across the entire genome of their children, detectable during adolescence (Cao-Lei et al., 2015).

3.3.7 Effect of psychoactive drugs

It appears that all psychoactive drugs, including prescription drugs, change the structure of the brain. The prefrontal regions typically are the most affected, although the nucleus accumbens and HPC also show significant changes (e.g., Crombag et al., 2005; Robinson & Kolb, 2004). Curiously, the effects in the mPFC and OFC regions of the prefrontal cortex are consistently different, and often in opposite directions (see Figure 4 from Crombag et al., 2005).

Less is known about the effects psychoactive drugs during development but there is now evidence that prenatal exposure to psychoactive drugs, including nicotine, valproic acid, amphetamine, morphine, fluoxetine, marijuana, and alcohol leave a clear footprint in the developing brain that can be seen in behavioral, neuronal structure, and epigenetic measures (e.g., Vassoler, Byrnes, & Pierce, 2014). There are several routes whereby the developing brain can be affected including directly by active drug metabolites that penetrate the fetal bloodstream to interfere with cell development, possibly through epigenetic changes, and indirectly through vasoconstriction that may interfere with oxygen supply (e.g., Minnes, Lang, & Singer, 2011).

Preliminary studies with generally small samples have done neuroimaging of children with prenatal exposure to both licit (e.g., alcohol and tobacco) and illicit drugs (e.g., cocaine, methamphetamine, marijuana). In general, dopamine-rich cortical (e.g., prefrontal cortex) and subcortical (e.g., basal ganglia) areas show reduced gray matter volume (see review by Derauf, Kekatpure, Neyzi, Lester, & Kosofsky, 2009) and suppressed activations in amygdala and the default mode network (a frontoparietal cortical network active when people are not engaged in specific tasks) (e.g., Li, Coles, Lynch, Luo, & Hu, 2016). The effects of different drugs are not identical, with the effects of tobacco generally being smaller than drugs like alcohol and cocaine. Although there are few MRI studies of late-adolescent or adults prenatally exposed to psychoactive drugs, the effects on cerebral morphology appear to reduce with time but there still are persistent differences in cortical thickness in mid-to-late adolescence (e.g., Gautam, Warner, Kan, & Sowell, 2015).

3.3.8 Effect of parent–infant and peer relationships

There is an extensive historical record of the effects of the importance of parent–child interactions (positive and negative) on the development of cognitive functions, and especially verbal abilities, in children, as well as many studies on the effects of early aversive experiences on behavior (see stress above). There are far fewer studies of the effect of parent–infant relations on neurobiological development, however, and virtually all have focused on the effects of maltreatment (but see Takeuchi et al., 2015; discussed below). Overall, childhood maltreatment is associated with consistent alterations in the development of prefrontal regions, anterior cingulate cortex, HPC, and the corpus callosum (Teicher & Samson, 2016). In addition, these authors conclude that maltreatment is consistently associated with enhanced amygdala response to threatening stimuli and diminished striatal response to anticipated reward.

Few studies have examined the neurobiological effects of early-life neglect such as institutional rearing. The Bucharest Early Intervention Project provides one of the best examples of the effects of neglect, in the absence of direct abuse. Because of high child to caregiver ratios, limited caregiver responsiveness, and an absence of typical emotional and cognitive stimulation, children are deprived of caregiver–child relationships that play a significant role in brain development. Romanian orphans were randomly assigned to high-level foster care or continued institutional care around 2 years of age (for details see Nelson, Fox, & Zeanah, 2014). Neuroimaging revealed that regardless of the location of the care, there was a 6.4% reduction in grey matter volume. In addition, all children with histories of institutional rearing showed abnormal development of the corpus callosum at 8–10 years but children moved to foster homes showed normalization of abnormalities in limbic connectivity that was still present in the care as usual children.

In contrast to the maltreatment studies, Taeuchi et al. (2015) examined the correlation of gray density and the amount of time children spent with their parents. There was a positive correlation between verbal comprehension scores and time with parents but a negative correlation between the superior temporal gyrus (STG) in both hemispheres and time with parents. There is previous evidence that reductions in STG gray matter reflect increased functional integrity of this area during development. For example, increased STG gray matter is found in children with a variety of maltreatments that are associated with poor verbal development (e.g., De Bellis et al., 2002). The changed STG gray matter in the right hemisphere is surprising, pointing to a role of the right STG in language.

Peer relationships also play a significant role in modifying brain development (e.g., Pellis & Pellis, 2009; see Chapter 12: Rough-and-Tumble Play and the Development of the Social Brain: What Do We Know, How Do We Know It and What Do We Need to Know? on Play). Although the details differ, all mammals play and the opportunity to play is rewarding. In addition, play has rules that govern it, including the reciprocal nature of roles, such as attacker and defender. It is possible to manipulate the amount of play that animals engage in by varying the number of potential playmates and the time provided for playing. For example, with rodents juveniles can be paired with varying numbers of potential partners from zero (living with adults who do not play with juveniles) to many. When Bell, Pellis, and Kolb (2010) varied the number of partners (0, 2, 4) they found that play behavior promotes the pruning of mPFC and nucleus accumbens. This pruning appears to make these regions, and associated behaviors, more plastic in adulthood (Burleson et al., 2016; Himmler, Pellis, & Kolb, 2013). Furthermore, perinatal experiences including tactile stimulation, gestational stress, and valproic acid all alter various aspects of play behavior, which presumably alters the development of the prefrontal cortex and nucleus accumbens (e.g., Muhammad & Kolb, 2011a; 2011b; Muhammad, Hossain, Pellis, & Kolb, 2011; Raza et al., 2015). We are unaware of any similar studies of brain development and play in humans.

3.3.9 Effect of diet

There is a large literature showing that nutrients influence brain development and behavior beginning in gestation and continuing into adult (see Moran & Lowe, 2016). The role of nutrients has largely been studied by examining the effects of nutrient deficiencies, rather than nutrient additives. Deficiencies related to protein energy, iron, zinc, copper, and choline have both global effects and brain circuit-specific on the developing brain, depending upon the precise timing of the nutrient deficit (see review by Georgieff, 2007).

Few studies have used neuroimaging to examine the effects of diet in the developing human brain, although there are some examples summarized by Isaacs (2013). For example, Isaacs et al. (2010) compared the effect of breast-feeding versus formula feeding on neural structure and cognition in adolescents who had been followed prospectively from infancy. MRI scans showed that breast-feeding was associated in adolescence by higher verbal IQs, and higher total brain volume and white matter volume, but not gray matter volume, the effects being larger in males.

Maternal diet during gestation significantly alters gene methylation in newborns (Dominguez-Salas et al., 2014). This study studied infants in rural Gambia who had been conceived in either the dry season or rainy season. Gambian's diets are dramatically different during these two seasons and so was gene methylation in the infants' blood. Although the authors did not study brain directly, it seems likely that a difference in global methylation in blood is a reasonable proxy for differences in gene methylation in neurons in the brain.

With this type of study in mind, researchers have administered a broad mixture of vitamins, minerals, and antioxidants and a blend of herbal supplements such as gingko biloba and amino acid precursors for neurotransmitters, to people with various mood disorders (e.g., Davison & Kaplan, 2012). Although there have not yet been human trials using such supplements for the developing brain, there are laboratory rat studies showing that feeding this supplemented diet to pregnant dams until the weaning of their infants increases cortical thickness, especially in frontal cortex measured in adulthood (Halliwell, 2004; 2011). More interesting, when the infant offspring were given perinatal frontal injuries, the animals showed significantly better outcomes, which was correlated with virtually complete recovery of both cognitive and motor behaviors, which was associated with increased brain weight, increased cortical thickness (Halliwell, Gibb, & Kolb, unpublished observations).

3.3.10 Effect of poverty

Childhood poverty is a major health problem worldwide. It is estimated that nearly 50 million children live below the national poverty level in industrial countries and it would be orders of magnitude greater in developing countries. Living in poverty is associated with poor cognitive development, including language, memory, socioemotional processing, and ultimately income and health in adulthood. Hanson et al. (2013) did repeated MRI scans on newborn to 3-year-old children demographically balanced to represent proportions defined by the US Census Bureau in terms of gender, race, ethnicity, and income distribution. The results showed that, although infants from different SES levels had similar gray matter volumes, by age 4 the lower SES children had lower gray matter volumes in the frontal and parietal cortex than more advantaged children. A subsequent study by Noble et al. (2015) found similar results when they examined the relationship between SES and cortical surface area in over 1000 participants between ages 3 and 20. Lower family income, independent of race or sex, was again associated with decreased cortical surface area in widespread regions of the frontal, temporal, and parietal cortex. This was associated with poorer cognitive performance on tests of attention, memory, vocabulary, and reading. Thus, lower SES is associated with smaller cortical surface area and poorer test outcomes (see Chapter 16, Socioeconomic Status for an expanded discussion of SES and brain development).

3.3.11 Effect of brain injury

Margaret Kennard was the first to do systematic studies on the effect of brain injury in development, beginning in the 1930s (e.g., Kennard, 1942). She studied the effects of unilateral motor cortex lesions in juvenile and adult monkeys and found milder impairments in the younger animals. Although she had no direct evidence, she suspected that there must be some change in cortical organization in the young animals to support more normal behavior (Kennard, 1942). Although she did not actually say "earlier is better," she is generally credited with this idea, a conclusion that Teuber (1975) dubbed the *Kennard Principle*. Hebb (1949) reached a rather different conclusion, however. His studies of Wilder Penfield's patients with early injuries to the frontal lobe showed that these children had worse outcomes than patients with similar injuries later in life. Hebb was studying the development of cognitive functions rather than motor functions as Kennard had done, and he concluded the early frontal injury was interfering with the development of neural networks needed to support many adult behaviors. Extensive research over the past 40 years, using children, monkeys, cats, and rodents as subjects, have shown that both Kennard and Hebb were partially correct. The outcomes depend upon the precise gestational age at injury, the age at behavioral assessment, the assessment instruments, the injury etiology, and whether the injury is unilateral or bilateral. Given that Chapter 15, Injury is devoted to the outcome of early brain injury in children, the focus here will be on a brief discussion of the laboratory animal literature (for a more extensive discussion, see Kolb, Mychasiuk, Muhammad, & Gibb, 2013).

Over the past 40 years, my colleagues and I have studied the effects of cerebral injuries in different rodent species (rats, hamsters, mice) and examined focal lesions to most neocortical regions at a variety of ages, comparing in many regions unilateral and bilateral injuries.

The general finding with rats and mice is that damage perinatally (Postnatal days 1–5; P1–5) has devastating effects on behavior, regardless of which cortical area is damaged (see Table 3.2). In stark contrast, similar damage as infants (P7–12) permits surprisingly normal functional outcomes, even though the brain is significantly smaller than normal. Injury in the late juvenile (P25) or early adolescence (P35) also led to unexpectedly normal behavioral outcomes, and again the brain is small. Damage at P55 (late adolescence) afforded no advantage, however, with the functional deficits appearing similar to those observed in adults with similar injuries. I am aware of only one study of prenatal cortical injury in rats, and there was remarkably normal behavior, in spite of a highly abnormal brain.

Results from studies of cats and monkeys present a similar pattern, although the dates vary because of differences in gestational rate. Villablanca, Hovda, Jackson,

Table 3.2 Summary of the effects of frontal cortical injury at different ages

Age at injury	Result	Basic references
E18 (prenatal)	Cortex develops with odd structure Functional recovery	Kolb, Cioe, and Muirhead (1998)
P1–6 (neonatal)	Small brain, dendritic atrophy Poor functional outcome	Kolb and Gibb (1993)
P7–12 (infant)	Dendrite and spine hypertrophy Cortical regrowth Functional recovery	Kolb and Gibb (1993) Kolb et al. (1998)
P25–P35 (juvenile)	Small brain Partial recovery Dendritic hypertrophy	Kolb and Whishaw (1981) Nemati and Kolb (2012)
P55 (late adolescence)	No recovery	Nemati and Kolb (2012)

& Infante (1993) conducted an extensive series of studies on the behavior of cats with frontal or prefrontal injuries. Cats are an interesting comparison to the rat and monkey because they are embryologically older than rats at birth with a gestation period of about 65 days, but they are much younger at birth than monkeys. Overall, Villablanca has found that although cats with prefrontal lesions shortly after birth show good recovery relative to animals with lesions later in life, cats with prenatal lesions have severe behavioral impairments. Thus, the newborn cats appear similar to P10 rats, whereas the prenatal cats are similar to P1–6 rats. Monkeys are different again. They are born much older than rats, cats, or even humans. Although Kennard reported better outcomes with infant lesions, as did Harlow, Akert, and Schiltz (1964), the bulk of the later evidence largely by Goldman and colleagues did not report this (see reviews by Goldman, 1974; Goldman, Isseroff, Schwrtz, & Bugbee, 1983). In contrast, however, prenatal lesions in monkeys allow substantial recovery (Goldman & Galkin, 1978). The prenatal lesions are more similar in embryological time to newborn cats and P10 rats. We can predict that if Goldman and Galkin had made their prenatal lesions even earlier, the outcome would be similar to the prenatal lesions in cats and lesions in newborn rats.

It is difficult to compare the effects of focal injuries in laboratory animals to those of children, in large part because of the very different etiologies and maturational rates as well as a host of environmental variables (e.g., Anderson, Spencer-Smith, & Wood, 2011). Nonetheless, the extensive studies of Vicki Anderson and her colleagues have consistently shown that the patterns of behavioral outcomes in children with early brain injuries varies with age at injury, just as it does in laboratory animals (Anderson et al., 2009; see Chapter 15: Injury).

As in children, many factors can influence the outcomes from early brain injury in rats (see Table 3.3). In fact, even animals with the worst spontaneous outcomes (i.e., P1–5 injury) can show remarkable functional recovery. It is important to note, however, that some experiences, especially psychoactive drugs (e.g., fluoxetine), gestational stress, and teratogens (e.g., bromodeoxyuridine), can act to effectively block recovery (see Table 3.3). As a rule of thumb, those experiences that enhance recovery are associated with synaptic proliferation and/or neurogenesis whereas those factors that impair recovery are associated with dendritic atrophy. We return to this in Chapter 13, Brain Plasticity and Experience.

3.4 INTERNAL EXPERIENCES AFFECTING BRAIN DEVELOPMENT

To this point we have emphasized the effects of external experiences on brain development. But internal factors are important as well, although much less well studied. We consider two as yet understudied factors.

Table 3.3 Summary of the effects of treatments enhancing recovery from frontal cortical injury at different ages

Treatment injury Outcome	Basic references
Positive outcomes:	
Complex housing (at weaning)	Kolb and Elliott (1987)
Gestational complex housing	Gibb et al. (2013)
Neonatal tactile stimulation	Kolb and Gibb (2010)
Gestational tactile stimulation	Gibb (2004)
Fibroblast growth factor-2	Comeau, Hastings, and Kolb (2007, 2008) Monfils et al. (2006)
Diazepam post injury	Kolb, Gibb, Pearce, and Tanguay (2008)
Negative outcomes:	
Gestational fluoxetine	Kolb et al. (2008)
Bromodeoxyuridine	Kolb, Pedersen, and Gibb, (2012)
Excessive exercise	Gibb (2004)
Neonatal noradrenaline depletion	Sutherland, Kolb, Becker, and Whishaw (1982)

3.4.1 Role of the microbiome

The enteric nervous system (ENS), which is sometimes considered part of the autonomic nervous system, functions largely independently to control digestion. The ENS is sometimes called the "second brain" because it contains such a diversity of neuron types, profusion of glial cells, and complex, integrated neural circuits. Its' estimated 200–500 million neurons is roughly equal to the number in the spinal cord. The ENS functions to control bowel motility, secretion, and blood flow to permit fluid and nutrient absorption and to support waste elimination (see Avetisyan, Schill, & Heuchkeroth, 2015). This is no simple task, given the number and balance of nutrients that are needed to support the body. In fact, it has been suggested that if bowel control required conscious thought, we could do little else. The gut responds to a range of hormones and other chemicals with exquisite neural responses.

The ENS interacts with gut bacteria, known collectively as the *microbiome*. About 10^{14} microbiota populate the adult gut, outnumbering the host cells by a factor of 10 (Farmer, Randall, & Azia, 2014), meaning that about 90% of the cells in and on the body are not human. The microbiota influence nutrient absorption and are a source of neurochemicals to regulate an array of physiological and psychological processes. This relationship is leading to the development of a class of compounds known as *psychobiotics* that can be used to treat behavioral disorders. Thus, the microbiota produce chemicals that can influence both the CNS and ENS, leading to changes in behavior.

Although the gut has no microbiota before birth, it is populated from the mother both from vaginal and anal fluids as well as the skin (especially breast) after birth.

It has been suggested that many neurodevelopmental disorders, including autism, may be related to microbial infections early in life (e.g., Finegold et al., 2002; Kohane et al., 2012). Hsiao et al. (2013) studied a mouse model that is known to display features of autism spectrum disorder. These mice have a very low production of social auditory vocalizations that measure about one-third of the normal levels. Manipulation of the gut bacteria toward strain-typical levels restored the vocalizations to normal, thus demonstrating that gut bacteria can alter behavior. To do this, it must be altering the brain.

There are few studies manipulating the microbiome in young animals but a study by Diaz Heijtz et al. (2011) is provocative. These authors manipulated gut bacteria in newborn mice and found that gut bacteria influence motor and anxiety-like behaviors, which were associated with changes in the production of synaptic-related proteins in cortex and striatum. This finding is important because it provides a mechanism whereby infections during development could influence brain development.

Finally, although there is only one study to date, a recent paper by Benakis, Brea, Caballero, and Faraco (2016) showed that manipulation of microbiota has an impact on ischemic stroke in adult mice. It remains to be seen if a similar manipulation could influence recovery from brain injury in development.

3.4.2 Role of the immune system

Although historically the brain was thought to be protected from the immune system, it is now clear that a large number of proteins originally discovered in the immune system are also found in the healthy, uninfected, nervous system, although the role may not always be immunological (Boulanger, 2009). Many immune proteins are found in the early postnatal brain, leading to speculation that they play a role in brain development. For example, immune proteins are expressed in neuronal stem cells, suggesting that immune signaling could influence neurogenesis (see review by Carpentier & Palmer, 2009). But the role of these proteins may be broader, possibly influencing the production and maturation of synapses (see Boulanger, 2009).

Immune signaling plays a key role in disease or injury in adults but the activation of the immune system during fetal or postnatal development could have significant functional consequences later in life. For example, cytokines coordinate the host response to infection but, in addition, they influence intercellular signaling including the nervous system. Thus, it is now known that cytokines influence all stages of brain development including neuronal proliferation and differentiation, migration, cell survival, and synapse modulation and elimination, as well as regulating gliogenesis including both astrocytes and oligodendroglia (see review by Deverman & Patterson, 2009).

Because many of the cytokines that are used for signaling in brain development also serve as immune modulators, normal cytokine-mediated developmental brain processes can be perturbed by maternal infection (see review by Patterson, 2009). In fact, maternal infection is a risk factor for several brain disorders including periventricular leukomalacia (PVL, white matter damage), autism (ASD), schizophrenia, and complications of Zika virus. PVL is a leading cause of cerebral palsy with associated motor and cognitive effects. There is a long history of a potential link between maternal infection and schizophrenia, which has been corroborated by sereological evidence. Estes and McAllister (2015, 2016) propose that activation of interlukin-17 (IL-17) in the maternal blood and IL-17 crosses the placental and increases expression of the IL-17 in fetal brain, leading to ASD-related cortical and behavioral abnormalities.

The link between infection by Zika virus and brain development is an emerging global health concern. Although research is just beginning, a study by Tang, Hammack, Ogden, and Wen (2016) in intriguing suggests a mechanism. The authors infected human neural progenitor cells in vitro. The Zika infection increased cell death and dysregulated cell-cycle progression, resulting in attenuated neural precursor, which could lead to abnormal brain development, including microcephaly.

3.5 CONCLUSIONS

The developing brain is remarkably sensitive to a wide range of factors that influence both brain and behavior, for better and worse. Two key questions remaining are: (1) what are the mechanisms underlying brain and behavioral changes; and (2) how does sex influence the effects of the different factors. There is little doubt that individual differences in gene expression provide a powerful mechanism but it is less clear how experiential factors modify gene expression, or how individual genes or groups of genes actually modify brain and ultimately behavior. Similarly, although all of the factors discussed here affect both sexes, the intensity of their effects vary by sex, as do the changes in gene expression. Finally, there is issue of how factors that influence brain development can interact with other factors (e.g., drugs, stress) throughout the rest of the life of an individual. We return to this issue in Chapter 13, Brain Plasticity and Experience.

REFERENCES

Als, H., Duffy, F. H., McAnulty, G., Butler, S. C., et al. (2012). NIDCAP improves brain function and structure in preterm infants with severe intrauterine growth restriction. *Journal of Perinatology, 32*, 797–803.

Anda, R. F., Felitti, V. J., Bremmer, J. D., Walker, J. D., Whitfield, C., ... Giles, W. H. (2006). The enduring effects of abuse and related adverse experiences in childhood. A convergence of evidence from neurobiology and epidemiology. *European Archives of Psychiatry and Clinical Neuroscience, 256*, 174–186.

Anderson, V., Spencer-Smith, M., & Wood, A. (2011). Do children really recover better? Neurobehavioural plasticity after early brain insult. *Brain, 134*, 2197–2221.

Anderson, V., Spencer-Smith, M., Leventer, R., Coleman, L., Anderson, P., ... Jacobs, R. (2009). Childhood brain insult: Can age at insult help us predict outcome? *Brain, 132*, 45–56.

Anvari, S. H., Trainor, L. J., Woodside, J., & Levy, B. A. (2002). Relations among musical skills, phonological processing, and early reading ability in preschool children. *Journal of Experimental Child Psychology*, *83*, 111–130.

Avetisyan, M., Schill, E. M., & Heuchkeroth, R. O. (2015). Building a second brain in the bowel. *Journal of Clinical Investigation*, *125*, 899–907.

Barac, R., & Bialystok, E. (2012). Bilingual effects on cognitive and linguistic development: role of language, cultural background, and education. *Child Development*, *83*, 413–422.

Bell, H. C., Pellis, S. M., & Kolb, B. (2010). Juvenile peer play experience and the development of the orbitofrontal and medial prefrontal cortex. *Behavioural Brain Research*, *207*, 7–13.

Benakis, C., Brea, D., Caballero, S., Faraco, G., et al. (2016). Commensal microbiota affects ischemic stroke outcome by regulating intestinal T cells. *Nature Medicine*, *22*, 516–523.

Berry, M., McConnell, P., & Sievers, J. (1980). Dendritic growth and the control of neuronal form. *Current Topics in Developmental Biology*, *15*(Pt. 1), 67–101.

Bialystok, E., Craik, F. I. M., & Luk, G. (2012). Bilingualism: Consequences for mind and brain. *Trends in Cognitive Science*, *16*, 240–250.

Boulanger, L. M. (2009). Immune proteins in brain development and synaptic plasticity. *Neuron*, *64*, 93–109.

Burleson, C. A., Pedersen, R. W., Seddighi, S., DeBusk, L. E., Burgharddt, G. M., & Cooper, M. A. (2016). Social play in juvenile hamsters alters dendritic morphology in the medial prefrontal cortex and attenuates effects of social stress in adulthood. *Behavioral Neuroscience*, in press.

Cao-Lei, L., Elgbeili, G., Massart, R., Laplante, D. P., Szyf, M., & King, S. (2015). Pregnant women's cognitive appraisal of a natural disaster affects DNA methylation in their children 13 years later: Project Ice Storm. *Translational Psychiatry*, *5*, e515. Available from http://dx.doi.org/10.1038/tp.2015.13.

Carpentier, P. A., & Palmer, T. D. (2009). Immune influence on adult neural stem cell regulation and function. *Neuron*, *64*, 79–92.

Cohen, S., Janicki-Deverts, D., & Miller, G. E. (2007). Psychological stress and disease. *JAMA*, *298*, 1685–1687.

Coleman, P. D., & Riesen, A. H. (1968). Environmental effects on cortical dendritic fields. *Journal of Anatomy*, *102*, 363–374.

Comeau, W., Gibb, R., Hastings, E., Cioe, J., & Kolb, B. (2008). Therapeutic effects of complex rearing or bFGF after perinatal frontal lesions. *Developmental Psychobiology*, *50*, 134–146.

Comeau, W., Hastings, E., & Kolb, B. (2007). Differential effect of pre and postnatal FGF-2 following medial prefrontal cortical injury. *Behavioural Brain Research*, *180*, 18–27.

Crombag, H. S., Gorny, G., Li, Y., Kolb, B., & Robinson, T. E. (2005). Opposite effects of amphetamine self-administration experience on dendritic spines in the medial and orbital prefrontal cortex. *Cerebral Cortex*, *15*, 341–348.

Davison, K. M., & Kaplan, B. J. (2012). Nutrient intakes are correlated with overall psychiatric functioning in adults with mood disorders. *Canadian Journal of Psychiatry*, *57*, 85–92.

De Bellis, M. D., Keschavan, M. S., Frustaci, K., Shifflett, H., Iyengar, S., ... Hall, J. (2002). Superior temporal gyrus volumes in maltreated children and adolescents with PTSD. *Biological Psychiatry*, *51*, 544–552.

Derauf, C., Kekatpure, M., Neyzi, N., Lester, B., & Kosofsky, B. (2009). Neuroimaging of children following prenatal drug exposure. *Seminars in Cell Developmental Biology*, *20*, 441–454.

Deverman, B. E., & Patterson, P. H. (2009). Cytokines and CNS development. *Neuron*, *64*, 61–78.

Diaz Heijtz, R., Wang, S., Anuar, F., Qian, Y., Bjorkholm, B., ... Pettersson, S. (2011). Normal gut microbiota modulates brain development and behavior. *Proceedings of the National Academy of Sciences of the United States of America*, *108*, 3047–3052.

Dominguez-Salas, P., Moore, S. E., Baker, M. S., Bergen, A. W., Cox, S. E., Dyer, R. A., et al. (2014). Maternal nutrition at conception modulates DNA methylation of human metastable epialles. *Nature Communications*, *5*, 3746. Available from http://dx.doi.org/10.1038/ncomms4746.

Elbert, T., Heim, S., & Rockstroh, B. (2001). Neural plasticity and development. In C. A. Nelson, & M. Luciana (Eds.), *Handbook of developmental cognitive neuroscience* (pp. 191–204). Cambridge, MA: MIT Press.

Estes, M. L., & McAllister, A. K. (2015). Immune mediators in the brain and peripheral tissues in autism spectrum disorder. *Nature Reviews Neuroscience*, *16*, 469–486.

Estes, M. L., & McAllister, A. K. (2016). Maternal TH17 cells take a toll on baby's brain. *Science*, *351*, 9189–9920.

Farmer, A. D., Randall, H. A., & Azia, Q. (2014). It's a gut feeling: How the gut microbiota affects the state of mind. *Journal of Physiology*, *592*, 2981–2988.

Field, T. (2014). Massage therapy research review. *Complementary Therapies in Clinical Practice*, *20*, 224–229.

Field, T., Diego, M., & Hernandez-Reif, M. (2010). Preterm infant massage therapy research: A review. *Infant Behavior and Development*, *33*, 115–124.

Finegold, S., Molitoris, D., Song, Y., Liu, C., Vaisanen, M., & Kaul, A. (2002). Gastrointestinal microflora studies in late onset autism. *Clinical and Infectious Diseases*, *35*, S6–S16.

Garbin, G., Sanjuan, A., Forn, C., Bustamanted, J. C., et al. (2010). Bridging language and attention: Brain basis of the impact of bilingualism on cognitive control. *NeuroImage*, *53*, 1272–1278.

Gautam, P., Warner, T. D., Kan, E. C., & Sowell, E. R. (2015). Executive function and cortical thickness in youths prenatally exposed to cocaine, alcohol and tobacco. *Developmental Cognitive Neuroscience*, *16*, 155–165.

Georgieff, M. K. (2007). Nutrition and the developing brain: nutrient priorities and measurement. *The American Journal of Clinical Nutrition*, *85S*(2007), 614–620.

Gibb et al., 2016 TS.

Gibb, R. (2004). Perinatal experience and recovery from brain injury. *Unpublished PhD Thesis*. University of Lethbridge.

Gibb, R., Gonzalez, C., & Kolb, B. (2014). Prenatal enrichment and recovery from perinatal cortical damage: Effects of maternal complex housing. *Frontiers in Behavioral Neuroscience*, *8*, 223. doi:10.3389/fnbeh.2014.00223.

Golding, J., Northtone, K., Gregory, S., Miller, L. L., & Pembrey, M. (2014). The anthropometry of children and adolescents may be influenced by the prenatal smoking habits of their grandmothers: A longitudinal cohort study. *American Journal of Human Biology*, *26*, 731–739.

Goldman, P. S. (1974). An alternative to developmental plasticity: Heterology of CNS structures in infants and adults. In D. G. Stein, J. Rosen, & N. Butters (Eds.), *Plasticity and recovery of function in the central nervous system* (pp. 149–174). New York, NY: Academic Press.

Goldman, P. S., & Galkin, T. W. (1978). Prenatal removal of frontal association neocortex in the fetal rhesus monkey: Anatomical and functional consequences. *Brain Research*, *152*, 451–485.

Goldman, P. S., Isseroff, A., Schwrtz, M., & Bugbee, N. (1983). The neurobiology of cognitive development. In P. H. Mussen (Ed.), *Handbook of child psychology: biology and infancy development* (pp. 311–344). New York, NY: Wiley.

Gooding, L. F., Abner, E. L., Jicha, G. A., Krylscio, R. J., & Schmitt, F. A. (2014). Musical training and late-life cognition. *American Journal of Alzheimer's Disease and Other Dementias*, *29*, 333–0343.

Groussard, M., Viader, F., Landeau, B., Desgranges, B., Eustache, F., & Platel, H. (2014). The effects of musical practice on structural plasticity: they dynamics of grey matter changes. *Brain and Cognition*, *90*, 174–180.

Guzzetta, A., Baldini, S., Bancale, Ad, Baroncelli, L., et al. (2009). Massage accelerates brain development and the maturation of visual function. *Journal of Neuroscience*, *219*, 6042–6051.

Halliwell, C. (2004). *Dietary factors and recovery from brain damage. Unpublished MSc thesis*. Lethbridge, Alberta, Canada: University of Lethbridge.

Halliwell, C. (2011). Treatment interventions following prenatal stress and neonatal cortical injury. *Unpublished PhD thesis*. Lethbridge, Alberta, Canada: University of Lethbridge.

Hanson, J. L., Hair, N., Shen, D. G., Shi, F., Gilmore, J. H., Wolfe, B. L., & Pollak, S. D. (2013). Family poverty affects the rate of human infant brain growth. *PLoS ONE*, *8*, e80954. Available from http://dx.doi.org/10.1371/journal.pone.0080954.

Harlow, H. F. (1959). Love in infant monkeys. *Scientific American*, *200*, 68–74.

Harlow, H. F., & Harlow, M. K. (1965). The effect of rearing conditions on behavior. *International Journal of Psychiatry*, *1*, 43–51.

Harlow, H., Akert, K., & Schiltz, K. (1964). The effects of bilateral prefrontal lesions on learned behavior of neonatal, infant, and preadolescent monkeys. In J. Warren, & K. Akert (Eds.), *The frontal granular cortex and behaviour* (pp. 126–148). New York, NY: McGraw-Hill.

Hart, B., & Risley, T. (1995). *Meaningful differences in the everyday experience of young American children*. Baltimore, MD: Brookes.

Hebb, D. O. (1947). The effects of early experience on problem solving at maturity. *American Pyschologist*, *2*, 737–745.

Hebb, D. O. (1949). *The organization of behavior*. New York: Wiley.

Hernandez, A. E., Martinez, A., & Kohnert, K. (2000). In search of the language switch: An fMRI study of picture naming in Spanish–English bilinguals. *Brain and Language*, *73*, 421–431.

Himmler, B. T., Pellis, S. M., & Kolb, B. (2013). Juvenile play experience primes neurons in the medial prefrontal cortex to be more responsive to later experiences. *Neuroscience Letters*, *556*, 42–45.

Holz, N. E., Boecker, R., Baumeister, S., Hohm, E., et al. (2014). Effect of prenatal exposure to tobacco smoke on inhibitory control: Neuroimaging results from a 25-year prospective study. *JAMA Psychiatry*, *71*, 786–796.

Hsiao, E. Y., McBride, S. W., Hsien, S., Sharon, G., Hyde, E. R., ... Petrosino, J. F., et al. (2013). Microbiota modulate behavioral and physiological abnormalities associated with neurodevelopmental disorders. *Cell*, *155*, 1451–1463.

Isaacs, E. B. (2013). Neuroimaging, a new tool for investigating the effects of early diet on cognitive and brain development. *Frontiers in Human Neuroscience*. Available from http://dx.doi.org/10.3389/fnhum.2013.00445.

Isaacs, E. B., Fischl, B. R., Quinn, B. T., Chong, W. K., Gadian, D. G., & Lucas, A. (2010). Impact of breast milk on intelligence quotient, brain size, and white matter development. *Peditaric Research*, *67*, 357–362.

James, C. E., Oechslin, M. S., van de Ville, D., Hauert, C. A., Descioux, C., & Lazeyras, F. (2014). Musical training intensity yields opposite effects on grey matter density in cognitive versus sensorimotor networks. *Brain Structure and Function*, *219*, 353–366.

Johnson, D. E., Guthrie, D., Smyke, A. T., Koga, S. F., Fox, N. A., Zeanah, C. H., & Nelson, C. A. (2010). Growth and associations between auxology, caregiving environment, and cognition in socially deprived Romanian children randomized to foster vs ongoing institutionalized care. *Archives of Pediatric Adolescent Medicine*, *164*, 507–516.

Joubert, B., Feliz, J. F., Yousefi, P., Bakulski, K. M., et al. (2016). DNA methylation in newborns and maternal smoking in pregnancy: Genome-wide consortium meta-analysis. *American Journal of Human Genetics*, *98*, 680–696.

Kaisser, A., Eppenberger, L. S., Smieskova, R., Borgwardt, S., et al. (2015). Age of second language acquisition in multilinguals has an impact on gray matter volume in language-associated brain areas. *Frontiers in Psychology*. Available from http://dx.doi.org/10.3399/psyg.2015.00638.

Kennard, M. (1942). Cortical reorganization of motor function. *Archives of Neurology*, *48*, 227–240.

King, S., Dancause, K., Turcotte-Tremblay, A. M., Veru, F., & Laplante, D. P. (2012). Using natural disasters to study the effects of prenatal maternal stress on child health and development. *Birth Defects Research C Embryo Today*, *96*, 273–288.

Klein, D., Mok, K., Chen, J.-K., & Watkins, K. E. (2014). Age of language learning shapes brain structure: A cortical thickness study of bilingual and monolingual individuals. *Brain and Language*, *131*, 20–24.

Kohane, I. S., McMurry, A., Weber, G., MacFadden, D., Rappaport, L., ... Murphy, S., et al. (2012). The co-morbidity burden of children and young adults with autism spectrum disorders. *PLoS ONE*, 7, e33224.

Kolb, B., & Elliott, W. (1987). Recovery from early cortical damage in rats. II. Effects of experience on anatomy and behavior following frontal lesions at 1 or 5 days of age. *Behavioural Brain Research*, *26*, 47–56.

Kolb, B., & Gibb, R. (1993). Possible anatomical basis of recovery of spatial learning after neonatal prefrontal lesions in rats. *Behavioral Neuroscience*, *107*, 799–811.

Kolb, B., Cioe, J., & Muirhead, D. (1998). Cerebral morphology and functional sparing after prenatal frontal cortex lesions in rats. *Behavioural Brain Research*, *91*, 143–155.

Kolb, B., & Gibb, R. (2010). Tactile stimulation facilitates functional recovery and dendritic change after neonatal medial frontal or posterior parietal lesions in rats. *Behavioural Brain Research, 214*, 115–120.

Kolb, B., & Whishaw, I. Q. (1981). Neonatal frontal lesions in the rat: sparing of learned but not species-typical behavior in the presence of reduced brain weight and cortical thickness. *Journal of Comparative and Physiological Psychology, 95*, 863–879.

Kolb, B., Gibb, R., Gorny, G., & Whishaw, I. Q. (1998). Possible brain regrowth after cortical lesions in rats. *Behavioural Brain Research, 91*, 127–141.

Kolb, B., Gibb, R., Pearce, S., & Tanguay, R. (2008). Prenatal exposure to prescription medications alters recovery following early brain injury in rats. *Society for Neuroscience Abstracts, 349*, 5.

Kolb, B., Mychasiuk, R., Muhammad, A., & Gibb, R. (2013). Brain plasticity in the developing brain. *Progress in Brain Research, 207*, 35–64.

Kolb, B., Pedersen, B., & Gibb, R. (2012). Embryonic pretreatment with bromodeoxyuridine blocks neurogenesis and functional recovery from perinatal frontal lesions in rats. *Developmental Neuroscience, 34*, 228–239.

Kolb, B., Whishaw, I. Q., & Teskey, G. C. (2016). *An Introduction to Brain and Behavior* (5th ed. New York: Worth.

Krech, D., Rosenzweig, M. R., & Bennett, E. L. (1960). Effects of environmental complexity and training on brain chemistry. *Journal of Comparative and Physiological Psychology, 53*, 509–519.

Lawler, J. M., Hostinar, C. E., Mliner, S. B., & Gunnar, M. R. (2014). Disinhibited social engagement in postinstitutionalized children: Differentiating normal from atypical behavior. *Developmental Psychopathology, 26*, 451–464.

Levi-Montalcini, R. (1949). The development of the acousticovestibular centers in the chick embryo in the absence of the afferent root fibers and descending fiber tracts. *Journal of Comparative Neurology, 91*, 209–242.

Li, Z., Coles, C. D., Lynch, M. E., Luo, Y., & Hu, X. (2016). Longitudinal changes of amygdala and default mode activation in adolescents prenatally exposed to cocaine. *Neurotoxicology and Teratology, 53*, 24–32.

McEwen, B. (2007). Physiology and neurobiology of stress and adaptation: central role of the brain. *Physiology Reviews, 87*, 873–904.

McEwen, B. S., & Morrison, J. H. (2013). The brain on stress: Vulnerability and plasticity of the prefrontal cortex over the life course. *Neuron, 79*, 16–29.

McEwen, B. S. (2008). Central effects of stress hormones in health and disease: Understanding the protective and damaging effects of stress and stress mediators. *European Journal of Pharmacology, 583*, 174–185.

Melzack, R. (1968). A neurophysiological approach to heredity–environmental interactions. In G. Newton, & S. Levine (Eds.), *Early experience and behavior: The psychobiology of development* (pp. 65–82). Springfield, Illinois: Thomas.

Metz, G. A. (2007). Stress as a modulator of motor system function and pathology. *Reviews in Neuroscience, 18*, 209–222.

Minnes, S., Lang, A., & Singer, L. (2011). Prenatal tobacco, marijuana, stimulant, and opiate exposure: Outcomes and practice implications. *Addiction Science and Clinical Practice, 6*, 57–70.

Monfils, M. H., Driscoll, I., Kamitakahara, H., Wilson, B., Flynn, C., ... Kolb, B. (2006). FGF-2 induced cell proliferation stimulates anatomical neurophysiological and functional recovery from neonatal motor cortex injury. *European Journal of Neuroscience, 24*, 739–749.

Moran, V. H., & Lowe, N. M. (2016). *Nutrition and the developing brain.* New York: Taylor and Francis.

Muhammad, A., & Kolb, B. (2011b). Mild prenatal stress modulated behaviour and Neuronal spine density without affecting amphetamine sensitization. *Developmental Neuroscience, 33*, 85–98.

Muhammad, A., & Kolb, B. (2011a). Maternal separation altered behavior and neuronal spine density without influencing amphetamine sensitization. *Behavioural Brain Research, 223*, 7–16.

Muhammad, A., Hossain, S., Pellis, S. M., & Kolb, B. (2011). Tactile stimulation during development attenuates amphetamine sensitization and structurally reorganizes prefrontal cortex and striatum in a sex-dependent manner. *Behavioral Neuroscience, 125*, 161–174.

Murmu, M., Salomon, S., Biala, Y., Weinstock, M., Braun, K., & Bock, J. (2006). Changes in spine density and dendritic complexity in the prefrontal cortex in offspring of mothers exposed to stress during pregnancy. *European Journal of Neuroscience, 24*, 1477–1487.

Nelson, C. A., Fox, N. A., & Zeanah, C. H. (2014). *Romania's abandoned children: deprivation, brain development and the struggle for recovery*. Cambridge, MA: Harvard University Press.

Nemati, F., & Kolb, B. (2012). Recovery from medial prefrontal cortex injury during adolescence: Implications for age-dependent plasticity. *Behavioural Brain Research, 229*, 168–175.

Nissen, H. W., Chow, K. L., & Semmes, J. (1951). Effects of restricted opportunity for tactual kinesthetic, and manipulative experience on the behavior of a chimpanzee. *American Journal of Psychology, 6*, 485–507.

Noble, K. G., Houston, S. M., Brito, N. H., Bartsch, H., Kan, E., ... Sowell, E. R. (2015). Family income, parental education and brain structure in children and adolescents. *Nature Neuroscience, 18*, 773–778.

Oers, H. J., de Kloet, E. R., Whelan, T., & Levine, S. (1998). Maternal deprivation effect on the infant's neural stress markers is reversed by tactile stimulation and feeding but not by suppressing corticosterone. *Journal of Neuroscience, 18*, 10171–10179.

Olsen, R. K., Pangelinan, M. M., Bogulski, C., Chakravarty, M. M., Luk, G., Grady, C. L., & Bialystok, E. (2015). The effect of lifelong bilingualism on regional grey and white matter volume. *Brain Research, 1612*, 128–139.

Patterson, P. H. (2009). Immune involvement in schizophrenia and autism: Etiology, pathology and animal models. *Behavioural Brain Research, 204*, 313–321.

Peal, E., & Lambert, W. (1962). The relation of bilingualism to intelligence. *Psychological Monographs, 76*, 1–23.

Pellis, S. M., & Pellis, V. C. (2009). *The playful brain: Venturing to the limits of neuroscience*. London: Oneworld Publications.

Pepino, V. C., & Mezzaccappa, M. A. (2015). Application of tactile/kinesthetic stimulation in preterm infants: a systematic review. *Jornal de Pediatria, 91*, 213–233.

Provencal, N., Suderman, M. J., Guillemin, C., Massart, R., et al. (2012). The signature of maternal rearing in the methylome in rhesus macaque prefrontal cortex and T Cells. *Journal of Neuroscience, 32*, 15626–15642.

Raza, S., Himmler, B. T., Harker, A., Kolb, B., Pellis, S. M., & Gibb, R. (2015). Effects of prenatal exposure to valproic acid on the development of juvenile-typical social play in rats. *Behavioral Pharmacology, 26*, 707–719.

Riesen, A. H. (Ed.), (1975). *The developmental neuropsychology of sensory deprivation* New York: Academic Press.

Riesen, A. H., & Zilbert, D. E. (1975). Behavioral consequences of variations in early sensory environments. In A. H. Riesen (Ed.), *The developmental neuropsychology of sensory deprivation* (pp. 211–252). New York: Academic Press.

Robinson, T. E., & Kolb, B. (2004). Structural plasticity associated with drugs of abuse. *Neuropharmacology, 47*(Suppl. 1), 33–46.

Rutter, M., O'Connor, T. G., & The English and Romanian Adoptees (ERA) Study Team (2004). Are there biological programming effects for psychological development? Findings from a study of Romanian adoptees. *Developmental Psychology, 40*, 81–94.

Schellenberg, E. G. (2011). Examining the association between music lessons and intelligence. *British Journal of Psychology, 102*, 283–302.

Schlaug, G. (2015). Musicians and music making as a model for the study of brain plasticity. *Progress in Brain Research, 217*, 37–55.

Schweizer, T. A., Ware, J., Fischer, C. E., Craik, F. I. M., & Bialystok, E. (2012). Bilingualism as a contributor to cognitive reserve: evidence from brain atrophy in Alzheimer's disease. *Cortex, 48*, 991–996.

Segerstrom, S., & Miller, G. (2004). Psychological stress and the human immune system: a meta-analytic study of 30 years of inquiry. *Psychological Bulletin, 130*, 601–630.

Sendon, M. Von (1932). *Raum-und Gestalt-auffassung bei operierten Blindgeborenen vor und nach der Operation*. Leipzig: Barth.

Suomi, S. J. (2002). How gene-environment interactions can shape the development of socioemotional regulation in rhesus monkeys. In B. S. Zuckerman, A. F. Lieberman, & N. A. Fox (Eds.), *Emotional regulation and developmental health: Infancy and early childhood* (pp. 5–26). New York: Johnson & Johnson Pediatric Institute.

Sutherland, R. J., Kolb, B., Becker, J. B., & Whishaw, I. Q. (1982). Cortical noradrenaline depletion eliminates sparing of spatial learning after neonatal frontal cortex damage in the rat. *Neuroscience Letters*, *32*, 125–130.

Takeuchi, H., Taki, Y., Hashizume, H., Asano, K., Asano, M., . . . Kawashima, R. (2015). The impact of parent–child interaction on brain structures: cross sectional and longitudinal analyses. *Journal of Neuroscience*, *35*, 2233–2245.

Tang, H., Hammack, C., Ogden, S. C., Wen, Z., et al. (2016). Zika virus infects human cortical neural progenitors and attenuates their growth. *Cell Stem Cell*, *18*. Available from http://dx.doi.org/10.1016/j.stem.2016.02.016.

Tees, R. C. (1967). Effects of early auditory restriction in the rat on adult pattern discrimination. *Journal of Comparative and Physiological Psychology*, *63*, 389–393.

Teicher, M. H., & Samson, J. A. (2016). Annual research review: Enduring neurobiological effects of childhood abuse and neglect. *Journal of Child Psychology and Psychiatry*, *57*, 241–266.

Teuber, H. (1975). *Recovery of function after brain injury inman. Outcome of severe damage to the nervous system. Ciba Foundation Symposium*. Amsterdam: Elsevier, North Holland.

Thompson, W. R., & Melzack, R. (1956). Early environment. *Scientific American*, *194*, 38–42.

Tieman, S. B., & Hirssch, H. V. B. (1982). Exposure to lines of only one orientation modifies dendritic morphology of cells in the visual cortex of the cat. *Journal of Comparative Neurology*, *211*, 353–362.

Trainor, L. J., Marie, C., Gerry, D., Whiskin, E., & Unrau, A. (2012). Becoming musically enculturated: effects of music classes for infants on brain and behavior. *Annals of the New York Academy of Sciences*, *1252*, 129–138.

Trehab, S., Schellenberg, E. G., & Ramenetsky, G. B. (1999). Infants' and adults' perception of scale structures. *Journal of Experimental Psychology: Human Perception and Performance*, *25*, 965–975.

van den Bergh, B. R., & Marcoen, A. (2004). High antenatal maternal anxiety is related to ADHD symptoms, externalizing problems, and anxiety in 8- and 9-year-olds. *Child Development*, *75*, 1085–1097.

Vassoler, F. M., Byrnes, E. M., & Pierce, R. C. (2014). The impact of exposure to addictive drugs on future generations: Physiological and behavioral effects. *Neuropharmacology*. 76(0 0):10.1016/j.neuropharm.2013.06.016. http://dx.doi.org/10.1016/j.neuropharm.2013.06.016.

Villablanca, J., Hovda, D., Jackson, G., & Infante, C. (1993). Neurological and behavioral effects of a unilateral frontal cortical lesion in fetal kittens, II. Visual system tests and proposing a 'critical period' for lesion effects. *Behavioural Brain Research*, *57*, 72–92.

Weaver, I. C., Cervoni, N., Champagne, F. A., et al. (2004). Epigenetic programming by maternal behavior. *Nature Neuroscience*, 7, 847–854.

Weinstock, M. (2008). The long-term behavioural consequences of prenatal stress. *Neuroscience and Biobehavioral Reviews*, *32*, 1073–1086.

Werker, J. F., & Tees, R. C. (1984). Cross-language speech perception: Evidence for perceptual reorganization during the first year of life. *Infant Behavior and Development*, 7, 49–63.

Wiseheart, M., Viswanathan, M., & Bialystok, E. (2016). Flexibility in task switching by monolinguals and bilinguals. *Biling (Cambridge, England)*, *19*, 141–146.

Yochum, C., Doherty-Lylon, S., Hoffman, C., Hossain, M. M., Zelikoff, J. T., & Richardson, J. R. (2014). Prenatal cigarette smoke exposure causes hyperactivity and aggressive behavior: role of altered catecholamines and BDNF. *Experimental Neurology*, *254*, 145–152.

FURTHER READING

Bick, J., Fox, N., Zeanah, C., & Nelson, C. A. (2015). Early deprivation, atypical brain development, and internalizing symptoms in late childhood. *Neuroscience*, in press. Neuroscience 342, 2017,140–153.

Bock, J., Poeschel, J., Schindler, J., Borner, F., Shachar-Dadon, A., ... Poeggel, G. (2016). Transgenerational sex-specific impact of preconception stress on the development of dendritic spines and dendritic length in the medial prefrontal cortex. *Brain Structure and Function*, *221*, 855–863.

Bock, J., Weinstock, T., Braun, K., & Segal, M. (2015). Stress in utero: Prenatal programming of brain plasticity and cognition. *Biological Psychiatry*, *78*, 315–326.

Georgieff, M. K., & Rao, R. (2001). The role of nutrition in cognitive development. In C. A. Nelson, & M. Luciana (Eds.), *Handbook in developmental cognitive neuroscience* (pp. 491–504). Cambridge, MA: MIT Press.

Levi-Montalcini, R. (1982). Developmental biology and the natural history of nerve growth factor. *Annual Review of Neuroscience*, *5*, 341–362.

Northstone, K., Golding, J., Smith, G. D., Miller, L. L., & Pembrey, M. (2014). Prepubertal start of father's smoking and increased body fat in his sons: further characterization of paternal transgenerational responses. *European Journal of Human Genetics*, *22*, 1382–1386.

CHAPTER 4

The Role of Animal Models in Developmental Brain Research

Celeste Halliwell
University of Lethbridge, Lethbridge, AB, Canada

4.1 INTRODUCTION

A challenge in directly studying human brain development has been in finding the appropriate study materials. In ancient times, human cadavers were not acceptable for study, and in more modern times, it has been largely carried out on postmortem tissues, when accessible. As a result, studies of brain development have relied on the study of brains of nonhuman species. It is possible to perturb the development of laboratory animals by harvesting their brains at specific time points during the developmental stages and the brains can be studied using a wide range of electrophysiological, morphological, and in vitro procedures. The brain provides a unique challenge however; although most of the body organs, such as the lungs and liver, show similarities across mammals, the brain is very different.

The most obvious difference about the brains across species is the overall size, but even the relative size of brain regions can vary considerably depending upon the ecological niche occupied by any given species. For example, primates have a large visual representation in the cortex, whereas laboratory rodents (mostly rats and mice) have a much smaller visual representation but have a large representation of facial whiskers that is not present in primates. Despite this problem, the overwhelming evidence is that the fundamental organization of most mammals is not much different from the human brain (for a more extensive discussion, see Kolb & Whishaw, 1983, 2015). Certainly, lab animals do not have as complex a cognitive life as do humans, and they do not talk as humans talk, but their cerebral organization is sufficiently similar to ours to allow generalizations to be made about development of the human brain. Laboratory animals (rodents, cats, monkeys) also have systems controlling emotional, attentional, and other cognitive processes that are remarkably similar to humans.

The processes of development are also similar, although we shall see that the timing of developmental events depends, in part, on the relative gestational length, which is related to the environmental niche that infants find themselves in at birth. In sum,

The Neurobiology of Brain and Behavioral Development
DOI: http://dx.doi.org/10.1016/B978-0-12-804036-2.00004-2

animal models have provided a sound basis for generalizing about principles of brain development in humans.

4.2 THE RISE OF COMPARATIVE NEUROBIOLOGY

The study of comparative neurobiology arose through the study of various brains of animals and has been reported to have begun with the Greek philosopher, Aristotle, at around 500 BC (Striedter, 2005). This was at a time when touching or looking at a human corpse was deemed forbidden and anyone who came into contact with the dying or deceased was required to undergo extensive rites and purification. Aristotle's anatomical work began as a systematic form of observations and comparisons among the biological study of animals. He examined approximately 500 different types of animals and at least 49 different animal species that are considered to have served as a foundation for technical insights still regarded valid today (Striedter, 2005; Swanson, 2003). His comparisons of body parts among all of the different species revealed what we now understand as homologous and analogous parts (Swanson, 2003). This concept did not become apparent as such until the embryological and evolutionary work in the 19th century (Striedter, 2005; Swanson, 2003).

During a brief period following Aristotle's time, public dissection of human cadavers became acceptable and was at a time when the Greek physician Herophilus described parts of the brain and the eye, nerves, and tendons (Kambi & Jain, 2012). Erastistratus followed with comparative studies across various species and made correlations about the size of the cerebellum with the ability to run fast and with having greater intellect with the larger cerebral convolutions. This brief period in history is considered to have laid the foundation for further advancements in brain anatomy and function. Scientific investigations into human anatomy again became banned during the great influence of the church until the post-Renaissance era (Kambi & Jain, 2012).

The 19th century marked a time of scientific exploration with advances in medicine and human anatomy, a general move away from church influences, and a time known for Darwinism, a theory of biological evolution that was developed by Charles Darwin. Darwin proposed his theory of natural selection through evolution, a theory that is considered to have transformed comparative neuroanatomy and furthered investigations into the ancestry of humans (Striedter, 2005).

A contemporary of Charles Darwin, Richard Owen, was a famous anatomist and known for his studies with ape brains (Striedter, 2005). Owen claimed that various species of animals, including humans, share similar traits by having the same organs to produce similar functions. Recognition of these common traits led to the classification of various species as having *homologous* organs; Owen thus proposed that mammalian species had common anatomical characteristics, including neural characteristics. Such theories led to a more general classification across all animal species and to the modern ideas of the development of the phylogenetic tree (Striedter, 2005).

The rejuvenation of interest in brain anatomy and function in humans led to ideas of localization of function beginning with the phrenology theory of Franz Joseph Gall who contended mental functions could be localized in certain areas of the brain (Kambi & Jain, 2012). Later, Broca and Wernicke proposed their theories of localization of speech production and speech comprehension based on postmortem studies of humans with specific neurological deficits. This was also during a time when electrical brain stimulation techniques in anesthetized dogs were revealing areas of localization of function in the brain (Kambi & Jain, 2012). Although comparative anatomy and brain stimulation techniques were informative about brain structure and function, neuroanatomical studies became transformed with Ramón y Cajal's discovery of the structure of the neuron, which now can be considered to have formed the basis of cellular and molecular neuroscience (Batista-Brito & Fishell, 2009; Kambi & Jain, 2012). Ramón y Cajal showed that the nervous system had anatomically distinct cells which lead to his theory of "neural atomism" (Kambi & Jain, 2012; Partsalis, Blazquez, &Triarhou, 2013). This theory was eventually replaced with the "neuron" by Heinrich Waldeyer in 1891 upon confirmation that neural cells were autonomous and not part of a continuous process that was supported by the nerve net theory (or the reticular theory) (Kambi & Jain, 2012). Waldeyer continued to systematize theories of the "ganglion cell" or "nerve cell" into the neuron doctrine theory, which describes the brain as being made up of polarized structures (cells) that contact each other through synaptic junctions (Partsalis et al., 2013). Additional staining techniques, such as those developed by Nissl and Marchi, led to methods for tracing neuronal connections that provided increased sensitivity and specificity for histological details of the brain (Jones, 2007). With the ongoing development of improved staining, histochemistry, and microscopy, theories continued to support a general homology of brain organization among mammalian species. This, in turn, has led to the development of animal models to approximate normal human brain development and for human neurodevelopmental disorders.

4.3 RESEARCH IN NEUROSCIENCES

Presently, the animals most often used to model development are rodents and these models account for approximately 90% of modern physiological medical research (Beery & Kauffer, 2015). The rat was the first species to be bred and domesticated for research and has been used to evaluate a wide range of psychological theories (Kolb, 2005). More recently, mice have also become popular as they are a close relative of the rat, but they are especially valuable when used with genetic manipulations to study various aspects of neurodevelopment—both normal and abnormal. Although rodent models have proven to be the workhorse for understanding brain development, there remains an important need to use species from other mammalian orders to ensure generality in principles. In 1929, Krogh asserted that "For many problems there will be some animal of choice or few such animals on which it can be most conveniently studied" (Krogh, 1929). For example, the conduction properties of neural impulses were demonstrated in the nervous system of the giant squid axon by Huxley and Hodgkin (Hodgkin & Huxley, 1952) who proposed a model that accurately described the action potential that is characteristic of all neural systems (Catterall, Raman, Robinson, Sejnowski, & Paulsen, 2012). The advantage to using the squid was that the size of its axon was large enough to insert a conduction electrode into the axon and one on the surface to measure electrical activity (Krebs, 1975). Similarly, the Tritonia fly was found to have large brain cells that were easier to study than those of mammals. Furthermore, the fly *Drosophila* continues to be a key species in genetics and behavior (Krebs, 1975; White, 2016; Edwards, 1986) as have other flies, worms, zebra fish, and mice, all for practical reasons of having short generation times and ease of husbandry in a laboratory setting (White, 2016).

Over the past 25 years, the expanding field of neurosciences has produced an abundance of studies of brain and behavior and their underlying mechanisms using a range of animal models. Naturally, there are inherent problems with building models for cross-species comparisons and making generalizations with regard to brain and behavior relationships (Kolb & Whishaw, 1983). Mammals that are used for research in behavioral neuroscience have at least analogous brain structures and cellular physiology, but there are obvious differences in the brain size relative to body size, even in animals that are evolutionarily close to humans such as the chimpanzee (Kolb, 2005; Krubizer, 1995; Rice & Barone, 2000). Nonetheless, aside from relative variations in brain volume among mammals, numerous similarities have been found in the fundamental processes of ontogeny and organization of the developing brain that are not much different from the human brain (Krubizer, 1995; Rice & Barone, 2000). Thus, the topography of mammalian brain systems of mammals ranging from rodents to carnivores and primates also show class common neural characteristics based on neurophysiological, anatomical, and lesion studies in addition to localization of function and cerebral laterality among mammalian cortices (Kolb & Whishaw, 1983; Kolb, 2005; Krubizer, 1995). For example, although laterality of function had once been considered an exclusive human trait, there is compelling evidence of lateralization of some structures and functions in the cortices of all mammals (Kolb & Whishaw, 2015; Rogers, 2014). Research of various vertebrate and invertebrate species has highlighted asymmetry in the brains of chicks, pigeons, and zebrafish, and there is even some evidence of neural lateralization in bees (Rogers, 2014). Hemispheric specialization for auditory analysis of conspecific vocalizations has also been observed in humans as well as nonhuman primates and other mammals, such as sea lions and mice (Poremba,

Bigelow, & Rossi, 2013). Abnormalities in laterality in some models have led to hypotheses about pathological human conditions such as schizophrenia, autism, and dyslexia (Rogers, 2014).

4.4 DEVELOPMENT OF THE CENTRAL NERVOUS SYSTEM

Our understanding of the development of the nervous system began with Aristotle's detailed descriptions of the sequences of embryological development in animals, which later contributed to the development of experimental embryology during the late 19th century (Lickliter & Honeycutt, 2010). Karl von Baer's work with the embryos of chicks provided a full description of the developmental stages that still define vertebrate biology today (Swanson, 2003). He succinctly illustrated the developmental stages of the formation of the neural plate that arises from the ectoderm, endoderm, and mesoderm, to the more differentiated stages that begins to define the central nervous system, the gut, and the various other tissues, such as bone and organs (Swanson, 2003). The rejuvenation of comparative neurobiology led to the discovery of the proliferative zones of the neural tube that would ultimately form the brain and the central nervous system and further clarify theories of homology (Striedter, 2005). During the latter decades of the 19th century, two Swiss scientists, Wilhelm His and Rudolf Albert von Kölliker, were very influential with their research studies of the development of the nervous system and the cerebral cortex (Bentivoglio & Mazzarello, 1999). He focused his research on the dividing germinal cells of the ventricular epithelium and migrating neuroblasts which came to define the foundations of human neuroembryology. He also thought that the epithelial cells could be instrumental with axon direction and guidance. Meanwhile, Kölliker, who is considered the father of modern histology, had published volumes of illustrations of histological results of the developing cortex. Both investigators are considered to have inspired the studies of Camillo Golgi, Ramón y Cajal, and Giuseppe Magini who further delineated the morphogenic events of the developing cortex using the Golgi-staining technique. It was this staining technique that revealed the detection of radial glial processes, how they emerged from the ventricular epithelium, and how they were identifiable as spherical cells. Some of these spherical cells emerged with the ependyma, whereas others were found higher up the very thin radial filaments. Numerous illustrations of the developing brain were revealed by using the Golgi-stain technique which was performed on human fetuses, calves, rabbits, dogs, guinea pigs, and invertebrates to arrive at a description of the neurodevelopmental stages in mammals. It was not until the heroic electron microscope studies of Pascko Rakic, however, that there was clear evidence of radially arranged glial processes that served as the foundation for migrating neuroblasts that ultimately form all of the neuronal components and the architecture of the brain (Bentivoglio & Mazzarello, 1999; Lui, Hansen, & Kriegstein, 2011).

Having described a basic neurodevelopmental scaffold for various species, it became possible for researchers to become more inquisitive about the processes of how the brain forms in a choreographed manner to form a network of functional parts (Krubizer, 1995). As advances in cellular and molecular technologies progress, animal studies have continued to reveal important findings about the developing brain. For example, cellular and genetic markers became increasingly seen as essential to maintaining the normal timing and expansion of the progenitor epithelial cells lining the ventricular zone of the neural tube that would eventually form the neocortex. Many investigations that currently define the markers critical for completion of the different developmental stages of the human brain come about largely come from rat and mouse models (Lui et al., 2011) in addition to parallel studies using cats, ferrets, and nonhuman primates (de Graff-Peters & Hadders-Algra, 2005; Hansen, Lui, Parker, & Kriegstein, 2010; Rakic, 2009). Evolutionary studies of diverse mammalian and nonmammalian species have also contributed greatly toward our understanding of brain development with the findings that mammals share many conserved genetic markers with other species as well as with humans (Krubizer, 1995; Martínez-Cerdeño et al., 2012; Poluch & Juliano, 2015).

In order to gain a better appreciation of the developmental events that ultimately produce a brain and an understanding of mechanisms involved, it is critical to understand the timing of these events with respect to birth and the maturation rate of the developing brain (Clancy, Finlay, Darlington, & Anand, 2007; Dobbing & Sands, 1973; Orr, Garbarino, Sallinas, & Buffentein, 2016; Rice & Barone, 2000; Workman, Charvet, Clancy, Darlington, & Finlay, 2013). This has been an enduring problem for comparative developmental research and the ability to extrapolate research findings from one species to another, particularly with that of humans is critical (Workman et al., 2013). The state of brain development in altricial mammals, which are born at an immature stage of nervous system development, requires more care for continued development; in contrast to a having a more mature brain at birth in precocial mammals. Thus, both the motor and cognitive capacity of newborns varies widely and is related of the length of gestational development. From the perspective of physical anthropology, nervous system development at birth can be defined by the timing of maturation at birth that forms a continuum of developmental stages in birds and mammals; however, this approach has proved difficult to use to classify human infants. Humans have actually been characterized as having "secondary altriciality" or as "exterogestate fetuses" to describe the prolonged gestation, but developmental immaturity at birth that is reflected by the helplessness of human newborns (Rosenberg & Trevathan, 2015). Thus, extrapolating research investigating the stages of brain development in animals to that of humans creates complications that require careful cross-species comparisons. It is vital to take into consideration differences in brain size and surface area, shape and in complexity between lissencephalic vs gyrencephalic brains of the various animal species.

Like the human brain, the rodent cortex also develops into roughly six layers with regional organizations in cortical topography of sensory, motor, and association areas (Lui et al., 2011), but unlike the human brain, it is smooth or lissencephalic. Modern biologists have suggested that some of the differences in gyration during cortical development can be attributed to changes in the length of the neurogenic period (Lewitus, Kelava, Kalinka, Tomancak, & Huttner, 2014) and in alterations in the timing of the neurogenic period that results in an increase in the number of progenitor cells (Caviness, Takahashi, & Nowakowski, 1995; Poluch & Juliano, 2015). This notable difference has been identified among various mammalian species and is considered to contribute to the number of proliferating cells in the subventricular zone that positively correlates with the degree of cortical gyrification (Lui et al., 2011). Brain development among mammals also share cytoarchitectonic features of having a cortex that emerges from a cellular sheet composed of pyramidal cells and interneurons arranged into horizontal layers that are interconnected by vertical radial processes and share an intrinsic connectivity (Rakic, 2009). The source of the cellular sheet is generated from proliferative and transient embryonic zones (the ventricular zone and subventricular zone) and the cortex is known to develop according to an "inside-out" gradient. Although humans have a larger brain with more extensive gyrification than other mammals, corticogenesis in most mammals is completed before or around the time of birth (Rakic, 2009).

Although the predominant species for studying brain development are rats and mice, other species can be considered a better "fit" for studying certain developmental processes. For example, the naked mole rat lives as long as 31 years of age and has a brain maturation rate that appears to be closer to that of primates at birth than that of mice or rats (Orr et al., 2016). This long-lived species has been suggested to provide many benefits for understanding brain developmental processes that are more compressed in the standard short-lived rodent models (Orr et al., 2016). The rodent *Octodon degus* is born with relatively mature sensory systems—their eyes are open and the auditory system is functional, and they are able to perceive and interact with their environment at a developmentally similar age as primates (Bock & Braun, 2011; Collonello, Iacobucci, Fuchs, Newberry, & Panksepp, 2011). Degus also share similarities with primates with respect to their complex social and family structures, their vocal communication systems, and in-play behavior. Like primates, and unlike other rodents, degu infants show patterns of attachments to caregivers (secure vs insecure) that resemble the pattern in primates. For this reason, degus have been proposed as a useful model for studying human attachment (Bock & Braun, 2011; Collonello et al., 2011). Degus also have an extended period of infancy with a dependence on close family members that could allow this species to be an animal model for the study of socioaffective behaviors through the developmental stages (Collonello et al., 2011; Nelson & Panksepp, 1998). Guinea pigs and spiny mice are precocial as they are born more mature than other rodents with their eyes open with fairly mature mobility. From this perspective, they could be useful models for studies of "toddler" behavior with the advantage that this phase of development begins without the effects of postnatal infant experience (Clancy et al., 2007).

The neuroinformatics model provides another approach to compare developmental events across species. Numerous research reports that measure developmental events in mammals have been generated. They include information regarding morphological comparisons, susceptibility patterns based on the "rule of thumb" method, and event-based analyses. All of these methodologies have contributed to the field of neuroinformatics that are comprised of neuroscience, evolutionary science, statistical modeling, and computer science (Clancy, Darlington, & Finlay, 2001). The initial model generated a timetable of 102 neural events across 10 mammalian species (that included humans) and showed that the timing and sequence of events that comprise brain development is remarkably conserved across mammalian species. Recent restructuring of this model surveyed 18 mammalian species and revealed 271 developmental events, in addition to describing some changes in the timing, or rate, of developmental events (so-called heterochronic changes) in brain evolution (Workman et al., 2013). Fig. 4.1 shows a comparison of brain development in five representative species over a selection from the 271 developmental events. Fig. 4.2 illustrates similar data relative to postconception days in the same species. It is clear that the date of birth shows a large variance across species relative to the developmental events. For example, the critical period for ocular dominance column plasticity is estimated to begin at event period 0.770 and ends at 0.901. Thus, the critical period is around birth for the macaque, but well after birth for cats and humans. Higher scores correlate with the later timing of events with respect to the duration of neurodevelopmental events. This comparison model allows cross species comparisons across developmental stages without considering the effects of postnatal life experiences that will alter brain development into adulthood (Workman et al., 2013). A notable advantage for developmental neuroscience to using this type of model with cross-species comparisons is that it suggests that the critical periods of perinatal development are the most accurate periods to make comparisons with humans (Clancy et al., 2007). The restructured informatics model includes an extension of developmental events to approximately 3 years postnatally in humans; an extension that allows for events during critical periods, such as early stress, and epigenetic modifications during brain development. The overall goal for this research is to determine how the different developmental rates and life histories can be incorporated into an evo–devo context of brain development; research that requires a large comparative database (Workman et al., 2013).

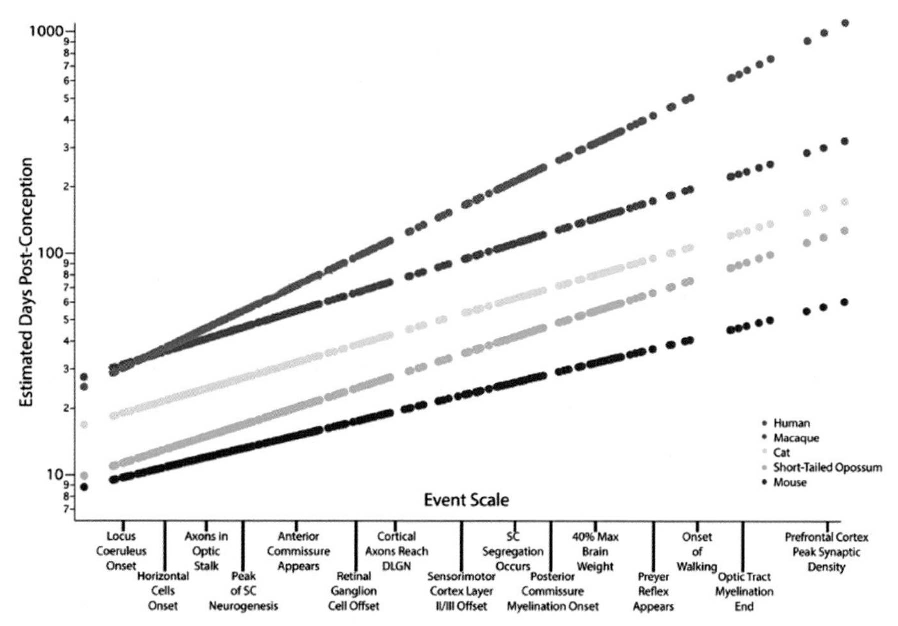

Figure 4.1 Predicted developmental schedules for five species of mammals. The dots represent 271 different developmental measurements in each species. The general pattern of development of the different regions is similar across species, although the postconceptual day of the different events varies. *Modified from Workman et al. (2013).*

Evo–devo is a subfield of biology that reflects the relationship of the two disciplines (evolution and development) that are considered to form the backbone of biological thinking (Innocenti, 2011). The realization that new biological features can arise through various inherited mutations in genetic expression and continue to be expressed among species through propagation by natural selection are now considered to be dynamically coupled (Innocenti, 2011; Rakic, 2009). It has been suggested that this theory should also be distinguished from Haeckel's law of ontogenetic recapitulation of phylogeny that was adopted in the 19th century (Rakic, 2009). Ernst Haeckel was a contemporary of Charles Darwin who compared embryological data to the building of phylogenetic trees and believed that an organisms evolutionary history could be read from its developmental history, hence the term "ontogeny recapitulates phylogeny" (Striedter, 2005).

4.5 THE MIXING OF GENES AND ENVIRONMENT IN DEVELOPMENT

For many years, developmental neurobiologists considered brain development processes to be deterministic, or unidirectional, such that these processes of development were a natural consequence of brain maturation (Stiles, 2011). The recent surge in

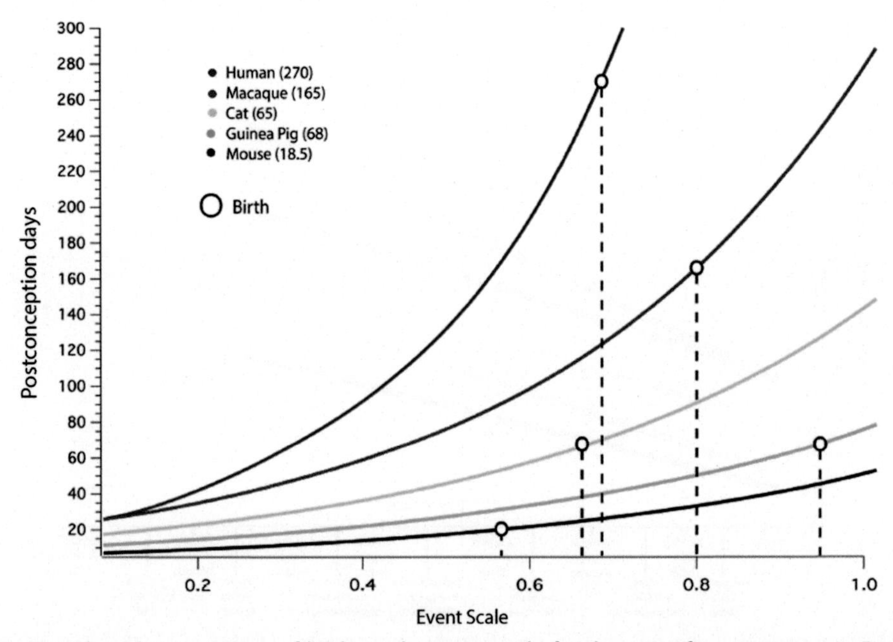

Figure 4.2 The relative position of birth on the event scale for the same five species as in Fig. 4.1. The mouse is born very early relative to brain events, whereas guinea pigs are born very late on the event scale. *From Workman et al. (2013).*

neuroscience research that encompasses both the biological intrinsic factors with the experiential factors has led to our understanding of the role of epigenetics in brain and behavioral development (see Chapter 7: Epigenetics and Genetics of Brain Development for an extensive discussion). Abnormalities in either patterns of gene expression or experience can lead to aberrant brain development. The process of brain development is dynamic, and it is thought that each individual will express a different balance on a continuum between inherited factors and environment (Stiles, 2011). Many insights into how the balance is achieved came from manipulations of early infant experiences. For example, Harlow showed that if infant monkeys were removed from their mothers at a few hours of age and raised in individual nurseries, the infants grew well and were healthy, but their behavior was not normal. They would also clasp themselves, rock, suck their thumbs and stare into space. When older and placed with other younger monkeys, they interacted poorly with others, if at all. If given a chance to soothe by hugging a diaper, the infants behaved much like Linus in the Peanuts cartoon strip. This was a pivotal observation leading Harlow to study the role of contact comfort in brain and behavioral development (for a recent review, see Suomi & Kolb, 2017). Similarly, the landmark studies of the effects of monocular deprivation of visual development by Wiesel and Hubel (1970) revealed that the removal of vision in one eye resulted in the elimination and rearrangement of connectivity in the brain that changed the course of ocular development as well as that of the brain (see Chapter 6: Critical Periods and Cortical Development for a detailed discussion). Finally, environmental manipulation, such as raising animals in complex environments or providing enhanced tactile stimulation produces marked changes in brain and behavioral development and gene expression (see Chapter 13: Brain Plasticity and Experience for more details).

4.6 ABERRANT BRAIN DEVELOPMENT

Previous research models of brain development have been successful for a close inspection of how the brain changes in response to various external or internal factors and how this affects behavior in the developing animal of choice. Models include manipulation of sensory and motor experiences, preconception stress and gestational stress, the effects of psychoactive drugs, inadequate diet, gene manipulations, and perinatal brain injury (see Chapter 13: Brain Plasticity and Experience for more details). Even paternal influences on brain development have been modeled using degus. Degus are bi-parental and offspring that are solely raised by the mother showed delays and/or impairments in the development of cortical synaptic circuits in the prefrontal cortex (Bock & Braun, 2011). Such studies have greatly contributed to our understanding of how the brain develops and what consequences can occur as a result of aberrant brain development or brain injury.

One challenge in using animal models is to find appropriate ways to replicate human psychiatric disorders. A major obstacle is how to identify and study—emotions such as sadness, guilt, delusions, or hallucinations. This is further complicated by the fact that diagnoses in humans are based upon behavioral signs rather than on the identification of the aberrant neural circuits and this likely occurs because aberrant neurological development has not been researched as thoroughly in humans as it has in laboratory animals (Nestler & Hyman, 2010). It is worth noting that although many emotions may be difficult to study in animals, there is clear evidence that nonhumans experience a wide range of emotions that develop much as those in humans (Montag & Panksepp, 2017). For example, aside from fear and anger, rhesus monkeys are born with multiple emotions that are necessary for the development of the mother–infant bond (Suomi, 2004). Once attachment and home base is established with the mother, infant monkeys can begin to explore their immediate environment and social surroundings. If they experience fear, they understand they can return to home base and their mother's protection. Through the social interactions and play activity into adolescence, both males and females learn to regulate their emotions based on their cognitive, physical, and social development. Infants that are deprived of this normal stage of development are known to exhibit socially inappropriate aggression later in life (Suomi, 2004). Panksepp and Lahvis (2011) reviewed a series of studies in mice and rats and concluded that rodents have the ability to experience empathy. He noted that rodents are capable of experiencing perceptions of social distress and shared experience of pain as "emotional contagion." Rodents are also capable of social modulation of associative learning through a cued-fear paradigm, and that rodents share the affect component of empathy. The advantage of this type of research is that if rodents do have the capacity to express empathy, this could be an opportunity to study the genetic substrates of empathy that could develop into pharmacotherapeutic intervention (Panksepp & Lahvis, 2011).

Until recently, the majority of published research was sex biased with the predominant use of males, in part to avoid the complication of the female estrous cycle. But, there is growing evidence that behavioral and neuroanatomical studies of males, including both human and nonhuman species do not represent changes in females, particularly with respect to drug responses, certain cognitive and emotional functions, and autoimmune function (Beery & Zucker, 2011).

Developing animal models for research in brain and behavioral development require evidence of face, predictive, and construct validity (Stewart & Kaleuff, 2015). Further arguments for supplementing biomedical neuroscience research with the use of comparative approaches are to consider using an appropriate species suited to specific research questions. These arguments arise from knowledge that not one species of animal is suitable for solving all the known medical diseases, neurodevelopmental disorders and basic developmental neurobiology (Bateson, 2010; Phelps, Campbell, Zheng, & Ophir, 2010; Nestler & Hyman, 2010).

Stewart and Kaleuff (2015) have recently proposed animal models to include a high validity for cross-species validation, or models of evolutionary validity, that target long-conserved behavioral domains and their underlying pathological mechanisms across various species as opposed to using species in closer taxa. Their argument is that interrelational validity would be able to capture the multiple overlapping domains found in comorbid disorders that comprise some well-defined neuropsychiatric disorders by using an animal model that can target the interplay among the domains of various human neuropsychiatric disorders. For example, anxiety is a physiological state that is known to co-occur with cognitive deficits and social phobias. There are three possible behavioral traits to address with overlapping domains. If an animal model could mimic all phenotypes of their disorder of interest, it would increase the interrelational validity of the model (Stewart & Kaleuff, 2015).

4.7 CONCLUSIONS

The use of nonhuman species to study brain development began in the early 20th century, in large part because of the challenge in studying brain development in the living human brain. Although there are clear limitations to the questions that animal models can address, animal models still provide the only means to investigate the underlying mechanisms of brain and behavioral development. Animal studies have allowed an understanding of the organizing principles of cortical development and function, as well as the constraints that limit or alter development. As noted forcefully by Leah Krubitzer (e.g., 1995), comparative neurobiology will remain "an integral part of attempts to understand the functional organization of the neocortex and ultimately, the evolution of more complex functions that are generated by the neocortex, such as perception, cognition, and consciousness."

Recently, advances in cell and molecular biology have contributed to the study of human brain development and have revealed quantitative and qualitative differences in gene composition and expression between humans with that of rodents and primates (Rakic, 2009). New techniques such as genetically modified rodents and techniques of optogenetics (using light to turn genes off and on) are allowing more precise control of brain activity and its relation to behavior. Although this type of work has mostly been used in adult animals, the application to developmental questions will certainly evolve over the next decade.

Finally, we must note that the growing use of noninvasive imaging is providing new insights into brain development in both humans and nonhuman animal models (see Chapter 1: Overview of Brain Development). Although it might be ideal to follow human brain development using various magnetic resonance imaging (MRI) techniques, there is a difficulty in ensuring that children between newborns and about the age of five remain still for extended periods of time in the MRI. Laboratory animals can be anesthetized repeatedly over time to follow the processes of brain development, but such techniques exclude the possibility of studying function using functional MRI. Nonetheless, we can foresee a time when both noninvasive imaging and more traditional neurobiological methods will be used in concert to study the developing brain.

REFERENCES

Bateson, P. (2010). The value of truly comparative and holistic approaches in the neurosciences. In M. S. Blumberg, J. H. Freeman, & S. R. Robinson (Eds.), *Oxford handbook of developmental behavioural neuroscience* (p. 7). New York, NY: Oxford University Press.

Batista-Brito, R., & Fishell, G. (2009). Developmental integration of cortical neurons. In O. Hobert (Ed.), *Development of neural circuitry* (p. 82). San Diego, CA, Burlingham, MA, USA, London, Oxford, UK: Academic Press, Elsevier Inc.

Beery, A. K., & Kauffer, D. (2015). Stress, social behavior, and resilience: Insights from rodents. *Neurobiology of Stress*, *1*, 116–127.

Beery, A. K., & Zucker, I. (2011). Sex bias in neuroscience and biomedical research. *Neuroscience and Behavioural Reviews*, *35*(2011), 565–572.

Bentivoglio, M., & Mazzarello, P. (1999). The history of radial glia. *Brain Research Bulletin*, *49*, 305–315.

Bock, J., & Braun, K. (2011). The impact of perinatal stress on the functional maturation of prefronto-cortical synaptic circuits: Implications for the pathophysiology of ADHD? In O. Braddick, J. Atkinson, & G. Innocenti (Eds.), *Progress in brain research* (vol. 189, pp. 155–169).

Catterall, W. A., Raman, I. M., Robinson, T. J., Sejnowski, T. J., & Paulsen, O. (2012). The Hodkin–Huxley heritage: From channels to circuits. *Journal of Neuroscience*, *32*(41), 14064–14073.

Caviness, V. S., Jr., Takahashi, T., & Nowakowski, R. S. (1995). Numbers, time, and neocortical neurogenesis: a general developmental and evolutionary model. *Trends in Neurosciences*, *18*(9), 379–382.

Clancy, B., Darlington, R. B., & Finlay, B. L. (2001). Translating developmental time across mammalian species. *Neuroscience*, *105*(1), 7–17.

Clancy, B., Finlay, B. L., Darlington, R. B., & Anand, K. J. S. (2007). Extrapolating brain development from experimental species to humans. *NeuroToxicology*, *28*, 931–937.

Collonello, V., Iacobucci, P., Fuchs, T., Newberry, R. C., & Pankseep, J. (2011). *Octodon degus*. A useful animal model for social-affective neuroscience research: Basic description of separate distress, social attachments, and play. *Neuroscience and Biobehavioral Reviews*, *35*, 1854–1863.

De Graff-Peters, V. B., & Hadders-Alga, M. (2005). Ontogeny of the human central nervous system. *Early Human Development*, *82*, 257–266.

Dobbing, J., & Sands, J. (1973). Quantitative growth and development of human brain. *Archives of Disease in Childhood*, *48*, 757–767.

Edwards, J. S. (1986). Pathways and changing connections in the developing insect nervous system. In W. T. Greenough, & J. J. Juraska (Eds.), *Developmental neuropsychobiology* (p. 73). Orlando FL: Academic Press Inc.

Hansen, D. V., Lui, J. H., Parker, P. R. L., & Kriegstein, A. (2010). Neurogenic radial glia in the outer subventricular zone of human neocortex. *Nature*, *464*, 554–563.

Hodgkin, A. L., & Huxley, A. F. (1952). Currents carried by sodium and potassium ions through the membrane of the giant axon of *Loligo*. *Journal of Physiology,*, *116*, 449–472.

Innocenti, G. M. (2011). Development and evolution: Two determinants of cortical connectivityIn O. Braddick, J. Atkinson, & G. Innocenti (Eds.), *Progress in Brain Research* (189, pp. 65–75).

Jones, E. G. (2007). Neuroanatomy: Cajal and after Cajal. *Brain Research Reviews*, *55*, 248–255.

Kambi, N., & Jain, N. (2012). Landmark discoveries in neurosciences. *Resonance*, *189*, 1054–1064.

Kolb, B. (2005). Neurological models. In I. Q. Whishaw, & B. Kolb (Eds.), *The behaviour of the laboratory rat* (pp. 449–459). New York, NY: Oxford University Press.

Kolb, B., & Whishaw, I. (1983). Problems and principles underlying interspecies comparisons. In T. E. Robinson (Ed.), *Behavioural approaches to brain research*. New York, NY: Oxford University Press, (pp. 237, 253).

Kolb, B., & Whishaw, I. Q. (2015). *Fundamental of human neuropsychology* (7th ed.New York, NY: Worth/MacMillan.

Krebs, H. A. (1975). The August Krogh principle: For many problems there is an animal on which it can be most conveniently studied. *Journal of Experimental Zoology*, *194*, 221–226.

Krogh, A. (1929). Progress of physiology. *American Journal of Physiology*, *90*, 243–251.

Krubizer, L. (1995). The organization of neocortex in mammals: Are species really so different? *Trends in Neuroscience*, *18*(9), 408–417.

Lewitus, E., Kelava, I., Kalinka, A. T., Tomancak, P., & Huttner, W. B. (2014). An adaptive threshold in mammalian neocortical evolution. *PLoS Biology*, *12*(11), 1–15 , e1002000.

Lickliter, R., & Honeycutt, H. (2010). Rethinking epigenesist and evolution inlight of developmental science. In M. S. Blumberg, J. H. Freeman, & S. R. Robinson (Eds.), *Oxford handbook of developmental behavioural neuroscience*. New York, NY: : Oxford University Press, (pp. 30, 31).

Lui, J. H., Hansen, D. V., & Kriegstein, A. R. (2011). Development and evolution of the human neocortex. *Cell*, *146*, 18–36.

Martínez-Cerdeño, V., Cunningham, C. L., Camacho, J., Antczak, J. L., Prakash, A. N., Cziep, M. E., et al. (2012). Comparative analysis of the subventricular zone in rat, ferret and macaque: Evidence for an outer subventricular zone in rodents. *PLoS One*, 7(1), e30178. Available from http://dx.doi.org/10.1371/journal.pone.0030178.

Montag, C., & Panksepp, J. (2017). Primary emotional systems and personality: An evolutionary perspective. *Frontiers in Psychology, 8*, 464. Available from http://dx.doi.org/10.3389/fpsyg.2017.00464.
Nelson, E. E., & Panksepp, J. (1998). Brain substrates of infant-mother attachment: contributions of opioid, oxytocin, and norepinephrine. *Neuroscience and Biobehavioural Reviews, 22*(3), 437–452.
Nestler, E. J., & Hyman, S. E. (2010). Animal models of neuropsychiatric disorders. *Nature Neuroscience, 13*(10), 1161–1169.
Orr, M., Garbarino, V. R., Sallinas, A., & Buffentein, R. (2016). Extended postnatal brain development in the longest-lived rodent: prolonged maintenance of neonotenous traits in the naked mole-rat brain. *Frontiers in Neuroscience, 10*, 1–17.
Panksepp, J., & Lahvis, G. P. (2011). Rodent empathy and affective neuroscience. *Neuroscience and Biobehavioural Reviews, 35*, 1864–1875.
Partsalis, A. M., Blazquez, P. M., & Triarhou, L. C. (2013). The renaissance of the neuron doctrine: Cajal rebutes the rector of Granada. *Translational Neuroscience, 4*(1), 104–114.
Phelps, S. M., Campbell, P., Zheng, D., & Ophir, A. G. (2010). Beating the boojum: comparative approaches to the neurobiology of social behavior. *Neuropharmacology, 58*, 17–28.
Poluch, S., & Juliano, S. L. (2015). Fine-tuning of neurogenesis is essential for the evolutionary expansion of the cerebral cortex. *Cerebral Cortex, 25*, 346–364.
Poremba, A., Bigelow, J., & Rossi, B. (2013). Processing of communication sounds: Contributions of learning, memory and experience. *Hearing Research, 305*, 31–44.
Rakic, P. (2009). Evolution of the neocortex: a perspective from developmental biology. *Nature, 10*, 724–735.
Rice, D., & Barone, S., Jr. (2000). Critical Periods of Vulnerability for the developing nervous system: Evidence from human and animal models. *Environmental Health Perspectives, 108*(Suppl 3), 511–533.
Rogers, L. J. (2014). Asymmetry of brain and behavior in animals: Its development function and human relevance. *Genesis, 52*, 555–571.
Rosenberg, K. R., & Trevathan, W. R. (2015). Are humans altricial? [Abstract]. The 84th Annual Meeting of the American Association of Physical Anthropologists. *American Journal of Physical Anthropology, 156*, 65.
Stewart, A. M., & Kaleuff, A. V. (2015). Developing better and more valid animal models of brain disorders. *Behavioural Brain Research, 276*, 28–31.
Stiles, J. (2011). Brain development and the nature versus nurture debateIn O. Braddick, J. Atkinson, & G. Innocenti (Eds.), *Progress in Brain Research* (189, pp. 3–22).
Striedter, G. F. (2005). *Principles of brain evolution* (pp. 20–49). Sunderland, MA: Sinauer Associates.
Suomi, S. J., & Kolb, B. (2017). Revisiting Harlow: Love in infant monkeys. In I. Q. Kolb,, & I. Q. Whishaw (Eds.), *Revisiting the classic studies in behavioural neuroscience*. London: Sage.
Suomi, S. J. (2004). How gene-environment interactions influence emotional development in Rhesus monkeys. In C. G. Cell, E. L. Bearer, & R. M. Lerner (Eds.), *Nature and nurture. The complex interplay of genetic and environmental influences on human brain development* (pp. 35–51). Mahwah, NJ: Lawrence Erlebaum Associates, Inc.
Swanson, L. W. (2003). *Brain architecture. Understanding the basic plan*. New York, NY: Oxford University Press, (pp. 5, 45).
White, B. J. (2016). What genetic model organisms offer the study of behavior and neural circuits. *Journal of Neurogenetics, 30*(2), 54–61. Available from http://dx.doi.org/10.1080/01677063.2016.1177049.
Wiesel, T. N., & Hubel, D. H. (1970). The period of susceptibility to the physiological effects of unilateral eye closure in kittens. *Journal of Physiology (London), 206*, 419–436.
Workman, A. D., Charvet, C. J., Clancy, B., Darlington, R. B., & Finlay, B. L. (2013). Modeling transformations of neurodevelopmental sequences across mammalian species. *The Journal of Neuroscience, 33*(17), 7368–7383.

PART II

Molecular Perspectives in Brain Development

CHAPTER 5

Stem Cells to Function

Maryam Faiz and Cindi M. Morshead
University of Toronto, Toronto, ON, Canada

Historically, it was believed that the central nervous system was incapable of regeneration. While most new neurons are produced in the embryonic period, it is now evident that NSCs are found in two regions of the adult mammalian brain. These two regions, the SE, which borders the lateral ventricles in the forebrain and the SGZ of the dentate gyrus in the hippocampus, produce new neurons into adulthood. Within these well-defined neurogenic niches, new cells are continuously generated from NSCs. Herein we will focus on forebrain neurogenesis derived from SE lining the lateral ventricles as they cells have been shown to contribute to neurogenesis in models of injury and disease following birth and into adulthood.

5.1 HISTORY OF NEUROGENESIS

In 1962, Joseph Altman proposed the idea of adult neurogenesis based on his observation that Thymidine-H^3 had incorporated into neurons after brain injury in rats (Altman, 1962). In 1965, Altman and Das provided the first evidence of postnatal neurogenesis in the rat using radioactively tagged thymidine to label dividing cells (Box 5.1) (Altman & Das, 1965) and in 1969, he proposed the idea that progenitors migrate from the SE through the rostral migratory stream (RMS) to the olfactory bulb, where they produce new neurons (Altman, 1969). These findings were largely ignored until research in the 1970s and 1980s by Kaplan and Hinds showed neurogenesis in the dentate gyrus (Kaplan & Bell, 1983) and olfactory bulb (Kaplan & Hinds, 1977) of adult rats. These findings however, were also met with skepticism and were thought to be irrelevant to humans as early studies in primates (Rakic, 1974) revealed no evidence to support ongoing neurogenesis. Interest in neurogenesis was revived in 1985 by work done by Fernando Nottebohm's group who showed functional integration of newly born neurons in neuronal circuits in songbirds (Nottebohm, 1985). This work in songbirds was quickly followed up by numerous studies in the 1990s that used bromodeoxyuridine (BrdU) to label newborn cells (Kuhn, Dickinson-Anson, & Gage, 1996), reinforcing the extent of neurogenesis in rodents (Box 5.1). Notably, it was the work of Eriksson et al. (1998) showing neurogenesis in humans that finally reversed the long-standing dogma that the adult brain was aneurogenic.

The Neurobiology of Brain and Behavioral Development
DOI: http://dx.doi.org/10.1016/B978-0-12-804036-2.00005-4

5.2 ADULT NEURAL STEM CELLS—DEFINITION, ORIGIN, & LOCATION

Even with the dogma dispelled and the belief that the adult brain continued to generate new neurons, the idea that resident neural stem cells (NSCs) were present in the adult central nervous system (CNS) was not a prominent view. This changed in 1992 when Reynolds and Weiss (Reynolds & Weiss, 1992) isolated cells from the adult CNS that displayed the cardinal properties of stem cells (Box 5.1). Irrespective of the tissue of origin, stem cells are minimally defined by their functional characteristics: proliferation, self-renewal, generation of progeny through transit-amplifying progenitors, and multilineage potential over time (Weiss et al., 1996b). With their isolation in vitro using a simple colony-forming assay (termed the neurosphere assay), the question of where the stem cells resided in the brain was an obvious one. Subsequent experiments determined that forebrain derived NSCs comprised a rare population of cells in a well-defined region known as the SE which lines the lateral ventricles (Morshead et al., 1994). Later experiments revealed that NSC derived neurospheres could be isolated from along the entire length of the ventricular neuroaxis (Weiss et al., 1996a; Vescovi, Reynolds, Fraser, & Weiss, 1993). Hence, while it is generally accepted that neurogenesis does not occur outside the two neurogenic regions in the brain, the SE and the subgranular zone (SGZ) in the hippocampus, NSCs are also found surrounding the central canal in the spinal cord.

BOX 5.1 Techniques that facilitated the discovery of neurogenesis

^{3}H-Thymidine and 5′ Bromodeoxyuridine (BrdU)

^{3}H-Thymidine is a radioactively labeled thymidine that is incorporated into the DNA of dividing cells in S phase and was used in the first experiments defining neurogenesis. BrdU is a newer, synthetic analogue of Thymidine. As with ^{3}H-Thymidine, BrdU labels dividing cells but it is preferred for experimental studies due to the lack of radioactivity. For experiments designed to study neurogenesis, BrdU is delivered to animals either by injection or in the drinking water (Wojtowicz & Kee, 2006) for various periods of time depending on the analysis to be performed. BrdU that is available when cells are synthesizing new DNA will be incorporated instead of Thymidine to produce the new strand of DNA. Notably, BrdU that is incorporated in the DNA will be diluted upon division as it is passed on to daughter cells hence rapidly dividing cells will lose their labeling before slowly dividing cells or those that enter mitotic quiescence. BrdU is detected using standard immunohistochemistry with an anti-BrdU antibody. A number of other analogues have been developed, including CLdU and IdU, which have been used in combination with BrdU in elegant investigations that detail the proliferation kinetics of progenitor cells (Cameron & McKay, 2001; Hayes & Nowakowski, 2002). By varying the pulsing paradigm and time points analyzed, this technique enables quantitative analysis of proliferation, differentiation, and survival of newborn cells (Kempermann, Kuhn, & Gage, 1997; Miller & Nowakowski, 1988).

(Continued)

BOX 5.1 Techniques that facilitated the discovery of neurogenesis (Continued)

Neurosphere Assay

Reynolds and Weiss (1992) were the first to isolate neural stem cells from the adult brain. When tissue was isolated using a relatively crude dissection of the adult forebrain, dissociated and cultured in serum free media in the presence of epidermal growth factor (EGF), a rare population of surviving cells proliferated to generate clonally derived free-floating colonies of cells over a period of 7 days. These free-floating colonies were termed neurospheres. The colonies were comprised of cells that expressed nestin, a marker of neuroepithelial stem cells. Neurospheres that were collected, dissociated into single cells and replated in the presence of mitogens, were able to generate multiple new neurospheres, revealing the-property of self-renewal. Moreover, when plated on a substrate suitable for differentiation, the cells gave rise to neurons and glia. In the field of stem cell biology, there is lack of specific markers for stem cells in general and neural stem cells are no exception. As such, neural stem cells have largely been assayed based on functional criteria making the neurosphere assay ((Fig. 5.1) Chojnacki & Weiss, 2008) a mainstay for assaying neural stem cells and the factors that modify their behavior. When used with rigor and with an understanding of it's limitations, the neurosphere assay remains a powerful tool (Pastrana, Silva-Vargas, & Doetsch, 2011; Jensen & Parmar, 2006). Considerations such as neurosphere size, cell density upon plating in culture, passaging techniques and mechanical manipulation of the cells must be taken into account (Coles-Takabe et al., 2008).

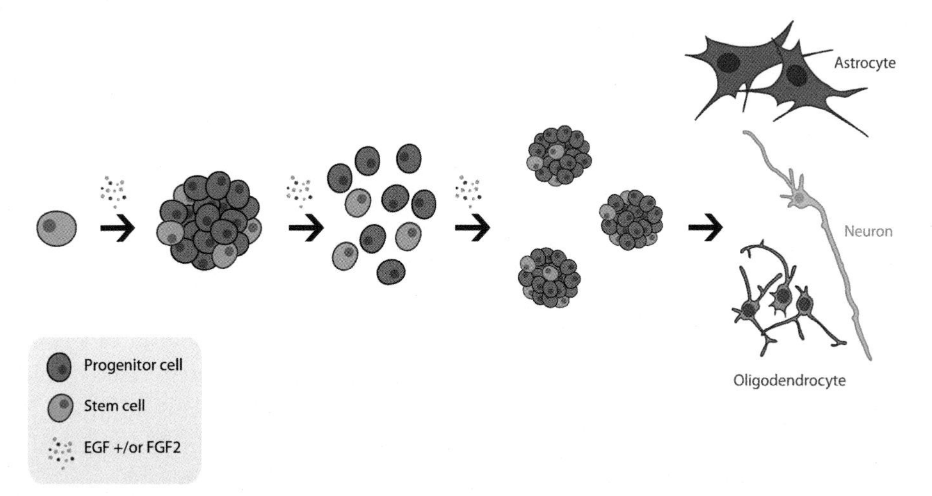

Figure 5.1 ***Neurosphere assay.*** Adult neural stem cells (NSCs) (blue) are isolated as single cells from the SE and grown in the presence of mitogens (FGF and EGF, orange and red dots). NSCs give rise to clonal colonies called neurospheres comprised of a mix of stem and progenitor cells (gray). Neurospheres can be dissociated and replated to form secondary neurospheres. When plated on substrates suitable for their differentiation, neurosphere derived cells give rise to all three neural cell types, astrocytes (purple), neurons (green) and oligodendrocytes (red). *Illustrated by Erina He.*

(*Continued*)

BOX 5.1 Techniques that facilitated the discovery of neurogenesis (Continued)

Carbon Birth Dating

Adult human neurogenesis has been studied using carbon birth dating. Nuclear bomb tests that occurred during the Cold War released a large amount of ^{14}C into the atmosphere. ^{14}C reacts with oxygen to form $^{14}CO_2$, which is taken up by plants during photosynthesis. The subsequent consumption of ^{14}C containing plant or animal matter results in similar concentrations of ^{14}C in the human body as are found in the atmosphere at any given time. When cells duplicate their DNA during cell division, ^{14}C will necessarily integrate into the genome. The concentration of ^{14}C remains stable within the cell after its last cell division. Hence, by correlating the measured genomic ^{14}C concentration to levels in the atmosphere, the birth date of cells can be determined (Bergmann, Spalding, & Frisen, 2015).

5.3 THE NSC CONTINUUM—FROM EMBRYONIC DEVELOPMENT TO ADULTHOOD

Neurodevelopment begins with neural plate induction at embryonic day 7.5 in the mouse (E7.5) followed by neural tube formation at E8.5. Neuroepithelial cells (NEPs) that make up the neural tube are thought to comprise the earliest NSC population (Fig. 5.2). At the onset of cortical development, NEPs divide symmetrically to produce two identical stem cells, contributing to the growth of the cortical primordium (for review see Conti & Cattaneo, 2010). When isolated and cultured in the presence of bFGF, NEPs demonstrate self-renewal and multipotency as revealed using the neurosphere assay. By E9.5, NEPs begin to transition into a distinct progenitor population known as radial glia (RG) (for reviews see Gotz & Huttner, 2005; Kriegstein & Alvarez-Buylla, 2009). By this point in development, the germinal zone is thought to have two regions, the ventricular zone (VZ), directly adjacent to the fluid filled ventricle, and the sub-VZ (SVZ) next to the VZ. The VZ contains the RG cells that divide asymmetrically to generate one self-renewing progenitor and one neuron or intermediate progenitor cell (Noctor, Martinez-Cerdeno, Ivic, & Kriegstein, 2004; Noctor, Martinez-Cerdeno, & Kriegstein, 2008; Haubensak, Attardo, Denk, & Huttner, 2004). These proliferating intermediate progenitor cells migrate away from the germinal zone to give rise to most of the postmitotic neurons in the cortex (Haubensak et al., 2004). As development continues, it is thought that the VZ decreases in size while the SVZ expands and becomes the major site of cell genesis (Brazel, Romanko, Rothstein, & Levison, 2003). It is hypothesized that the VZ transitions into a nonmitotic ependymal cell layer, while the SVZ remains mitotically active and comprises the SE. In the postnatal period, radial glia either transform into astrocytes that are found in the parenchyma or remain in the SE in the adult CNS. Notably, the SE is often referred to as the adult SVZ or VZ–SVZ in the literature (Merkle, Tramontin, Garcia-Verdugo, & Alvarez-Buylla, 2004; Gallagher et al., 2013). These findings support a continuum in the lineage of NEPs-RG-adult NSCs through development and into adulthood.

5.4 ADULT NEURAL STEM CELLS—THE SE NICHE

5.4.1 Cellular composition

The SE contains four basic cell types that were defined by ultrastructural and immunocytochemical studies (Garcia-Verdugo, Doetsch, Wichterle, Lim, & Alvarez-Buylla, 1998; Doetsch, Garcia-Verdugo, & Alvarez-Buylla, 1997); (1) astrocyte-like (type B) NSCs, (2) transit-amplifying progenitors (type C), (3) migrating neuroblasts (type A), and (4) ependymal cells (type E) (Fig. 5.2). Type E cells line the ventricle wall. The adjacent SE is a few cell layers thick and contains Type B cell bodies (NSCs) that extend a primary cilium into the ventricles though the center of a rosette of ependymal cells, giving the appearance of "pinwheel structures." The cell bodies of Type C and Type A cells reside within the SE. The Type B cells make contact with the vasculature found in the lateral portion of the SE (Mirzadeh, Merkle, Soriano-Navarro, Garcia-Verdugo, & Alvarez-Buylla, 2008) (Fig. 5.2).

The ependymal cells (type E cells) express the markers S100β and CD24 and are postmitotic, multiciliated cells in direct contact with the cerebral spinal fluid (CSF) in the ventricles. Through coordinated beating of their cilia they control the flow of CSF within the ventricles, which is important for the guidance of newly born neurons that are generated by the NSCs in this region (Sawamoto et al., 2006). It has been proposed that Type E cells act as self-renewing, multipotent stem cells in the adult brain (Johansson et al., 1999) however the findings from these studies have not been substantiated (Chiasson, Tropepe, Morshead, & Van Der Kooy, 1999; Laywell, Rakic, Kukekov, Holland, & Steindler, 2000).

In the adult brain, two types of B cells have been identified—type B1 and B2 cells. Slowly dividing B1 cells are generally accepted to be bona fide NSCs. B1 cells contact the ependymal layer and have a nonmotile primary cilium that extends into the lateral ventricle (Mirzadeh et al., 2008), while type B2 cells also reside within the SE, surrounding Type A cells, but do not contact the ventricle (Doetsch et al., 1997). B1 cells express markers known to be expressed by astrocytes including GFAP, vimentin, GLAST, nestin, connexin 30, Aldh1 and BLBP, and as such, are sometimes referred to as SE astrocytes. In addition, B1 cells express "stem cell markers" such as Lex1 and CD133 (reviewed in Basak & Taylor, 2009). To date, however, there are no markers that can be used to exclusively identify B1 cells. Proof for the existence of an adult GFAP+ NSC in vivo came from a series of studies using ablation paradigms and transgenic mouse models to demonstrate that a slowly dividing cell was the source of neurosphere formation (Morshead et al., 1994). The original work in 1994 used high

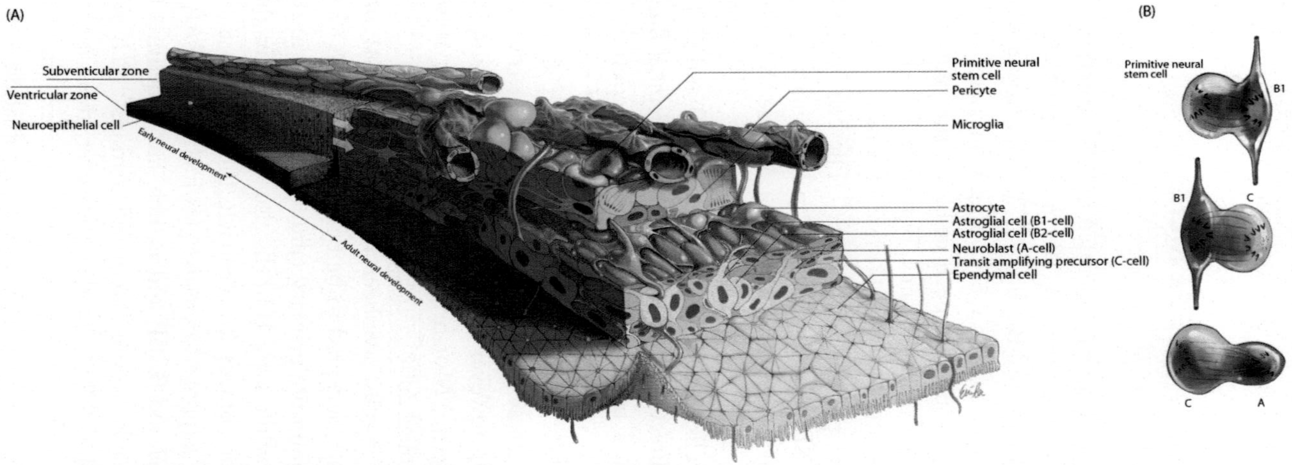

Figure 5.2 ***Neurogenesis in the subependyma.*** (A) The subependyma (SE) is a specialized niche that harbors NSCs in the adult brain. In development, neuroepithelial cells that behave as multipotent neural stem cells are found in the ventricular zone. A second germinative zone, the subventricular zone, is adjacent to the ventricular zone. As development progresses, it is hypothesized that the ventricular zone transitions into the ependymal cell layer found in the adulthood and the subventricular zone becomes the main germinative zone known as the SE. In the adult brain a single layer of multiciliated ependymal cells lines the lateral ventricle, which contains cerebral spinal fluid. The SE, a few cell layers thick and adjacent to the ependyma, harbors primitive neural stem cells, astrocyte-like B1 and B2 cells, transit-amplifying cells, and neuroblasts, as well as a specialized vasculature and various support cells (microglia and pericytes). Within the SE, B1 cell bodies (NSCs) extend a primary cilium into the ventricles through the center of a rosette of ependymal cells, giving the appearance of "pinwheel structures." (B) Image depicting the neural stem cell lineage. Primitive neural stem cells (pink) give rise to Type B1 cells (blue), which in turn generate Type C cells (transit-amplifying progenitors, green). These cells give rise to Type A cells (neuroblasts, red). *Illustrated by Erina He.*

doses of ^{3}H-thymidine to ablate actively proliferating cells in the SE. In this paradigm, a loss of progenitor cells was seen but there was no effect on the numbers of NSCs that formed neurospheres in vitro. Indeed, this study not only revealed that NSCs were relatively mitotically quiescent in vivo but further, that the spared NSC could repopulate the SE with rapidly dividing progeny (type A and type C cells) (Morshead et al., 1994). This was the first evidence that an adult NSC with the capacity to self-renew and regenerate lost cells, was present in the adult brain. This work was later confirmed and extended by using cytosine β-D arabinofuranoside (AraC), an antimitotic drug that incorporates into the DNA of dividing cells and induces apoptosis to kill the proliferating progeny of NSCs (Doetsch, Caille, Lim, Garcia-Verdugo, & Alvarez-Buylla, 1999a). Similar to the 3H-Thymidine kill, NSCs were activated and gave rise to progenitors that generated new neuroblasts. The evidence that NSCs expressed GFAP came from elegant studies by Doetsch et al. (1999a) using viral agents to specifically label GFAP expressing SE cells and showing that these cells gave rise to neuroblast progeny. Finally, two studies used GFAP-TK transgenic mice that express tyrosine kinase (TK) under the control of the GFAP promoter, to confirm that NSCs express GFAP in late embryogenesis and into adulthood. Administration of the antiviral agent ganciclovir (GCV) led to the loss of TK expressing, GFAP + NSC cells and the subsequent loss of neurospheres (Morshead, Garcia, Sofroniew, & Van Der Kooy, 2003; Imura, Kornblum, & Sofroniew, 2003). Together, these seminal studies revealed that Type B1, GFAP expressing cells are NSCs.

The direct progeny of NSCs are Type C, transit-amplifying progenitors (TAPs) that express EGF receptor and the transcription factors Ascl1 and dlx2. It has been suggested that Type C cells, in addition to type B cells, can give rise to neurospheres in the presence of EGF in vitro (Belluzzi, Benedusi, Ackman, & Loturco, 2003). However, the complete loss of neurosphere formation from GFAP-TK mice reveals that neurosphere forming cells are GFAP expressing B cells (Morshead et al., 2003). In vivo, type C cells divide and give rise to neuroblasts (Type A cells) that are characterized by their expression of doublecortin (DCX) and polysialylated neural cell adhesion molecule (PSA-NCAM). Type A cells migrate rostrally in chains along the well-defined RMS towards the olfactory bulb where they differentiate into interneurons (Lois & Alvarez-Buylla, 1994; Belluzzi et al., 2003).

Recently, studies have demonstrated regional heterogeneity within the SE surrounding the lateral ventricles. The number of neural stem/progenitor cells varies along the dorsoventral and rostrocaudal axes (Azim et al., 2012). There are regions with concentrated numbers of NSC cells termed neurogenic hot spots (Mirzadeh et al., 2008). In addition it has been shown that different types of olfactory bulb interneurons come from specific subregions of the SE surrounding the ventricles (Merkle, Mirzadeh, & Alvarez-Buylla, 2007) and that this patterning is defined in early embryogenesis (Fuentealba et al., 2015).

The SE also contains a specialized vasculature. Vessels in this region are devoid of astrocyte endfeet and pericyte coverage, which is in contrast to most vessels in the brain (Tavazoie et al., 2008). NSCs and TAPs are in contact with blood vessels within the SE (Tavazoie et al., 2008; Montanari, Tute, Beezer, & Mitchell, 1996), which permits the cells to have direct access to blood borne factors (Tavazoie et al., 2008). The SE also harbors a specialized type of microglial cells, which are distinct from their parenchymal counterparts and seem to have adapted to support adult neurogenesis. They are characterized by low expression of purinoreceptors and lack of ATP-inducible chemotaxis. Interestingly, depletion of these cells impairs the survival and migration of SE neuroblasts (Ribeiro Xavier, Kress, Goldman, Lacerda De Menezes, & Nedergaard, 2015).

5.5 DEFINING THE LINEAGE

It is generally assumed that relatively quiescent NSCs (B1 cells) give rise to transit-amplifying progenitors (C cells) that give rise to neuroblasts (A cells). As mentioned above, early studies using ablation paradigms, and later work using transgenic reporter systems to label the progeny of GFAP, Nestin or GLAST expressing NSCs (Dhaliwal & Lagace, 2011) have provided evidence for a B1 to C to A lineage relationship (Fig. 5.2). However, more recent work has uncovered a novel population of NSCs in the SE niche that lies upstream of the current lineage: a "primitive NSC" that generates B1 cells and functions as a reserve pool of NSCs in the adult brain.

5.5.1 Primitive NSCs

Neural specification is thought to occur either via instructive cues (Spemann & Mangold, 2001; Bachiller et al., 2000) or through a default pathway (Tropepe et al., 2001; Smukler, Runciman, Xu, & Van Der Kooy, 2006; Hitoshi et al., 2004). Default neural specification is supported by the fact that pluripotent embryonic stem cells will transition to a neural state in the absence of exogenous cues (Smukler et al., 2006). These early NSCs have been called "primitive NSCs" (pNSCs); and are leukemia inhibitory factor (LIF) dependent, express neural precursor markers, generate neurons and glia in vitro and can contribute to both neural and nonneural tissues in vivo (Tropepe et al., 2001). Previously, pNSCs, isolated from the embryonic day E5.5–E8.5, were thought to be a transient population that gave rise to definitive NSCs (type B1 cells) through development (Hitoshi et al., 2004). Most recently, however, pNSCs have been isolated from the adult mammalian brain (Sachewsky et al., 2014b). Similar to pNSCs in the developing embryo, the adult-derived pNSCs are LIF responsive and have a similar gene expression profile. In addition, pNSCs generate multipotent, self-renewing colonies that give rise to the GFAP expressing NSCs in vitro and repopulate the SE after neural stem and progenitor cell ablation in vivo.

Together, these findings indicate that this rare population of largely quiescent pNSCs persist in the adult SE and lie upstream of the well-described GFAP expressing Type B1 NSCs.

5.5.2 Quiescent versus active B1 cells

Recently, two stem cell states were characterized in the adult SE: quiescent NSCS (qNSCs) and activated NSCs (aNSCs) (Codega et al., 2014). Both activated and quiescent NSC are found to have processes that contact the ventricle but aNSCs express EGFR and Nestin and higher levels of Sox2 relative to qNSCs, which may contribute to the observed faster cycling times in the activated subpopulation (Codega et al., 2014). In contrast qNSCs are largely dormant, demonstrate a slower production of OB neurons compared to aNSCs, and show less frequent neurosphere formation in vitro (Codega et al., 2014). The current literature suggests that qNSCs lie upstream of aNSCs in the NSC lineage.

5.6 OLFACTORY BULB NEUROGENESIS

Neuroblasts born in the SE migrate through the RMS to the OB and then turn radially towards the granular and periglomerular layers, where they differentiate into interneurons (Fig. 5.3) (Altman, 1969; Luskin, 1993; Belluzzi et al., 2003) for review see Pignatelli and Belluzzi (2010).

5.6.1 Migration of cells to the OB

More than 30,000 neuroblasts are born in the SE each day (Alvarez-Buylla, Garcia-Verdugo, & Tramontin, 2001). Migration extends rostrally along the RMS towards the OB. Neuroblasts migrate in a "chain formation," whereby cells move alongside each other without direction from glial extensions or axonal fibers unlike what is found during neural development (Wichterle, Garcia-Verdugo, & Alvarez-Buylla, 1997). Instead, migratory cells are ensheathed by mature astrocytes in structures known as glial tubes. The glial tubes are thought to play an important role in physically separating neuroblasts from the surrounding parenchyma (Lois, Garcia-Verdugo, & Alvarez-Buylla, 1996; Doetsch et al., 1997). Neuroblasts move rapidly (about 122 μm/hour) and traverse long distances of up to 5 mm in rodents and 20 mm in monkeys, to reach the OB (Lois et al., 1996; Kornack & Rakic, 2001).

The directional migration of neuroblasts is regulated by a number of factors including chemorepulsive slit proteins that are expressed in the septum (Hu & Rutishauser, 1996; Wu et al., 1999) and choroid plexus (Hu, 1999). Despite the separation of these structures from the SE, the flow of CSF within the ventricle permits these factors to influence neuroblast migration (Sawamoto et al., 2006). The

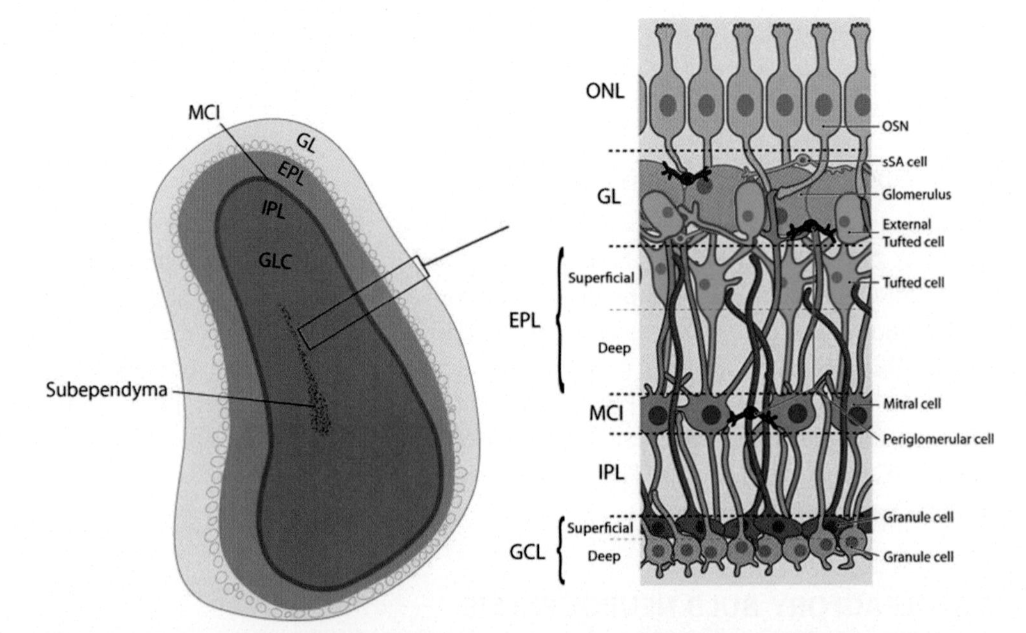

Figure 5.3 ***Structure of the olfactory bulb.*** The cross-section of the olfactory bulb shows its' laminar structure. The outermost olfactory nerve layer (ONL) consists of axons of olfactory sensory neurons that enter the glomeruli and form synapses with the dendrites of mitral tufted and periglomerular cells in the glomerular layer (GL). Juxtaglomerular neurons surround the glomeruli and consist of three cell types: periglomerular cells (red), external tufted cells (green) and superficial short-axon (sSA) cells. The principal projection neurons of the olfactory bulb, tufted (green) and mitral (blue) cells are found in the external plexiform layer (EPL) and mitral cell layer (MCL), respectively. These cells make dendrodendritic synapses with glomerular interneurons and granule cells. The internal plexiform layer (IPL) contains axons and deep collaterals of mitral and tufted cells and granule cell bodies. The innermost layer of olfactory bulb, the granule cell layer (GCL), is largely composed of granule cells (purple) that extend their dendrites to the EPL and also contains migrating interneurons. *Illustrated by Erina He.*

coordinated beating of ependymal cilia directs CSF and sets up a concentration gradient that provides directional information to newborn neurons. Expression of Sonic Hedgehog, a secreted morphogen with varied roles in neural development (Hor & Tang, 2010), and Eph/Ephrin, cell surface proteins that regulate cell adhesion (Laussu, Khuong, Gautrais, & Davy, 2014), in the SE determine the proportion of neuroblasts that migrate to the OB (Angot et al., 2008). Notably, studies that involve removal of the rostral OB reveal a significant reduction in neuroblast migration, which can be recovered by an OB transplant (Liu & Rao, 2003), suggesting that the OB itself provides chemoattractant signals for migrating neuroblasts.

Once neuroblasts arrive in the OB, they must detach from chains and migrate radially to populate the OB without using radial glial scaffolds that are typically used during development. The mechanism underlying the radial migration is less well understood. The initial detachment of neural precursor cells from chains appears to be mediated by the expression of Tenascin-R in the granule cell and internal plexiform layers (Saghatelyan, De Chevigny, Schachner, & Lledo, 2004) and Reelin expression mainly in the mitral cell (Hack, Bancila, Loulier, Carroll, & Cremer, 2002). Indeed, ectopic expression of Tenascin-R in vivo reroutes migrating neuroblasts (Saghatelyan et al., 2004) and exogenous application of Reelin in vitro induces cell dispersal of germinal zone explants (Hack et al., 2002). There is also evidence to suggest to suggest that the radial migration involves vascular scaffolds since migrating cells are seen extending leading processes along, or twisted around, vessels (Bovetti et al., 2007).

5.6.2 Survival of cells in the RMS/OB

Estimates of the number of newly born OB cells per day are in the tens of thousands (Lois & Alvarez-Buylla, 1994; Winner, Cooper-Kuhn, Aigner, Winkler, & Kuhn, 2002) which translates into about 1% of the total OB granule cell population being generated daily with limited change in the size of the OB (Kaplan, Mcnelly, & Hinds, 1985; Biebl, Cooper, Winkler, & Kuhn, 2000; Pomeroy, Lamantia, & Purves, 1990). Thus, neurogenesis must be compensated by cell death (Biebl et al., 2000). Estimates range from approximately 15% turnover of interneurons in the granule cell layer (Ninkovic, Mori, & Gotz, 2007; Lagace et al., 2007) to studies that report massive cell death within the OB. Pulse labeling studies using BrdU estimate that within a 19-month period, reflecting 75% of the lifespan of the animal, only 50% of the newborn cells survive (Winner et al., 2002). Further studies using transgenic mouse models to examine the rate of replacement of granule cells suggest that almost all the deep granule cells and half of the superficial granule cells are replaced in a year (Imayoshi et al., 2008). These studies utilized NestinCreER^{T2} mice crossed with reporter mice, to specifically and permanently label Nestin expressing neural precursor cells following tamoxifen administration. Interestingly, this same study showed that programed cell death is not dependent on the incoming supply of new neurons; rather, death and neurogenesis are two independent processes (Imayoshi et al., 2008) that together determine the numbers of new cells in the OB. A recent study highlighted a novel role for neuroblasts in the clearance of dead cells from the neurogenic niche. Migrating neuroblasts expressing DCX were shown to express the intracellular engulfment protein ELM01 enabling them to act as phagocytes. Moreover, genetic or pharmacological disruption of engulfment led to decreased phagocytosis by DCX expressing cells leading to accumulation of apoptotic nuclei in the neurogenic niches and impaired neuronal differentiation (Lu et al., 2011).

5.6.3 Differentiation of cells in OB

When neuroblasts begin their radial trajectory in the OB, they also start the process of neuronal differentiation and maturation (Carleton, Petreanu, Lansford, Alvarez-Buylla, A. & Lledo, 2003; Lledo & Saghatelyan, 2005). The majority of the neuroblasts differentiate into GABAergic granule cells (3:1 granule: periglomerular cells), which lack axons and form dendrodendritic synapses with mitral and tufted cells (reviewed in Lledo, Alonso, & Grubb, 2006) (Fig. 5.3). A minority become GABAergic periglomerular neurons, a small percentage of which are also dopaminergic (Kato et al., 2000; Luskin, 1993) reviewed in Lledo et al. (2006). In addition, a small percentage of neuroblasts form glutamatergic juxtaglomerular cells, found in the glomerular layer of the olfactory bulb (Brill et al., 2009) (Fig. 5.3).

Newly born granule cells are found in the deep region of the granule cell layer (GCL) and preferentially synapse with mitral cells in the deep lamina of the external plexiform layer, while older granule cells are located in the more superficial GCL whose dendrites mainly synapse with tufted cells. One hypothesis is that these two populations (deep and superficial) modulate distinct neural circuits (reviewed in Sakamoto, Kageyama, & Imayoshi, 2014). It has therefore been suggested that tufted cells would be regulated by the earlier born granule cells while newly born cells, that preferentially occupy the deep GCL, would supply inhibition to mitral and tufted cells (Sakamoto et al., 2014). It takes about two weeks for a newborn neuron to integrate into the existing circuitry. Interestingly, spiking, or action potential acquisition is the last neuronal characteristic acquired by these cells and it has been suggested that this delayed excitability prevents uncontrolled neurotransmitter release and disruption of the existing neural network (Carleton et al., 2003). These new interneurons play an important role in regulating the output of mitral and tufted cells from the OB to the cortex (Greer, 1987; Lledo, Merkle, & Alvarez-Buylla, 2008).

5.6.3.1 Wiring newborn neurons into the preexisting circuitry

What triggers the differentiation of cells in the olfactory bulb? One hypothesis is that cells are replaced when needed, for example, in response to a novel external stimulus. However, given that it takes approximately 2 weeks for newly born neurons to functionally integrate into the circuitry (Ortega-Perez, Murray, & Lledo, 2007) new cells would not likely respond to the novel stimulus trigger in an appropriate time to serve in this role. However, if the adult production of OB neurons recapitulates development and neurons are produced in excess prior to pruning then one might predict that new cells are produced, migrate to the olfactory bulb and freeze in the mitral layer where they wait for a signal to differentiate (Pignatelli & Belluzzi, 2010). There is little evidence in the literature that specifically supports either of these hypotheses.

5.6.4 Regional and temporal specification of OB neurons

While there is little more than speculation regarding the differentiation cues for OB neurons, recent research on OB neurogenesis has provided significant understanding of the various subtypes of OB bulb neurons that are generated. Initially, it was thought that OB neurons originated from a restricted region of the anterior SE, close to the RMS (Luskin, 1993; Lois & Alvarez-Buylla, 1994). However, recent reports demonstrate that olfactory neuron subtypes are derived by specific subpopulations of progenitors that reside along the entire SE lining the LV (Merkle et al., 2007). Labeling and tracing the fate of specific subsets of radial glia cells, which become NSCs in the adult, with viral vectors that enable to the specific and permanent labeling of SE cells and their neural progeny, has demonstrated well delineated regionalization. Progenitor cells in specific regions of the SE give rise to specific subsets of OB neurons in the adult (Merkle et al., 2007). NSCs in the dorsal wall of the SE produce primarily TH expressing periglomerular OB neurons and a small population of Calretinin expressing cells. Conversely, labeling of precursors in the ventrolateral SE reveals the production of Calbindin expressing OB cells (Merkle et al., 2007).

BOX 5.2 Human neurogenesis

The human SE differs in architecture from the adult rodent SE. The human SE has four well-defined layers (Quinones-Hinojosa, Sanai, Gonzalez-Perez, & Garcia-Verdugo, 2007). Layer 1 is adjacent to the lateral ventricle and is comprised of multiciliated ependymal cells. Layer 2 is a hypocellular layer with few cell bodies but numerous interdigitated ependymal and astrocyte cell projections. In Layer 3 there is a ribbon of proliferating astrocyte cell bodies and it is this layer that contains the multipotent neurosphere forming cells (Sanai et al., 2004). Layer 4 is adjacent to the brain parenchyma, and is primarily comprised of neuronal bodies and myelin tracts (Quinones-Hinojosa et al., 2007).

Human neurogenesis was first documented in postmortem human brain samples from cancer patients that had received a single dose of BrdU for diagnostic purposes (Eriksson et al., 1998) (Box 5.1). Analysis of cell proliferation showed BrdU labeling in neurogenic regions, the hippocampus and the SE, supporting the presence of adult neurogenesis in the adult brain. It is generally accepted that the adult human brain contains neural stem cells, but whether or not the SE generates cells that migrate through a human RMS to the OB into adulthood is a matter of debate (Sanai, Berger, Garcia-Verdugo, & Alvarez-Buylla, 2007). In infancy, DCX + /PSA-NCAM + neuroblasts have been identified (Wang et al., 2011; Sanai et al., 2004, 2011). These cells are highly proliferative and migrate both in the RMS and into the prefrontal cortex (Sanai et al., 2011) but are significantly reduced by 6 months of age. These findings are contrasted by a study by Curtis et al. (2007b), which showed significant proliferation and migration in a human RMS in adults. The development of a ^{14}C birth dating technique in Jonas Frisen's laboratory has shed some light on the existence and function

(Continued)

BOX 5.2 Human neurogenesis (Continued)

of human neurogenesis (Spalding, Bhardwaj, Buchholz, Druid, & Frisen, 2005; Box 5.1). In a study by Bergman et al., the age of olfactory neurons corresponded to the time of birth, indicating that there is no postnatal adult neurogenesis and it has been suggested that the loss of olfactory neurogenesis in humans may be due to an evolutionary deterioration of our olfactory epithelium. This is correlated with humans being less able detect new odors (Bergmann et al., 2015). Notably however, this carbon dating technique is not sensitive enough to detect single cells hence a low level of neurogenesis in the OB remains a distinct possibility (Spalding et al., 2005).

This carbon dating technique has provided evidence of striatal neurogenesis in the adult brain. The estimated rate of adult born interneurons is 2.7% per year (Ernst et al., 2014). Interestingly, these newly born cells are depleted in Huntington's disease (Ernst et al., 2014). This technique has also been extended to studies of the injured brain. In a study of carbon dating after stroke, no new neurogenesis was detected in the cortex (Huttner et al., 2014).

5.7 FUNCTION OF OB NEUROGENESIS UNDER PHYSIOLOGICAL CONDITIONS

Understanding the functional relevance of adult neurogenesis in the olfactory bulb has been challenging. The olfactory bulb is responsible for olfaction that influences food seeking, social, danger detecting, maternal and sexual behaviors (Breton-Provencher & Saghatelyan, 2012; Lazarini & Lledo, 2011). Odor information is first processed in the OB and then transferred to the olfactory cortex via mitral and tufted cells and it is becoming increasing clear that the newly born neurons are integral to the proper functioning of this system. To investigate the function of olfactory bulbs, many studies have used approaches to inhibit various aspects of neurogenesis (proliferation, migration, survival) and examined subsequent changes in the processing of olfactory information (i.e. changes in behavior).

5.7.1 Olfaction related behaviors

5.7.1.1 Odor discrimination and memory

Odor discrimination is most simply assessed by repeated presentations of the same odor followed by presentation of a novel odor. Control animals that are able to discriminate between two odors will display reduced sniffing time with repeated presentations of the same odor and this will increase upon presentation of a novel odor. Further, this discrimination task can be used to assess odor memory by varying the time intervals between repeated presentations (reviewed in Breton-Provencher & Saghatelyan, 2012). Many studies have assessed the effects of reducing neurogenesis on odor discrimination and have yielded conflicting results.

In studies where neurogenesis was reduced, for example, by infusing the antimitotic agent AraC to kill neuroblasts, mice showed a 75% reduction in cell migration to the OB. Interestingly, no reduction in odor discrimination was detected when examined after 28 days of AraC treatment. Mice behaved similar to controls showing decreased sniffing time with repeated odors and increased time when presented with a new odor. However, the authors did observe poor odor memory as revealed by a lack of recognition (increased sniffing time) when the time interval between odor presentations was increased (Breton-Provencher, Lemasson, Peralta, & Saghatelyan et al., 2009). In a separate set of studies, neurogenesis was reduced by irradiation and again, no change in odor discrimination was observed but long-term memory was impaired. Similarly, transgenic mice that permitted the specific ablation of nestin expressing cells (NSCs and progenitor cells) with diphtheria toxin revealed no deficit in odor discrimination however, in this study, in contrast to the findings using other ablation paradigms, no impairment was seen in odor memory. A number of other mouse models have been used to explore the functional relevance of newborn neurons. For instance, EGFR knockout mice (Enwere et al., 2004) have reduced SE cell proliferation and BDNF knockout mice (Bath et al., 2008) have impaired survival of cells in the OB and in both cases, decreased odor discrimination is observed in adult mice (Enwere et al., 2004). In NCAM deficient mice there is a dramatic 40% reduction in the size of the OB due to reduced neuroblast migration and mice display impaired odor discrimination. However, these same knockout mice did not display deficits in short-term odor memory. The conflicting results may in part, be explained by differences in the difficulty of odor discrimination tasks; newborn cells may play a role in discrimination between highly similar odors and fail to distinguish odors that are too distinct (Akers, Sakaguchi, & Arruda-Carvalho, 2010; Breton-Provencher et al., 2009). The age of the animal could also be a factor in olfactory outcomes. In this regard, developmental deficits in neurogenesis or those that occur during aging could differentially affect odor discrimination (Akers et al., 2010). Moreover, preexisting olfactory bulb cells (specifically granule cells) may compensate for the loss of neurogenesis and mask deficits. In an elegant study by Arruda-Carvalho et al. (2014), an ablation paradigm was used that specifically eliminated newly generated neurons after integration into an odor related memory trace. Post training ablation of newly generated neurons impaired the formation of subsequent odor-reward memories while the ablation of the same cells after training is ineffective. These findings suggest that these newly born neurons play a time-limited role in the formation of odor memories (Arruda-Carvalho et al., 2014).

5.7.1.2 Odor associative learning/fear conditioning

Odor associative learning and fear conditioning behavioral tasks have also been used to evaluate the function of adult neurogenesis. Animals learn to associate a reward or noxious stimulus with a particular odor. In odor cued fear conditioning, mice receive a noxious stimulus, such as an electrical foot shock, that normally elicits freezing behavior, after exposure to a particular odor. Mice with irradiation-impaired neurogenesis show impairments in this behavior and freeze less in response to the cued odor (Valley, Mullen, Schultz, Sagdullaev, & Firestein, 2009). In odor-associated learning, mice are attracted to an odor when it is associated with a reward, for example, sugar water. When neurogenesis is disrupted using ablation paradigms (either by irradiation or AraC ablation) mice continue to show associative learning but have deficits in long-term memory (Lazarini et al., 2009; Sultan et al., 2010). Other studies report long-term memory retention of more than a week after AraC ablation (Breton-Provencher et al., 2009) and more than two months after ablation of newborn neurons using transgenic mouse models (Imayoshi et al., 2008). Discrepancies in the findings of these studies have been attributed to differences in experimental design (Breton-Provencher & Saghatelyan, 2012; Lazarini & Lledo, 2011; Sakamoto et al., 2014) given that the ablation paradigms target different cells types (progenitors versus newborn neurons).

5.7.2 Reproductive, maternal, and social behaviors

Olfaction is necessary for a number of behaviors crucial to reproduction and maternal behavior, including mate selection and offspring recognition. These functions are mediated by prolactin-induced neurogenesis. During pregnancy and lactation, prolactin increases the number and integration glomerular and periglomerular neurons (Shingo et al., 2003). A reduction in prolactin results in the rejection of offspring by the mother (Shingo et al., 2003) and increased postpartum anxiety and impaired maternal behavior (Larsen & Grattan, 2010). In males, recognition of their own offspring is also related to prolactin mediated increases in neurogenesis (Mak et al., 2007). Interestingly, prolactin also plays a role in mate preference; male pheromones cause an increase in prolactin in females, which increases olfactory bulb neurogenesis and mediates female mate preference for dominant over subordinate males (Mak et al., 2007). Blocking neurogenesis with AraC ablation of proliferating progenitors eliminates this preference, suggesting that neurogenesis is required for female mate selection.

Genetic ablation of adult neurogenesis showed a similar disruption in sex-specific behaviors. As predicted from the prolactin studies, female mice show deficits in fertility and nurturing (Sakamoto et al., 2011) with ablation induced loss of neurogenesis. Males showed deficits in male—male aggression and sexual behaviors toward females but interestingly, male and female mice displayed normal social recognition behaviors towards mice of the opposite sex. In contrast, a study, of irradiation-impaired neurogenesis showed no deficit in maternal behaviors or young recognition, but rather irradiated mice displayed increased interaction with male conspecifics, suggesting a deficit in female recognition of male odors (Feierstein et al., 2010).

Interestingly, neurogenesis is also important for recognizing predator odors. Mice will demonstrate freezing behaviors in response to predator odors. However, when neurogenesis is disrupted using transgenic mouse models, mice approach predator odors that they have been trained to associate with a reward (Sakamoto et al., 2011), suggesting that continuous neurogenesis is required for predator avoidance (Breton-Provencher & Saghatelyan, 2012).

5.7.3 Nonneurogenic functions of neural precursor cells

In addition to olfactory bulb related functions, recent evidence shows support for other roles of SE-derived cells, whereby NSCs may contribute to CNS homeostasis through nonneurogenic functions (Martino, Butti, & Bacigaluppi, 2014). Two novel roles of SE-derived cells have recently been described. First, neural precursor cells (neural stem and progenitor cells; NPCs) can modulate the activation, proliferation and phagocytosis of microglia (referred to later in "glia" role after injury). NPCs have a distinct secretory profile and it has been shown that NPC derived VEGF in particular, may regulate microglial function and activity in the brain. Indeed injections of NPCs into the striatum cause increased activation of surrounding microglia. Second, NSCs play a role in regulating systemic immune responses directed against the brain. It has been proposed that dendritic cells (antigen presenting cells of the immune system) are recruited to the SE via factors from the choroid plexus and/or the associated vasculature. Dendritic cells use the RMS to move from the CNS to the systemic immune compartment. Once the cells reach the OB they access the cervical lymph nodes where they can present antigens, acquired while migrating along the RMS, to the systemic immune system (Mohammad et al., 2014). AraC ablation of SE progenitors leads to the accumulation of dendritic cells within the RMS (Mohammad et al., 2014). While the interaction between dendritic cells and SE cells has been demonstrated, it remains to be determined how NPCs interact with dendritic cells to regulate CNS immunity.

BOX 5.3 Aging and the SE

Aging causes a change in the organization of the SE niche. Neuroblasts and transit-amplifying cells decrease with age while astrocytes and ependymal cells remain relatively constant (Luo, Daniels, Lennington, Notti, & Conover, 2006) reviewed in Capilla-Gonzalez, Herranz-Perez, and Garcia-Verdugo (2015). These changes are associated with an age-related decline in neurogenesis (Enwere et al., 2004; Ahlenius, Visan, Kokaia, Lindvall, & Kokaia, 2009; Tropepe, Craig, Morshead, & Van Der Kooy, 1997). There is a 50% decrease in proliferation (Tropepe et al., 1997; Maslov, Barone, Plunkett, & Pruitt, 2004; Jin et al., 2003) fewer newly generated neurons in the olfactory bulb (Luo et al., 2006; Enwere et al., 2004; Tropepe et al., 1997) and reduced odor discrimination in old versus young mice (Enwere et al., 2004).

(Continued)

BOX 5.3 Aging and the SE (Continued)

While the age-induced reduction in neurogenesis is generally accepted, there is controversy as to whether this change is due to differences in the stem cell niche or changes in NSCs themselves. Some evidence supports a reduction in the numbers of stem cells in the brain (Ahlenius et al., 2009; Enwere et al., 2004; Maslov et al., 2004), while other studies have shown equivalent numbers of stem cells in the aged and young brain, suggesting changes in the microenvironment (Tropepe et al., 1997; Piccin, Tufford, & Morshead, 2014).

Decreased numbers of NSCs in the aged SE could be explained by changes in cell cycle kinetics (reviewed in Conover & Shook, 2011). Aging cells accumulate DNA damage, which can affect their proliferation and may decrease the cycling time of cells to allow for DNA repair (reviewed in Conover & Shook, 2011). Indeed, p16Ink-4a, a cell cycle regulator that mediates cell cycle entry in response to DNA damage, is increased in aged NSCs and leads to decreased proliferation of NSCs and their progeny (Molofsky et al., 2006). Telomeres are also important for maintaining proliferation. Telomerase reverse transcriptase, implicated in telomere elongation, and telomere activity decreases over time (Ferron et al., 2004). Moreover, if the accumulated damage is too great, NSCs could exit the cell cycle and undergo cell death (Conover & Shook, 2011).

Changes in extrinsic or environmental factors have also been described. There is an age-related decrease in growth factors such as EGF (Enwere et al., 2004) IGF, FGF-2, and VEGF (Shetty et al., 2005), and morphogens, such as Wnt (Piccin & Morshead, 2011; Okamoto et al., 2011). Intriguingly, coculture experiments show that factors from the young SE can rescue this age-related decline and transplants of old cells into a young niche show no difference in the behavior compared to young cells (Piccin et al., 2014), suggesting that niche factors may play a role in determining the levels of adult neurogenesis. Strong evidence for the regulation of neurogenesis by the microenvironment comes from experiments using heterochronic parabiosis. For example, GDF11, a circulating member of the BMP/TGF-β family, is found in higher concentrations in young blood and induces vascular remodeling, culminating in increased neurogenesis and improved olfactory discrimination in aging mice (Katsimpardi et al., 2014).

5.8 THE RESPONSE OF SE-DERIVED NEURAL PRECURSOR CELLS AFTER INJURY

Brain damage, including stroke, traumatic brain injury, seizures, hypoxia-ischemia and neurodegenerative disease (Arvidsson, Collin, Kirik, Kokaia, & Lindvall, 2002; Parent, Vexler, Gong, Derugin, & Ferriero, 2002; Zhang, Zhang, Zhang, & Chopp, 2001; Jin et al., 2001, 2004; Szele & Chesselet, 1996; Tattersfield et al., 2004; Fallon et al., 2000) all have been reported to influence the behavior of neural precursor cells. In animal models, a change in SE proliferation and production of newborn neurons and glial cells in areas of damage is often seen, indicating a potential for the SE to contribute to brain repair. However, the neurogenic response in and of itself is not sufficient to induce a significant functional improvement.

5.8.1 SE regeneration following NPC ablation

The SE shows a remarkable response to injury, responding quickly to generate a large number of cells following ablation of NPCs (Morshead et al., 1994; Doetsch, Garcia-Verdugo, & Alvarez-Buylla, 1999b, Piccin & Morshead, 2011; Sachewsky et al., 2014b). Following radiation induced or AraC ablation of proliferating progenitors, NSCs replenish the niche. This NSC activation effectively combats the loss of migrating cells, which contribute to injury-induced neurogenesis (see below). The signals important for activation of NPCs and concomitant repopulation of the SE are not entirely clear however a plethora of factors, many in response to injury, have been shown to activate endogenous NPCs and promote their proliferation (Chojnacki, Shimazaki, Gregg, Weinmaster, & Weiss, 2003; Piccin & Morshead, 2011; Shimazaki, Shingo, & Weiss, 2001). Most interesting, recent work has proposed that NSCs themselves, and not just the neural progeny, will migrate away from the SE and contribute to the production of new cells at the site of injury (Faiz et al., 2015). Finally, in the event of the loss of NSCs, the rare, upstream primitive NSCs are able to repopulate the NSC (Sachewsky et al., 2014b).

5.8.2 Increased SE neurogenesis after acute injury

5.8.2.1 Neurogenesis in the striatum

The response of the SE to ischemic damage has been particularly well characterized and it is clear that stroke increases SE proliferation and neurogenesis. Early studies showed that ischemic damage, both by middle cerebral artery occlusion (MCAO) and global ischemia (Jin et al., 2001; Parent et al., 2002; Arvidsson et al., 2002) results in a transient increase in the number of proliferating cells in the SE that peaks one week after injury. This increase is primarily due to an increased number of rapidly dividing transit-amplifying progenitors (type A) and neuroblasts (type C) cells after stroke (Chen et al., 2004). However, an increase in the size of the stem cell pools has also been demonstrated poststroke (Sachewsky et al., 2014a; Erlandsson, Lin, Yu, & Morshead, 2011). Many of the newly generated neuroblasts migrate to the damaged striatum (Arvidsson et al., 2002; Jin et al., 2001; Ohab, Fleming, Blesch, & Carmichael, 2006; Parent et al., 2002; Yamashita et al., 2006), migrating along reactive astrocytes and blood vessels (Yamashita et al., 2006; Thored et al., 2006) in chain-like structures, akin to their normal migration to the RMS (Lois et al., 1996; Peretto, Merighi, Fasolo, & Bonfanti, 1997; Parent et al., 2002; Arvidsson et al., 2002). This migration occurs at the expense of olfactory bulb migration such that a reduction in olfactory bulb neurogenesis is seen that is concomitant with the increased migration to the injured parenchyma (Goings, Sahni, & Szele, 2004). In the striatum the cells differentiate into regionally appropriate neurons; they express DARPP-32 a marker of striatal projection neurons and Calbindin, a marker of striatal medium spiny neurons (Arvidsson et al., 2002; Parent et al., 2002). Unfortunately, few new neurons have been shown to persist for more than a month poststroke, suggesting that there is limited survival and integration of newly born cells (Arvidsson et al., 2002; Parent et al., 2002). More recently however, studies have shown that increased neurogenesis in the striatum persists for at least four months following MCAO induced ischemia (Thored et al., 2006) and up to one year after photothrombotic cortical stroke (Osman, Porritt, Nilsson, & Kuhn, 2011a). The functional significance of the transient and long-term survival of the newborn neurons is not clear.

5.8.2.2 Neurogenesis in the cortex

Similar to what is observed in the striatum, neurogenesis has been observed after cortical stroke (reviewed in Saha, Jaber, & Gaillard, 2012), as well as after other types of injury such as aspiration lesions, inflammatory demyelination and fluid percussion injury (Szele & Chesselet, 1996; Calza et al., 1998; Chirumamilla, Sun, Bullock, & Colello, 2002). Interestingly, both tangential and radial migration is seen after demyelinating lesions and cortical injury (Szele & Chesselet, 1996; Szele et al., 1994; Ong, Plane, Parent, & Silverstein, 2005). In addition, it has been shown that a neurovascular "niche" is established in the poststroke cortex, whereby neuroblasts may migrate using vasculature when traveling long distances (Ohab et al., 2006). Indeed, in areas of poststroke angiogenesis, neuroblasts are found ensheathing and interdigitating with vascular endothelial cells. This interaction is meditated by the vascular ligands such as SDF1β and Ang1 that act on neuroblast receptors, CXCR4 and Tie2 respectively, to guide migrating neuroblasts (Ohab et al., 2006). Hence, in instances of neuroblast migration to injury sites, the relationship between angiogenesis and neurogenesis is observed. However, different between cortical and striatal injury, the degree of neurogenesis is significantly less in the injured cortex.

5.8.3 SE-derived neurogenesis in response to (chronic) neurodegeneration

Changes in neurogenesis are also seen in a number of neurodegenerative conditions (reviewed in Curtis, Faull, & Eriksson, 2007a; Winner & Winkler, 2015). In Huntington's disease there is increased SE proliferation and an associated increase in migratory neuroblasts to the OB. In contrast, experimental models of Parkinson's disease and Alzheimer's disease generally report impaired OB neurogenesis (Curtis et al., 2007a; Demars, Hu, Gadadhar, & Lazarov, 2010; Young, Taylor, & Bordey, 2011). Unlike the NPC response to acute injury that is often considered a "first response" to brain damage, it is not yet understood why various chronic neurodegenerative conditions exhibit different neurogenic responses. It is intriguing that symptoms observed in the early stages of a number of different neurodegenerative diseases are linked to olfactory (or hippocampal) function and it has been suggested that deficits in neurogenesis may be responsible for the early symptomology observed (Winner & Winkler, 2015).

5.9 THE ROLE OF SE-DERIVED CELLS AFTER INJURY

While there is evidence showing limited survival and functional integration of newborn neurons into the existing neuronal circuitry following injury, ablation of neurogenesis has been shown to be detrimental to injury recovery. Ablation of SE proliferation using AraC infusion for 7 days after stroke caused an increase in infarct size and a greater neurological deficit (Li et al., 2010). Similarly, targeted ablation of nestin expressing precursors result in larger infarcts and more severe neurological deficits as early as one day after stroke (Wang et al., 2012a). This finding strongly suggests that the early effects of injury on brain repair are not mediated by the production of new neurons. Indeed, recent studies have shown that SE activation plays an important role in facilitating brain repair through mechanisms such as trophic support for damaged cells and tissue. Additional roles of the SE that have recently been identified postinjury include the production of glial cells that contribute to reactive gliosis and immunomodulatory effects (for examine regulating immune cells and microglia) thereby highlighting novel **nonneurogenic** roles for the SE postinjury (reviewed in Martino et al., 2014).

5.9.1 Trophic support

The SE-derived newborn cells generated after injury have been shown to provide trophic support through release of growth factors and neurotrophins that serve to enhance cell survival and mediate neuroplasticity, thus facilitating functional recovery from injury (reviewed in Kernie & Parent, 2010). Much information on trophic factor release has been garnered from studies of transplanted cells that improve cell survival and angiogenesis at the injury site (Rafuse, Soundararajan, Leopold, & Robertson, 2005; Pluchino et al., 2003; Ourednik, Ourednik, Lynch, Schachner, & Snyder, 2002; Lindvall & Kokaia, 2011; Horie et al., 2011; Minnerup et al., 2011; Lee et al., 2007; Lee, Kim, Kim, & Kim, 2009). More recently, studies have shown direct evidence of trophic factor release by endogenous neural stem and progenitor cells after injury. Using an in vitro assay of neuronal excitotoxic cell death, it was shown that NPCs release trophic factors such as BDNF and VEGF, which are responsible for NPC-mediated neuroprotection (Li et al., 2010). SE-derived progenitors have also been shown to protect striatal neurons from excitotoxicity through secretion of endocannabinoids, which bind to cannabinoid receptors on medium spiny striatal neurons and modulate glutamatergic transmission (Butti et al., 2012). Interestingly, restoring cannabinoid levels reverses the increased morbidity and mortality seen following the ablation of precursors after injury (Butti et al., 2012), highlighting a neuroprotective role for cannabinoids on precursors residing within the SE. Moreover, a study of the role of Wnt signaling in stroke repair showed that although injection of Wnt ligands into the postischemic striatum improved functional outcome after 28 days, a similar injection directly into the SE resulted in improvement in just 2 days. This effect was attributed to a neuroprotective effect of immature neurons as no increase in the number of mature neurons was observed following the SE injection (Shruster, Ben-Zur, Melamed, & Offen, 2012). There are many roles for signaling molecules that can enhance the ability of the SE to contribute to recovery including those that regulate the behavior of NSCs, progenitors and the neuroblasts themselves (Choe, Pleasure, & Mira, 2015; Faigle & Song, 2013; Mu, Lee, & Gage, 2010; Riquelme, Drapeau, & Doetsch, 2008).

5.9.2 Glia

Recent studies have shown the importance of glia derived from the SE after an insult. Reactive astrocytes are produced from SE-derived cells after cortical injury both in the adult and neonate, and are critical for glial scar formation and blood brain barrier repair (Faiz et al., 2015; Benner et al., 2013) postinjury. In addition, in demyelinating lesions, SE-derived NG2 expressing glial precursors facilitate remyelination by forming functional glutamatergic synapses with demyelinated axons (Etxeberria, Mangin, Aguirre, & Gallo, 2010). Postinjury neurogenesis has been extensively studied (perhaps due to a neurocentric view of the brain) but the production of glia and generation of trophic factors may turn out to be the more interesting response of the SE to damage and the basis of neural regeneration postinjury.

5.10 FUTURE DIRECTIONS

5.10.1 Mobilizing endogenous neural stem cells for repair

Since the discovery of injury-induced neurogenesis, endogenous NSCs have been purported as a means for brain repair/regeneration. The biggest hurdle currently faced is the inability of these cells to mature and integrate into functional neurons. Thus increasing the survival and integration of NPCs could be a promising strategy for brain repair. Several factors have been shown to increase production/survival of new neurons, including FGF-2, EGF, stem cell factor, erthyropietin (EPO), BDNF, caspase inhibitors and antiinflammatory drugs (Lindvall & Kokaia, 2011; Kokaia & Lindvall, 2003). Some factors have been shown to have multiple effects including EPO which is able to promote neurogenesis and angiogenesis to facilitate stroke recovery (Wang, Zhang, Wang, Zhang, & Chopp, 2004) and EPO receptor (EpoR) knockdowns show impairment of poststroke neurogenesis by reducing proliferation and migration of newly born cells (Tsai et al., 2006). EPO has also been used in combination with other factors, known to promote neurogenesis including EGF, and this combination was shown to reverse motor impairments after stroke (Kolb et al., 2007). After success in a many preclinical animal models, EPO was used and deemed safe in a pilot trial for stroke treatment but a later clinical trial was suspended due to safety concerns regarding hemorrhaging and mortality in patients (Minnerup, Schmidt, Albert-Weissenberger, & Kleinshnitz, 2013). To circumvent these issues, EPO derivatives are being designed to maximize neuroprotective effects while minimizing off target effects in other organs (Carmichael, 2010). Similarly, analogues of other NPC activation molecules, such as Cyclosporin A (CsA), have been shown to be effective in promoting behavioral recovery in preclinical models of stroke (Osman et al., 2011b; Sachewsky et al., 2014a). These studies are built on preclinical animal model paradigms demonstrating that CsA, an immunosuppressant used in the clinic, promotes NPC survival and functional recovery following stroke (Sachewsky et al., 2014a; Hunt, Cheng, Hoyles, Jervis, & Morshead, 2010). Although clinical trials have demonstrated its safety in traumatic brain injury (Minnerup et al., 2013) there are remaining concerns of the safety of CsA, including nephrotoxicity and neurotoxicity (Osman et al., 2011b) that limit its potential for use in regenerative medicine strategies. These two examples highlight one of the drawbacks surrounding molecules used for therapies that activate endogenous repair mechanisms, which is the unwanted secondary/systemic effects (Carmichael, 2010). Along these lines, clinical trials of Trafermin (bFGF) treatment after stroke were unsuccessful due to undesirable secondary effects of the treatment (Carmichael, 2010; Bogousslavsky et al., 2002) and EGF administration, although shown to boost neurogenesis, has been shown to promote tumorigenesis (Herbst, 2004; Lindberg et al., 2012). One approach to circumvent the systemic effects of drug treatment is to develop local drug delivery strategies (Vulic & Shoichet, 2014). Local release delivery of EPO (Wang, Cooke, Morshead, & Shoichet, 2012b) and CsA (Tuladhar, Morshead, & Shoichet, 2015) in hydrogels have both shown to reduce stroke infarct volume and promote functional recovery and in the case of local CsA delivery to the brain, the concentration of CsA that was found in tissues within the body was negligible compared to the systemic delivery of the drug. In support of this approach it is thought that stereotaxic access to the stroke cavity will soon be common in neurosurgical suites (Carmichael, 2010).

5.10.2 Reprograming cells to an earlier NSC like state

Cellular reprograming has become an intense field of research since the discovery of induced pluripotent stem (iPS) cells (Takahashi et al., 2007). It has been shown that the expression of specific transcription factors can not only drive somatic cells to pluripotency but can also directly convert somatic cells between cell fates. In theory, both neurons and NSCs derived from iPS cells or directly converted from patient-specific somatic cells (Faiz & Nagy, 2013) would overcome rejection issues when developing transplantation interventions to promote repair. The idea of reprograming nonneuronal resident brain cells has also been suggested as an avenue for repair; in vivo reprograming would theoretically provide a local source of cells critical for brain repair. Indeed, a number of groups have reported the successful in vivo conversion of astrocytes to neuroblasts or neurons (Torper et al., 2013; Niu et al., 2013; Heinrich et al., 2010; Guo et al., 2014; Grande et al., 2013), and of NG2 + (glial) progenitors (Heinrich et al., 2014) and pericytes (Karow et al., 2012) into neurons. While the reprograming strategies continue to be developed and optimized, it is entirely unclear whether in vivo reprograming is a reasonable approach to promote tissue repair and functional recovery after brain injury. For instance, given the important roles of reactive astrocytes after injury (Myer, Gurkoff, Lee, Hovda, & Sofroniew, 2006), it is not obvious that generating neurons at the expense of reactive astrocytes or other nonneuronal cell types, is the key to recovery.

In considering the future of regenerative medicine, it seems increasingly clear that no single approach will be sufficient to promote recovery. Indeed, combinatorial therapies directed at maintaining whole brain homeostasis instead of single pathogenic mechanisms (Iadecola & Anrather, 2011) already seem to be the most effective. We envision reprogramed NSCs transplanted into the stroke site in hydrogels that are genetically engineered to locally secrete molecules important for plasticity, antiinflammation and mobilization of endogenous precursors in the future.

5.11 CONCLUDING REMARKS

Adult neurogenesis is now well established in the adult mammalian brain. Many of the structural and cellular components of the SE niche have been identified. However, questions regarding the NSC lineage remain. Investigation of the degree of lineage commitment of NSCs and their progeny and the mode of division within the lineage remain to be determined as they relate to cell production and ultimately the capacity for these cells to contribute to neural function and neurorepair. In addition, the discovery of novel nonneurogenic roles of neural precursor cells is adding further complexity to the SE niche. To address outstanding questions in the field, further research is needed on the function of SE neurogenesis in both the normal and injured brain. This will provide insight to design better strategies for neurorepair and regeneration.

REFERENCES

Ahlenius, H., Visan, V., Kokaia, M., Lindvall, O., & Kokaia, Z. (2009). Neural stem and progenitor cells retain their potential for proliferation and differentiation into functional neurons despite lower number in aged brain. *The Journal of Neuroscience*, *29*, 4408–4419.

Akers, K. G., Sakaguchi, M., & Arruda-Carvalho, M. (2010). Functional contribution of adult-generated olfactory bulb interneurons: odor discrimination versus odor memory. *The Journal of Neuroscience, 30*, 4523–4525.

Altman, J. (1962). Are new neurons formed in the brains of adult mammals? *Science (New York, N.Y.), 135*, 1127–1128.

Altman, J. (1969). Autoradiographic and histological studies of postnatal neurogenesis. IV. Cell proliferation and migration in the anterior forebrain, with special reference to persisting neurogenesis in the olfactory bulb. *The Journal of Comparative Neurology, 137*, 433–457.

Altman, J., & Das, G. D. (1965). Autoradiographic and histological evidence of postnatal hippocampal neurogenesis in rats. *The Journal of Comparative Neurology, 124*, 319–335.

Alvarez-Buylla, A., Garcia-Verdugo, J. M., & Tramontin, A. D. (2001). A unified hypothesis on the lineage of neural stem cells. *Nature Reviews. Neuroscience, 2*, 287–293.

Angot, E., Loulier, K., Nguyen-Ba-Charvet, K. T., Gadeau, A. P., Ruat, M., & Traiffort, E. (2008). Chemoattractive activity of sonic hedgehog in the adult subventricular zone modulates the number of neural precursors reaching the olfactory bulb. *Stem Cells (Dayton, Ohio), 26*, 2311–2320.

Arruda-Carvalho, M., Akers, K. G., Guskjolen, A., Sakaguchi, M., Josselyn, S. A., & Frankland, P. W. (2014). Posttraining ablation of adult-generated olfactory granule cells degrades odor-reward memories. *The Journal of Neuroscience, 34*, 15793–15803.

Arvidsson, A., Collin, T., Kirik, D., Kokaia, Z., & Lindvall, O. (2002). Neuronal replacement from endogenous precursors in the adult brain after stroke. *Nature Medicine, 8*, 963–970.

Azim, K., Fiorelli, R., Zweifel, S., Hurtado-Chong, A., Yoshikawa, K., Slomianka, L., & Raineteau, O. (2012). 3-dimensional examination of the adult mouse subventricular zone reveals lineage-specific microdomains. *PLoS ONE, 7*, e49087.

Bachiller, D., Klingensmith, J., Kemp, C., Belo, J. A., Anderson, R. M., May, S. R., ... De Robertis, E. M. (2000). The organizer factors Chordin and Noggin are required for mouse forebrain development. *Nature, 403*, 658–661.

Basak, O., & Taylor, V. (2009). Stem cells of the adult mammalian brain and their niche. *Cellular and Molecular Life Sciences: CMLS, 66*, 1057–1072.

Bath, K. G., Mandairon, N., Jing, D., Rajagopal, R., Kapoor, R., Chen, Z. Y., ... Lee, F. S. (2008). Variant brain-derived neurotrophic factor (Val66Met) alters adult olfactory bulb neurogenesis and spontaneous olfactory discrimination. *The Journal of Neuroscience, 28*, 2383–2393.

Belluzzi, O., Benedusi, M., Ackman, J., & Loturco, J. J. (2003). Electrophysiological differentiation of new neurons in the olfactory bulb. *The Journal of Neuroscience, 23*, 10411–10418.

Benner, E. J., Luciano, D., Jo, R., Abdi, K., Paez-Gonzalez, P., Sheng, H., Warner, D. S., ... Kuo, C. T. (2013). Protective astrogenesis from the SVZ niche after injury is controlled by Notch modulator Thbs4. *Nature, 497*, 369–373.

Bergmann, O., Spalding, K. L., & Frisen, J. (2015). Adult Neurogenesis in Humans. *Cold Spring Harbor Perspectives in Biology*, 7, a018994.

Biebl, M., Cooper, C. M., Winkler, J., & Kuhn, H. G. (2000). Analysis of neurogenesis and programmed cell death reveals a self-renewing capacity in the adult rat brain. *Neuroscience Letters, 291*, 17–20.

Bogousslavsky, J., Victor, S. J., Salinas, E. O., Pallay, A., Donnan, G. A., Fieschi, C., ... European-Australian Fiblast In Acute Stroke, G. (2002). Fiblast (trafermin) in acute stroke: results of the European-Australian phase II/III safety and efficacy trial. *Cerebrovascular Diseases (Basel, Switzerland), 14*, 239–251.

Bovetti, S., Hsieh, Y. C., Bovolin, P., Perroteau, I., Kazunori, T., & Puche, A. C. (2007). Blood vessels form a scaffold for neuroblast migration in the adult olfactory bulb. *The Journal of Neuroscience, 27*, 5976–5980.

Brazel, C. Y., Romanko, M. J., Rothstein, R. P., & Levison, S. W. (2003). Roles of the mammalian subventricular zone in brain development. *Progress in Neurobiology, 69*, 49–69.

Breton-Provencher, V., Lemasson, M., Peralta, M. R., III, & Saghatelyan, A. (2009). Interneurons produced in adulthood are required for the normal functioning of the olfactory bulb network and for the execution of selected olfactory behaviors. *The Journal of Neuroscience, 29*, 15245–15257.

Breton-Provencher, V., & Saghatelyan, A. (2012). Newborn neurons in the adult olfactory bulb: unique properties for specific odor behavior. *Behavioural Brain Research, 227*, 480–489.

Brill, M. S., Ninkovic, J., Winpenny, E., Hodge, R. D., Ozen, I., Yang, R., ... Gotz, M. (2009). Adult generation of glutamatergic olfactory bulb interneurons. *Nature Neuroscience*, *12*, 1524–1533.

Butti, E., Bacigaluppi, M., Rossi, S., Cambiaghi, M., Bari, M., Cebrian Silla, A., ... Martino, G. (2012). Subventricular zone neural progenitors protect striatal neurons from glutamatergic excitotoxicity. *Brain*, *135*, 3320–3335.

Calza, L., Giardino, L., Pozza, M., Bettelli, C., Micera, A., & Aloe, L. (1998). Proliferation and phenotype regulation in the subventricular zone during experimental allergic encephalomyelitis: in vivo evidence of a role for nerve growth factor. *Proceedings of the National Academy of Sciences of the United States of America*, *95*, 3209–3214.

Cameron, H. A., & Mckay, R. D. (2001). Adult neurogenesis produces a large pool of new granule cells in the dentate gyrus. *The Journal of Comparative Neurology*, *435*, 406–417.

Capilla-Gonzalez, V., Herranz-Perez, V., & Garcia-Verdugo, J. M. (2015). The aged brain: genesis and fate of residual progenitor cells in the subventricular zone. *Frontiers in Cellular Neuroscience*, *9*, 365.

Carleton, A., Petreanu, L. T., Lansford, R., Alvarez-Buylla, A., & Lledo, P. M. (2003). Becoming a new neuron in the adult olfactory bulb. *Nature Neuroscience*, *6*, 507–518.

Carmichael, S. T. (2010). Translating the frontiers of brain repair to treatments: starting not to break the rules. *Neurobiology of Disease*, *37*, 237–242.

Chen, J., Li, Y., Zhang, R., Katakowski, M., Gautam, S. C., Xu, Y., Lu, M., Zhang, Z., & Chopp, M. (2004). Combination therapy of stroke in rats with a nitric oxide donor and human bone marrow stromal cells enhances angiogenesis and neurogenesis. *Brain Research*, *1005*, 21–28.

Chiasson, B. J., Tropepe, V., Morshead, C. M., & Van Der Kooy, D. (1999). Adult mammalian forebrain ependymal and subependymal cells demonstrate proliferative potential, but only subependymal cells have neural stem cell characteristics. *The Journal of Neuroscience*, *19*, 4462–4471.

Chirumamilla, S., Sun, D., Bullock, M. R., & Colello, R. J. (2002). Traumatic brain injury induced cell proliferation in the adult mammalian central nervous system. *Journal of Neurotrauma*, *19*, 693–703.

Choe, Y., Pleasure, S. J., & Mira, H. (2015). Control of adult neurogenesis by short-range morphogenic-signaling molecules. *Cold Spring Harbor Perspectives in Biology*, *8*, 1–18.

Chojnacki, A., Shimazaki, T., Gregg, C., Weinmaster, G., & Weiss, S. (2003). Glycoprotein 130 signaling regulates Notch1 expression and activation in the self-renewal of mammalian forebrain neural stem cells. *The Journal of Neuroscience*, *23*, 1730–1741.

Chojnacki, A., & Weiss, S. (2008). Production of neurons, astrocytes and oligodendrocytes from mammalian CNS stem cells. *Nature Protocols*, *3*, 935–940.

Codega, P., Silva-Vargas, V., Paul, A., Maldonado-Soto, A. R., Deleo, A. M., Pastrana, E., & Doetsch, F. (2014). Prospective identification and purification of quiescent adult neural stem cells from their in vivo niche. *Neuron*, *82*, 545–559.

Coles-Takabe, B. L., Brain, I., Purpura, K. A., Karpowicz, P., Zandstra, P. W., Morshead, C. M., & Van Der Kooy, D. (2008). Don't look: growing clonal versus nonclonal neural stem cell colonies. *Stem Cells (Dayton, Ohio)*, *26*, 2938–2944.

Conover, J. C., & Shook, B. A. (2011). Aging of the subventricular zone neural stem cell niche. *Aging and Disease*, *2*, 49–63.

Conti, L., & Cattaneo, E. (2010). Neural stem cell systems: physiological players or in vitro entities? *Nature Reviews. Neuroscience*, *11*, 176–187.

Curtis, M. A., Faull, R. L., & Eriksson, P. S. (2007a). The effect of neurodegenerative diseases on the subventricular zone. *Nature Reviews. Neuroscience*, *8*, 712–723.

Curtis, M. A., Kam, M., Nannmark, U., Anderson, M. F., Axell, M. Z., Wikkelso, C., ... Eriksson, P. S. (2007b). Human neuroblasts migrate to the olfactory bulb via a lateral ventricular extension. *Science (New York, N.Y.)*, *315*, 1243–1249.

Demars, M., Hu, Y. S., Gadadhar, A., & Lazarov, O. (2010). Impaired neurogenesis is an early event in the etiology of familial Alzheimer's disease in transgenic mice. *Journal of Neuroscience Research*, *88*, 2103–2117.

Dhaliwal, J., & Lagace, D. C. (2011). Visualization and genetic manipulation of adult neurogenesis using transgenic mice. *The European Journal of Neuroscience*, *33*, 1025–1036.

Doetsch, F., Caille, I., Lim, D. A., Garcia-Verdugo, J. M., & Alvarez-Buylla, A. (1999a). Subventricular zone astrocytes are neural stem cells in the adult mammalian brain. *Cell*, *97*, 703–716.
Doetsch, F., Garcia-Verdugo, J. M., & Alvarez-Buylla, A. (1997). Cellular composition and three-dimensional organization of the subventricular germinal zone in the adult mammalian brain. *The Journal of Neuroscience*, *17*, 5046–5061.
Doetsch, F., Garcia-Verdugo, J. M., & Alvarez-Buylla, A. (1999b). Regeneration of a germinal layer in the adult mammalian brain. *Proceedings of the National Academy of Sciences of the United States of America*, *96*, 11619–11624.
Enwere, E., Shingo, T., Gregg, C., Fujikawa, H., Ohta, S., & Weiss, S. (2004). Aging results in reduced epidermal growth factor receptor signaling, diminished olfactory neurogenesis, and deficits in fine olfactory discrimination. *The Journal of Neuroscience*, *24*, 8354–8365.
Eriksson, P. S., Perfilieva, E., Bjork-Eriksson, T., Alborn, A. M., Nordborg, C., Peterson, D. A., & Gage, F. H. (1998). Neurogenesis in the adult human hippocampus. *Nature Medicine*, *4*, 1313–1317.
Erlandsson, A., Lin, C. H., Yu, F., & Morshead, C. M. (2011). Immunosuppression promotes endogenous neural stem and progenitor cell migration and tissue regeneration after ischemic injury. *Experimental Neurology*, *230*, 48–57.
Ernst, A., Alkass, K., Bernard, S., Salehpour, M., Perl, S., Tisdale, J., ... Frisen, J. (2014). Neurogenesis in the striatum of the adult human brain. *Cell*, *156*, 1072–1083.
Etxeberria, A., Mangin, J. M., Aguirre, A., & Gallo, V. (2010). Adult-born SVZ progenitors receive transient synapses during remyelination in corpus callosum. *Nature Neuroscience*, *13*, 287–289.
Faigle, R., & Song, H. (2013). Signaling mechanisms regulating adult neural stem cells and neurogenesis. *Biochimica et Biophysica Acta*, *1830*, 2435–2448.
Faiz, M., & Nagy, A. (2013). Induced pluripotent stem cells and disorders of the nervous system: progress, problems, and prospects. *The Neuroscientist: A Review Journal Bringing Neurobiology, Neurology and Psychiatry*, *19*, 567–577.
Faiz, M., Sachewsky, N., Gascon, S., Bang, K. W., Morshead, C. M., & Nagy, A. (2015). Adult neural stem cells from the subventricular zone give rise to reactive astrocytes in the cortex after stroke. *Cell Stem Cell*, *17*, 624–634.
Fallon, J., Reid, S., Kinyamu, R., Opole, I., Opole, R., Baratta, J., ... Loughlin, S. (2000). In vivo induction of massive proliferation, directed migration, and differentiation of neural cells in the adult mammalian brain. *Proceedings of the National Academy of Sciences of the United States of America*, *97*, 14686–14691.
Feierstein, C. E., Lazarini, F., Wagner, S., Gabellec, M. M., De Chaumont, F., Olivo-Marin, J. C., ... Gheusi, G. (2010). Disruption of adult neurogenesis in the olfactory bulb affects social interaction but not maternal behavior. *Frontiers in Behavioral Neuroscience*, *4*, 176.
Ferron, S., Mira, H., Franco, S., Cano-Jaimez, M., Bellmunt, E., Ramirez, C., ... Blasco, M. A. (2004). Telomere shortening and chromosomal instability abrogates proliferation of adult but not embryonic neural stem cells. *Development (Cambridge, England)*, *131*, 4059–4070.
Fuentealba, L. C., Rompani, S. B., Parraguez, J. I., Obernier, K., Romero, R., Cepko, C. L., & Alvarez-Buylla, A. (2015). Embryonic origin of postnatal neural stem cells. *Cell*, *161*, 1644–1655.
Gallagher, D., Norman, A. A., Woodard, C. L., Yang, G., Gauthier-Fisher, A., Fujitani, M., ... Miller, F. D. (2013). Transient maternal IL-6 mediates long-lasting changes in neural stem cell pools by deregulating an endogenous self-renewal pathway. *Cell Stem Cell*, *13*, 564–576.
Garcia-Verdugo, J. M., Doetsch, F., Wichterle, H., Lim, D. A., & Alvarez-Buylla, A. (1998). Architecture and cell types of the adult subventricular zone: in search of the stem cells. *Journal of Neurobiology*, *36*, 234–248.
Goings, G. E., Sahni, V., & Szele, F. G. (2004). Migration patterns of subventricular zone cells in adult mice change after cerebral cortex injury. *Brain Research*, *996*, 213–226.
Gotz, M., & Huttner, W. B. (2005). The cell biology of neurogenesis. *Nature Reviews. Molecular Cell Biology*, *6*, 777–788.
Grande, A., Sumiyoshi, K., Lopez-Juarez, A., Howard, J., Sakthivel, B., Aronow, B., ... Nakafuku, M. (2013). Environmental impact on direct neuronal reprogramming in vivo in the adult brain. *Nature Communications*, *4*, 2373.

Greer, C. A. (1987). Golgi analyses of dendritic organization among denervated olfactory bulb granule cells. *The Journal of Comparative Neurology, 257*, 442–452.

Guo, Z., Zhang, L., Wu, Z., Chen, Y., Wang, F., & Chen, G. (2014). In vivo direct reprogramming of reactive glial cells into functional neurons after brain injury and in an Alzheimer's disease model. *Cell Stem Cell, 14*, 188–202.

Hack, I., Bancila, M., Loulier, K., Carroll, P., & Cremer, H. (2002). Reelin is a detachment signal in tangential chain-migration during postnatal neurogenesis. *Nature Neuroscience, 5*, 939–945.

Haubensak, W., Attardo, A., Denk, W., & Huttner, W. B. (2004). Neurons arise in the basal neuroepithelium of the early mammalian telencephalon: a major site of neurogenesis. *Proceedings of the National Academy of Sciences of the United States of America, 101*, 3196–3201.

Hayes, N. L., & Nowakowski, R. S. (2002). Dynamics of cell proliferation in the adult dentate gyrus of two inbred strains of mice. *Brain Research. Developmental Brain Research, 134*, 77–85.

Heinrich, C., Bergami, M., Gascon, S., Lepier, A., Vigano, F., Dimou, L., ... Gotz, M. (2014). Sox2-mediated conversion of NG2 glia into induced neurons in the injured adult cerebral cortex. *Stem Cell Reports, 3*, 1000–1014.

Heinrich, C., Blum, R., Gascon, S., Masserdotti, G., Tripathi, P., Sanchez, R., ... Berninger, B. (2010). Directing astroglia from the cerebral cortex into subtype specific functional neurons. *PLoS Biology, 8*, e1000373.

Herbst, R. S. (2004). EGFR inhibition in NSCLC: the emerging role of cetuximab. *Journal of the National Comprehensive Cancer Network: JNCCN, 2*(Suppl 2), S41–S51.

Hitoshi, S., Seaberg, R. M., Koscik, C., Alexson, T., Kusunoki, S., Kanazawa, I., ... Van Der Kooy, D. (2004). Primitive neural stem cells from the mammalian epiblast differentiate to definitive neural stem cells under the control of Notch signaling. *Genes & Development, 18*, 1806–1811.

Hor, C. H., & Tang, B. L. (2010). Sonic hedgehog as a chemoattractant for adult NPCs. *Cell Adhesion & Migration, 4*, 1–3.

Horie, N., Pereira, M. P., Niizuma, K., Sun, G., Keren-Gill, H., Encarnacion, A., ... Steinberg, G. K. (2011). Transplanted stem cell-secreted vascular endothelial growth factor effects poststroke recovery, inflammation, and vascular repair. *Stem Cells (Dayton, Ohio), 29*, 274–285.

Hu, H. (1999). Chemorepulsion of neuronal migration by Slit2 in the developing mammalian forebrain. *Neuron, 23*, 703–711.

Hu, H., & Rutishauser, U. (1996). A septum-derived chemorepulsive factor for migrating olfactory interneuron precursors. *Neuron, 16*, 933–940.

Hunt, J., Cheng, A., Hoyles, A., Jervis, E., & Morshead, C. M. (2010). Cyclosporin A has direct effects on adult neural precursor cells. *The Journal of Neuroscience, 30*, 2888–2896.

Huttner, H. B., Bergmann, O., Salehpour, M., Racz, A., Tatarishvili, J., Lindgren, E., ... Frisen, J. (2014). The age and genomic integrity of neurons after cortical stroke in humans. *Nature Neuroscience, 17*, 801–803.

Iadecola, C., & Anrather, J. (2011). Stroke research at a crossroad: asking the brain for directions. *Nature Neuroscience, 14*, 1363–1368.

Imayoshi, I., Sakamoto, M., Ohtsuka, T., Takao, K., Miyakawa, T., Yamaguchi, M., ... Kageyama, R. (2008). Roles of continuous neurogenesis in the structural and functional integrity of the adult forebrain. *Nature Neuroscience, 11*, 1153–1161.

Imura, T., Kornblum, H. I., & Sofroniew, M. V. (2003). The predominant neural stem cell isolated from postnatal and adult forebrain but not early embryonic forebrain expresses GFAP. *The Journal of Neuroscience, 23*, 2824–2832.

Jensen, J. B., & Parmar, M. (2006). Strengths and limitations of the neurosphere culture system. *Molecular Neurobiology, 34*, 153–161.

Jin, K., Galvan, V., Xie, L., Mao, X. O., Gorostiza, O. F., Bredesen, D. E., & Greenberg, D. A. (2004). Enhanced neurogenesis in Alzheimer's disease transgenic (PDGF-APPSw, Ind) mice. *Proceedings of the National Academy of Sciences of the United States of America, 101*, 13363–13367.

Jin, K., Minami, M., Lan, J. Q., Mao, X. O., Batteur, S., Simon, R. P., & Greenberg, D. A. (2001). Neurogenesis in dentate subgranular zone and rostral subventricular zone after focal cerebral ischemia in the rat. *Proceedings of the National Academy of Sciences of the United States of America, 98*, 4710–4715.

Jin, K., Sun, Y., Xie, L., Batteur, S., Mao, X. O., Smelick, C., ... Greenberg, D. A. (2003). Neurogenesis and aging: FGF-2 and HB-EGF restore neurogenesis in hippocampus and subventricular zone of aged mice. *Aging Cell, 2*, 175–183.

Johansson, C. B., Momma, S., Clarke, D. L., Risling, M., Lendahl, U., & Frisen, J. (1999). Identification of a neural stem cell in the adult mammalian central nervous system. *Cell, 96*, 25–34.

Kaplan, M. S., & Bell, D. H. (1983). Neuronal proliferation in the 9-month-old rodent-radioautographic study of granule cells in the hippocampus. *Experimental Brain Research, 52*, 1–5.

Kaplan, M. S., & Hinds, J. W. (1977). Neurogenesis in the adult rat: electron microscopic analysis of light radioautographs. *Science (New York, N.Y.), 197*, 1092–1094.

Kaplan, M. S., Mcnelly, N. A., & Hinds, J. W. (1985). Population dynamics of adult-formed granule neurons of the rat olfactory bulb. *The Journal of Comparative Neurology, 239*, 117–125.

Karow, M., Sanchez, R., Schichor, C., Masserdotti, G., Ortega, F., Heinrich, C., ... Berninger, B. (2012). Reprogramming of pericyte-derived cells of the adult human brain into induced neuronal cells. *Cell Stem Cell, 11*, 471–476.

Kato, T., Yokouchi, K., Kawagishi, K., Fukushima, N., Miwa, T., & Moriizumi, T. (2000). Fate of newly formed periglomerular cells in the olfactory bulb. *Acta Oto-laryngologica, 120*, 876–879.

Katsimpardi, L., Litterman, N. K., Schein, P. A., Miller, C. M., Loffredo, F. S., Wojtkiewicz, G. R., ... Rubin, L. L. (2014). Vascular and neurogenic rejuvenation of the aging mouse brain by young systemic factors. *Science (New York, N.Y.), 344*, 630–634.

Kempermann, G., Kuhn, H. G., & Gage, F. H. (1997). More hippocampal neurons in adult mice living in an enriched environment. *Nature, 386*, 493–495.

Kernie, S. G., & Parent, J. M. (2010). Forebrain neurogenesis after focal Ischemic and traumatic brain injury. *Neurobiology of Disease, 37*, 267–274.

Kokaia, Z., & Lindvall, O. (2003). Neurogenesis after ischaemic brain insults. *Current Opinion in Neurobiology, 13*, 127–132.

Kolb, B., Morshead, C., Gonzalez, C., Kim, M., Gregg, C., Shingo, T., & Weiss, S. (2007). Growth factor-stimulated generation of new cortical tissue and functional recovery after stroke damage to the motor cortex of rats. *Journal of Cerebral Blood Flow and Metabolism: Official Journal of the International Society of Cerebral Blood Flow and Metabolism, 27*, 983–997.

Kornack, D. R., & Rakic, P. (2001). The generation, migration, and differentiation of olfactory neurons in the adult primate brain. *Proceedings of the National Academy of Sciences of the United States of America, 98*, 4752–4757.

Kriegstein, A., & Alvarez-Buylla, A. (2009). The glial nature of embryonic and adult neural stem cells. *Annual Review of Neuroscience, 32*, 149–184.

Kuhn, H. G., Dickinson-Anson, H., & Gage, F. H. (1996). Neurogenesis in the dentate gyrus of the adult rat: age-related decrease of neuronal progenitor proliferation. *The Journal of Neuroscience, 16*, 2027–2033.

Lagace, D. C., Whitman, M. C., Noonan, M. A., Ables, J. L., Decarolis, N. A., Arguello, A. A., ... Eisch, A. J. (2007). Dynamic contribution of nestin-expressing stem cells to adult neurogenesis. *The Journal of Neuroscience, 27*, 12623–12629.

Larsen, C. M., & Grattan, D. R. (2010). Prolactin-induced mitogenesis in the subventricular zone of the maternal brain during early pregnancy is essential for normal postpartum behavioral responses in the mother. *Endocrinology, 151*, 3805–3814.

Laussu, J., Khuong, A., Gautrais, J., & Davy, A. (2014). Beyond boundaries—Eph:ephrin signaling in neurogenesis. *Cell Adhesion & Migration, 8*, 349–359.

Laywell, E. D., Rakic, P., Kukekov, V. G., Holland, E. C., & Steindler, D. A. (2000). Identification of a multipotent astrocytic stem cell in the immature and adult mouse brain. *Proceedings of the National Academy of Sciences of the United States of America, 97*, 13883–13888.

Lazarini, F., & Lledo, P. M. (2011). Is adult neurogenesis essential for olfaction? *Trends in Neurosciences, 34*, 20–30.

Lazarini, F., Mouthon, M. A., Gheusi, G., Chaumont, D. E., Olivo-Marin, F., Lamarque, J. C., ... Lledo, P. M. (2009). Cellular and behavioral effects of cranial irradiation of the subventricular zone in adult mice. *PLoS ONE, 4*, e7017.

Lee, H. J., Kim, K. S., Kim, E. J., Choi, H. B., Lee, K. H., Park, I. H., ... Kim, S. U. (2007). Brain transplantation of immortalized human neural stem cells promotes functional recovery in mouse intracerebral hemorrhage stroke model. *Stem Cells (Dayton, Ohio)*, *25*, 1204–1212.

Lee, H. J., Kim, M. K., Kim, H. J., & Kim, S. U. (2009). Human neural stem cells genetically modified to overexpress Akt1 provide neuroprotection and functional improvement in mouse stroke model. *PLoS ONE*, *4*, e5586.

Li, B., Piao, C. S., Liu, X. Y., Guo, W. P., Xue, Y. Q., Duan, W. M., ... Zhao, L. R. (2010). Brain self-protection: the role of endogenous neural progenitor cells in adult brain after cerebral cortical ischemia. *Brain Research*, *1327*, 91–102.

Lindberg, O. R., Brederlau, A., Jansson, A., Nannmark, U., Cooper-Kuhn, C., & Kuhn, H. G. (2012). Characterization of epidermal growth factor-induced dysplasia in the adult rat subventricular zone. *Stem Cells and Development*, *21*, 1356–1366.

Lindvall, O., & Kokaia, Z. (2011). Stem cell research in stroke: how far from the clinic? *Stroke; A Journal of Cerebral Circulation*, *42*, 2369–2375.

Liu, G., & Rao, Y. (2003). Neuronal migration from the forebrain to the olfactory bulb requires a new attractant persistent in the olfactory bulb. *The Journal of Neuroscience*, *23*, 6651–6659.

Lledo, P. M., Alonso, M., & Grubb, M. S. (2006). Adult neurogenesis and functional plasticity in neuronal circuits. *Nature Reviews. Neuroscience*, *7*, 179–193.

Lledo, P. M., Merkle, F. T., & Alvarez-Buylla, A. (2008). Origin and function of olfactory bulb interneuron diversity. *Trends in Neurosciences*, *31*, 392–400.

Lledo, P. M., & Saghatelyan, A. (2005). Integrating new neurons into the adult olfactory bulb: joining the network, life-death decisions, and the effects of sensory experience. *Trends in Neurosciences*, *28*, 248–254.

Lois, C., & Alvarez-Buylla, A. (1994). Long-distance neuronal migration in the adult mammalian brain. *Science (New York, N.Y.)*, *264*, 1145–1148.

Lois, C., Garcia-Verdugo, J. M., & Alvarez-Buylla, A. (1996). Chain migration of neuronal precursors. *Science (New York, N.Y.)*, *271*, 978–981.

Lu, Z., Elliott, M. R., Chen, Y., Walsh, J. T., Klibanov, A. L., Ravichandran, K. S., & Kipnis, J. (2011). Phagocytic activity of neuronal progenitors regulates adult neurogenesis. *Nature Cell Biology*, *13*, 1076–1083.

Luo, J., Daniels, S. B., Lennington, J. B., Notti, R. Q., & Conover, J. C. (2006). The aging neurogenic subventricular zone. *Aging Cell*, *5*, 139–152.

Luskin, M. B. (1993). Restricted proliferation and migration of postnatally generated neurons derived from the forebrain subventricular zone. *Neuron*, *11*, 173–189.

Mak, G. K., Enwere, E. K., Gregg, C., Pakarainen, T., Poutanen, M., Huhtaniemi, I., & Weiss, S. (2007). Male pheromone-stimulated neurogenesis in the adult female brain: possible role in mating behavior. *Nature Neuroscience*, *10*, 1003–1011.

Martino, G., Butti, E., & Bacigaluppi, M. (2014). Neurogenesis or non-neurogenesis: that is the question. *The Journal of Clinical Investigation*, *124*, 970–973.

Maslov, A. Y., Barone, T. A., Plunkett, R. J., & Pruitt, S. C. (2004). Neural stem cell detection, characterization, and age-related changes in the subventricular zone of mice. *The Journal of Neuroscience*, *24*, 1726–1733.

Merkle, F. T., Mirzadeh, Z., & Alvarez-Buylla, A. (2007). Mosaic organization of neural stem cells in the adult brain. *Science (New York, N.Y.)*, *317*, 381–384.

Merkle, F. T., Tramontin, A. D., Garcia-Verdugo, J. M., & Alvarez-Buylla, A. (2004). Radial glia give rise to adult neural stem cells in the subventricular zone. *Proceedings of the National Academy of Sciences of the United States of America*, *101*, 17528–17532.

Miller, M. W., & Nowakowski, R. S. (1988). Use of bromodeoxyuridine-immunohistochemistry to examine the proliferation, migration and time of origin of cells in the central nervous system. *Brain Research*, *457*, 44–52.

Minnerup, J., Kim, J. B., Schmidt, A., Diederich, K., Bauer, H., Schilling, M., ... Schabitz, W. R. (2011). *Effects of neural progenitor cells on sensorimotor recovery and endogenous repair mechanisms after photothrombotic stroke, . Stroke* (42, pp. 1757–1763).

Minnerup, J., Schmidt, A., Albert-Weissenberger, C., & Kleinshnitz, C. (2013). Stroke: Pathophysiology and therapy. In D. N. Granger, & J. P. Granger (Eds.), *Colloquium series on integrated systems physiology: From molecule to function to disease*. San Rafael (CA): Morgan & Claypool Life Sciences.

Mirzadeh, Z., Merkle, F. T., Soriano-Navarro, M., Garcia-Verdugo, J. M., & Alvarez-Buylla, A. (2008). Neural stem cells confer unique pinwheel architecture to the ventricular surface in neurogenic regions of the adult brain. *Cell Stem Cell*, *3*, 265–278.

Mohammad, M. G., Tsai, V. W., Ruitenberg, M. J., Hassanpour, M., Li, H., Hart, P. H., … Brown, D. A. (2014). Immune cell trafficking from the brain maintains CNS immune tolerance. *The Journal of Clinical Investigation*, *124*, 1228–1241.

Molofsky, A. V., Slutsky, S. G., Joseph, N. M., He, S., Pardal, R., Krishnamurthy, J., … Morrison, S. J. (2006). Increasing p16INK4a expression decreases forebrain progenitors and neurogenesis during ageing. *Nature*, *443*, 448–452.

Montanari, C. A., Tute, M. S., Beezer, A. E., & Mitchell, J. C. (1996). Determination of receptor-bound drug conformations by QSAR using flexible fitting to derive a molecular similarity index. *Journal of Computer-Aided Molecular Design*, *10*, 67–73.

Morshead, C. M., Garcia, A. D., Sofroniew, M. V., & Van Der Kooy, D. (2003). The ablation of glial fibrillary acidic protein-positive cells from the adult central nervous system results in the loss of forebrain neural stem cells but not retinal stem cells. *The European Journal of Neuroscience*, *18*, 76–84.

Morshead, C. M., Reynolds, B. A., Craig, C. G., Mcburney, M. W., Staines, W. A., Morassutti, D., … Van Der Kooy, D. (1994). Neural stem cells in the adult mammalian forebrain: a relatively quiescent subpopulation of subependymal cells. *Neuron*, *13*, 1071–1082.

Mu, Y., Lee, S. W., & Gage, F. H. (2010). Signaling in adult neurogenesis. *Current Opinion in Neurobiology*, *20*, 416–423.

Myer, D. J., Gurkoff, G. G., Lee, S. M., Hovda, D. A., & Sofroniew, M. V. (2006). Essential protective roles of reactive astrocytes in traumatic brain injury. *Brain*, *129*, 2761–2772.

Ninkovic, J., Mori, T., & Gotz, M. (2007). Distinct modes of neuron addition in adult mouse neurogenesis. *The Journal of Neuroscience*, *27*, 10906–10911.

Niu, W., Zang, T., Zou, Y., Fang, S., Smith, D. K., Bachoo, R., & Zhang, C. L. (2013). In vivo reprogramming of astrocytes to neuroblasts in the adult brain. *Nature Cell Biology*, *15*, 1164–1175.

Noctor, S. C., Martinez-Cerdeno, V., Ivic, L., & Kriegstein, A. R. (2004). Cortical neurons arise in symmetric and asymmetric division zones and migrate through specific phases. *Nature Neuroscience*, *7*, 136–144.

Noctor, S. C., Martinez-Cerdeno, V., & Kriegstein, A. R. (2008). Distinct behaviors of neural stem and progenitor cells underlie cortical neurogenesis. *The Journal of Comparative Neurology*, *508*, 28–44.

Nottebohm, F. (1985). Neuronal replacement in adulthood. *Annals of the New York Academy of Sciences*, *457*, 143–161.

Ohab, J. J., Fleming, S., Blesch, A., & Carmichael, S. T. (2006). A neurovascular niche for neurogenesis after stroke. *The Journal of Neuroscience*, *26*, 13007–13016.

Okamoto, M., Inoue, K., Iwamura, H., Terashima, K., Soya, H., Asashima, M., & Kuwabara, T. (2011). Reduction in paracrine Wnt3 factors during aging causes impaired adult neurogenesis. *The FASEB Journal*, *25*, 3570–3582.

Ong, J., Plane, J. M., Parent, J. M., & Silverstein, F. S. (2005). Hypoxic-ischemic injury stimulates subventricular zone proliferation and neurogenesis in the neonatal rat. *Pediatric Research*, *58*, 600–606.

Ortega-Perez, I., Murray, K., & Lledo, P. M. (2007). The how and why of adult neurogenesis. *Journal of Molecular Histology*, *38*, 555–562.

Osman, A. M., Porritt, M. J., Nilsson, M., & Kuhn, H. G. (2011a). Long-term stimulation of neural progenitor cell migration after cortical ischemia in mice. *Stroke; A Journal of Cerebral Circulation*, *42*, 3559–3565.

Osman, M. M., Lulic, D., Glover, L., Stahl, C. E., Lau, T., Van Loveren, H., & Borlongan, C. V. (2011b). Cyclosporine-A as a neuroprotective agent against stroke: its translation from laboratory research to clinical application. *Neuropeptides*, *45*, 359–368.

Ourednik, J., Ourednik, V., Lynch, W. P., Schachner, M., & Snyder, E. Y. (2002). Neural stem cells display an inherent mechanism for rescuing dysfunctional neurons. *Nature Biotechnology*, *20*, 1103–1110.

Parent, J. M., Vexler, Z. S., Gong, C., Derugin, N., & Ferriero, D. M. (2002). Rat forebrain neurogenesis and striatal neuron replacement after focal stroke. *Annals of Neurology, 52*, 802–813.

Pastrana, E., Silva-Vargas, V., & Doetsch, F. (2011). Eyes wide open: A critical review of sphere-formation as an assay for stem cells. *Cell Stem Cell, 8*, 486–498.

Peretto, P., Merighi, A., Fasolo, A., & Bonfanti, L. (1997). Glial tubes in the rostral migratory stream of the adult rat. *Brain Research Bulletin, 42*, 9–21.

Piccin, D., & Morshead, C. M. (2011). Wnt signaling regulates symmetry of division of neural stem cells in the adult brain and in response to injury. *Stem Cells (Dayton, Ohio), 29*, 528–538.

Piccin, D., Tufford, A., & Morshead, C. M. (2014). Neural stem and progenitor cells in the aged subependyma are activated by the young niche. *Neurobiology of Aging, 35*, 1669–1679.

Pignatelli, A., & Belluzzi, O. (2010). Neurogenesis in the adult olfactory bulb. In A. Menini (Ed.), *The Neurobiology of Olfaction*. Boca Raton, FL: CRC Press.

Pluchino, S., Quattrini, A., Brambilla, E., Gritti, A., Salani, G., Dina, G., ... Martino, G. (2003). Injection of adult neurospheres induces recovery in a chronic model of multiple sclerosis. *Nature, 422*, 688–694.

Pomeroy, S. L., Lamantia, A. S., & Purves, D. (1990). Postnatal construction of neural circuitry in the mouse olfactory bulb. *The Journal of Neuroscience, 10*, 1952–1966.

Quinones-Hinojosa, A., Sanai, N., Gonzalez-Perez, O., & Garcia-Verdugo, J. M. (2007). The human brain subventricular zone: Stem cells in this niche and its organization. *Neurosurgery Clinics of North America, 18*, 15–20, vii.

Rafuse, V. F., Soundararajan, P., Leopold, C., & Robertson, H. A. (2005). Neuroprotective properties of cultured neural progenitor cells are associated with the production of sonic hedgehog. *Neuroscience, 131*, 899–916.

Rakic, P. (1974). Neurons in rhesus monkey visual cortex: systematic relation between time of origin and eventual disposition. *Science (New York, N.Y.), 183*, 425–427.

Reynolds, B. A., & Weiss, S. (1992). Generation of neurons and astrocytes from isolated cells of the adult mammalian central nervous system. *Science (New York, N.Y.), 255*, 1707–1710.

Ribeiro Xavier, A. L., Kress, B. T., Goldman, S. A., Lacerda De Menezes, J. R., & Nedergaard, M. (2015). A distinct population of microglia supports adult neurogenesis in the subventricular zone. *The Journal of Neuroscience, 35*, 11848–11861.

Riquelme, P. A., Drapeau, E., & Doetsch, F. (2008). Brain micro-ecologies: neural stem cell niches in the adult mammalian brain. *Philosophical Transactions of the Royal Society of London. Series B, Biological Sciences, 363*, 123–137.

Sachewsky, N., Hunt, J., Cooke, M. J., Azimi, A., Zarin, T., Miu, C., ... Morshead, C. M. (2014a). Cyclosporin A enhances neural precursor cell survival in mice through a calcineurin-independent pathway. *Disease Models & Mechanisms, 7*, 953–961.

Sachewsky, N., Leeder, R., Xu, W., Rose, K. L., Yu, F., Van Der Kooy, D., & Morshead, C. M. (2014b). Primitive neural stem cells in the adult mammalian brain give rise to GFAP-expressing neural stem cells. *Stem Cell Reports, 2*, 810–824.

Saghatelyan, A., De Chevigny, A., Schachner, M., & Lledo, P. M. (2004). Tenascin-R mediates activity-dependent recruitment of neuroblasts in the adult mouse forebrain. *Nature Neuroscience, 7*, 347–356.

Saha, B., Jaber, M., & Gaillard, A. (2012). Potentials of endogenous neural stem cells in cortical repair. *Frontiers in Cellular Neuroscience, 6*, 14.

Sakamoto, M., Imayoshi, I., Ohtsuka, T., Yamaguchi, M., Mori, K., & Kageyama, R. (2011). Continuous neurogenesis in the adult forebrain is required for innate olfactory responses. *Proceedings of the National Academy of Sciences of the United States of America, 108*, 8479–8484.

Sakamoto, M., Kageyama, R., & Imayoshi, I. (2014). The functional significance of newly born neurons integrated into olfactory bulb circuits. *Frontiers in Neuroscience, 8*, 121.

Sanai, N., Berger, M. S., Garcia-Verdugo, J. M., & Alvarez-Buylla, A. (2007). Comment on "Human neuroblasts migrate to the olfactory bulb via a lateral ventricular extension". *Science (New York, N.Y.), 318*, 393 , author reply 393.

Sanai, N., Nguyen, T., Ihrie, R. A., Mirzadeh, Z., Tsai, H. H., Wong, M., ... Alvarez-Buylla, A. (2011). Corridors of migrating neurons in the human brain and their decline during infancy. *Nature, 478*, 382–386.

Sanai, N., Tramontin, A. D., Quinones-Hinojosa, A., Barbaro, N. M., Gupta, N., Kunwar, S., ... Alvarez-Buylla, A. (2004). Unique astrocyte ribbon in adult human brain contains neural stem cells but lacks chain migration. *Nature*, *427*, 740–744.

Sawamoto, K., Wichterle, H., Gonzalez-Perez, O., Cholfin, J. A., Yamada, M., Spassky, N., ... Alvarez-Buylla, A. (2006). New neurons follow the flow of cerebrospinal fluid in the adult brain. *Science (New York, N.Y.)*, *311*, 629–632.

Shetty, A. K., Hattiangady, B., & Shetty, G. A. (2005). Stem/progenitor cell proliferation factors FGF-2, IGF-1, and VEGF exhibit early decline during the course of aging in the hippocampus: role of astrocytes. *Glia*, *51*, 173–186.

Shimazaki, T., Shingo, T., & Weiss, S. (2001). The ciliary neurotrophic factor/leukemia inhibitory factor/gp130 receptor complex operates in the maintenance of mammalian forebrain neural stem cells. *The Journal of Neuroscience*, *21*, 7642–7653.

Shingo, T., Gregg, C., Enwere, E., Fujikawa, H., Hassam, R., Geary, C., ... Weiss, S. (2003). Pregnancy-stimulated neurogenesis in the adult female forebrain mediated by prolactin. *Science (New York, N.Y.)*, *299*, 117–120.

Shruster, A., Ben-Zur, T., Melamed, E., & Offen, D. (2012). Wnt signaling enhances neurogenesis and improves neurological function after focal ischemic injury. *PLoS ONE*, *7*, e40843.

Smukler, S. R., Runciman, S. B., Xu, S., & Van Der Kooy, D. (2006). Embryonic stem cells assume a primitive neural stem cell fate in the absence of extrinsic influences. *The Journal of Cell Biology*, *172*, 79–90.

Spalding, K. L., Bhardwaj, R. D., Buchholz, B. A., Druid, H., & Frisen, J. (2005). Retrospective birth dating of cells in humans. *Cell*, *122*, 133–143.

Spemann, H., & Mangold, H. (2001). Induction of embryonic primordia by implantation of organizers from a different species. 1923. *The International Journal of Developmental Biology*, *45*, 13–38.

Sultan, S., Mandairon, N., Kermen, F., Garcia, S., Sacquet, J., & Didier, A. (2010). Learning-dependent neurogenesis in the olfactory bulb determines long-term olfactory memory. *The FASEB Journal*, *24*, 2355–2363.

Szele, F. G., & Chesselet, M. F. (1996). Cortical lesions induce an increase in cell number and PSA-NCAM expression in the subventricular zone of adult rats. *The Journal of Comparative Neurology*, *368*, 439–454.

Szele, F. G., Dowling, J. J., Gonzales, C., Theveniau, M., Rougon, G., & Chesselet, M. F. (1994). Pattern of expression of highly polysialylated neural cell adhesion molecule in the developing and adult rat striatum. *Neuroscience*, *60*, 133–144.

Takahashi, K., Tanabe, K., Ohnuki, M., Narita, M., Ichisaka, T., Tomoda, K., & Yamanaka, S. (2007). Induction of pluripotent stem cells from adult human fibroblasts by defined factors. *Cell*, *131*, 861–872.

Tattersfield, A. S., Croon, R. J., Liu, Y. W., Kells, A. P., Faull, R. L., & Connor, B. (2004). Neurogenesis in the striatum of the quinolinic acid lesion model of Huntington's disease. *Neuroscience*, *127*, 319–332.

Tavazoie, M., Van Der Veken, L., Silva-Vargas, V., Louissaint, M., Colonna, L., Zaidi, B., ... Doetsch, F. (2008). A specialized vascular niche for adult neural stem cells. *Cell Stem Cell*, *3*, 279–288.

Thored, P., Arvidsson, A., Cacci, E., Ahlenius, H., Kallur, T., Darsalia, V., ... Lindvall, O. (2006). Persistent production of neurons from adult brain stem cells during recovery after stroke. *Stem Cells (Dayton, Ohio)*, *24*, 739–747.

Torper, O., Pfisterer, U., Wolf, D. A., Pereira, M., Lau, S., Jakobsson, J., ... Parmar, M. (2013). Generation of induced neurons via direct conversion in vivo. *Proceedings of the National Academy of Sciences of the United States of America*, *110*, 7038–7043.

Tropepe, V., Craig, C. G., Morshead, C. M., & Van Der Kooy, D. (1997). Transforming growth factor-alpha null and senescent mice show decreased neural progenitor cell proliferation in the forebrain subependyma. *The Journal of Neuroscience*, *17*, 7850–7859.

Tropepe, V., Hitoshi, S., Sirard, C., Mak, T. W., Rossant, J., & Van Der Kooy, D. (2001). Direct neural fate specification from embryonic stem cells: A primitive mammalian neural stem cell stage acquired through a default mechanism. *Neuron*, *30*, 65–78.

Tsai, P. T., Ohab, J. J., Kertesz, N., Groszer, M., Matter, C., Gao, J., . . . Carmichael, S. T. (2006). A critical role of erythropoietin receptor in neurogenesis and post-stroke recovery. *The Journal of Neuroscience*, *26*, 1269–1274.

Tuladhar, A., Morshead, C. M., & Shoichet, M. S. (2015). Circumventing the blood–brain barrier: Local delivery of cyclosporin A stimulates stem cells in stroke-injured rat brain. *Journal of Controlled Release: Official Journal of the Controlled Release Society*, *215*, 1–11.

Valley, M. T., Mullen, T. R., Schultz, L. C., Sagdullaev, B. T., & Firestein, S. (2009). Ablation of mouse adult neurogenesis alters olfactory bulb structure and olfactory fear conditioning. *Frontiers in Neuroscience*, *3*, 51.

Vescovi, A. L., Reynolds, B. A., Fraser, D. D., & Weiss, S. (1993). bFGF regulates the proliferative fate of unipotent (neuronal) and bipotent (neuronal/astroglial) EGF-generated CNS progenitor cells. *Neuron*, *11*, 951–966.

Vulic, K., & Shoichet, M. S. (2014). Affinity-based drug delivery systems for tissue repair and regeneration. *Biomacromolecules*, *15*, 3867–3880.

Wang, C., Liu, F., Liu, Y. Y., Zhao, C. H., You, Y., Wang, L., . . . Yang, Z. (2011). Identification and characterization of neuroblasts in the subventricular zone and rostral migratory stream of the adult human brain. *Cell research*, *21*, 1534–1550.

Wang, L., Zhang, Z., Wang, Y., Zhang, R., & Chopp, M. (2004). Treatment of stroke with erythropoietin enhances neurogenesis and angiogenesis and improves neurological function in rats. *Stroke; a Journal of Cerebral Circulation*, *35*, 1732–1737.

Wang, X., Mao, X., Xie, L., Sun, F., Greenberg, D. A., & Jin, K. (2012a). Conditional depletion of neurogenesis inhibits long-term recovery after experimental stroke in mice. *PLoS ONE*, *7*, e38932.

Wang, Y., Cooke, M. J., Morshead, C. M., & Shoichet, M. S. (2012b). Hydrogel delivery of erythropoietin to the brain for endogenous stem cell stimulation after stroke injury. *Biomaterials*, *33*, 2681–2692.

Weiss, S., Dunne, C., Hewson, J., Wohl, C., Wheatley, M., Peterson, A. C., & Reynolds, B. A. (1996a). Multipotent CNS stem cells are present in the adult mammalian spinal cord and ventricular neuroaxis. *The Journal of Neuroscience*, *16*, 7599–7609.

Weiss, S., Reynolds, B. A., Vescovi, A. L., Morshead, C., Craig, C. G., & Van Der Kooy, D. (1996b). Is there a neural stem cell in the mammalian forebrain? *Trends in Neurosciences*, *19*, 387–393.

Wichterle, H., Garcia-Verdugo, J. M., & Alvarez-Buylla, A. (1997). Direct evidence for homotypic, glia-independent neuronal migration. *Neuron*, *18*, 779–791.

Winner, B., Cooper-Kuhn, C. M., Aigner, R., Winkler, J., & Kuhn, H. G. (2002). Long-term survival and cell death of newly generated neurons in the adult rat olfactory bulb. *The European Journal of Neuroscience*, *16*, 1681–1689.

Winner, B., & Winkler, J. (2015). Adult neurogenesis in neurodegenerative diseases. *Cold Spring Harbor Perspectives in Biology*, 7, a021287.

Wojtowicz, J. M., & Kee, N. (2006). BrdU assay for neurogenesis in rodents. *Nature Protocols*, *1*, 1399–1405.

Wu, W., Wong, K., Chen, J., Jiang, Z., Dupuis, S., Wu, J. Y., & Rao, Y. (1999). Directional guidance of neuronal migration in the olfactory system by the protein Slit. *Nature*, *400*, 331–336.

Yamashita, T., Ninomiya, M., Hernandez Acosta, P., Garcia-Verdugo, J. M., Sunabori, T., Sakaguchi, M., . . . Sawamoto, K. (2006). Subventricular zone-derived neuroblasts migrate and differentiate into mature neurons in the post-stroke adult striatum. *The Journal of Neuroscience*, *26*, 6627–6636.

Young, S. Z., Taylor, M. M., & Bordey, A. (2011). Neurotransmitters couple brain activity to subventricular zone neurogenesis. *The European Journal of Neuroscience*, *33*, 1123–1132.

Zhang, R. L., Zhang, Z. G., Zhang, L., & Chopp, M. (2001). Proliferation and differentiation of progenitor cells in the cortex and the subventricular zone in the adult rat after focal cerebral ischemia. *Neuroscience*, *105*, 33–41.

CHAPTER 6

Critical Periods in Cortical Development

Takao K. Hensch[1,2]
[1]Harvard University, Cambridge, MA, United States
[2]Harvard Medical School, Boston, MA, United States

6.1 CRITICAL PERIODS IN EARLY BRAIN DEVELOPMENT

Critical periods, times at which the organism is especially sensitive to environmental stimuli, have been observed across species in various neural systems (see review by Hensch, 2004). Primary sensory areas exhibit striking examples of experience-dependent neural plasticity in specific windows in early life. Critical periods allow the developing nervous system to establish an optimal neural representation of the environment that the organism finds itself in and thus to guide future action. Ensuring that the neural circuitry is wired to construct accurate, immutable maps early in life for later use is a far more efficient biological strategy than attempting to rewire circuitry later. But environments can change so the closure of early critical periods can compromise future revisions to the circuitry. There is growing evidence, however, that although adaptive changes are set in place early, there are mechanisms to ensure that later plasticity is still possible.

Although in principle the consequences of sensory deprivation during critical periods could be found throughout cortical regions, the effects are most obvious for sensory cortical regions such as visual, auditory, or somatosensory cortex. The shift in ocular dominance related to monocular visual deprivation is perhaps the canonical model of synaptic plasticity confined to a postnatal critical period so I will focus on it here as a model of critical periods in development.

6.2 THE AMBLYOPIA MODEL

When one eye is occluded in early development there is an enduring loss of responsiveness to later input to this eye in primary visual cortex (V1), a condition referred to as amblyopia (poor visual acuity). Following this initial discovery by Wiesel and Hubel (1963), a picture has emerged that during development, input from the two eyes competes with one another for influence on individual neurons in V1. This interocular competition has revealed a postnatal period of enhanced synaptic plasticity

The Neurobiology of Brain and Behavioral Development
DOI: http://dx.doi.org/10.1016/B978-0-12-804036-2.00006-6

that is confined to a critical period. This enhanced plasticity appears related to a peak phase of physical growth of the head and likely allows for constant perception as the distance between the eyes increases during the critical period (Wang, Sarnaik, & Cang, 2010). If there is an asymmetry in the quality of input from the two eyes at this time there will be a reduction in visual acuity in the affected eye with no obvious pathology in the eye, thalamus, or visual cortex. The severity of the amblyopia varies considerably depending upon the age at initiation and type of asymmetry. Types of causes include form deprivation (e.g., cataract), unequal alignment of the eyes (strabismus) or unequal refractive error of the eyes (anisometropia). In children the critical period for developing amblyopia is up to 8 years and is relatively easy to correct until that age by improving the quality of input to the affected eye (see reviews by Daw, 1998; Mitchell & MacKinnon, 2002; Simons, 2005). Beyond this age the condition becomes increasingly resistant to reversal. The emerging constraints on plasticity in this system have the advantage of lending stability to mature visual cortical circuitry, but have the disadvantage of impeding later recovery from amblyopia.

Amblyopia is most often produced in animal models by eyelid suture (monocular deprivation, MD), which significantly occludes the pattern visual input to one eye. Research using various laboratory animals species have shown that MD leads to functional and structural changes in V1 that result in a reduction in acuity in the deprived eye because of a shift in ocular dominance of binocular neurons away from the deprived eye toward the open eye (e.g., Hensch et al., 1998; LeVay, Stryker, & Shatz, 1978; Mataga, Nagai, & Hensch, 2002; Sato & Stryker, 2008; Taha & Stryker, 2002; Wiesel & Hubel, 1963, 1970).

Although the plasticity of ocular dominance peaks during the early postnatal critical period, it remains to some extent in many species, including rodents and cats, beyond sexual maturity. Thus the older visual cortex can retain some forms of neural plasticity but it may be expressed differently from the plasticity observed during the critical period. It is important to note that the original studies of Wiesel & Hubel measured ocular dominance plasticity by recording the spiking output of V1 neurons but today there are many measures including visually-evoked synaptic potentials, intrinsic hemodynamic signals, immediate early gene activation, thalamocortical axon or dendritic spine morphology and motility, and calcium responses in individual cell types. Each of these methods are differently sensitive to subthreshold visual inputs and yield different levels of resolution (Morishita & Hensch, 2008), which are important considerations in measuring the effectiveness of therapies for recovery of visual function.

6.3 REORGANIZING CONNECTIONS

A reduction in functional strength and selectivity of deprived eye visual responses is the initial effect of MD during the critical period (Frenkel & Bear, 2004;

Gordon & Stryker, 1996; Hensch et al., 1998; Trachtenberg, Trepel, & Stryker, 2000). Reduction of deprived-eye responses may occur by synaptic depression at both thalamocortical and intracortical connections. Thus, although the historical emphasis has been on excitatory pyramidal neurons, recent studies have demonstrated that inhibitory GABAergic synapses show robust activity-dependent plasticity. Inhibitory synapses in the cortex are formed by a diverse group of GABAergic interneurons that differentially respond to early sensory experience. The most rapid shifts in visual response are seen in parvalbumin (PV)-expressing inhibitory interneurons which may enable further functional changes within V1 (Aton, Broussard, Dumoulin et al., 2013; Kuhlman, Olivas, Tring et al., 2013; Yazaki-Sugiyama, Kang, Câteau, Fukai, & Hensch, 2009). Depression is then followed by a relatively slower, homeostatic strengthening of open eye responses (Frenkel & Bear, 2004; Kaneko, Stellwagen, Malenka, & Stryker, 2008; Sawtell, Frenkel, Philpot et al., 2003).

The changes in functional strength of synapses during MD is reflected in robust morphological plasticity. Cells are surrounded by an extracellular matrix (ECM), which provides scaffolding for the cells but in addition initiates biochemical and biomechanical cues related to tissue morphogenesis, differentiation, and homeostasis. Not much is known about the structure and function of the brain's ECM although it constitutes nearly 20% of the brain's volume. There is an emerging evidence that the ECM plays a significant role in neurogenesis, synaptogenesis, and synaptic and homeostatic plasticity, as well as in neurodegenerative and neuropsychiatric diseases (e.g., Dityatev, Wehrle-Ahller, & Pitkanen, 2014). The ECM shows an initial degradation by the upregulation of proteases occurs within the first 2 days after MD in the mouse, and may elevate spine motility (Mataga, Mizuguchi, & Hensch, 2004; Oray, Majewska, & Sur, 2004). One component of the ECM is the neurofilament light protein, which can be reduced in V1, resulting in structural plasticity by destabilizing the cytoskeleton (Duffy & Livingstone, 2005; Duffy & Mitchell, 2013; Duffy, Murphy, Frosch, & Livingstone, 2007). Brief MD during the critical period alters spine density on pyramidal neurons (Djurisic, Vidal, Mann et al., 2013; Mataga et al., 2004; Tropea, Majewska, Garcia, & Sur, 2010; Yu, Majewska, & Sur, 2011), leading to a transient decrease in synaptic density formed by thalamocortical axons originating from the lateral geniculate nucleus (LGN) (Coleman, Nahmani, Gavornik et al., 2010). Longer MD leads to enduring alterations in the extent and length of thalamocortical arbors serving the two eyes (Antonini, Fagiolini, & Stryker, 1999; Hubel, Wiesel, & LeVay, 1977; Shatz & Stryker, 1978) and a significant reduction in dendritic spine density (Montey & Quinlan, 2011).

Studies in nonhuman primates and humans show that rather than an abrupt closure of the critical period in development, there appears to be a protracted decline in visual plasticity that extends into adulthood. The residual plasticity during the period of decline differs from critical period plasticity in various important ways.

First, the effect of MD on ocular dominance in adults is slower and likely requires longer periods of deprivation to engage. Second, mature plasticity may not require depression of deprived eye responses for subsequent strengthening of responses to the nondeprived eye. Third, because plasticity in layer 4 has been shown to be constrained early in postnatal development, the later form of plasticity may be restricted to synapses in layers 1/2/3 and 5/6. Fourth, saturated synapses (i.e., the receptors are saturated by a quantum of transmitter) may restrict the amount of recovery of visual function that can be accomplished. Furthermore, the structural alterations (pruning and spine motility) accompanying the plasticity in the critical period do not result from MD in adults (Lee, Huang, Feng et al., 2006; Mataga et al., 2004; Oray et al., 2004). Indeed, one of the hallmarks of the termination of the critical period is a general decline in structural plasticity. Nonetheless, plasticity may persist in the adult layer 1 after adult MD (Hofer, Mrsic-Flogel, Bonhoeffer, & Hübener, 2009).

6.4 MECHANISMS CONTROLLING CRITICAL PERIODS

The emergence of powerful new molecular genetic techniques available in mice are providing new insights into the mechanisms that may initiate and terminate critical periods. Research using rodents has shown that the critical period for ocular dominance plasticity peaks in the third postnatal week, which demonstrates that elevated plasticity is not the initial state of immature circuits. Indeed, the initiation of the critical period requires the maturation of specific inhibitory circuitry. Maturation of this circuit can be accelerated by activating inhibitory $GABA_A$ receptors with allosteric modulators such as benzodiazepines (Fagiolini & Hensch, 2000; Fagiolini, Fritschy, Löw et al., 2004; Hensch et al., 1998; Iwai, Fagiolini, Obata, & Hensch, 2003). Inhibitory interneurons containing the calcium binding protein PV can be induced to mature early, and thus prompt premature initiation of the critical period by increasing levels of growth factors (brain derived neurotrophic factor: Hanover, Huang, Tonegawa, & Stryker, 1999; Huang, Kirkwood, Pizzorusso et al., 1999; orthodenticle homeobox 2(Otx2): Spatazza, Lee, Di Nardo, et al., 2013; Sugiyama, Di Nardo, Aizawa et al., 2008), or removing cell-adhesion (polysialic acid: Di Cristo, Chattopadhyaya, Kuhlman et al., 2007) or DNA binding proteins (methyl CP binding protein 2: Durand, Patrizi, Quast et al., 2012; Krishnan, Wang, Lu et al., 2015).

Inhibition mediated by fast-spiking PV-interneurons exerts powerful control over the excitability and plasticity of downstream pyramidal neurons, potentially sharpening the spike-timing required for synaptic plasticity (Katagiri, Fagiolini, & Hensch, 2007; Kuhlman, Olivas, Tring et al., 2013; Toyoizumi, Miyamoto, Yazaki-Sugiyama et al., 2013) (see Fig. 6.1). Synaptic strength and/or number of synapses is regulated by several proteins acting at excitatory synapses onto PV interneurons, and impact the timing of the critical period (neuro activity regulated pentaxin (NARP): Gu, Huang,

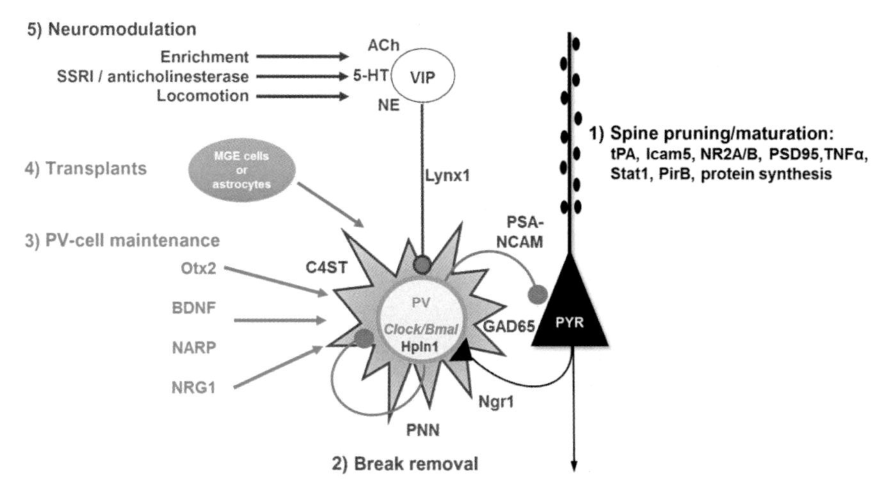

Figure 6.1 Summary of molecular factors serving critical period plasticity in primary visual cortex. (1) rewiring is a sequential process of synaptic pruning and maturation; (2) molecular brakes limiting plasticity accumulate to close the critical period; (3) the perineuronal net (PNN) attracts maturation/maintenance factors for pivotal parvalbumin (PV) cells; (4) transplantation of PV precursor cells or immature astrocytes reopens plasticity in adulthood; (5) noninvasive neuromodulation also resets PV-cells via a second, disinhibitory circuit.

Chang et al., 2013; neuregulin 1 (NRG1): Gu, Tran, Murase et al., 2016; Sun, Ikrar, Davis, et al., 2016). The critical period fails initiation in NARP-deficient mice unless there is an enhancement of inhibitory output or excitatory drive onto PV interneurons (Gu, Huang, Chang et al., 2013; Gu, Tran, Murase et al., 2016). Thus, the dynamic balance of excitation–inhibition is central to the timing of the critical period, partly by driving oscillatory activity in the gamma frequency (30–80 Hz) range (Bartos, Vida, Jonas, 2007). Such shifts in electroencephalogram oscillations appear to be relevant for plasticity and various cognitive functions (e.g., Baltus & Herrmann, 2016).

Just as the maturation of inhibition regulates the timing of the critical period, so too is a further increase in inhibition thought to terminate the critical period. It follows that the critical period can be reopened in adulthood by reduction of inhibition, such as by pharmacological manipulations (Harazov, Spolidoro, DiCristo et al., 2010) or the knockdown of Otx2 (Beurdeley, Spatazza, Lee et al., 2012; Spatazza, Lee Di Nardo et al., 2013). Whereas treatment with an NRG1 peptide induces a precocious termination of the critical period, the critical period can be reactivated in adults by inhibition of the activity of the NRG receptor (ErbB) (Kameyama, Sohya, Ebina et al., 2010). Taken together these studies reveal that PV inhibitory cells exert bidirectional control over ocular dominance plasticity (van Versendaal & Levelt, 2016), acting as a type of "molecular brake."

Other classes of inhibitory neurons may influence the expression of plasticity, either through the regulation of inhibitory PV neurons or through independent actions. For example, layer 2/3 neurons expressing vasoactive intestinal peptide (VIP) and layer 1 inhibitory neurons in V1 can be strongly activated and exert cortical effects by disinhibition of pyramidal neurons (Donato, Rompani, & Caroni, 2013; Fu, Kaneko, Tang, Alvarez-Buylla, & Stryker, 2015; Letzkus, Wolff, Meyer et al., 2011; Pfeffer, Xue, He, Huang, & Scanziani, 2013; Pi, Hangya, Kvistsiani et al., 2013). Hence, VIP neurons can be activated by locomotion resulting in enhanced neural activity in V1 (Niell & Stryker, 2010), which can enhance adult plasticity by increasing inhibition onto other interneuron subtypes that target pyramidal neurons (Fu, Kaneko, Tang et al., 2015; Fu, Tucciarone, Espinosa, et al., 2014). Similarly, when animals learn an auditory discrimination task the reinforcement signals (reward and punishment) activate VIP neurons in auditory cortex, which in turn increase the gain of a functional subpopulation of pyramidal neurons by disinhibition (Pi, Hangya, Kvistsiani et al., 2013). Thus, the inhibition of inhibitory neurons provides a mechanism for enabling plasticity in the adult cortex.

6.5 MOLECULAR CONSTRAINTS ON CRITICAL PERIOD PLASTICITY

If the induction of molecular brakes acts to terminate the critical period, then one solution to promoting recovery from amblyopia would be to remove the molecular brakes in adulthood. This could be accomplished via epigenetic mechanisms (see Chapter 7: Epigenetics and Genetics of Brain Development). For example, because increased histone deacetylase (HDAC) activity may down-regulate expression of genes promoting plasticity in development and thus reduce plasticity, inhibition of HDAC could enhance plasticity in adult V1, allowing for recovery from amblyopia (Putignano, Lonetti, Cancedda et al., 2007; Silingardi, Scali, Belluomini, & Pizzorusso, 2010). The targets for histone acetylation remain to be identified, however.

Molecular brakes could also reflect the increased expression of specific genes over development to actively limit plastic rewiring. A key candidate gene is *Lynx1* which is an endogenous inhibitor of nicotinic acetylcholine receptors. The expression of *Lynx1* emerges in V1 as the critical period closes and would act to dampen neuromodulatory actions of acetylcholine (Miwa, Ibanez-Tallon, Crabtree et al., 1999; Morishita, Miwa, Heintz, & Hensch, 2010). Both genetic deletion of *lynx1* and administration of acetylcholinesterase inhibitors enhance spine motility and the morphological plasticity induced by MD (Sajo, Ellis-Davies, & Morishita, 2016) and enables recovery of visual acuity following MD throughout life (Morishita, Miwa, Heintz et al., 2010). Another molecular brake is PirB (paired Ig-like receptor B) which is the major histocompatibility complex class I receptor. Disruption of PirB signaling facilitates recovery from amblyopia in adults and enhances Ocular dominance plasticity throughout life (Bochner, Sapp, Adelson et al., 2014; Syken, Grandpre, Kanold, & Shatz, 2006). Stat1 (signal transducer and activator of transcription 1), an immune system molecule, restricts the increase of open eye responses following MD. Genetic deletion of Stat1 enhances plasticity (Nagakura, Van Wart, Petravicz, Tropea, & Sur, 2014). The reopening of the critical period may therefore be enhanced by the identification of specific molecules that act to suppress plasticity in the adult cortex, which could lead to pharmacological interventions to reopen the critical period.

Perineuronal nets are a specialized structure in the ECM that play a key role in stabilizing synapses in the adult brain, and especially PV neurons. Perineuronal nets reach maturity at the end of the critical period and can act as a physical barrier to morphological plasticity. Disruption of the molecular latticework of the perineuronal net (Carulli, Pizzorusso, Kwok et al., 2010; Pizzorusso, Medini, Berardi et al., 2002; Pizzorusso, Medini, Landi et al., 2006) or the molecules which bind to it (Otx2: Beurdeley, Spatazza, Lee et al., 2012) permits recovery from amblyopia in adults. Mice lacking the Nogo receptor (NGR1), either globally or only from PV-cells spontaneously recover visual acuity following chronic MD because they retain their critical period plasticity (McGee, Yang, Fischer, Daw, & Strittmatter, 2005; Stephany, Chan, Parivash et al., 2014). NGR1 may work in concept with PirB to reduce spine plasticity in adults (Bochner, Sapp, Adelson et al., 2014).

Another molecular brake may be within the dendritic spine. An intracellular scaffold that is highly enriched at excitatory synapses is composed of postsynaptic protein 95 (PSD-95), which binds to N-Methyl-D-aspartate (NMDA) receptors, AMPA receptors, and potassium channels. PSD-95 is believed to accelerate maturation of excitatory synapses and acts to stabilize synaptic changes, partly by promoting AMPA receptors to incorporate into synapses containing only NMDA receptors, leading to functionally "silent" synapses. The removal of the AMPA receptors is promoted by the immediate early gene Arc, which blocks visual plasticity when the AMPA receptors are deleted (McCurry, Shepherd, Tropea et al., 2010). If PSD-95 is genetically reduced in adulthood there is an increase in the number of silent synapses, reactivating the juvenile form of ocular dominance plasticity (Huang, Stodieck, Goetze et al., 2015). It has been proposed that converting silent synapses to functional synapses could be a general mechanism for constraining neural plasticity (Greifzu, Pielecka-Fortuna, Kalogeraki et al., 2013; Huang, Stodieck, Goetze et al., 2015).

6.6 ENVIRONMENTAL REACTIVATION OF CRITICAL PERIODS IN ADULTHOOD

The nervous system is strongly impacted by the external physical or sensory environment throughout life, leading to alterations in the function and plasticity of neural circuits. We have reviewed some of the determinants of the time course of the critical period for ocular dominance plasticity but most studies have examined animals in relatively impoverished laboratory conditions. Surprisingly, enriching the housing environments with sensory, motor, or social experiences alters the time course of ocular dominance plasticity. For example, whereas standard laboratory housing closes the critical period for ocular dominance plasticity in the juvenile brain, plasticity persists into adulthood when mice are raised in large complex cages with multisensory and motor enrichment (Greifzu, Pielecka-Fortuna, Kalogeraki et al., 2013; Sale, Maya Vetencourt, Medini et al., 2007). This type of housing is not only more representative of "normal" conditions for laboratory rodents but may provide a better model for identifying mechanisms and treatments of ocular dominance plasticity in humans.

The effects of enriching experience can be seen at the molecular level through increases in H3 acetylation (Baroncelli, Scali, Sansevero et al., 2016), reductions in PV and GAD67 expression in visual cortex inhibitory neurons, and weakened GABA signaling. This leads to enhanced plasticity in cortex and hippocampus (Donato et al., 2013; Greifzu, Pielecka-Fortuna, Kalogeraki et al., 2013; Sale, Maya Vetencourt, Medini et al., 2007).

Curiously, plasticity can be reactivated in adult V1 by total visual deprivation, which can also promote recovery from chronic MD (Duffy & Mitchell, 2013; Eaton, Sheehan, & Quinlan, 2016; He, Ray, Dennis, & Quinlan, 2007; Mitchell, MacNeill, Crowder, Holman, & Duffy, 2016; Montey & Quinlan, 2011; Stodieck, Greifzu, Goetze, Schmidt, & Löwel, 2014). Cooper and Bear (2012) have proposed several mechanisms whereby dark exposure could increase plasticity in pyramidal neurons. For instance, the NMDA type glutamate receptors can be returned to a "juvenile" form (containing the NR2B subunit) that shows enhanced temporal summation (He, Hodos, & Quinlan, 2006; Yashiro, Corlew, Philpot, 2005). Furthermore, a typically juvenile form of plasticity returns (Huang, Gu, Quinlan, & Kirkwood, 2010; Montey & Quinlan, 2011). Thus, pyramidal neuron spines show immature structure and dynamics (Tropea, Majewska, Garcia et al., 2010) and there is a strengthening of immature excitatory synapses on pyramidal neurons. These changes increase neuronal excitability and expand the integration window for spike-timing dependent plasticity (Goel & Lee, 2007; Guo, Huang, de Pasquale et al., 2012; He, Ray, Dennis et al., 2007).

Total visual deprivation also decreases the excitability of PV interneurons, thus reactivating plasticity (Gu, Tran, Murase et al., 2016). In addition, dark exposure leads to a loss of cytoskelatal stability in the LGN, possibly contributing to the reactivation ocular dominance plasticity (Duffy, Lingley, Holman, & Mitchell, 2016; O'Leary, Kutcher, Mitchell et al., 2012). Taken together, the seemingly opposite interventions of environmental enrichment and total visual deprivation may both act to remove constraints on plasticity that normally act to stabilize V1 circuitry.

Although both enrichment and deprivation can lead to enhanced visual acuity, this is only the first step of a two-step process that requires the experience be followed by repetitive visual experience (Eaton, Sheehan, Quinlan et al., 2016; Greifzu, Kalogeraki, & Löwel, 2016; Kaneko & Stryker, 2014; Montey, Eaton, & Quinlan, 2013). Thus, the global experiences act to reactivate the machinery of plasticity (essentially a permissive step) whereas the focused sensory experience is necessary to stimulate perceptual learning (acting as an instructive step). The challenge is to identify the optimal combination of experiences that lead to change.

6.7 REACTIVATING PLASTICITY TO ENHANCE FUNCTIONAL RECOVERY

As discussed above, the pioneering work of Hubel and Wiesel led to the idea that there is a critical period during which MD induces structural rearrangements in V1 (Hubel et al., 1977; Shatz & Stryker, 1978; Wiesel & Hubel, 1963). The growing evidence that there can be reactivation of plasticity in V1 has led to the conclusion that critical periods are not limited to early postnatal development and that full recovery of visual acuity is possible (Bavelier, Levi, Li, Dan, & Hensch, 2010; Sengpiel, 2014; Takesian & Hensch, 2013). Given the growing understanding of mechanisms of the postnatal critical period (see above), novel therapies have been identified that have the potential to reverse the effects of MD. Several commonly prescribed drugs, such as selective serotonin reuptake inhibitors (SSRIs) (Maya Vetencourt, Sale, Viegi et al., 2008), valproate (Gervain, Vines, Chen et al., 2013; Lennartsson, Arner, Fagiolini et al., 2015), or cholinesterase inhibitors (Morishita, Miwa, Heintz et al., 2010), could be repurposed to rescue adult amblyopic patients. As we have seen, PV interneurons are key to V1 plasticity and reduced PV function may provide a common mechanism to many of these interventions. Total visual deprivation may rejuvenate intracortical inhibition by reducing the excitatory drive onto PV neurons (Gu, Tran, Murase et al., 2016). Similarly, fluoxetine (an SSRI) acts to reduce both basal levels of extracellular GABA (Maya Vetencourt, Sale, Viegi et al., 2008) and the number of PV interneurons surrounded by dense perineuronal nets (Guirado, Perez-Rando, Sanchez-Matarredona, Castrén, & Nacher, 2014). The use of action video games or vagal nerve stimulation may also be effective in treating sensory abnormalities by recruiting neuromodulatory pathways that engage attention and motivation (Hess & Thompson, 2015; Levi, Knill, & Bavelier, 2015; Mitchell & Duffy, 2014; Murphy, Roumeliotis, Williams, Beston, & Jones, 2015; Tsirlin, Colpa, Goltz, & Wong, 2015; Engineer et al., 2011).

6.8 GENERALIZING BEYOND AMBLYOPIA

The amblyopia model has generated extensive research and provided important insights at the mechanistic level. Critical periods are observed across many modalities (see Hensch, 2004 for review) and it is reasonable to wonder how well the model generalizes to other systems. Once believed to be primarily a feature of primary sensory regions, novel windows of plasticity have been shown for functions such as fear extinction in the basolateral amygdala (Gogolla, Caroni, Lüthi, & Herry, 2009), multisensory integration in the insular cortex (Gogolla, Takesian, Feng, Fagiolini, & Hensch, 2014), and the acquisition of preference behaviors in the medial prefrontal cortex (Yang, Lin, Hensch, 2012). These examples are united by the common principles of PV cell maturation observed in the visual system. A challenge is to identify electrophysiological signatures of shifting excitatory–inhibitory balance reflecting the timing of critical periods and potentially to translate noninvasively to humans.

There are also many unstudied visual deficits beyond the impaired acuity associated with amblyopia. These include impairments in shape discrimination, loss of stereoscopic depth perception, and deficits in object tracking and motion and direction perception (reviewed in Daw, 2013). Neurons in V1 have separable response properties showing distinct overlapping critical periods (reviewed in Kiorpes, 2015). For instance, direction selectivity precedes ocular dominance in kittens (Daw & Wyatt, 1976). Similarly, rhesus monkeys show an early critical period for basic spectral sensitivity that ends at about 6 months but another longer lasting critical period extending to 25 months for complex representations such as contrast sensitivity and binocularity (Harwerth, Smith, Duncan, Crawfor, & von Noorden, 1986). Different critical periods likely depend on different underlying mechanisms and thus may require multiple interventions to restore all features of V1 responses. An example can be seen in mouse visual cortex where a genetic deletion of PSD-95 does not affect the plasticity of ocular dominance in juveniles but disrupts the development of orientation preference in visual cortex (Fagiolini, Katagiri, Miyamoto et al., 2003).

Moreover, amblyopia is also associated with abnormalities in extrastriate regions (reviewed in Kiorpes, 2015). Changes in extrastriate cortex may explain why the magnitude of compromised vision demonstrated in psychophysical experiments is much greater than the changes seen in the function of V1 neurons (Shooner, Hallum, Kumbhani et al., 2015). For example, the aberrant development of extrastriate areas MT/V5 in amblyopic monkeys may explain the deficits in higher order visual functions such as motion perception (Kiorpes, Tang, & Movshon, 2006). Such extrastriate changes may suggest that higher cortical regions might provide additional targets to facilitate recovery from the effects of amblyopia on V1 (Masuda, Dumoulin, Nakadomari, & Wandell, 2008), especially because higher cortical areas are believed to express prolonged windows of plasticity that extend well beyond that of V1. But much is still to be learned about plasticity and critical periods with higher order cortical regions.

Finally, there is a large literature showing that there are critical periods in the responsiveness of cortical regions to early injury (for reviews see Kolb, 1995; Kolb, Mychasiuk, Muhammad, & Gibb, 2013). Nothing is known about the role of interneurons in recovery nor about the mechanisms underlying the effectiveness of certain interventions in stimulating recovery. Future studies will be needed to determine the extent to which the ocular dominance model will inform these research questions and potentially lead to translation to humans.

6.9 THE FUTURE

Neural circuitry is potently shaped by experience during "critical periods" in development but the brain retains the capacity to be rewired by experiences beyond early life. Importantly, although there are similarities in functional plasticity in early life and adulthood, each period may utilize distinct underlying mechanisms. The development of new molecular tools in neuroscience has led to a new understanding of the "triggers" and "brakes" that control the onset and offset of plastic processes during critical periods. Understanding the differences and similarities between early and adult plasticity, including differences in how they are measured, will lead to new therapies for remediating the effects of experiential perturbations in both children and adults.

Future work should include the development of better models for critical period plasticity across animal species and humans. Experiences in early life often have delayed effects on brain and behavior and many aversive experiences are known risk factors for behavioral disorders including schizophrenia and depression. The mechanisms for these effects are poorly understood but molecules implicated in regulating the timing of the critical periods in development as well constraining adult plasticity, are known risk factors for neurodevelopmental disorders such as schizophrenia (Do, Cuenod, & Hensch, 2015; Penzes, Buonanno, Passafaro, Sala, & Sweet, 2013; Rico & Marín, 2011). Curiously, the prevalence of visual refractive errors (amblyopia, myopia, hyperopia) is two times less common in male schizophrenics (Caspi, Vishne, Reichenberg et al., 2009).

Finally, capitalizing on genetic diversity in mice and humans should provide new insights into the individual variability that influences the timing of critical periods. Furthermore, understanding electrophysiological and biochemical correlates of critical periods will allow us to compare developmental trajectories more readily across species and disease states and to generalize beyond models such as amblyopia.

REFERENCES

Antonini, A., Fagiolini, M., & Stryker, M. P. (1999). Anatomical correlates of functional plasticity in mouse visual cortex. *Journal of Neuroscience*, *19*, 4388–4406.

Aton, S. J., Broussard, C., Dumoulin, M., Seibt, J., Watson, A., Coleman, T., & Frank, M. G. (2013). Visual experience and subsequent sleep induce sequential plastic changes in putative inhibitory and excitatory cortical neurons. *Proceedings of the National Academy of Sciences of the United States of America*, *110*(8), 3101–3106.

Baltus, A., & Herrmann, C. S. (2016). The importance of individual frequencies of endogenous brain oscillations for auditory cognition. *Brain Research*, *1640*, 243–250.

Baroncelli, L., Scali, M., Sansevero, G., Olimpico, F., Manno, I., Costa, M., & Sale, A. (2016). Experience affects critical period plasticity in the visual cortex through an epigenetic regulation of histone post-translational modifications. *Journal of Neuroscience*, *36*(12), 3430–3440.

Bartos, M., Vida, I., & Jonas, P. (2007). Synaptic mechanisms of synchronized gamma oscillations in inhibitory interneuron networks. *Nature Reviews Neuroscience*, *8*(1), 45–56.

Bavelier, D., Levi, D. M., Li, R. W., Dan, Y., & Hensch, T. K. (2010). Removing brakes on adult brain plasticity: From molecular to behavioral interventions. *Journal of Neuroscience*, *30*(45), 14964–14971.

Beurdeley, M., Spatazza, J., Lee, H. H., Sugiyama, S., Bernard, C., Di Nardo, A. A., & Prochiantz, A. (2012). Otx2 binding to perineuronal nets persistently regulates plasticity in the mature visual cortex. *Journal of Neuroscience*, *32*(27), 9429–9437.

Bochner, D. N., Sapp, R. W., Adelson, J. D., Zhang, S., Lee, H., Djurisic, M., ... Shatz, C. J. (2014). Blocking PirB up-regulates spines and functional synapses to unlock visual cortical plasticity and facilitate recovery from amblyopia. *Science Translational Medicine*, *6*(258), 258ra140.

Carulli, D., Pizzorusso, T., Kwok, J. C., Putignano, E., Poli, A., Forostyak, S., ... Fawcett, J. W. (2010). Animals lacking link protein have attenuated perineuronal nets and persistent plasticity. *Brain*, *133* (Pt 8), 2331–2347.

Caspi, A., Vishne, T., Reichenberg, A., Weiser, M., Dishon, A., Lubin, G., ... Davidson, M. (2009). Refractive errors and schizophrenia. *Schizophrenia Research*, *107*(2–3), 238–241.

Coleman, J. E., Nahmani, M., Gavornik, J. P., Haslinger, R., Heynen, A. J., Erisir, A., & Bear, M. F. (2010). Rapid structural remodeling of synapses parallels experience-dependent functional plasticity in mouse primary visual cortex. *Journal of Neuroscience*, *30*(29), 9670–9682.

Cooper, L. N., & Bear, M. F. (2012). The BCM theory of synapse modification at 30: Interaction of theory with experiment. *Nature Reviews Neuroscience*, *13*(11), 798–810.

Daw, N. W. (1998). Critical periods and amblyopia. *Archives Ophthalmology*, *116*, 502–505.

Daw, N. W. (2013). *Visual Development* (2d ed.,). New York: Springer Verlag, 282pp.

Daw, N. W., & Wyatt, H. J. (1976). Kittens reared in a unidirectional environment: Evidence for a critical period. *The Journal of Physiology*, *257*(1), 155–170.

Di Cristo, G., Chattopadhyaya, B., Kuhlman, S. J., Fu, Y., Bélanger, M. C., Wu, C. Z., ... Huang, Z. J. (2007). Activity-dependent PSA expression regulates inhibitory maturation and onset of critical period plasticity. *Nature Neuroscience*, *10*(12), 1569–1577.

Dityatev, A., Wehrle-Ahller, B., & Pitkanen, A. (2014). Brain extracellular matrix in health and disease. *Progress in Brain Research*, *214*, 1–506.

Djurisic, M., Vidal, G. S., Mann, M., Aharon, A., Kim, T., Ferrao Santos, A., ... Shatz, C. J. (2013). PirB regulates a structural substrate for cortical plasticity. *Proceedings of the National Academy of Sciences of the United States of America*, *110*(51), 20771–20776.

Do, K. Q., Cuenod, M., & Hensch, T. K. (2015). Targeting oxidative stress and aberrant critical period plasticity in the developmental trajectory to schizophrenia. *Schizophrenia Bulletin*, *41*, 835–846.

Donato, F., Rompani, S. B., & Caroni, P. (2013). Parvalbumin-expressing basket-cell network plasticity induced by experience regulates adult learning. *Nature*, *504*(7479), 272–276.

Duffy, K. R., Lingley, A. J., Holman, K. D., & Mitchell, D. E. (2016). Susceptibility to monocular deprivation following immersion in darkness either late into or beyond the critical period. *Journal of Comparative Neurology*, *524*(13), 2643–2653.

Duffy, K. R., & Livingstone, M. S. (2005). Loss of neurofilament labeling in the primary visual cortex of monocularly deprived monkeys. *Cerebral Cortex*, *15*(8), 1146–1154.

Duffy, K. R., & Mitchell, D. E. (2013). Darkness alters maturation of visual cortex and promotes fast recovery from monocular deprivation. *Current Biology*, *23*(5), 382–386.

Duffy, K. R., Murphy, K. M., Frosch, M. P., & Livingstone, M. S. (2007). Cytochrome oxidase and neurofilament reactivity in monocularly deprived human primary visual cortex. *Cerebral Cortex*, *17*(6), 1283–1291.

Durand, S., Patrizi, A., Quast, K. B., Hachigian, L., Pavlyuk, R., Saxena, A., ... Fagiolini, M. (2012). NMDA receptor regulation prevents regression of visual cortical function in the absence of Mecp2. *Neuron*, *76*(6), 1078–1090.

Eaton, N. C., Sheehan, H. M., & Quinlan, E. M. (2016). Optimization of visual training for full recovery from severe amblyopia in adults. *Learning & Memory, 23*(2), 99–103.

Engineer, N. D., Riley, J. R., Seale, J. D., Vrana, W. A., Shetake, J. A., Sudanagunta, S. P., & Kilgard, M. P. (2011). Reversing pathological neural activity using targeted plasticity. *Nature, 470*(7332), 101–104.

Fagiolini, M., Fritschy, J. M., Löw, K., Möhler, H., Rudolph, U., & Hensch, T. K. (2004). Specific GABA-A circuits for visual cortical plasticity. *Science, 303*, 1681–1683.

Fagiolini, M., & Hensch, T. K. (2000). Inhibitory threshold for critical period activation in primary visual cortex. *Nature, 404*(6774), 183–186.

Fagiolini, M., Katagiri, H., Miyamoto, H., Mori, H., Grant, S. G., Mishina, M., & Hensch, T. K. (2003). Separable features of visual cortical plasticity revealed by *N*-methyl-D-aspartate receptor 2A signaling. *Proceedings of the National Academy of Sciences of the United States of America, 100*(5), 2854–2859.

Frenkel, M. Y., & Bear, M. F. (2004). How monocular deprivation shifts ocular dominance in visual cortex of young mice. *Neuron, 44*(6), 917–923.

Fu, Y., Kaneko, M., Tang, Y., Alvarez-Buylla, A., & Stryker, M. P. (2015). A cortical disinhibitory circuit for enhancing adult plasticity. *Elife*, e05558.

Fu, Y., Tucciarone, J. M., Espinosa, J. S., Sheng, N., Daracy, D. P., Nicoll, R. A., ... Stryker, M. P. (2014). A cortical circuit for gain control by behavioral state. *Cell, 156*, 1139–1152.

Gervain, J., Vines, B. W., Chen, L. M., Seo, R. J., Hensch, T. K., Werker, J. F., & Young, A. H. (2013). Valproate reopens critical-period learning of absolute pitch. *Frontiers in Systems Neuroscience*, 7, 102.

Goel, A., & Lee, H. K. (2007). Persistence of experience-induced homeostatic synaptic plasticity through adulthood in superficial layers of mouse visual cortex. *Journal of Neuroscience, 27*(25), 6692–6700.

Gogolla, N., Caroni, P., Lüthi, A., & Herry, C. (2009). Perineuronal nets protect fear memories from erasure. *Science, 325*, 1258–1261.

Gogolla, N., Takesian, A. E., Feng, G., Fagiolini, M., & Hensch, T. K. (2014). Sensory integration in mouse insular cortex reflects GABA circuit maturation. *Neuron, 83*, 894–905.

Gordon, J. A., & Stryker, M. P. (1996). Experience-dependent plasticity of binocular responses in the primary visual cortex of the mouse. *Journal of Neuroscience, 16*(10), 3274–3286.

Greifzu, F., Kalogeraki, E., & Löwel, S. (2016). Environmental enrichment preserved lifelong ocular dominance plasticity, but did not improve visual abilities. *Neurobiology of Aging, 41*, 130–137.

Greifzu, F., Pielecka-Fortuna, J., Kalogeraki, E., Kremplar, K., Favaro, P. D., Schlüter, O. M., & Löwel, S. (2013). Environmental enrichment extends ocular dominance plasticity into adulthood and protects from stroke-induced impairments of plasticity. *Proceedings of the National Academy of Sciences of the United States of America, 111*(3), 1150–1155.

Gu, Y., Huang, S., Chang, M. C., Worley, P., Kirkwood, A., & Quinlan, E. M. (2013). Obligatory role for the immediate early gene NARP in critical period plasticity. *Neuron, 79*(2), 335–346.

Gu, Y., Tran, T., Murase, S., Borrell, A., Kirkwood, A., & Quinlan, E. M. (2016). Neuregulin-dependent regulation of fast-spiking interneuron excitability controls the timing of the critical period. *Journal of Neuroscience, 36*(40), 10285–10295.

Guirado, R., Perez-Rando, M., Sanchez-Matarredona, D., Castrén, E., & Nacher, J. (2014). Chronic fluoxetine treatment alters the structure, connectivity and plasticity of cortical interneurons. *International Journal of Neuropsychopharmacology, 17*(10), 1635–1646.

Guo, Y., Huang, S., de Pasquale, R., McGehrin, K., Lee, H. K., Zhao, K., & Kirkwood, A. (2012). Dark exposure extends the integration window for spike-timing-dependent plasticity. *Journal of Neuroscience, 32*(43), 15027–15035.

Hanover, J. L., Huang, Z. J., Tonegawa, S., & Stryker, M. P. (1999). Brain-derived neurotrophic factor overexpression induces precocious critical period in mouse visual cortex. *Journal of Neuroscience, 19* (22), RC40.

Harauzov, A., Spolidoro, M., DiCristo, G., Pasquale, R. D., Cancedda, L., Pizzorusso, T., ... Maffei, L. (2010). Reducing intracortical inhibition in the adult visual cortex promotes ocular dominance plasticity. *Journal of Neuroscience, 30*(1), 361–371.

Harwerth, R. S., Smith, E. L., Duncan, G. C., Crawfor, M. L., & von Noorden, G. K. (1986). Multiple sensitive periods in the development of the primate visual system. *Science, 232*(4747), 235–238.

He, H. Y., Hodos, W., & Quinlan, E. M. (2006). Visual deprivation reactivates rapid ocular dominance plasticity in adult visual cortex. *Journal of Neuroscience, 26*(11), 2951–2955.

He, H. Y., Ray, B., Dennis, K., & Quinlan, E. M. (2007). Experience-dependent recovery of vision following chronic deprivation amblyopia. *Nature Neuroscience, 10*(9), 1134–1136.

Hensch, T. K. (2004). Critical period regulation. *Annual Review of Neuroscience, 27*, 549–579.

Hensch, T. K., Fagiolini, M., Mataga, N., Stryker, M. P., Baekkeskov, S., & Kash, S. F. (1998). Local GABA circuit control of experience-dependent plasticity in developing visual cortex. *Science, 282*, 1504–1508.

Hess, R. F., & Thompson, B. (2015). Amblyopia and the binocular approach to its therapy. *Vision Research, 114*, 4–16.

Hofer, S. B., Mrsic-Flogel, T. D., Bonhoeffer, T., & Hübener, M. (2009). Experience leaves a lasting structural trace in cortical circuits. *Nature, 457*(7227), 3137.

Huang, S., Gu, Y., Quinlan, E. M., & Kirkwood, A. (2010). A refractory period for rejuvenating GABAergic synaptic transmission and ocular dominance plasticity with dark exposure. *Journal of Neuroscience, 30*(49), 16636–16642.

Huang, X., Stodieck, S. K., Goetze, B., Cui, L., Wong, M. H., Wenzel, C., ... Schlüter, O. M. (2015). Progressive maturation of silent synapses governs the duration of a critical period. *Proceedings of the National Academy of Sciences of the United States of America, 112*(24), E3131–E3140.

Huang, Z. J., Kirkwood, A., Pizzorusso, T., Porcialtti, V., Morales, B., Bear, M. F., & Maffei, L. (1999). BDNF regulates the maturation of inhibition and the critical period of plasticity in mouse visual cortex. *Cell, 98*, 7965–7980.

Hubel, D. H., Wiesel, T. N., & LeVay, S. (1977). Plasticity of ocular dominance columns in monkey striate cortex. *Philosophical Transactions of the Royal Society of London. Series B, Biological Sciences, 278*(961), 377–409.

Iwai, Y., Fagiolini, M., Obata, K., & Hensch, T. K. (2003). Rapid critical period induction by tonic inhibition in visual cortex. *Journal of Neuroscience, 23*(17), 6695–6702.

Kameyama, K., Sohya, K., Ebina, T., Fukuda, A., Yanagawa, Y., & Tsumoto, T. (2010). Difference in binocularity and ocular dominance plasticity between GABAergic and excitatory cortical neurons. *Journal of Neuroscience, 30*(4), 1551–1559.

Kaneko, M., Stellwagen, D., Malenka, R. C., & Stryker, M. P. (2008). Tumor necrosis factor-alpha mediates one component of competitive, experience-dependent plasticity in developing visual cortex. *Neuron, 58*(5), 673–680.

Kaneko, M., & Stryker, M. (2014). Sensory experience during locomotion promotes recovery of function in adult visual cortex. *eLife, 3*, e02798.

Katagiri, H., Fagiolini, M., & Hensch, T. K. (2007). Optimization of somatic inhibition at critical period onset in mouse visual cortex. *Neuron, 53*(6), 805–812.

Kiorpes, L. (2015). Visual development in primates: Neural mechanisms and critical periods. *Developmental Neurobiology, 75*(10), 1080–1090.

Kiorpes, L., Tang, C., & Movshon, J. A. (2006). Sensitivity to visual motion in amblyopic macaque monkeys. *Visual Neuroscience, 23*, 247–256.

Kolb, B. (1995). *Brain plasticity and behavior.* Hillsdale, NJ: Lawrence Erlbaum.

Kolb, B., Mychasiuk, R., Muhammad, A., & Gibb, R. (2013). Brain plasticity in the developing brain. *Progress in Brain Research, 207*, 35–64.

Krishnan, K., Wang, B. S., Lu, J., Wang, L., Maffei, A., Cang, J., & Huang, Z. J. (2015). MeCP2 regulates the timing of critical period plasticity that shapes functional connectivity in primary visual cortex. *Proceedings of the National Academy of Sciences of the United States of America, 112*(34), E4782–E4791.

Kuhlman, S. J., Olivas, N. D., Tring, E., Ikrar, T., Xu, T., & Trachtenberg, J. T. (2013). A disinhibitory microcircuit initiates critical-period plasticity in the visual cortex. *Nature, 56*(5), 908–923.

Lee, W. C., Huang, H., Feng, G., Sanes, J. R., Brown, E. N., So, P. T., & Nedivi, E. (2006). Dynamic remodeling of dendritic arbors in GABAergic interneurons of adult visual cortex. *PLoS Biology, 4*, e29.

Lennartsson, A., Arner, E., Fagiolini, M., Saxena, A., Andersson, R., Takahashi, H., ... Carninci, P. (2015). Remodeling of retrotransposon elements during epigenetic induction of adult visual cortical plasticity by HDAC inhibitors. *Epigenetics & Chromatin, 8*, 55.

Letzkus, J. J., Wolff, S. B., Meyer, E. M., Tovote, P., Courtin, J., Herry, C., & Lüthi, A. (2011). A disinhibitory microcircuit for associative fear learning in the auditory cortex. *Nature, 480*, 331.

LeVay, S., Stryker, M. P., & Shatz, C. J. (1978). Ocular dominance columns and their development in layer IV of the cat's visual cortex: A quantitative study. *Journal of Comparative Neurology, 179*, 223–244.

Levi, D. M., Knill, D. C., & Bavelier, D. (2015). Stereopsis and amblyopia: A mini-review. *Vision Research, 114*, 17–30.

Masuda, Y., Dumoulin, S. O., Nakadomari, S., & Wandell, B. A. (2008). V1 Projection zone signals in human macular degeneration depend on task, not stimulus. *Cerebral Cortex, 18*(11), 2483–2493.

Mataga, N., Mizaguchi, Y., & Hensch, T. K. (2004). Experience-dependent pruning of dendritic spines in visual cortex by tissue plasminogen activator. *Neuron, 44*(6), 1031–1041.

Mataga, N., Nagai, N., & Hensch, T. K. (2002). Permissive proteolytic activity for visual cortical plasticity. *Proceedings of the National Academy of Sciences of the United States of America, 99*(11), 7717–7721.

Maya Vetencourt, J. F., Sale, A., Viegi, A., Baroncelli, L., De Pasquale, R., O'Leary, O. F., ... Maffei, L. (2008). The antidepressant fluoxetine restores plasticity in the adult visual cortex. *Science, 18*, 385–388.

McCurry, C. L., Shepherd, J. D., Tropea, D., Wang, K. H., Bear, M. F., & Sur, M. (2010). Loss of Arc renders the visual cortex impervious to the effects of sensory deprivation or experience. *Nature Neuroscience, 13*, 450–457.

McGee, A. W., Yang, Y., Fischer, Q. S., Daw, N. W., & Strittmatter, S. M. (2005). Experience-driven plasticity of visual cortex limited by myelin and Nogo receptor. *Science, 309*(5744), 2222–2226.

Mitchell, D. E., & Duffy, K. R. (2014). The case from animal studies for balanced binocular treatment strategies for human amblyopia. *Ophthalmic & Physiological Optics, 34*(2), 129–145.

Mitchell, D. E., & MacKinnon, S. (2002). The present and potential impact of research on animal models for clinical treatment of stimulus deprivation amblyopia. *Clinical and Experimental Optometry, 85*(1), 5–18.

Mitchell, D. E., MacNeill, K., Crowder, N. A., Holman, K., & Duffy, K. R. (2016). Recovery of visual functions in amblyopic animals following brief exposure to total darkness. *The Journal of Physiology, 594*(1), 149–167.

Miwa, J. M., Ibanez-Tallon, I., Crabtree, G. W., Sanchez, R., Sali, A., Role, L. W., & Heintz, N. (1999). Lynx1, an endogenous toxin-like modulator of nicotinic acetylcholine receptors in the mammalian CNS. *Neuron, 23*(1), 105–114.

Montey, K. L., Eaton, N. C., & Quinlan, E. M. (2013). Repetitive visual stimulation enhances recovery from severe amblyopia. *Learning & Memory, 20*(6), 311–317.

Montey, K. L., & Quinlan, E. M. (2011). Recovery from chronic monocular deprivation following reactivation of thalamocortical plasticity by dark exposure. *Nature Communications, 2*, 317.

Morishita, H., & Hensch, T. K. (2008). Critical period revisited: Impact on vision. *Current Opinion in Neurobiology, 18*(1), 101–107.

Morishita, H., Miwa, J. M., Heintz, N., & Hensch, T. K. (2010). Lynx1, a cholinergic brake, limits plasticity in adult visual cortex. *Science, 330*, 1238–1240.

Murphy, K. M., Roumeliotis, G., Williams, K., Beston, B. R., & Jones, D. G. (2015). Binocular visual training to promote recovery from monocular deprivation. *Vision Research, 114*, 68–78.

Nagakura, I., Van Wart, A., Petravicz, J., Tropea, D., & Sur, M. (2014). STAT1 regulates the homeostatic component of visual cortical plasticity via an AMPA receptor-mediated mechanism. *Journal of Neuroscience, 34*, 10256–10263.

Niell, C. M., & Stryker, M. P. (2010). Modulation of visual responses by behavioral state in mouse visual cortex. *Neuron, 65*(4), 472–479.

Oray, S., Majewska, A., & Sur, M. (2004). Dendritic spine dynamics are regulated by monocular deprivation and extracellular matrix degradation. *Neuron, 44*, 1021–1030.

Penzes, P., Buonanno, A., Passafaro, M., Sala, C., & Sweet, R. A. (2013). Developmental vulnerability of synapses and circuits associated with neuropsychiatric disorders. *Journal of Neurochemistry, 126*(2), 165–182.

Pfeffer, C., Xue, M., He, M., Huang, Z., & Scanziani, M. (2013). Inhibition of inhibition in visual cortex: The logic of connections between molecularly distinct interneurons. *Nature Neuroscience, 16*, 1068–1076.

Pi, H. J., Hangya, B., Kvitsiani, D., Sanders, J. I., Huang, Z. J., & Kepecs, A. (2013). Cortical interneurons that specialize in disinhibitory control. *Nature, 503*, 521–524.

Pizzorusso, T., Medini, P., Berardi, N., Chierzi, S., Fawcett, J. W., & Maffei, L. (2002). Reactivation of ocular dominance plasticity in the adult visual cortex. *Science, 298*(5596), 1248–1251.

Pizzorusso, T., Medini, P., Landi, S., Baldini, S., Berardi, N., & Maffei, L. (2006). Structural and functional recovery from early monocular deprivation in adult rats. *Proceedings of the National Academy of Sciences of the United States of America, 103*(22), 8517–8522.

Putignano, E., Lonetti, G., Cancedda, L., Ratto, G., Costa, M., Maffei, L., & Pizzorusso, T. (2007). Developmental downregulation of histone posttranslational modifications regulates visual cortical plasticity. *Neuron, 53*, 747–759.

Rico, B., & Marín, O. (2011). Neuregulin signaling, cortical circuitry development and schizophrenia. *Current Opinion in Genetics & Development, 21*(3), 262–270.

Sajo, M., Ellis-Davies, G., & Morishita, H. (2016). Lynx1 Limits Dendritic Spine Turnover in the Adult Visual Cortex. *Journal of Neuroscience, 36*(36), 9472–9478.

Sale, A., Maya Vetencourt, J. F., Medini, P., Cenni, M. C., Baroncelli, L., De Pasquale, R., & Maffei, L. (2007). Environmental enrichment in adulthood promotes amblyopia recovery through a reduction of intracortical inhibition. *Nature Neuroscience, 10*(6), 679–681.

Sato, M., & Stryker, M. P. (2008). Distinctive features of adult ocular dominance plasticity. *Journal of Neuroscience, 28*(41), 10278–10286.

Sawtell, N. B., Frenkel, M. Y., Philpot, B. D., Nakazawa, K., Tonegawa, S., & Bear, M. F. (2003). NMDA receptor-dependent ocular dominance plasticity in adult visual cortex. *Neuron, 38*(6), 977–985.

Sengpiel, F. (2014). Plasticity of the visual cortex and treatment of amblyopia. *Current Biology, 24*(18), R936–R940.

Shatz, C. J., & Stryker, M. P. (1978). Ocular dominance in layer IV of the cat's visual cortex and the effects of monocular deprivation. *The Journal of Physiology, 281*, 267–283.

Shooner, C., Hallum, L. E., Kumbhani, R. D., Ziemba, C. M., Garcia-Marin, V., Kelly, J. G., ... Kiorpes, L. (2015). Population representation of visual information in areas V1 and V2 of amblyopic macaques. *Vision Research, 114*, 56–67.

Silingardi, D., Scali, M., Belluomini, G., & Pizzorusso, T. (2010). Epigenetic treatments of adult rats promote recovery from visual acuity deficits induced by long-term monocular deprivation. *European Journal of Neuroscience, 31*, 2185–2192.

Simons, K. (2005). Amblyopia characterization, treatment, and prophylaxis. *SUR Ophthalmology, 50*, 123–166.

Spatazza, J., Lee, H. H., Di Nardo, A. A., Tibaldi, L., Joliot, A., Hensch, T. K., & Prochiantz, A. (2013). Choroid-plexus-derived Otx2 homeoprotein constrains adult cortical plasticity. *Cell Reports, 3*(6), 1815–1823.

Stephany, C. É., Chan, L. L., Parivash, S. N., Dorton, H. M., Piechowicz, M., Qiu, S., & McGee, A. W. (2014). Plasticity of binocularity and visual acuity are differentially limited by nogo receptor. *Journal of Neuroscience, 34*(35), 11631–11640.

Stodieck, S. K., Greifzu, F., Goetze, B., Schmidt, K. F., & Löwel, S. (2014). Brief dark exposure restored ocular dominance plasticity in aging mice and after a cortical stroke. *Experimental Gerontology, 60*, 1–11.

Sugiyama, S., Di Nardo, A. A., Aizawa, S., Matsuo, I., Volovitch, M., Prochiantz, A., & Hensch, T. K. (2008). Experience-dependent transfer of Otx2 homeoprotein into the visual cortex activates postnatal plasticity. *Cell, 134*(3), 508–520.

Sun, Y., Ikrar, T., Davis, M. F., Gong, N., Zheng, Z., Luo, Z. D., ... Xu, X. (2016). Neurogulin-1/ErbB4 signaling regulates visual cortical plasticity. *Neuron, 92*(1), 160–173.

Syken, J., Grandpre, T., Kanold, P. O., & Shatz, C. J. (2006). PirB restricts ocular-dominance plasticity in visual cortex. *Science, 313*(5794), 1795–1800.

Taha, S., & Stryker, M. P. (2002). Rapid ocular dominance plasticity requires cortical but not geniculate protein synthesis. *Neuron, 34*, 425–436.

Takesian, A. E., & Hensch, T. K. (2013). Balancing plasticity/stability across brain development. *Progress in Brain Research, 207*, 3–34.

Toyoizumi, T., Miyamoto, H., Yazaki-Sugiyama, Y., Atapour, N., Hensch, T. K., & Miller, K. D. (2013). A theory of the transition to critical period plasticity: Inhibition selectively suppresses spontaneous activity. *Neuron*, *80*(1), 51–63.

Trachtenberg, J. T., Trepel, C., & Stryker, M. P. (2000). Rapid extragranular plasticity in the absence of thalamocortical plasticity in the developing primary visual cortex. *Science*, *287*, 2029–2032.

Tropea, D., Majewska, A. K., Garcia, R., & Sur, M. (2010). Structural dynamics of synapses in vivo correlate with functional changes during experience-dependent plasticity in visual cortex. *Journal of Neuroscience*, *30*, 11086–11095.

Tsirlin, I., Colpa, L., Goltz, H. C., & Wong, A. M. (2015). Behavioral training as new treatment for adult amblyopia: A meta-analysis and systemic review. *Investigative Ophthalmology & Visual Science*, *56*(6), 4061–4075.

van Versendaal, D., & Levelt, C. N. (2016). Inhibitory interneurons in visual cortical plasticity. *Cellular and Molecular Life*, *73*(19), 3677–3691.

Wang, B. S., Sarnaik, R., & Cang, J. (2010). Critical period plasticity matches binocular orientation preference in the visual cortex. *Neuron*, *65*(2), 246–256.

Wiesel, T. N., & Hubel, D. H. (1963). Single-cell responses in striate cortex of kittens deprived of vision in one eye. *Journal of Neurophysiology*, *26*, 1003–1017.

Wiesel, T. N., & Hubel, D. H. (1970). The period of susceptibility to the physiological effects of unilateral eye closure in kittens. *The Journal of Physiology (London)*, *206*, 419–436.

Yang, E. J., Lin, E. W., & Hensch, T. K. (2012). Critical period for acoustic preference in mice. *Proceedings of the National Academy of Sciences of the United States of America*, *109*(Suppl 2), 17213–17220.

Yashiro, K., Corlew, R., & Philpot, B. D. (2005). Visual deprivation modifies both presynaptic glutamate release and the composition of perisynaptic/extrasynaptic NMDA receptors in adult visual cortex. *Journal of Neuroscience*, *25*(50), 11684–11692.

Yazaki-Sugiyama, Y., Kang, S., Câteau, H., Fukai, T., & Hensch, T. K. (2009). Bidirectional plasticity in fast-spiking GABA circuits by visual experience. *Nature*, *462*, 218–221.

Yu, H., Majewska, A. K., & Sur, M. (2011). Rapid experience-dependent plasticity of synapse function and structure in ferret visual cortex in vivo. *Proceedings of the National Academy of Sciences of the United States of America*, *108*, 21235–21240.

FURTHER READING

Antonini, A., & Stryker, M. P. (1993). Rapid remodeling of axonal arbors in the visual cortex. *Science*, *260*(5115), 1819–1821.

Atwal, J. K., Pinkston-Gosse, J., Syken, J., Stawicki, S., Wu, Y., Shatz, C., & Tessier-Lavigne, M. (2008). PirB is a functional receptor for myelin inhibitors of axonal regeneration. *Science*, *322*(5903), 967–970.

Baroncelli, L., Sale, A., Viegi, A., Maya Vetencourt, J. F., De Pasquale, R., Baldini, S., & Maffei, L. (2010). Experience-dependent reactivation of ocular dominance plasticity in the adult visual cortex. *Experimental Neurology*, *226*(1), 100–109.

Blakemore, C., Garey, L. J., & Vital-Durand, F. (1978). The physiological effects of monocular deprivation and their reversal in the monkey's visual cortex. *The Journal of Physiology(London)*, *283*, 223–262.

Bonaccorsi, J., Berardi, N., & Sale, A. (2014). Treatment of amblyopia in the adult: Insights from a new rodent model of visual perceptual learning. *Frontiers in Neural Circuits*, *8*, 82.

Chang, M. C., Park, J. M., Pelkey, K. A., Grabenstatter, H. L., Xu, D., Linden, D. J., ... Worley, P. F. (2010). Narp regulates homeostatic scaling of excitatory synapses on parvalbumin-expressing interneurons. *Nature Neuroscience*, *13*(9), 1090–1097.

Cornelissen, L., Kim, S. E., Purdon, P. L., Brown, E. N., & Berde, C. B. (2015). Age-dependent electro-encephalogram (EEG) patterns during sevoflurane general anesthesia in infants. *Elife*, *4*, e06513.

Dickendesher, T. L., Baldwin, K. T., Mironova, Y. A., Koriyama, Y., Raiker, S. J., Askew, K. L., ... Giger, R. J. (2012). NgR1 and NgR3 are receptors for chondroitin sulfate proteoglycans. *Nature Neuroscience*, *15*(5), 703–712.

El-Shamaylah, Y., Kiorpes, L., Kohn, A., & Movshon, J. A. (2010). Visual motion processing by neurons in area MT of macque monkeys with experimental amblyopia. *Journal of Neuroscience*, *30*(36), 12198–12209.

Fagiolini, M., Pizzorusso, T., Berardi, N., Domenici, L., & Maffei, L. (1994). Functional postnatal development of the rat primary visual cortex and the role of visual experience: Dark rearing and monocular lid suture. *Vision Research*, *34*, 709–720.

Gil-Pagés, M., Stiles, R. J., Parks, C. A., Neier, S. C., Radulovic, M., Oliveros, A., ... Schrum, A. G. (2013). Slow angled-descent forepaw grasping (SLAG) could be incorporated into psy: An innate behavioral task for identification of individual experimental mice possessing functional vision. *Behavioral and Brain Functions*, *23;9*(1), 35.

Heimel, J. A., Hermans, J. M., & Sommeijer, J. P. (2008). Neuro-Bsik Mouse Phenomics consortium, Levelt CN (2008) Genetic control of experience-dependent plasticity in the visual cortex. *Genes, Brain and Behavior*, 7(8), 915–923.

Hensch, T. K. (2005). Critical period plasticity in local cortical circuits. *Nature Reviews Neuroscience*, *6* (11), 877–888.

Kaplan, E. S., Cooke, S. F., Komorowski, R. W., Chubykin, A. A., Thomazeau, A., Khibnik, L. A., ... Bear, M. R. (2016). Contrasting roles for parvalbumin expressing inhibitory neurons in two forms of adult visual cortical plasticity. *Elife*, *5*, e11450.

Kawato, M., Lu, Z. L., Sagi, D., Sasaki, Y., Yu, C., & Watanabe, T. (2014). Perceptual learning—the past, present, and future. *Vision Research*, *99*, 1–4.

Kiorpes, L., & McKee, S. P. (1999). Neural mechanisms underlying amblyopia. *Current Opinion in Neurobiology*, *9*(4), 4880–4886.

Lehmann, K., & Löwel, S. (2008). Age-dependent ocular dominance plasticity in adult mice. *PLoS ONE*, *3*(9), e3120.

Levi, D. M., & Li, R. W. (2009). Perceptual learning as a potential treatment for amblyopia: A mini-review. *Vision Research*, *49*(21), 2535–2549.

Li, B., Woo, R. S., Mei, L., & Malinow, R. (2007). The neuregulin-1 receptor erbB4 controls glutamatergic synapse maturation and plasticity. *Neuron*, *54*(4), 583–597.

Liao, D. S., Krahe, T. E., Prusky, G. T., Medina, A. E., & Ramoa, A. S. (2004). Recovery of cortical binocularity and orientation selectivity after the critical period for ocular dominance plasticity. *Journal of Neurophysiology*, *92*(4), 2113–2121.

Maruko, I., Zhang, B., Tao, X., Tong, J., Smith, E. L., 3rd, & Chino, Y. M. (2008). Postnatal development of disparity sensitivity in visual area 2 (v2) of macaque monkeys. *Journal of Neurophysiology*, *100*, 2486–2495.

Movshon, J. A., & Dürsteler, M. R. (1977). Effects of brief periods of unilateral eye closure on the kitten's visual system. *Journal of Neurophysiology*, *40*, 1255–1265.

O'Brien, R. J., Xu, D., Petralia, R. S., Steward, O., Huganir, R. L., & Worley, P. (1999). Synaptic clustering of AMPA receptors by the extracellular immediate-early gene NARP. *Neuron*, *23*(2), 309–323.

O'Leary, T. P., Kutcher, M. R., Mitchell, D. E., & Duffy, K. R. (2012). Recovery of neurofilament following early monocular deprivation. *Frontiers in Systems Neuroscience*, *6*, 22.

Olson, C. R., & Freeman, R. D. (1975). Progressive changes in kitten striate cortex during monocular vision. *Journal of Neurophysiology*, *38*, 26–32.

Pelkey, K. A., Barksdale, E., Craig, M. T., Yuan, X., Sukumaran, M., Vargish, G. A., ... McBain, C. J. (2015). Pentraxins coordinate excitatory synapse maturation and circuit integration of parvalbumin interneurons. *Neuron*, *85*(6), 1257–1272.

Pham, T. A., Graham, S. J., Suzuki, S., Barco, A., Kandel, E. R., Gordon, B., & Lickey, M. E. (2004). *Learning & Memory*, *11*(6), 738–747.

Prusky, G. T., & Douglas, R. M. (2003). Developmental plasticity of mouse visual acuity. *European Journal of Neuroscience*, *17*, 167–173.

Quinlan, E. M., Philpot, B. D., Huganir, R. L., & Bear, M. F. (1999). Rapid, experience-dependentexpression of synaptic NMDA receptors in visual cortex in vivo. *Nature Neuroscience*, *2*(4), 352–357.

Saiepour, M. H., Rajendran, R., Omrani, A., Ma, W. P., Tao, H. W., Heimel, J. A., & Levelt, C. N. (2015). Ocular dominance plasticity disrupts binocular inhibition–excitation matching in visual cortex. *Current Biology, 25*(6), 713–721.

Scali, M., Baroncelli, L., Cenni, M. C., Sale, A., & Maffei, L. (2012). A rich environmental experience reactivates visual cortex plasticity in aged rats. *Experimental Gerontology, 47*(4), 337–341.

Scholl, B., Burge, J., & Priebe, N. J. (2013). Binocular integration and disparity selectivity in mouse primary visual cortex. *Journal of Neurophysiology, 109*(12), 3013–3024.

Schwarzkopf, D. S., Vorobyov, V., Mitchell, D. E., & Sengpiel, F. (2007). Brief daily binocular vision prevents monocular deprivation effects in visual cortex. *European Journal of Neuroscience, 25*(1), 270–280.

Smith, G. B., & Bear, M. F. (2010). Bidirectional ocular dominance plasticity of inhibitory networks: Recent advances and unresolved questions. *Frontiers in Cellular Neuroscience, 4*, 21.

Sugiyama, S., Prochiantz, A., & Hensch, T. K. (2009). From brain formation to plasticity: Insights on Otx2 homeoprotein. *Development, Growth & Differentiation, 51*(3), 369–377.

Ting, A. K., Chen, Y., Wen, L., Yin, D. M., Shen, C., Tao, Y., ... Mei, L. (2011). Neuregulin 1 promotes excitatory synapse development and function in GABAergic interneurons. *Journal of Neuroscience, 31*(1), 15–25.

Trachtenberg, J. T., & Stryker, M. P. (2001). Rapid anatomical plasticity of horizontal connections in the developing visual cortex. *Journal of Neuroscience, 21*, 3476–3482.

Xu, D., Hopf, C., Reddy, R., Cho, R. W., Guo, L., Lanahan, A., ... Worley, P. (2003). Narp and NP1 form heterocomplexes that function in developmental and activity-dependent synaptic plasticity. *Neuron, 39*(3), 513–528.

CHAPTER 7

Epigenetics and Genetics of Development

Alexandre A. Lussier*, Sumaiya A. Islam* and Michael S. Kobor
University of British Columbia, Vancouver, BC, Canada

ABBREVIATIONS

5-HTT	serotonin transporter gene
ADP	adenosine diphosphate
AHRR	aryl hydrocarbon receptor repressor
ARC	arcuate nucleus
AS	Angelman syndrome
ATP	adenosine triphosphate
BDNF	brain-derived neurotrophic factor
BWS	Beckwith–Wiedemann syndrome
CpG	cytosine-guanine dinucleotide
CpH	cytosine-adenosine/cytosine/thymine dinucleotide
CGI	cytosine-guanine dinucleotide island
COMT	catechol-*O*-methyltransferase
DNA	deoxyribonucleic acid
DNAhm	DNA hydroxymethylation
DNMT	DNA methyltransferase
DOHaD	developmental origins of health and disease
ESC	embryonic stem cell
F0/F1/F2/F3	initial, first, second, or third generation
FASD	fetal alcohol spectrum disorder
FGF	fibroblast growth factor
FKBP5	FK506 Binding Protein 5
G×E	gene by environment
GnRH	gonadotropin-releasing hormone
GR	glucocorticoid receptor
H3K4me	histone 3 lysine 4 methylation
H3K9me	histone 3 lysine 9 methylation
H3K27me	histone 3 lysine 27 methylation
HAR1A	highly accelerated region 1A
HAT	histone acetyltransferases
HDAC	histone deacetylase

*Authors contributed equally.

The Neurobiology of Brain and Behavioral Development
DOI: http://dx.doi.org/10.1016/B978-0-12-804036-2.00007-8

HDM	histone demethylases
HMT	histone methyltransferases
HTT	Huntington gene
HP1	heterochromatin protein 1
IAP	intracisternal A particles
ICM	inner cell mass
IGF2	insulin growth factor 2
IQ	intellectual quotient
Kiss1	kisspeptin gene
LEARn	latent early-life associated regulation
LINE-1	long interspersed nuclear element 1
MAF	minor allele frequency
MAOA	monoamine oxidase-A
MeCP2	methyl CpG binding protein 2
NFR	nucleosome-free regions
NPC	neural progenitor cell
NR3C1	nuclear receptor subfamily 3 group C (GR-encoding gene)
PCDH	protocadherin
PcG	polycomb group
PD	Parkinson's disorder
PGC	primordial germ cell
PTM	posttranslational modifications
PTSD	posttraumatic stress disorder
PWS	Prader Willi syndrome
QTL	quantitative trait locus
mQTL	methylation quantitative trait locus
RNA	ribonucleic acid
RNAi	RNA interference
dsRNA	double-stranded RNA
lncRNA	long noncoding RNA
mRNA	messenger RNA
miRNA	microRNA
ncRNA	noncoding RNA
piRNA	PIWI-interacting RNA
pri-miRNA	primary miRNA
siRNA	short interfering RNA
snoRNA	small nucleolar RNA
snRNA	small nuclear RNA
spliRNA	splice-site RNA
ssRNA	single-stranded RNA
tRNA	transfer RNA
tiRNA	transcription initiation RNA
RSS	Russell–Silver syndrome
SNCA	alpha-synuclein gene
SNP	single nucleotide polymorphism
SNV	single nucleotide variant
TE	trophectoderm
TET	ten–eleven-translocation

TDG	thymidine DNA glycosylase
TrxG	trithorax group
Xi	inactivated X chromosome
Xic	X-inactivation center
Xist	X-inactive specific transcript
XCI	X chromosome inactivation

7.1 FOREWORD

Brain development consists of a series of complex and dynamic events, integrating contributions from genetic, epigenetic, and environmental factors to mediate the establishment and maintenance of neural networks throughout the life course. As such, characterizing the molecular mechanisms underpinning trajectories of brain development across the lifespan constitutes a fundamental cornerstone of neurobiological research. Although genetic factors comprise the inherited basis of developmental patterning, epigenetic factors are thought to form the regulatory overlay of the genome, playing crucial roles in the global shaping and maintenance of these patterns. Importantly, the same epigenetic processes that maintain ontogenetic stability can also exhibit environmentally influenced dynamic variation. Such responsiveness is critical to neural development in the brain and may be particularly critical for brain plasticity during sensitive periods of development. Furthermore, the dysregulation of epigenetic mechanisms can often lead to pathological consequences, contributing to a spectrum of disease phenotypes and neurodevelopmental disorders. Accordingly, delineating the key functions of both genetic and epigenetic factors throughout different developmental periods is crucial to understanding the complex molecular processes that underlie brain development and plasticity. Here, we provide a general framework for such genetic and epigenetic variation, detailing their roles in the context of different developmental stages across the life course, as well as neurodevelopmental and psychiatric disorders (Figure 7.1).

7.2 INTRODUCTION TO GENETICS

The mechanisms underlying the inherited basis of human health and disease have been a longstanding scientific challenge, particularly in the context of developmental research. Our understanding of the factors driving such inheritance only truly blossomed in the late 19th century, when Gregor Mendel ushered in the new era of genetics through his groundbreaking experiments in pea plants. Now considered the father of modern genetics, Mendel identified the basic principles of hereditary transmission, illustrating the predictable action of discrete units to produce inherited traits known as phenotypes. Further research in the field later identified DNA as the main hereditary molecule of the cell and the inherited units as specific genomic sequences known as

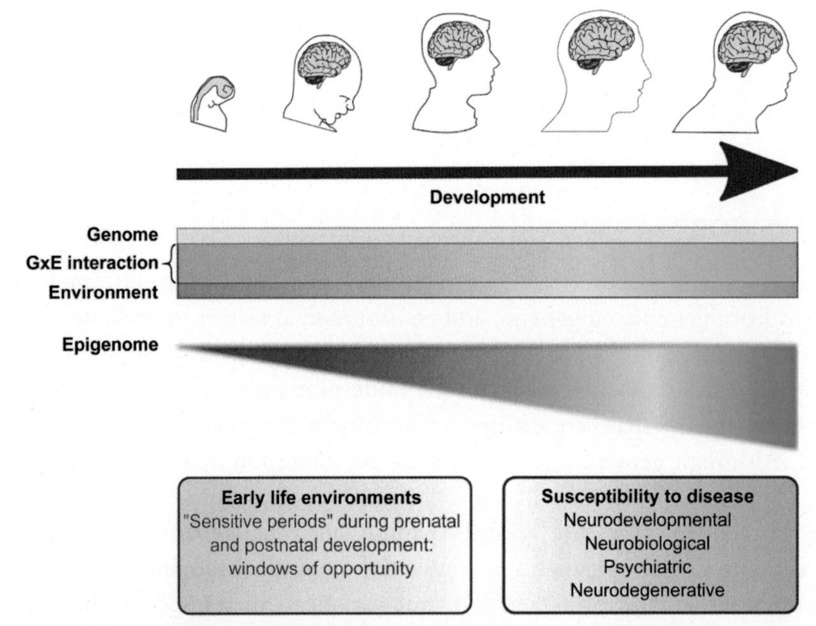

Figure 7.1 Biological variation and brain-related phenotypes are influenced by multiple sources throughout brain development. Genetic variation (blue bar) is inherited at birth and remains mostly stable throughout the lifetime. By contrast, the environment presents shifting conditions that can influence long-term health and behavior (rainbow bar). The intersection of these two influences represents gene by environment (G × E) interactions (overlap). Together, these factors are reflected in the epigenome, which is highly malleable in response to environmental conditions and strongly influenced by genetic variation. Epigenetic variability increases across the life course, with different developmental windows conferring differential sensitivity to G × E and environmental influences (widening triangle gradient). These windows of opportunity for developmental programing of epigenetic patterns and subsequent health are more vulnerable during early life, which ranges from preconception and prenatal life to postnatal environments. In turn, these effects can also influence vulnerability to disease, including neurodevelopmental disorders, neurobiological dysregulation, psychiatric disease, and neurodegeneration later in life. Taken together, the influences of both the genome and shifting environments are reflected in the epigenome, which can shape development and vulnerability to disease.

genes, which in turn encode different cellular products that interact to produce various phenotypes. Early research into the genetic underpinnings of disease focused mainly on monogenic, or Mendelian, diseases, which are caused by mutations in a single gene and show a defined inheritance pattern (dominant or recessive, and autosomal or sex-linked). For example, Huntington's disease is an autosomal dominant Mendelian disorder caused by trinucleotide repeat expansion mutations in the huntingtin gene (*HTT*) and results in age-related neurodegeneration, loss of motor control, and progressive cognitive decline. Although relatively few genes produce such drastic phenotypes, the study of Mendelian disorders has provided vital insight into the genes required for the typical development of different physiological systems, especially in the context of the brain. While disruption of only a single gene's function is primarily involved in Mendelian disease pathogenesis, common complex diseases (e.g., type 2 diabetes, schizophrenia, and coronary heart disease), and quantitative biological traits (e.g., height, weight, and IQ) display multifactorial etiology involving variation from multiple genes and environmental conditions (Hirschhorn & Daly, 2005). The spectrum of genetic variation underlying such complex traits and common diseases ranges from single nucleotide variants (SNVs) to structural variants, such as insertions, deletions, duplications, inversions, and copy number differences (Frazer, Murray, Schork, & Topol, 2009). The former are further classified according to the frequency of the minor allele (i.e., less common allele in a given population). Specifically, single nucleotide polymorphisms (SNPs) refer to SNVs with a minor allele frequency (MAF) of at least 1% in the population, whereas low frequency and rare variants have a MAF of less than 1% or 0.05%, respectively (Frazer et al., 2009). Most common variants have relatively small effect sizes on a given phenotype when analyzed individually or in combination and can therefore only explain a small proportion of disease or trait heritability defined as the proportion of phenotypic variance attributed to additive genetic factors (Manolio et al., 2009). For example, while the human height shows an estimated heritability of 80%, its associated variants only account for 5% of the observed phenotypic variance (Visscher, 2008). This suggests that other undetected factors likely contribute to phenotypic variance and therefore constitute an appreciable proportion of the "missing heritability" in these complex diseases or traits (Manolio et al., 2009). Such unaccounted factors may arise from an inadequate ability to detect rare variants with large effect sizes, poor detection of structural variations using commercial SNP arrays, and low power to account for nonadditive (i.e., epistatic) effects or gene–gene interactions (Visscher, Hill, & Wray, 2008). Importantly, the contribution of environmental influences, particularly during dynamic periods of developmental susceptibility, and their interplay with allelic variation (known as gene × environment interactions) represent additional sources of heritable variability, which are thought to inflate estimates of total heritability (Boyce, 2016; Zannas, Provençal, & Binder, 2015; Zuk, Hechter, Sunyaev, & Lander, 2012). Accordingly, a functional characterization of the genetic architecture underlying complex phenotypes requires a thorough analysis of the diverse regulatory layers coordinating the precise and timely expression of the ~25,000 genes in the human genome.

7.3 INTRODUCTION TO EPIGENETICS

7.3.1 What is epigenetics?

Although genetics may be considered the inscribed "blueprint" underlying the central dogma of molecular biology (i.e., DNA → RNA → protein), epigenetics can be thought of as the regulatory overlay of genetic sequence that fine-tunes gene activity during development and in response to external signals (Boyce & Kobor, 2015). From a historical perspective, the term "epigenetics" was first introduced by Conrad Waddington in the early 1940s to describe "the branch of biology which studies the causal interactions between genes and their products which bring the phenotype into being" (Waddington, 1968). Waddington argued that epigenetics play a critical role in the development of multicellular organisms by creating "epigenetic landscapes" that drive cellular differentiation along a programmed trajectory toward a specific cell-type lineage (Waddington, 1968). Since the first introduction of this concept, the field of epigenetics has flourished into a highly active area of study aimed at characterizing the molecular mechanisms underlying gene regulation and biological programing. Today, epigenetics is operationally defined as modifications of DNA and its regulatory components, including chromatin and noncoding RNA, to potentially modulate gene transcription without changing the DNA sequence itself (Bird, 2007; Henikoff & Greally, 2016; Meaney, 2010). Notably, Waddington's initial hypothesis still holds true: The ontogeny of the ~200 different cell types in the human body is largely shaped by the unique epigenomic profiles and transcriptional activity of each cellular subtype (Domcke et al., 2015; Schuebeler, 2015). In addition, epigenetic regulation is now becoming increasingly recognized as a potential biological mediator of environmental influences, which can contribute to sculpting the epigenome (Feil & Fraga, 2012). Accordingly, epigenetic regulation involves both dynamic tissue- and cell-type-specific variation during development, as well as the preservation of the cellular memory required for developmental stability. The orchestration of such epigenetic control involves a number of distinct mechanisms, namely, (1) DNA modifications, (2) regulation of chromatin structure, and (3) noncoding RNA activity.

7.3.2 DNA modifications

Covalent modifications on DNA nucleotides, primarily cytosine, have long been an established form of epigenetic regulation. Specifically, DNA modifications are comprised of DNA methylation (which can occur in the context of cytosine-guanine (CpG) dinucleotides or at non-CpG positions) as well as oxidized derivatives of DNA methylation such as DNA hydroxymethylation (DNAhm). In this section, we review these variant forms of DNA modifications in the context of brain development.

7.3.2.1 DNA methylation

DNA methylation is arguably the most studied epigenetic mark and involves the covalent attachment of a methyl group to the 5′ position of cytosine, typically at CpG dinucleotide sites (Jones & Takai, 2001). These CpG dinucleotides occur relatively infrequently in the genome in order to minimize the potential for DNA methylation-induced sequence mutability, as methylated cytosines can undergo spontaneous deamination to thymine (Gardiner-Garden & Frommer, 1987; Illingworth & Bird, 2009; Weber et al., 2007). Areas with comparatively high CpG content in the genome have been termed "CpG islands" (CGIs) and these CGIs are thought to exist as regions that were either never methylated or only transiently methylated in the germline while the rest of the genome experienced a loss of CpGs at methylated sequences (Gardiner-Garden & Frommer, 1987; Illingworth & Bird, 2009; Weber et al., 2007). Importantly, the DNA methylation status of the ~28 million CpG sites in the human genome is often dependent on genomic context (Jones, 2012; Ulahannan & Greally, 2015). For example, CGIs, which are associated with approximately 50%–70% of known promoters, tend to contain low levels of methylation in somatic cells, while nonisland CpGs exhibit generally higher methylation levels (Illingworth & Bird, 2009; Saxonov, Berg, & Brutlag, 2006; Weber et al., 2007). Moreover, DNA methylation is associated with the regulation of gene expression, although its effects on transcription are highly dependent on genomic context (Edgar, Tan, Portales-Casamar, & Pavlidis, 2014; Jones & Baylin, 2007; Lam et al., 2012). For example, DNA methylation at gene promoters is generally associated with gene expression silencing, although its role may be more variable within gene bodies (Jones & Baylin, 2007; Schuebeler, 2015). Conversely, in regions of lower CpG density which flank CGIs, known as "island shores," high DNA methylation levels are generally associated with highly expressed genes, especially if the associated CGI is lowly methylated (Baubec & Schuebeler, 2014; Edgar et al., 2014; Irizarry et al., 2008). While the exact mechanisms remain mostly unknown, transcriptional silencing by DNA methylation may potentially occur through the direct blocking of transcription factor binding or the recruitment of transcriptional repressors to promoter, enhancers, or insulator regions (Tate & Bird, 1993). Although DNA methylation patterns in promoter regions tend to negatively correlate with gene expression within an individual, emerging evidence shows that when comparing a single gene across a population, the association between DNA methylation and gene expression can be negative, positive, or nonexistent, highlighting the complex relationship between DNA methylation and transcription (Gutierrez-Arcelus et al., 2013; Jones, Fejes, & Kobor, 2013; Lam et al., 2012). Moreover, DNA methylation can be both active, by being a likely cause of gene expression variation, or passive, by being a consequence or an independent mark of gene expression levels (Gutierrez-Arcelus et al., 2013; Jones et al., 2013). In addition to its role in transcriptional control, DNA methylation within introns has been associated with altered messenger RNA (mRNA) splicing, and its presence within certain exons potentially regulates alternative transcriptional start sites (Maunakea et al., 2010; Maunakea, Chepelev, Cui, & Zhao, 2013; Shukla et al., 2011). Finally, DNA methylation in repetitive elements, which comprise more than half of the human genome including intergenic sequences, tends to occur at relatively high levels and is associated with maintenance of chromosome structure and genomic integrity (Cordaux & Batzer, 2009; Donnelly, Hawkins, & Moss, 1999).

The establishment and maintenance of DNA methylation patterns are carried out by a highly conserved family of enzymes known as DNA methyltransferases (DNMTs). In mammals, three major DNMTs have been identified, DNMT1, DNMT3a, and DNMT3b, which are characterized by a conserved stretch of amino acids in the C-terminal catalytic domain that target the 5′ carbon of cytosine (Bestor, 2000). As the most abundant form in adult cells, DNMT1 maintains DNA methylation patterns during cell division by binding hemi-methylated CpG sites following DNA replication and methylating the cytosine on the newly synthesized daughter strand (Robertson et al., 1999). In contrast to DNMT1's role in DNA methylation maintenance, DNMT3a and DNMT3b establish de novo genome-wide DNA methylation patterns following embryo implantation (discussed in Section 7.5) (Okano, Xie, & Li, 1998). These enzymes show equal affinity for hemi-methylated or nonmethylated DNA, and are essential for early development, as deleting their encoding genes causes embryonic lethality in mice (Okano et al., 1998). DNMTs are vital components of epigenetic regulation, modulating the expression of different genes and suppressing repetitive elements throughout the genome to prevent transposition events and chromosomal instability, which could have severe consequences on the organism. For instance, missense mutations in DNMT3b can result in immunodeficiency-centromeric instability-facial anomalies syndrome, caused by chromatin dysregulation and increased genome instability due to recombination events between homologous repetitive elements (Ehrlich et al., 2009). Of note, the maintenance and de novo roles of these enzymes are not clear-cut, as DNMT1 can function as a *de novo* DNMT when overexpressed in cultured human fibroblasts (Vertino, Yen, Gao, & Baylin, 1996). Conversely, DNMT3a and DNMT3b can also play a role in DNA methylation maintenance, as loss of DNMT1 in human colorectal cancer cells results in a loss of DNA methylation at only 20% of CpG sites (Rhee et al., 2000). Finally, two additional DNMTs exist in vertebrates, DNMT3L and DNMT2. Although catalytically inactive, the former binds to DNMT3a/b to increase their enzymatic activity, a function required for reproduction, but not development (Gowher, Liebert, Hermann, Xu, & Jeltsch, 2005). Interestingly, the male germ cells of mice lacking this enzyme display increased activity of transposable elements, including LINE-1 (long interspersed nuclear element 1) and IAP (intracisternal A particles) classes, highlighting the role of DNMTs in suppressing transposition events (Bourc'his & Bestor, 2004; Webster et al., 2005). By contrast, DNMT2 has barely any detectable CpG methylation activity, but has been linked to transfer RNA (tRNA) methylation (Goll et al., 2006; Jeltsch, Nellen, & Lyko, 2006; Schaefer et al., 2010).

7.3.2.2 Non-CpG DNA methylation

While DNA methylation primarily occurs in the context of CpG dinucleotides, it can also occur at CpH (where H = A/C/T) sites. Indeed, both the maintenance DNMT1 and *de novo* DNMT3a/b enzymes have been shown to methylate non-CpG cytosines in vitro (Guo et al., 2014; Yokochi & Robertson, 2002). Previous studies have shown that methylated CH dinucleotides (mCH) occur in cultured embryonic stem cells (ESCs) and induced pluripotent stem cells (Laurent et al., 2010; Lister et al., 2009; Lister et al., 2012; Ramsahoye et al., 2000; Ziller et al., 2011). Moreover, analysis of adult human and mouse CNS neurons found that mCH is specifically enriched in neurons compared to other cell types, as non-CpG methylation is nearly absent in nonneuronal adult somatic cells, but can reach up to ~25% of all cytosines in neurons of the adult mouse dentate gyrus (Guo et al., 2014; Lister et al., 2013; Ziller et al., 2011). Levels of mCH increase rapidly during early postnatal brain development (mouse ~2–4 weeks; human 0–2 years), suggesting that mCH potentially plays an important role in the regulation of postnatal brain development. These changes are associated with a transient rise in DNMT3a levels, as knockdown of this enzyme results in significant loss of mCH, but not methylcytosine levels (Guo et al., 2014). Genome-wide profiling also showed that in neurons, mCH is present throughout the 5′ upstream, gene–body, and 3′ downstream regions of genes, where it is negatively correlated with gene expression (Guo et al., 2014; Lister et al., 2013). Furthermore, in vitro plasmid reporter gene analyses have shown that CH methylation is associated with transcriptional repression in mouse neurons (Guo et al., 2014). However, mCH is not associated with gene silencing in all cell types, as non-CpG methylation in ESCs positively correlates with gene expression (Lister et al., 2009). It is thought that the distinct distribution and role in gene expression of mCH in different cell types relates to differences in the relative abundance and activity of specific "readers" and "writers" of non-CpG methylation (Kinde, Gabel, Gilbert, Griffith, & Greenberg, 2015). Furthermore, in addition to CH methylation, very recent research has detected the presence of methylated adenosine nucleotides in vertebrates, suggesting that that DNA modification variants may be more diverse than previously thought (Dominissini et al., 2013; Koziol et al., 2015; Meyer & Jaffrey, 2016; Meyer et al., 2012).

7.3.2.3 DNA hydroxymethylation

Although the mechanisms underlying the establishment and maintenance of DNA methylation by DNMTs have been well characterized, the process of DNA demethylation remains unclear. Thought to involve both active and passive pathways, this phenomenon is vital for typical development and genetic regulation, particularly in the brain (Ooi & Bestor, 2008; Tognini, Napoli, & Pizzorusso, 2015; Wu & Zhang, 2014). For example, neuronal activity-induced DNA demethylation of specific promoters and expression of corresponding genes such as brain-derived neurotrophic factor (*BDNF*) and fibroblast growth factors (*FGF*) occurs through the action of Gadd45b and represents an activity-dependent form of modulating neurogenesis in the adult brain (Ma et al., 2009). Passive DNA demethylation can occur due to a lack of DNMT1 activity, resulting in a gradual loss of DNA methylation over several rounds of replication (Bhutani, Burns, & Blau, 2011). By contrast, active DNA demethylation may potentially occur through the oxidation of 5-methylcytosine (5mC), catalyzed by the ten–eleven-translocation (TET) family of enzymes (Santiago, Antunes, Guedes, Sousa, & Marques, 2014; Tahiliani et al., 2009). This process generates a series of oxidized cytosine base variants, including hydroxymethylcytosine (hmC), formylcytosine (fC), and carboxycytosine (caC) (Ito et al., 2010; Tahiliani et al., 2009; Ulahannan & Greally, 2015). The oxidized site can then be removed by thymine DNA glycosylase to create an abasic site, which undergoes base excision repair to yield an unmodified cytosine (He et al., 2011). Alternatively, hmC can be converted to hydroxymethyluracil by activation-induced deaminase prior to base excision repair (Nabel et al., 2012). Although the exact details of active DNA demethylation remain unclear, the emerging evidence points to a process involving the coordinated activity of a number of key enzymatic players and intermediate modified cytosine species. In addition to their potential role in DNA demethylation, these cytosine variants may also play a role in modulating chromatin structure or recruiting various factors to key regions of the genome (Sadakierska-Chudy, Kostrzewa, Filip, & Comprehensive View, 2014). For instance, various members of the methyl-CpG-binding domain (MBD) protein family display different affinities for hmC, and given their role in recruiting different chromatin-modifying complexes, hmC could potentially alter chromatin landscapes throughout the genome (Pfeifer, Kadam, & Jin, 2013). Interestingly, DNAhm is present at high levels in pluripotent cells and the brain, where it has been implicated in neural stem cell functions, although its exact functional role remains to be uncovered (Ito et al., 2010; Kriaucionis & Heintz, 2009; Santiago et al., 2014). Genome-wide mapping of DNAhm in various brain regions, including the frontal cortex, hippocampus, and cerebellum, identified an enrichment of hmC in gene bodies, which was positively associated with gene transcription, particularly at developmentally activated genes (Lister et al., 2013; Wang et al., 2012). Finally, active DNA demethylation and TET activity is associated with memory formation and addiction in mice, further supporting its functional role in neural activity (Alaghband, Bredy, & Wood, 2016).

7.3.3 Regulation of chromatin structure

In eukaryotic genomes, DNA does not exist as a naked template but rather as a dynamic, nucleoprotein polymer. Known as chromatin, this structure is regulated by a multitude of factors to modulate access to genetic material and allow for DNA compaction within the nucleus. The fundamental repeating unit of chromatin is the nucleosome, consisting of 147 base pairs of DNA wrapped around a protein octamer containing two molecules each of core histone proteins, H2A, H2B, H3, and H4 (Kornberg, 1974; Luger, Mäder, Richmond, & Sargent, 1997). Nucleosomes are interconnected by short sections of DNA associated with the linker histone H1 to form an 11-nm-wide "beads-on-a-string" structure (Kornberg, 1974; Olins & Olins, 1974). In turn, this structure is thought to coil into a 30-nm chromatin fiber, although the elucidation of this structure remains widely debated (Joti et al., 2014; Li & Reinberg, 2011). The chromatin fiber structure can subsequently undergo additional levels of condensation to ultimately form the typical chromosome structure seen during mitosis (Li & Reinberg, 2011; Robinson, Fairall, Huynh, & Rhodes, 2006). Transitions between higher order chromatin conformations are dependent on several factors, such as cell cycle phase, transcriptional activity, and cell type (Ma, Kanakousaki, & Buttitta, 2015). For example, the beads-on-a-string structure represents the active and largely unfolded configuration that occurs during interphase to permit access of transcriptional machinery to DNA, while condensed chromosomes occur during mitosis to permit segregation of replicated DNA during anaphase (Ma et al., 2015). Eukaryotic nuclei display two main types of chromatin structure, tightly packed regions known as heterochromatin, and relatively less condensed regions known as euchromatin (Allis & Jenuwein, 2016). In general, euchromatin assumes a relatively open and more accessible conformation that contains actively transcribed genes. By contrast, heterochromatin is highly condensed and tends to contain largely transcriptionally silenced element, such as repetitive sequences in pericentromeric and telomeric regions (Allis & Jenuwein, 2016; Grewal & Jia, 2007). Given that chromatin structure and transcriptional activity are tightly linked processes, their relationship requires a dynamic interplay between histone modifications, histone variants, and chromatin remodeling complexes (Deal & Henikoff, 2010).

7.3.3.1 Histone modifications

Histones are small and highly basic proteins possessing a flexible N-terminal tail that protrudes from the nucleosome core particle. This structure is the predominant target of posttranslational modifications (PTMs) in which chemical groups or short protein peptides are covalently attached to specific amino acid residues to confer an additional level of regulatory control on the chromatin structure. Different types of histone PTMs include acetylation, methylation, phosphorylation, ubiquitination, sumoylation, and adenosine diphosphate-ribosylation (Table 7.1) (Bannister & Kouzarides, 2011; Bowman & Poirier, 2015; Kouzarides, 2007). These influence nucleosome stability and positioning by altering the chemical interactions within nucleosomes, between neighboring nucleosomes, or between histone-DNA contacts (Shogren-Knaak et al., 2006;

Table 7.1 Summary of histone posttranslational modifications[a]

Histone	Modified residue	Modification	Modifying enzymes	Proposed function
H1	Lys26	Methylation	EZH2	Transcriptional silencing
	Ser27	Phosphorylation	Unknown	Transcriptional activation, chromatin decondensation
H2A	Lys5	Acetylation	Tip60, p300/CBP	Transcriptional activation
	Ser1	Phosphorylation	Unknown, MSK1	Mitosis, chromatin assembly, transcriptional repression
	Ser139 (mammalian H2AX)	Phosphorylation	ATR, ATM, DNA-PK	DNA repair
	Lys119	Ubiquitylation	Ring2	Spermatogenesis
	Lys9	Biotinylation	Biotinidase	Unknown
	Lys13	Biotinylation	Biotinidase	Unknown
H2B	Lys5	Acetylation	p300, ATF2	Transcriptional activation
	Lys12	Acetylation	p300/CBP, ATF2	Transcriptional activation
	Lys15	Acetylation	p300/CBP, ATF2	Transcriptional activation
	Lys20	Acetylation	p300	Transcriptional activation
	Ser14	Phosphorylation	Mst1, unknown	Apoptosis, DNA repair
	Lys120	Ubiquitylation	UbcH6	Meiosis
H3	Lys9	Acetylation	Gcn5, SRC-1, unknown	Transcriptional activation, histone deposition
	Lys14	Acetylation	Unknown, Gcn5, PCAF, Esal, Tip60, SRC-1, Elp3, Hpa2, hTFIIIC90, TAF1, Sas2, Sas3, p300	Histone deposition, transcriptional activation, DNA repair, RNA polymerase II & III transcription, euchromatin
	Lys18	Acetylation	Gcn5, p300/CBP	Transcriptional activation, DNA repair and replication
	Lys23	Acetylation	Unknown, Gcn5, SAs3, p300/CBP	Histone deposition, transcriptional activation, DNA repair
	Lys27	Acetylation	Gcn5	Transcriptional activation
	Lys4	Methylation	Set7/9 (vertebrates), MLL, ALL-1	Permissive euchromatin (di-Me), transcriptional activation
	Arg8	Methylation	PRMT5	Transcriptional repression
	Lys9	Methylation	Suv39h,Clr4, G9a, SETDB1	Transcriptional silencing (tri-Me), transcriptional repression, genomic imprinting, DNA methylation (tri-Me), transcriptional activation

	Arg17	Methylation	CARM1	Transcriptional activation
	Lys27	Methylation	EZH2, G9a	Transcriptional silencing, X inactivation (tri Me)
	Lys36	Methylation	Set2	Transcriptional activation (elongation)
	Lys79	Methylation	Dot1	Eurchromatin, transcriptional activation (elongation), checkpoint response
	Thr3	Phosphorylation	Haspin/Gsg2	Mitosis
	Ser10	Phosphorylation	Aurora-B kinase, MSK1, MSK2, IKK-α, Snf1	Mitosis, meiosis, immediate-early gene activation, transcriptional activation
	Thr11	Phosphorylation	Dlk/Zip	Mitosis
	Ser28	Phosphorylation	Aurora-B kinase, MSK1, MSK2	Mitosis, immediate-early activation
	Lys4	Biotinylation	Biotinidase	Gene expression
	Lys9	Biotinylation	Biotinidase	Gene expression
	Lys18	Biotinylation	Biotinidase	Gene expression
H4	Lys5	Acetylation	Hat1, Esal, Tip60, ATF2, Hpa2, p300	Histone deposition, transcriptional activation, DNA repair
	Lys8	Acetylation	Gcn5, PCAF, Esal, Tip60, ATF2, Elp3, p300	Transcriptional activation, DNA repair
	Lys12	Acetylation	Hat1, Esal, Tip60, Hpa2, p300	Histone deposition, telomeric silencing, transcriptional activation, DNA repair
	Lys16	Acetylation	Gcn5, Esal, Tip60, ATF2, Sas2	Transcriptional activation, DNA repair, euchromatin
	Arg3	Methylation	PRMT1/PRMT5	Transcriptional activation
	Lys20	Methylation	PR-Set7, Suv4-20	Transcriptional silencing (mono-Me), heterochromatin (tri-Me), transcriptional activation, checkpoint response
	Lys59	Methylation	Unknown	Transcriptional silencing
	Ser1	Phosphorylation	Unknown, CK2	Mitosis, chromatin assembly, DNA repair
	Lys12	Biotinylation	Biotinidase	DNA damage response

[a]Note only listed histone modifications pertaining to vertebrates and mammal.

Tropberger et al., 2013; Williams, Truong, & Tyler, 2008). These changes modulate the accessibility of the local chromatin structure, resulting in either open or closed chromatin states (Berger, Kouzarides, Shiekhattar, & Shilatifard, 2009; Venkatesh & Workman, 2015). For example, acetylation of lysine neutralizes its positive charge, thereby weakening interactions with the negatively charged DNA and promoting a more accessible chromatin structure (Grunstein, 1997). Moreover, different PTMs may regulate transcription in their vicinity by serving as a docking platform to recruit various effector proteins, such as chromatin remodeling complexes, or by preventing various proteins from binding to chromatin (Venkatesh et al., 2013). For example, heterochromatin protein 1 (HP1) binds to the methylated lysine 9 of histone H3, subsequently leading to the formation of heterochromatin and transcriptional silencing (Talbert & Henikoff, 2006). In general, most histone PTM marks can be classified as activating or repressive in relation to transcriptional status (Smolle & Workman, 2013). Importantly, histone modifications, along with their corresponding "readers," play key roles in mediating crosstalk between transcriptional regulators and chromatin-modifying complexes to dynamically regulate chromatin structure and function (Bannister & Kouzarides, 2011).

Covalent modifications of histones are reversible, requiring the action of histone-modifying enzymes. Specifically, separate enzymes that add and remove histone PTMs are referred to as "writers" and "erasers," respectively. Distinct classes of "writers" and "erasers" exist in eukaryotes, each specialized to act on a specific histone to catalyze a different PTM. For example, acetylation of specific lysine residues is catalyzed by histone acetyltransferases, while methylation is catalyzed by histone methyltransferases, phosphorylation is performed by kinases, and ubiquitination is carried out by ubiquitinases (Table 7.1) (Marmorstein & Trievel, 2009). Reciprocally, "erasers" include histone deacetyltransferases (HDACs) for deacetylation, histone demethylases (HDMs) for demethylation, phosphatases for dephosphorylation, and deubiquitinases for deubiquitination (Table 7.1) (Marmorstein & Trievel, 2009). The targeted activity of these enzymes often occurs on the same histone in a manner that can influence the transcriptional potential of the chromatin region. For example, Trithorax Group (TrxG) and Polycomb Group (PcG) proteins catalyze the trimethylation of lysine 4 and lysine 27 of histone 3 (H3K4me3 and H3K27me3), respectively, and the colocalization of these marks occur in regions known as bivalent domains (Bernstein et al., 2006; Voigt, Tee, & Reinberg, 2013). These bivalent domains are often found in ESCs at the promoters of developmental genes and are believed to poise the expression of developmental genes, keeping them in a reversibly silenced state, which allows for rapid activation or stable silencing upon differentiation (Mikkelsen et al., 2007; Pan et al., 2007). Given the immense diversity and biological specificity associated with distinct patterns of histone PTMs, it has been proposed that a "histone code" exists in which the combination of histone modifications on a single nucleosome would result in a unique downstream effect (Strahl & Allis, 2000). However, the idea of a "histone code" remains a debated topic, which requires more defined biochemical, epigenomic, and functional characterizations of its potential combinatorial complexity (Rando, 2012).

7.3.3.2 Histone variants

An additional feature in the regulatory landscape of chromatin is the presence of histone variants, which are nonallelic variants of the core histones encoded by unique genes (Weber & Henikoff, 2014). These variants are highly conserved across species and exist for all canonical histones, except for histone H4 (Weber & Henikoff, 2014). The H2A and H2B families of variants exhibit substantial sequence variations, while the H3 family of variants tends to be less diverse (Talbert & Henikoff, 2010). While most eukaryotes express the core histones during the S phase of the cell cycle to allow for replication-coupled deposition, histone variants are largely expressed in a replication-independent manner (Talbert & Henikoff, 2010). This allows for varying levels of deposition that is dependent on the degree of exchange with its corresponding canonical histone (Weber & Henikoff, 2014). For example, variant H3.3 comprises ~90% of the histone 3 in terminally differentiated neurons, but only ~20% in dividing cells (McKittrick, Gaften, Ahmad, & Henikoff, 2004; Piña & Suau, 1987; Szenker, Ray-Gallet, & Almouzni, 2011). Moreover, histone variants exhibit specific genomic localization and distribution patterns, creating distinct chromatin neighborhoods. For instance, studies in yeast have shown that the H2A.Z is localized in genomic areas flanking the nucleosome-free regions, which occur at many gene promoters (Albert et al., 2007; Ranjan et al., 2013; Yen, Vinayachandran, & Pugh, 2013). In relation to the brain, H2A.Z exchange in the hippocampus and prefrontal cortex of the adult brain has been shown to have a restrictive effect on memory formation (Schauer et al., 2013; Zovkic, Paulukaitis, Day, Etikala, & Sweatt, 2015). Accordingly, histone variants can have a profound effect on nucleosome structure, stability, dynamics, and ultimately, gene expression and cellular function.

7.3.3.3 Chromatin remodeling complexes

Chromatin remodeling complexes play a critical role in regulating the chromatin landscape, often using the energy from ATP hydrolysis to disrupt histone-DNA contacts to assemble nucleosomes, alter nucleosome composition, or modulate DNA accessibility (Cairns, 2007; Narlikar, Sundaramoorthy, & Owen-Hughes, 2013). The assembly class of chromatin remodeling complexes facilitates the formation and positioning of nucleosomes, often working in concert with histone chaperones, specialized proteins involved in histone storage, transport, and nucleosome assembly and disassembly (Venkatesh & Workman, 2015). The second class of chromatin remodelers edits the composition of nucleosomes, exchanging canonical histones with histone variants. For example, the human SRCAP complex alters the composition of conventional nucleosomes by removing H2A/H2B dimers and incorporating histone variant H2A.Z/H2B dimers, forming unique chromatin neighborhoods characterized by H2A.Z occupancy (Ruhl et al., 2006; Wong, Cox, & Chrivia, 2007). The final class of chromatin remodeling complexes facilitates access to DNA by sliding or ejecting the histone octamer (Cairns, 2007). Notably, chromatin remodelers possess "reading" domains, allowing interactions with histone PTMs and other chromatin factors to target their functions. Accordingly, the collective function and crosstalk between histone modifications, histone variants, and chromatin remodeling complexes regulates the chromatin structure in a dynamic and responsive manner (Bannister & Kouzarides, 2011). Importantly, disruption in the expression and/or activity of these factors involved in chromatin regulation can lead to a number of neurobiological disorders (**see Section 7.8**).

7.3.4 Noncoding RNA

The final layer of epigenetic regulation is mediated through noncoding RNAs (ncRNAs), which are distinct from mRNA in that they are not translated into protein. This category includes several different species of RNA, which widely differ both in terms of length and cellular roles. These mediate a wide variety of regulatory functions, ranging from the regulation of mRNA and protein levels to the repression of repetitive and transposable elements. This additional level of epigenetic control appears to play an essential role in the central nervous system, as neural cells express very high levels of ncRNAs (Gustincich et al., 2006; Kapranov et al., 2010; Spadaro & Bredy, 2012). Furthermore, ncRNAs may have played a key role in the evolution of the human brain, given that noncoding regions encoding RNA transcripts involved in neural development display the quickest evolution rates in primate genomes (Pollard et al., 2006). By contrast, several ncRNAs are also found in evolutionarily constrained elements, displaying highly conserved spatiotemporal expression profiles across different species (Chodroff et al., 2010). Given that the majority of these are expressed from genomic regions associated with neurodevelopmental genes displaying correlated expression patterns, they likely play important roles in brain development. Here, we focus on several species of short ncRNAs, long noncoding RNAs (lncRNA), as well as other ncRNAs with functions related to neural development.

7.3.4.1 Short noncoding RNA

Some of the best characterized species of ncRNA are short, ranging from 20 to 200 base pairs. These include several different forms of ncRNA, which play crucial roles in the regulation of cellular functions. For instance, RNA interference (RNAi) pathways involve several different short ncRNA species, including microRNA (miRNA), PIWI-interacting RNA (piRNA), and short interfering (siRNA), and mainly inhibit mRNA translation or suppress transposition events in the genome. These ncRNAs act through different, although similar mechanisms, and are typically generated from longer precursor transcripts, which are enzymatically processed to create several small ncRNAs. For instance, miRNA are typically expressed from primary miRNA or introns (mirtrons), which are then processed and exported from the nucleus as precursor miRNA. These are then processed by DICER1 into mature miRNA, composed of 20–23 base pairs of single-stranded RNA (ssRNA), and loaded into the RISC complex where they repress target mRNA through imperfect complementarity of the 5′ seed sequence to induce subsequent translational repression. Of note, a single miRNA can target multiple different mRNAs or ncRNAs and multiple miRNAs can repress the same RNA target, providing a wide range of target sensitivity and regulatory specificity. Furthermore, miRNA also appear to have considerable importance in alterations to neural gene expression patterns, as knockdown of DICER1 in mice produces a wide range of neurodevelopmental defects and several miRNA have been directly implicated in brain development to date (Fiore, Khudayberdiev, Saba, & Schratt, 2011). Moreover, different miRNA variants (isomiRs) display tissue-specific and developmental stage-dependent profiles, suggesting that miRNA are likely integrated into the complex transcriptional and epigenetic networks involved in brain development (Martí et al., 2010). In contrast to miRNA, endogenous siRNA (endo-siRNA) are expressed from cis- or trans-acting sense–antisense pairs, inverted repeats, and transposons. Although they play similar functional roles to miRNA, these are processed by DICER2 to produce double-stranded RNA 21–26 nucleotides in length (Okamura et al., 2008). Moreover, once incorporated into the RISC complex as single strands, they silence both mRNA and transposon-derived RNA through endonucleic cleavage of perfectly complementary sequences to the 5′ seed sequence. Finally, piRNAs are a distinct class of RNAi, 26–30 nucleotides in length. Generated from piRNA loci or transposons, they ultimately regulate mRNA expression levels and transposon activity, respectively, following a series of complex processing mechanisms known as the "ping-pong" cycle. While these ncRNAs were initially identified in germ cells, their expression has been recently discovered in neurons, suggesting a key role not only for genome maintenance and regulation, but also in brain function and development (Yan et al., 2011). Moreover, the ncRNAs involved in retrotransposons regulation, endo-siRNAs and piRNA, may potentially mediate the high frequency of transposition events in the human brain necessary for neural development and plasticity. Given their role in the regulation of transposable elements, these small ncRNA may control the establishment and maintenance of the brain's transcriptional architecture by increasing neuronal cellular diversity (Baillie et al., 2011; Coufal et al., 2009; Faunes et al., 2011; Muotri, Zhao, Marchetto, & Gage, 2009). Additional species of short ncRNA also play important roles in the posttranscriptional control of mRNA, although through distinct mechanisms than RNAi pathways. In particular, small nuclear RNA and small nucleolar (snoRNA) are involved in RNA splicing and modification, mediating alternative splicing events and guiding RNA-modifying enzymes to promote posttranscriptional modifications of different RNA species. Given the wide variety of isoforms and protein variants in the brain, it is possible that these molecular pathways are involved in neurodevelopment and neural function through life. For instance, brain-specific snoRNAs expressed from the C/D box 115 locus promote alternative splicing of the serotonin receptor 2C, suggesting that these ncRNAs play an important role in regulating key isoforms within neural cells (Kishore & Stamm, 2006; Soeno et al., 2010). Furthermore, the majority of genomic regions containing snoRNA clusters can also express miRNA-like RNAs, which suggests a potential role for snoRNA loci in RNAi pathways (Ender et al., 2008; Taft et al., 2009).

7.3.4.2 Long noncoding RNA

Defined as any noncoding transcript longer than 200 nucleotides in length, lncRNA are critical regulators of the cell, providing an essential layer of epigenetic control for a number of different processes. Mainly transcribed from intergenetic regions, telomeres, antisense to protein-coding genes, or regulatory regions such as promoters and enhancers, lncRNA serve a wide variety of cellular functions (Guttman et al., 2009; Hung et al., 2011; Khalil et al., 2009; Mercer et al., 2011; Orom et al., 2010). Several of these regulate the genomic regions from which they are expressed, either through enhancer-like functions or the promotion of repressive epigenetic states (Orom et al., 2010; Rinn et al., 2007; Wutz, 2011). Others play key roles in transcriptional and epigenetic regulation of different genes, recruiting transcription factors or chromatin remodeling complexes to specific genetic loci. In addition, they can act as decoys for different chromatin modifiers and transcription factors, preventing their accumulation within certain regions of the genome. Furthermore, some lncRNAs contribute to the regulation of chromatin states across broad genomic regions and even an entire chromosome in the case of *XIST* (**see Section 7.5.2**). LncRNA also appear to play roles in alternative splicing, posttranscriptional RNA modifications, nuclear-cytoplasmic shuttling, and translational control (Wang & Chang, 2011). Indeed, they sometimes serve as precursors for other small ncRNAs, including snoRNA, and miRNA, suggesting an important role in RNAi pathways. Furthermore, they can act as buffers for miRNA, particularly in the case of lncRNA expressed antisense to protein-coding genes, modulating the levels of RNAi transcripts through complementary binding sites. Multiple lines of evidence also point to the crucial role of lncRNAs in brain function and development. In particular, the highly accelerated region 1 A (*HAR1A)* lncRNA evolved rapidly following the divergence of humans from apes, and its expression highly correlates with that of reelin, a crucial gene in forebrain organization, which suggests that it helps coordinate brain development (Pollard et al., 2006).

Furthermore, the lncRNA antisense to DLX6, *Evf2*, is expressed from an ultraconserved enhancer and plays an essential role in regulating the population of GABAergic interneurons in the hippocampus and dentate gyrus. Several hundreds of additional lncRNAs are also dynamically expressed during GABAergic neurogenesis and oligodendrocyte lineage specification, underscoring the importance of these ncRNA species in brain development (Mercer et al., 2010). Finally, aberrant expression of several different lncRNAs has been associated with different neurodevelopmental disorders, including autism, Fragile X syndrome, Rett syndrome, schizophrenia, and anxiety-like disorder (Barry et al., 2013; Millar et al., 2000; Pastori et al., 2014; Petazzi et al., 2013; Spadaro et al., 2015; Williams et al., 2009; Ziats & Rennert, 2013). Overall, lncRNA appear to play important roles in the integration of developmental, spatial, temporal, and stimulus-specific cues by regulating a wide variety of cellular processes, necessary for the complex epigenetic and transcriptional patterns required during brain development (Wang & Chang, 2011).

7.3.4.3 Additional noncoding RNA classes and RNA modifications

Perhaps the most important RNA species for basic cellular functions are tRNA and rRNA, as they are essential for mRNA translation into protein. However, they do not perform any regulatory functions *per se*, and as such might not be considered epigenetic regulators, but instead represent vital mediators of translation. Furthermore, additional subclasses of ncRNA are now emerging, including, among many others, antisense termini-associated short RNAs, splice-site RNA (spliRNA), transcription initiation RNAs (tiRNAs), termini-associated short RNAs, centromere-associated RNAs, telomere small RNAs, mitochondrial ncRNAs, and miRNA-offset RNAs (Cao et al., 2009; Carone et al., 2009; Kapranov et al., 2010; Landerer et al., 2011; Lung et al., 2006; Taft et al., 2009; Taft et al., 2010). Although some work has characterized the functions of these transcripts, such as research demonstrating a role for tiRNA in transcription factor binding, their regulatory roles remain poorly understood and their implication in cellular functions unclear (Taft, Hawkins, Mattick, & Morris, 2011). As such, these are important research avenues in which further investigation may identify important roles for these transcripts in mediating developmental and tissue-specific epigenetic variation. Finally, covalent modifications to mRNA and different ncRNA species also play important regulatory roles, including in modulating RNA expression, stability, and subsequent translation, interactions with different protein complexes, and cell-specific developmental programs. Although still a nascent field, several different chemical groups have been identified in RNA, including N^6-methyladenosine (m^6A), 5-methylcytidine, and 5-hydroxymethylcytidine (Shafik, Schumann, Evers, Sibbritt, & Preiss, 2016). In particular, levels of m^6A increase in the prefrontal cortex of mice following behavioral training, suggesting a potential role in key cognitive functions such experience-dependent learning (Widagdo et al., 2016).

In addition to their role in mRNA stability, these modifications can regulate the interaction profiles of lncRNA and miRNA, modulating their ability to bind other RNA molecules, DNA, or proteins (Liu et al., 2015). Although the functional implications of these modifications remain unclear, they are relatively abundant in the brain and evidence is emerging to support their vital importance in regulating RNA species in neural tissues.

7.4 GENE–ENVIRONMENT INTERACTIONS

Although epigenetic processes play a fundamental role in maintaining ontogenetic stability in cell differentiation during development, these same epigenetic mechanisms exhibit dynamic variation to allow for finely tuned gene regulation in response to environmental influences. Such a phenomenon represents an "epigenetic paradox," whereby both cellular stability and environmentally influenced plasticity is conferred by the same epigenetic processes (Boyce & Kobor, 2015). Related to its latter role, there is growing interest in epigenetics as a potential mediator of gene–environment (G × E) interactions, defined as genetic or environmental effects on phenotype or outcome that are dependent on each other. More specifically, certain genes can moderate an environment's influence on a particular individual, or environmental influences can only be revealed among individuals of a particular genotype (Rutter, 2010). Classic examples of G × E interaction include the effect of serotonin transporter gene (*5-HTT*) variants in moderating the influence of stressful life events on depression and common regulatory variants of the monoamine oxidase-A (*MAOA*) gene moderating the effects of childhood maltreatment on male antisocial behaviors (Caspi et al., 2003; Caspi, Hariri, Holmes, Uher, & Moffitt, 2010; Haberstick et al., 2014; Kim-Cohen, Caspi, Taylor, & Williams, 2006; Sharpley, Palanisamy, Glyde, Dillingham, & Agnew, 2014; Taylor & Kim-Cohen, 2007). A burgeoning body of work has implicated epigenetic processes as putative molecular mechanisms underlying G × E interactions in various contexts. For example, a link between childhood maltreatment and an allelic risk variant for posttraumatic stress disorder (PTSD) was established in the *FKBP5* gene, which encodes a chaperone of the glucocorticoid receptor (GR), a key mediator of the stress response (Klengel et al., 2012). This association is potentially mediated through a decrease in methylation of a CpG located in the intron of the *FKBP5* risk allele, leading to the suppression of GR function, dysregulation of stress responsivity, and increased risk for PTSD (Klengel et al., 2012). This work not only provided one of the first demonstrations of epigenetics as a molecular mediator in G × E interplay, but also pointed to the functional effects of a methylation quantitative trait locus (mQTL), defined as an allelic variant that correlates with CpG methylation levels in its vicinity (Jones et al., 2013). A number of studies have explored the occurrence of mQTLs in the context of the human brain, finding that mQTLs tend to occur as *cis* associations in different brain regions and may underlie risk loci of various neuropsychiatric diseases, such as schizophrenia and bipolar disorder (Gamazon et al., 2013; Gibbs et al., 2010; Hannon et al., 2015; Jaffe et al., 2015; Zhang et al., 2010). Indeed, a recent study reported brain-specific differential susceptibility of G × E effects on the epigenome in which maternal anxiety × BDNF Val66Met polymorphism interact with DNA methylation as measured in neonatal cord tissue to predict differential volumes of the amygdala and the hippocampus, two brain structures that have been associated with the risk for psychopathology (Chen et al., 2015). Moreover, when modeling variability in human neonatal methylomes, the inclusion G × E interaction terms account for up to 75% of the variably methylated regions between individuals over models containing only G or E terms, suggesting that G × E interactions play an important role in mediating the genomic response to external stimuli and potentially shaping developmental trajectories in early life (Teh et al., 2014). Finally, more recent demonstration of epigenetic mediation of G × E interaction in the context of substance use intervention programs in youth has highlighted the potential positive impact of considering G × E effects in the design of intervention schemes and prevention strategies (Sharpley et al., 2014). Taken together, these observations provide a compelling framework for further investigating the role of epigenetics as a potential molecular mediator of G × E interactions. Future studies aimed at addressing if such epigenetic mediation occurs across many or select tissue types or what critical periods of development are particularly vulnerable to epigenetically mediated G × E interactions would help clarify the biological implications of gene–environment interplay.

7.5 MAJOR EPIGENETIC EVENTS DURING DEVELOPMENT

The epigenome undergoes a number of fundamental changes during development, which play crucial roles in preparing the genome for the activation and implementation of developmental programs. In particular, DNA methylation patterns in the genome undergo widespread changes throughout the course of gametogenesis, fertilization, and subsequent embryonic development (Smith & Meissner, 2013). Prior to gametogenesis, genome-wide DNA demethylation occurs in primordial germ cells (PGCs), erasing epigenetic marks throughout the vast majority of the genome. Epigenetic patterns are reestablished de novo during gametogenesis to restore developmental potency and establish imprinted loci. However, following fertilization of the oocyte, the genome undergoes a second wave of global DNA demethylation, though imprinted genes, repetitive elements, and transposons appear to be largely protected from this epigenetic remodeling. DNA methylation patterns are subsequently reestablished in the blastocyst during cellular differentiation (**see Section 7.6 for more details**). Finally, two additional processes play important roles in development and have laid the foundation for the study of developmental epigenetics—imprinting and

X-inactivation. Highly conserved among mammals, imprinting and X-inactivation events provide substantial insight into the molecular mechanisms underlying epigenetic regulation, both in general and in the context of developmental programing. Furthermore, given that imprinting and X-inactivation both play important roles in brain development, alterations to the typical unfolding of these processes may potentially beget the emergence of certain neurodevelopmental disorders.

7.5.1 Imprinting

The vast majority of genes in diploid cells are expressed from both the maternally and paternally contributed chromosomes. However, a small but very important subset of genes, known as imprinted genes, display monoallelic expression from either the maternal or paternal chromosome, where epigenetic mechanisms repress the other parental allele. In general, paternally expressed imprinted genes function as growth promoters, while maternally derived monoallelic genes act as growth repressors (Barlow & Bartolomei, 2014). Though the evolutionary aspects of this phenomenon remain unclear, the leading theory for the presence of imprinted genes proposes that they arose in response to a "parental conflict," stemming from opposing goals of the parental genomes (Moore & Haig, 1991). Here, paternal genes would seek to maximize fitness and growth in their offspring, while maternal genes would strive to maximize the female's own reproductive fitness and allocate equal resources to each offspring, which may have originated from different fathers.

Nearly 100 imprinted genes have been confirmed in humans, with nearly double that number predicted as potential genes with monoallelic expression. In the mouse genome, which shows considerable overlap with human imprinting, 150 imprinted genes have been identified, with more than 80% mapping to one of 16 genomic clusters containing two or more genes (Wan & Bartolomei, 2008). Furthermore, the mechanisms regulating imprinting are relatively well conserved between species, and animal models provide a useful framework to assess their molecular underpinnings and functional implications. Thus far, it appears that several different epigenetic mechanisms determine which allele becomes inactivated, with DNA methylation and lncRNA playing integral parts in the suppression of gene expression from one parental allele (for a complete review of the mechanisms underlying imprinting, please see the review by Massah, Beischlag, & Prefontaine, 2015). While the majority of imprinted genes are conserved between rodents and humans, some genes display species-specific imprinting, where some imprinted genes in mice display biallelic expression in human tissues. Furthermore, expression appears to be dependent on both the stage of development and tissue, as certain genes that are imprinted in one tissue display biallelic expression in another (Davies, 1994). Others are imprinted in a tissue-specific manner, such as *UBE3A*, which is maternally expressed in the brain and likely plays a key role in neurodevelopment. Interestingly, in mice, the maternal genome appears to contribute considerably to embryonic brain development, particularly within the forebrain, while the paternal genome is more important in adult limbic structures, such as the cortex and hypothalamus (Gregg et al., 2010).

As imprinted genes are essential for proper mammalian growth and development, their disruption can often lead to severe consequences. Indeed, several disorders are associated with alterations of imprinted loci, with mutations or deletions within the same region presenting different phenotypes depending on whether the maternal or paternal chromosome was affected (Cattanach & Kirk, 1985). For instance, classic examples of imprinting disorders are Russell–Silver (RSS) and Beckwith–Wiedemann (BWS) syndromes, which are caused by epigenetic or genetic abnormalities of 11p15.5 on the paternal and maternal chromosomes, respectively. These syndromes display opposite phenotypes, where individuals with RSS show undergrowth and dwarfism, whereas patients with BWS display abnormal overgrowth. Furthermore, Prader Willi Syndrome (PWS) and Angelman Syndrome (AS), caused by deletions within the paternal or maternal 15q11–13 chromosomal region, respectively, also present with distinct symptoms. Individuals with PWS (paternal deletion) display hypotonia and intellectual disability, while those with AS (maternal deletion) present with speech impediments and developmental impairments (Kernohan & Bérubé, 2010). Of note, transcription of the maternal lncRNA from the PWS region rescues the growth retardation phenotype associated with this disorder in mice, highlighting the vital importance of ncRNA in the regulation of imprinted genes (Rozhdestvensky et al., 2016). Furthermore, mouse models with various alterations to imprinted genes often display deficits in higher brain functions, such as social behavior, learning, and memory (Agis-Balboa et al., 2011; Champagne, Curley, Swaney, Hasen, & Keverne, 2009; Chen et al., 2011; Drake, Devito, Cleland, & Soloway, 2011; Lefebvre et al., 1998; Li et al., 1999). Deviations from typical imprinted gene expression may also be linked to psychosis and autistic spectrum disorders (Badcock, 2011; Crespi, 2008). Given that many imprinting-related disorders cause varying degrees of mental retardation, imprinted genes likely play a critical part in the central nervous system.

7.5.2 X-inactivation

Early in development, the second X chromosome of females is inactivated through epigenetic mechanisms in a process known as dosage compensation or X chromosome inactivation (XCI) (Lyon, 1961; Lyon, 1962). This ensures equal levels of X-linked gene expression between males (XY) and females (XX), and perhaps not surprisingly, its failure results in embryonic lethality in females (Takagi & Abe, 1990). From the 2-cell stage and up to the morula stage, the paternal X chromosome of the preimplantation embryo is gradually inactivated through paternal imprinting XCI (Okamoto, Otte, Allis, Reinberg, & Heard, 2004). However, upon implantation of the early blastocyst into the uterus, this phenomenon is reversed, with both chromosomes becoming active once again (Mak et al., 2004). At this stage, several epigenetic mechanisms within each cell come into play to randomly inactivate one copy of the X chromosome, an irreversible process resulting in mosaicism that is passed on to subsequent daughter cells. Nevertheless, some exceptions to this rule have been observed. In the mouse placenta, the paternal X chromosome is always inactivated, and XCI is reversed in oocytes of the female germline prior to meiosis, given that oocytes are haploid and must contain an active X chromosome (Ohhata & Wutz, 2013; Takagi & Sasaki, 1975).

Several epigenetic mechanisms act in concert to regulate XCI, which is triggered through a unique locus, the X-inactivation center (Xic), expressing several genes involved in XCI (Massah et al., 2015; Rastan, 1983). In particular, the X-inactive specific transcript (Xist), a 17-kilobase lncRNA expressed from the inactivated X (Xi), plays an essential role in this process by coating the chromosome from which it is expressed and recruiting epigenetic remodelers to catalyze XCI (Borsani et al., 1991; Brown et al., 1991; Penny, Kay, Sheardown, Rastan, & Brockdorff, 1996). Following the initiation of XCI, Xist RNA spreads over the X chromosome, recruiting chromatin remodeling factors such as PRC2 to increase the proportion of repressive histone modifications, while decreasing activating marks (Peters et al., 2002; Boggs et al., 2002; Boggs, Connors, Sobel, Chinault, & Allis, 1996; Heard et al., 2001; Jeppesen & Turner, 1993; Plath et al., 2003; Silva et al., 2003). Furthermore, different macroH2A isoforms such as macroH2A1.1, macroH2A1.2, and macroH2A2 are incorporated into chromatin to create a tightly compacted chromatin structure, which condenses into a structure known as the Barr body (Chadwick & Willard, 2001; Csankovszki, Panning, Bates, Pehrson, & Jaenisch, 1999). In combination with alterations to the chromatin landscape, the association of DNMT1 with the Xi catalyzes the accumulation of DNA methylation, which preserves the inactive state of the X chromosome and potentially acts as the final "lock" on the Xi (Lock, Takagi, & Martin, 1987; Sado et al., 2000). A number of additional genes also come into play during X inactivation, such as *Tsix*, a lncRNA maintaining the activation of Xa and a number of additional proteins and ncRNA species mediating the finely tuned recruitment of epigenetic regulators to the X chromosome. For a complete review of the factors involved in XCI, please refer to the following reviews (Gendrel & Heard, 2014; Massah et al., 2015).

Although most genes on the Xi are fully inactive, nearly 15% remain somewhat expressed. Known as escape genes, many of these play important roles in brain development, regulating a number of key neuronal processes, such as cellular differentiation, dendritic outgrowth, and cell survival (reviewed in Berletch, Yang, Xu, Carrel,

& Disteche, 2011). Moreover, a number of escape genes have also been associated with intellectual disability syndromes, further supporting their crucial role in typical brain development (Ropers, 2010). These genes display different chromatin structure than their inactivated counterparts, showing a depletion of repressive histone marks, loss of macroH2A, and generally feature of open and active chromatin. Furthermore, unlike CpGs within the promoters of inactivated genes on the X chromosome, which are heavily methylated, those in XCI-escaping genes are generally hypomethylated (Weber et al., 2005; Cotton et al., 2015). From a genetic standpoint, escape genes appear to contain intrinsic genomic elements allowing them to escape XCI, including Alu elements, short ACG/CGT motifs, and the L1 subclass of LINEs (Carrel & Willard, 2005; Cotton et al., 2014; Tannan et al., 2014). Nevertheless, the exact mechanisms underlying the escape of these genes from X inactivation remains mostly unknown, though a considerable amount of research is currently being devoted to uncovering the molecular mechanisms and evolutionary advantages conferred by this process. For a more complete review of genes that escape XCI, please see the following review (Peeters, Cotton, & Brown, 2014).

7.6 PRECONCEPTION EPIGENETICS

Characterized by the unique transcriptional and epigenetic landscapes of gametes, the preconception period has the power to shape developmental trajectories in offspring. As such, environmental conditions, nutritional factors, and stress during this time could influence the maturation of gametes and reprogram subsequent epigenetic patterns, potentially leading to altered brain development or disease susceptibility in subsequent generations. Nevertheless, epigenetic programing in gametes is crucial to prime developmental potency and pass on genetic and potentially epigenetic information to the offspring.

7.6.1 Spermatozoa

The development of the male gamete lasts approximately four months, during which time immature sperm cells are generated from PCGs and differentiate into fertile spermatozoa. Throughout this period, the epigenome undergoes widespread epigenetic alterations, crucial for the highly specialized cellular functions of sperm cells and the subsequent programming of zygotic development. Environmental factors present during this time, such as nutrition, exposure to toxins, or stress, can alter these dynamic epigenetic patterns, potentially resulting in long-term changes to developmental trajectories in the offspring. For instance, in mice, paternal exposure to a high-fat diet leads to deficits in glucose tolerance and insulin sensitivity in offspring, as well as alterations to epigenetic and transcriptomic profiles (Ng et al., 2010; Wei et al., 2014). Similarly, human studies have shown that paternal obesity results in altered DNA methylation patterns of imprinted genes, including the *IGF2* locus (Soubry et al., 2013; Soubry et al., 2013). As such, epigenetic inheritance through the father may represent a vital aspect of long-term health outcomes in children, adults, and potentially subsequent generations as well (reviewed in Soubry, Hoyo, Jirtle, & Murphy, 2014).

To facilitate nuclear compaction of the spermatozoa, the vast majority (90%–95%) of histones within chromatin are replaced with highly basic proteins known as protamines (Corzett, Mazrimas, & Balhorn, 2002; Oliva, 2006). These may play an important role in genome protection from oxidative damage caused by the female reproductive tract, as elevated DNA damage in sperm appears to directly correlate with increased histone retention (Aoki et al., 2005; Garcia-Peiro et al., 2011; Torregrosa et al., 2006). Once inserted into chromatin, several PTMs to protamines are required for higher order chromatin condensation, including phosphorylation at a number of sites (Aoki & Carrell, 2003). Furthermore, decreased fertility has been linked to abnormal protamine levels and ratios, including decreased sperm counts and function, as well as diminished embryo quality during in vitro fertilization, suggesting that protamines are essential for fertilization and subsequent development (Carrell, Emery, & Hammoud, 2007; de Mateo et al., 2009; Simon, Castillo, Oliva, & Lewis, 2011; Torregrosa et al., 2006). In addition to their role in DNA compaction and protection, these proteins also effectively inhibit transcription, leading to repression of most genes within the sperm genome. However, while the majority of the sperm genome is repressed within protamine-induced compaction, some regions retain histones, which are typically located in the promoters of genes involved in development, miRNA, and imprinted genes (Arpanahi et al., 2009; Hammoud et al., 2009). Furthermore, histones in developmental genes are generally marked with transcription-activating H3K4me2 or H3K4me3 and repressive H3K27me3, creating a poised bivalent state similar to ESCs. By contrast, H3K4me3-specific regions are mostly located within genes involved in spermatogenesis, HOX gene clusters, noncoding RNAs, and paternally expressed imprinted loci (Hammoud et al., 2009). Taken together, these findings suggest that histone modifications play a critical role in poising developmental programs within the spermatozoal genome to ensure proper development following fertilization. In addition to canonical nucleosomal subunits, spermatozoal chromatin also contains a unique histone variant, testes-specific histone H2B, which is abundant in mature sperm cells (Churikov et al., 2004; Gatewood, Cook, Balhorn, Schmid, & Bradbury, 1990). This variant shows enrichment in the promoters of ion channel and spermatogenesis genes, which may reflect a remnant/history of the conditions found during sperm maturation (Hammoud et al., 2009). Additional histone variants are also involved in the regulation of the sperm genomes, including H2A.Z, which plays a role in poising essential developmental genes (Rangasamy, Berven, Ridgway, & Tremethick, 2003). As a whole, histones within spermatozoal chromatin appear to potentially play a role in the retention of a historical record of sperm maturation and preparation for later developmental programs (Carrell & Hammoud, 2009). In addition to the contribution of 50% of genetic material and some specialized epigenetic patterns to the zygote, the paternal gamete also appears to transmit a number of ncRNA species to the zygote, including piRNA, lncNRA, miRNA, and tRNA, which may also influence subsequent development (Hamatani, 2012; Jodar, Selvaraju, Sendler, Diamond, & Krawetz, 2013; Liebers, Rassoulzadegan, & Lyko, 2014).

7.6.2 Oocytes

In contrast to male germ cells, which are mitotically arrested until spermatogenesis begins during puberty, female germ cells remain in the diplotene stage of the first meiotic division from birth until ovulation. As no new oocytes are produced in mammals following birth, the growth and release of each cell is tightly regulated. During the growth stage of oocytes, lasting approximately 4 months, the cell accumulates a large supply of mRNA and organelles to regulate and direct embryogenesis (Ma et al., 2013; Oktem & Oktay, 2008; Sánchez & Smitz, 2012). However, in the late stages of growth, transcriptional activity within the oocyte decreases drastically, with little to no transcription occurring once it reaches full size. At this time, the chromatin structure of the oocyte partially condenses around the nucleolus, coinciding with the arrest of gene expression (Andreu-Vieyra et al., 2010; Bouniol-Baly et al., 1999; De La Fuente et al., 2004; Tan et al., 2009; Wickramasinghe, Ebert, & Albertini, 1991; Zuccotti, Garagna, Merico, Monti, & Redi, 2005). Once fully grown, luteinizing hormone, secreted by the hypothalamus, stimulates oocyte maturation and reentry into the cell cycle. Here, maturing oocytes complete meiosis I and advance to the metaphase of meiosis II, where they arrest again until fertilization (Kang & Han, 2011). While no active gene expression is present in the maturing oocyte, the large stockpile of mRNA generated during the growth phase regulates this process to produce a fertile oocyte, highlighting the importance of the pre- and periconceptional periods in developmental programing.

Female gametes contain a highly specialized and dynamic epigenome, which exerts tight regulatory control of their unique physiological status and development, not to mention the potential biological embedding of environmental factors. At birth, oocytes contain almost no DNA methylation, having been reset as PGCs to erase and then reestablish the marks necessary for later development. However, retrotransposons from the IAP family, repetitive elements, and a number of additional CpG islands within the genome retain DNA methylation patterns in the oocyte (Guibert & Weber, 2013; Popp et al., 2010; Seisenberger et al., 2012). This state persists until the growth phases of the oocytes, when DNA methylation begins to accumulate in the genome, resulting in approximately 15% of CpG and 15% CH sites becoming fully methylated by the end of growth, making oocytes one of the few cell types with appreciable levels of non-CpG methylation (Kobayashi et al., 2012; Shirane et al., 2013; Smallwood et al., 2011; Tomizawa et al., 2011). Although the absolute levels of DNA methylation in oocytes are only half those found in spermatozoa, sperm cells do not contain any CH methylation (Kobayashi et al., 2012). DNA methylation also accumulates in imprinted loci, although the various loci acquire epigenetic marks during different stages of oocyte development, which may reflect transcriptional activity within or near those regions or sensitive period of oocyte development and programing (Chotalia et al., 2009; Obata & Kono, 2002). During the growth phase, oocytes also show increasing methylation levels of histone H3 on K4 (me2/me3) and K9 (me2), as well as high levels of acetylation on histones H3 and H4 (Clarke & Vieux, 2015; Kageyama et al., 2007). As such, the chromatin state of growing oocyte is quite permissive to transcription, consistent with the very high levels of mRNA accumulation. Furthermore, oocyte chromatin includes the specific histone H1 variant H1FOO, an essential variant for oocyte maturation that maintains the expression of pluripotent genes during early development (Hayakawa, Ohgane, Tanaka, Yagi, & Shiota, 2012). Finally, endo-siRNAs appear to play a crucial role in postnatal oocyte development. While their exact roles remain to be determined, they likely act to repress transposition events and modulate mRNA translation during oocyte growth and maturation (Clarke & Vieux, 2015).

Given the protracted nature of oocyte development (oocytes may be arrested for decades in humans), environmental factors, potentially throughout the lifetime, could affect the growth and maturation of the female gamete. While robust DNA damage repair mechanisms prevent the majority of potential genetic mutations, which could decrease offspring fitness, epigenetic marks within the genome are more responsive to environmental factors and may mediate their long-term effects (Bock, 2009; Carroll & Marangos, 2013). As such, stressors or environmental conditions affecting the mother may have the power to shape the developmental trajectories of the offspring through epigenetic mechanisms, potentially altering cognitive and behavioral functions later in life, or predisposing the infant to disease throughout life. For a thorough review of epigenetic dynamics in the oocyte, please refer to the following reviews (Clarke & Vieux, 2015; Dean, 2013).

7.6.3 Trans and intergenerational inheritance

Putative evidence is now emerging to support the possibility that environmentally induced phenotypic variation persists over multiple generations, which potentially occurs through epigenetic alterations retained during germ-cell development and passed on to following generations. Although this is an intriguing avenue of research, it remains a young field requiring moderated interpretation and further consideration before it is taken at face value (van Otterdijk & Michels, 2016). This is especially true considering the limitations imposed by biological processes during gamete development and fertilization outlined in the previous and following sections. To date, two different, though similar, types of transmission have been proposed, intergenerational and transgenerational effects. The former manifest themselves in any organisms directly exposed to the factors influencing these changes, while the latter are defined by alterations persisting in generations that have not been directly exposed to the condition. For example, environmental conditions affecting a pregnant mother (F0) could affect the PGCs of her direct offspring (F1), which in turn give rise to a second generation (F2) manifesting intergenerational effects, as the gametes of F1 were directly exposed to the altering factor. However, if the third generation (F3) were to display the same environmentally induced phenotype, these effects would be considered a transgenerational effect, given that no cells were directly exposed to the conditions. By contrast, transgenerational inheritance through males or the maternal prior to gestation only requires transmission to F2, as the reproductive cells of F1 will not be directly affected. While the evidence for transgenerational effects in mammals remains elusive, some early examples are emerging, notably in rodent models. For example, odor fear conditioning in male mice induces behavioral sensitivity in the F1 and F2 generations, as well as CpG hypomethylation in the *Olfr151* gene of F0 and F1 sperm (Dias & Ressler, 2014). In addition to paternal transmission, prenatal maternal stress during the first week of pregnancy alters the transcriptional profiles of male rats, and more specifically, miRNA expression patterns in the brain, with the progeny of these animals displaying similar alterations (Morgan & Bale, 2011). Although few examples of intergenerational epigenetic inheritance in humans exist to date, some lines of evidence are beginning to emerge. For instance, children conceived during the Dutch Hunger Winter of 1944–45 displayed lower DNA methylation levels in the imprinted gene *IGF2* compared to unexposed siblings, even 60 years after the exposure (Heijmans et al., 2008). Nevertheless, transgenerational inheritance in humans may difficult to confirm for a number of reasons. In contrast to the controlled environments of animal models, environmental conditions differ between individuals, introducing additional stochastic variation. Moreover, the heterogeneity of human populations and influence of genetic background on epigenetics limit our ability to distinguish effects caused by genetics or epigenetic transmission. Finally, longitudinal studies combining several generations are required to prove that epigenetic patterns are transmitted, and a cohort of this type is not yet available (van Otterdijk & Michels, 2016).

Although the exact mechanisms underlying these effects remain unclear, several different hypotheses have been put forth to explain their persistence over time, and most require an incomplete erasure of epigenetic patterns during gametogenesis and early development. Thus far, DNA methylation has been an attractive candidate for inter/transgenerational inheritance, given its potential role in the regulation of gene expression, relative stability over time, and response to environmental cues (Bock, 2009). While the majority of the DNA methylome is wiped clean in PGCs, certain regions of the genome appear to resist the initial wave of demethylation, particularly those containing repetitive elements (Sakashita et al., 2014). A number of studies using the Agouti mouse model have demonstrated that environmental conditions can alter the DNA methylation status of an IAP element upstream of the agouti gene to produce variable phenotypes persisting across multiple generations (Lane et al., 2003; Morgan, Sutherland, Martin, & Whitelaw, 1999; Wolff, Kodell, Moore, & Cooney, 1998). Imprinted loci may also play a role in transgenerational inheritance, as they resist the second wave of DNA demethylation following fertilization and the gametes of mice treated with streptozocin display altered DNA methylation of the Peg3 imprinted gene (Ge et al., 2013; Stouder & Paoloni-Giacobino, 2010). However, in all likelihood, several epigenetic mechanisms probably act in concert to pass on epigenetic programing to the subsequent generation. For instance, some histone modifications originating from the paternal gametes are stable until the blastocyst stage of development, suggesting that histone modifications may also play a role in intergenerational epigenetic inheritance (Sarmento et al., 2004). Furthermore, ncRNA are also likely to play a role in inter/transgenerational inheritance, with potentially great importance for male transmission, given their abundance in spermatozoa. In particular, injection of certain miRNA into fertilized eggs can cause effects across at least three generations, indiscriminate of the parent of origin (Wagner et al., 2008). Furthermore, the exposure of male mice to chronic stress prior to breeding alters the miRNA profile of their spermatozoa, leading to a blunted stress response in offspring and altered gene expression programs within the hypothalamus (Rodgers, Morgan, Leu, & Bale, 2015). Although the authenticity of epigenetically induced inheritance has not yet been fully established, this remains a fascinating opportunity to explain the missing heritability of different factors or lasting programing effects of environmental conditions. For balanced reviews of the current literature on transgenerational epigenetic inheritance, please refer to Bohacek, Gapp, Saab, and Mansuy (2012) and van Otterdijk and Michels (2016).

7.7 PRENATAL DEVELOPMENT

Prenatal development begins upon fertilization of the oocyte by a sperm cell, where the two haploid parental genomes combine to produce a diploid zygote. This event initiates a number of widespread genetic and epigenetic events to effectively sculpt the new organism's development. The epigenome of the zygote must be reprogrammed to achieve a totipotent state capable of generating the wide variety of cellular subtypes found in a fully developed organism.

7.7.1 Periconceptional development

The preimplantation period leads to the creation of two distinct cell lineages, the inner cell mass (ICM) and trophectoderm, which result in the fetus and placenta, respectively, as well as several distinct tissues, including the umbilical cord, chorion, and amnion. During the first step of this process, known as the maternal-to-zygotic transition, oocyte-specific transcripts are degraded and replaced with zygotic transcripts, facilitating epigenetic reprogramming of the early embryo (Latham, Solter, & Schultz, 1991). In parallel, chromatin takes on a transcriptionally repressive state throughout the majority of the genome, though the promoters of key totipotency genes become active to establish gene expression patterns required for early development (Li, Lu, & Dean, 2013).

Given that both the sperm and oocytes contain highly specific and diverse DNA methylation patterns, necessary for their specialized functions, these must be reset following fertilization to create the totipotent state required for cell lineage generation. Sperm DNA is heavily methylated (80%–90% of all CpGs) relative to maternal haploid DNA, which contains half the levels of sperm DNA methylation (Kobayashi et al., 2012). As such, both parental pronuclei must undergo radical global DNA demethylation before the zygote can reach a totipotent state, though they occur through different mechanisms (Marcho, Cui, Mager, 2015; Smith & Meissner, 2013). By the morula stage of development, DNA methylation within the genome is nearly absent (Lane et al., 2003; Santos, Hendrich, Reik, & Dean, 2002).

The paternal genome is quickly and completely demethylated through active DNA demethylation during the first several hours postfertilization, independent of replication. This process begins in the paternal pronucleus immediately following fertilization and prior to the first cell division through the action of TET3, which mediates the conversion of 5-methylcytosine (5mC) to 5-hydromethylcytosine (5hmC) (Gu et al., 2011; Wossidlo et al., 2011). This oxidized version of cytosine is subsequently removed through thymidine DNA glycosylase (TDG)-mediated base excision repair (Kohli & Zhang, 2013). However, the TDG repair machinery is not required for loss of DNA methylation in the paternal genome, suggesting that additional mechanisms come into play (Guo et al., 2014). Indeed, as the maintenance DNMT, DNMT1, has low affinity for 5hmC and is absent from the preimplantation nucleus, replication-dependent dilution of 5hmC also appears to plays a vital role in this process (Hashimoto et al., 2012; Howell et al., 2001).

In contrast to the paternal genome, the maternal genome mostly undergoes passive demethylation through the progressive dilution of DNA methylation levels over the course of cell divisions throughout the preimplantation period. This process is facilitated by its relatively low initial DNA methylation levels (Guo et al., 2014). Furthermore, while the maternal pronucleus contains TET3, the maternal genome is protected from active DNA demethylation by the protein STELLA, which is expressed in oocytes and gonads, as well as during germ-cell specification (Saitou, Barton, & Surani, 2002; Sato et al., 2002). Through its binding to H3K9me2, a feature enriched in the maternal genome and protamine-free regions of the paternal genome, STELLA alters the chromatin structure to prevent TET3 binding and activity (Nakamura et al., 2012). The protection of imprinted genes from the wave of global DNA demethylation is thought to partially occur through STELLA-dependent mechanisms as well (Nakamura et al., 2007).

Considerable alterations to chromatin structure also begin immediately following fertilization, with several epigenetic modifications taking place in both haploid genomes. In the paternal genome, protamines are replaced by maternal histones in a replication-independent fashion, decondensing the highly compacted chromatin and causing the sperm nucleus to mature into the paternal pronucleus (Marcho et al., 2015). More specifically, protamines are replaced with nucleosomes containing the H3 variant H3.3, which is included in place of H3 when nucleosome assembly occurs outside transcription (Torres-Padilla, Bannister, Hurd, Kouzarides, & Zernicka-Goetz, 2006). While this variant is typically associated with transcriptionally active chromatin states, the zygotic genome is not transcriptionally active immediately upon fertilization. As such, H3.3 inclusion likely poises the paternal genome for subsequent activation of developmental programs (Lin, Conti, & Ramalho-Santos, 2013). Histone modification patterns also undergo widespread changes following fertilization, displaying asymmetric patterns between the maternal and paternal haploid genomes. While the exact functions and genomic localization of such marks remain mostly unknown, they highlight the complexity of the distinct programming required for proper embryonic gene activation or repression (Marcho et al., 2015). Most importantly, they likely serve to establish the totipotency programs required for early zygote development, while also providing a framework for the later divergence of epigenomic landscapes and generation of the wide variety of cellular subtypes.

7.7.2 Embryonic development and neurogenesis

Following implantation of the zygote at the blastocyst stage, DNA methylation patterns become reestablished in the ICM of the blastocyst, rising rapidly in the primitive ectoderm, which eventually matures into the entire embryo (Borgel et al., 2010; Santos et al., 2002; Smith et al., 2012). These changes correlate with the resumption of DNMT expression in the embryo, which establish lineage-specific DNA methylation patterns to guide cells along their specified developmental trajectories. These eventually culminate into the wide diversity of cellular subtypes arising from the different germ layers of the blastocyst (Santos et al., 2002). Given that the cells present in the blastocyst are precursors to all cells in the body, disruption of epigenetic programs by different teratogens during this fundamental stage of development can result in widespread defects in the organism. Cells arising from the same lineage display more similar epigenetic patterns than those from other lineages. However, as cells differentiate, their epigenetic profiles become increasingly adapted to their particular functions. As such, differences in cell type remain the major drivers of epigenetic modifications (Smith & Meissner, 2013).

Of particular importance to the present chapter, several different layers of epigenetic regulation modulate the differentiation of neural cells, regulating neurogenesis in the developing brain. Epigenetic processes at this stage ensure adequate proportions of neurons, glia, astrocytes, and their specialized subtypes, as well as the proper encoding of the gene expression programs controlling brain patterning and maturation (Imamura, Uesaka, & Nakashima, 2014; Lilja, Heldring, & Hermanson, 2012; Shen, Ji, & Jiao, 2015). These act in concert to drive neurogenesis in the embryo, mediating the differentiation of neural progenitor cells (NPC) and patterning of the developing brain (LaSalle, Powell, & Yasui, 2013). In NPCs, chromatin remodeling complexes interact with repressive transcription factors to inhibit the expression of neuronal-specific genes, thus preventing differentiation. Upon cell division, repression of these genes is released and the daughter cells enter into the neuronal lineage, which is characterized by widespread alterations to epigenetic patterns throughout the cell and the activation of neuronal-specific transcription programs.

A complex interplay between different epigenetic factors is necessary during neural development, as each factor plays a fundamental role in regulating neuronal differentiation and brain patterning. For one, DNA methylation is vital for neuronal differentiation, as NPCs lacking DNMT1 give rise to dysfunctional neurons (Fan et al., 2001; LaSalle et al., 2013). By contrast, DNAhm is associated with neuronal differentiation, accumulating in intragenic regions during neurogenesis, concomitant with decreases in H3K27me3 and increased gene expression (Hahn et al., 2013). Several ncRNA species are also involved in neuronal differentiation and patterning, with miRNA and lncRNA playing key roles in NPC proliferation, lineage commitment, and spatiotemporal regulation of brain development (Fatica & Bozzoni, 2014; Lv, Jiang, Liu, Lei, & Jiao, 2014; Mercer et al., 2010; Pollard et al., 2006; Wang & Chang, 2011; Zhao, Sun, Li, & Shi, 2009). Finally, monoallelic gene expression, which shows similar mechanisms to imprinting, may also play a key role in generating neuronal cell surface diversity through the stochastic expression of different protocadherin (*PCDH*) genes, which encode cell-surface adhesion molecules (Massah et al., 2015). As such, various neuronal subpopulations might express vastly different subsets of these genes from a single allele, likely resulting in their wide variety of functions and localizations (Esumi et al., 2005; Frank et al., 2005).

Furthermore, several disorders associated with mutations in epigenetic mediators and regulators are apparent even at a young age, supporting a vital role for epigenetic mechanisms in typical early brain development and function (Kramer & van Bokhoven, 2009; Vissers, Gilissen, & Veltman, 2015). For instance, a number of severe diseases marked by defects in brain development and function result from mutations in epigenetic regulators, including Fragile X, Rett, Rubinstein-Taybi, Sotos, and Weaver syndromes, as well as several additional types of X-linked mental retardation. Interestingly, the disruption of distinct epigenetic mechanisms, which likely regulate different gene sets, gives rise to various forms of mental retardation with similar phenotypes, suggesting that global epigenetic patterns are crucial for normal brain function and that most epigenetic marks are closely interconnected (Kramer & van Bokhoven, 2009; Vissers et al., 2015). In particular, X-linked intellectual disability is associated with various mutations on the X chromosome, with the majority falling within key epigenetic factors such as MeCP2, the histone kinase RSK2, and the H3K4-specific HDM JARID1C. Furthermore, minor alterations to MeCP2 or other methyl-binding protein expression and genome-wide changes to DNA methylation patterns have been associated with cases of autism spectrum disorder, further supporting the crucial role of epigenetic regulators in brain organization and development (Berko et al., 2014; Cukier et al., 2010; Gonzales & LaSalle, 2010). While the majority of severe disorders resulting from mutations in epigenetic regulators manifest during childhood, more complex disorders such as schizophrenia, depression, or Alzheimer's disease often manifest later in life, potentially requiring additional environmental factor or brain maturation to fully emerge (Boyce & Kobor, 2015). Nevertheless, epigenetic mechanisms likely play a vital role in shaping brain development and subsequent biological trajectories during early life and can potentially influence health outcomes throughout the lifetime.

7.8 DEVELOPMENTAL VULNERABILITY THROUGHOUT THE LIFETIME

From a biological standpoint, development is an ongoing, lifelong process involving the intersection of genetic programming and environmental cues (Boyce & Kobor, 2015). For instance, prenatal environments may shape the vulnerability to disease, leading to long-term effects on health and behavior. Childhood is a highly sensitive period of life, given the high neuronal plasticity of the neonatal and child brain and the high rates of neurogenesis and synapse formation. Adolescence represents an important window of vulnerability for brain development, with critical neuronal changes occurring parallel to the physical changes of puberty. Given the heightened plasticity of the brain during these periods, neurobiological development may be particularly sensitive to environmental factors, leading to long-term effects of these exposures on cognitive abilities, behavioral patterns, emotional functioning, and susceptibility to mental illnesses.

7.8.1 Prenatal environments

The developmental origins of health and disease (DOHaD) hypothesis suggest that environmental factors during early life development can influence future health outcomes (Barker, 1990; Barker, 1995). Specifically, signals received during development, such as nutritional and hormonal status, may preemptively lead the organism toward a phenotype best adapted for the anticipated external environment (Hanson, Low, & Gluckman, 2011). This early-life programming is a manifestation of developmental plasticity, where a single genotype can lead to multiple phenotypic outcomes due to differing environmental conditions (Barker, 2007). However, in the event of a mismatch between early and later life environments, this adaptive response may no longer confer a fitness advantage, but instead, lead to deleterious phenotypes (Godfrey, Lillycrop, Burdge, Gluckman, & Hanson, 2007). As such, adverse early-life conditions, including maternal undernutrition, obesity, stress, or exposure to teratogens, such as lead, ethanol, and nicotine, have the potential to permanently imprint physiological and behavioral systems during development, leading to long-term consequences in offspring (Godfrey & Robinson, 1998; Hanson & Gluckman, 2008). Given the relationship of epigenetic factors, biological programs, and environmental responsivity, epigenetics may represent a potential mechanism for the biological embedding of prenatal exposure. Thus, a considerably body of research has investigated the relationship between the alterations to epigenetic patterns and prenatal exposures.

Fetal alcohol spectrum disorder (FASD) is a prime example of the programming of physiological systems by early-life teratogens. Caused by gestational exposure to alcohol, this disorder manifests through both immediate and long-term alterations to cognition, behavior, immune function, and vulnerability to mental health disorders (Hellemans, Sliwowska, Verma, & Weinberg, 2010). In turn, these have been associated with changes in epigenetic patterns in the brain, including alterations to ncRNA, histone modifications, and DNA methylation (Lussier et al., 2017). Importantly, several findings from animal models were replicated through a study of DNA methylation patterns in the buccal epithelial cells of children with FASD, suggesting that alcohol may leave a lasting signature on the epigenome (Portales-Casamar et al., 2016). Several other studies have also investigated DNA methylation patterns following prenatal exposures in human populations. For instance, maternal undernutrition during pregnancy can also cause changes in the DNA methylation status of *IGF2*, which may influence growth and vulnerability to disease later in life (Heijmans et al., 2008). Finally, perhaps the best replicated association in human populations, prenatal exposure to cigarette smoke is associated with persistent alterations to DNA methylation of the *AHRR* gene (Joubert et al., 2012).

Although these findings suggest robust effects of prenatal exposures on the epigenome, and potentially long-term health and behavioral outcomes, they remain correlative and cannot yet be interpreted as causative alterations. By virtue of fundamental biological properties, alterations to the epigenetic patterns of central tissues are not easily measurable in humans, given the difficulty of obtaining brain samples. Thus, the vast majority of epigenome-wide association studies are performed in peripheral tissues such as blood and buccal epithelial cells in the hope that they might reflect epigenomic variation in the brain. As epigenetic patterns are highly dependent on cell types that may respond differently in the face of the same exposures, these surrogate tissues may not fully portray the true changes driving disease. As such, the establishment of common epigenetic profiles between central and peripheral tissues is an ongoing and essential topic of research.

7.8.2 Childhood

The brain is particularly vulnerable during childhood, given its plasticity in response to the different stimuli shaping its development. Given the plasticity of the brain during early childhood, environmental factors during this critical developmental period have the potential to shape the trajectories of cognitive and behavioral functioning. Indeed, associations between later life disease and numerous environmental conditions have been identified throughout these windows of vulnerability, including, but not limited to socio-economic status, parental stress, abuse, and exposure to toxins such as alcohol and tobacco (Godfrey & Robinson, 1998; Hanson & Gluckman, 2008). In line with the DOHaD hypothesis, these factors may potentially become biologically embedded into the epigenome, underlying lifelong vulnerability to stressors and chronic disorders (Kubota, Miyake, Hariya, & Mochizuki, 2015). As epigenetic mechanisms play an integral role in mediating experience-dependent processes and brain plasticity to regulate the establishment, refinement, and maintenance of neural circuits at this stage, it is possible that they could mediate the biological integration of external cues during this sensitive period of development, when these mechanisms are the most responsive to stimuli (Tognini et al., 2015).

An emerging number of epigenome-wide association studies are beginning to uncover potentially lasting relationships between adverse childhood environments and altered DNA methylation patterns later in life (Borghol et al., 2012; Essex et al., 2013; Lam et al., 2012; Powell et al., 2013). In turn, these changes have been associated with altered neurobehavioral profiles, supporting a potential link between the epigenome and long-term health outcomes. For example, hippocampal cells of suicide victims abused as children display increased DNA methylation levels in the *NR3C1* gene promoter (McGowan et al., 2009). Given that physically abused children also manifest these altered DNA methylation patterns in blood, it is possible that adverse childhood experiences integrate biological circuitry during childhood and potentially lead to long-term consequences (Romens, Mcdonald, Svaren, & Pollak, 2015).

7.8.3 Puberty

Puberty is characterized by the activation of the gonadal hormone systems, rapid growth, sexual maturation, and widespread changes in brain function. Several brain regions undergo extensive maturation and reorganization during this period, including sexually dimorphic alterations to the prefrontal cortex, hippocampus, and amygdala, among many others. Given that these sex-specific differences last throughout the lifetime, as well as the complex systems and timing involved, epigenetic mechanisms likely play a key role in pubertal maturation. Indeed, similar to early development, the epigenome is poised during puberty to modulate brain development and potentially influence disease risk (Morrison, Rodgers, Morgan, & Bale, 2014).

Brain maturation upon the onset of puberty is driven by the surge of gonadal hormones to which the brain had not previously been exposed. Given that steroid hormones readily cross the lipid bilayer and that their receptors also act as transcription factors, the long-lasting alterations to brain function are potentially mediated through epigenetic mechanisms. Studies in females have uncovered that the onset of puberty is associated with increased kisspeptin (*kiss1*) expression in the arcuate nucleus (ARC) of the medial basal hypothalamus (Oakley, Clifton, & Steiner, 2009). This derepression is linked to increased DNA methylation in the promoters of key epigenetic regulators, which decreases their expression and leads to a more transcriptionally permissive chromatin state of *kiss1*. Neurons expressing this neuropeptide then directly stimulate gonadotropin-releasing hormone (GnRH) neurons to initiate estrous cyclicity and pubertal onset. Furthermore, DNA methylation patterns and histone modifications within the promoter of GnRH modulate its expression levels in the Rhesus monkey brain, suggesting that this key regulator of puberty is also under epigenetic control (Kurian & Terasawa, 2013).

Several studies have investigated the role of global genetic and epigenetic factors during pubertal onset, identifying key roles for epigenetic mechanisms in the regulation of both pubertal onset and pubertal brain plasticity. For instance, disruption of global DNA methylation patterns delays the beginning of puberty in female rats, while exposure to high levels of methyl donors results in earlier puberty, highlighting a potential role for DNA methylation in pubertal onset (Almstrup et al., 2016; Morrison et al., 2014). Although little genome-wide data of epigenetic alteration during puberty have been collected, a genome-wide analysis of DNA methylation in the rat hypothalamus identified an increase in DNA methylation within the promoters of chromatin modifiers, such as the Polycomb group genes and their interaction partners (Lomniczi et al., 2013). These alterations could potentially underlie the plasticity of the pubescent brain, and mediate vulnerability to the programming effects of different environmental stressors, which may lead to deleterious long-term effects on mental and physical health.

While these findings certainly present an attractive mechanism for the biological embedding of early life experiences in the genome and their manifestation in cognitive and behavioral profiles, the vast majority of epigenetic studies in humans are performed on peripheral tissues, such as buccal epithelial cells and blood, rather than central tissues. As environmental stimuli may not be associated with the same alterations in the brain, these results may not represent causal mechanisms and must be interpreted with caution. However, a number of studies have begun to investigate the concordance between peripheral and central tissues, assessing the relevance of different biological samples to neurobiological deficits and their potential use in biomarker discovery for neurological disorders (Farré et al., 2015; Marzi et al., 2016; Walton et al., 2015; Yu et al., 2016, Edgar et al., 2017). Furthermore, epigenetic studies have also begun to incorporate potential confounding factors, such as allelic variation, ethnicity, age, socio-economic status, and cellular heterogeneity, providing a solid framework for the study of epigenetic mechanisms in neurobiological disorders (Farré et al., 2015; Lister et al., 2013).

7.8.4 Later life manifestation of early life environments

In contrast to genetically linked neurodevelopmental disorders, which generally present early in life, alterations to the epigenome occurring during development may be more likely to manifest during later life. The concept of allostatic load suggests that the "wear and tear" caused by chronic stressors throughout the life course can contribute to the dysregulation of physiological systems and subsequent risk of maladaptive outcomes (McEwen & Wingfield, 2003). In the context of neurobiological diseases, it has been proposed by the "latent early-life associated regulation" (LEARn) model that the accumulation of epigenetic alterations across the life course, particularly beginning at early developmental stages, can lead to pathological outcomes that manifest much later in life (Delgado-Morales & Esteller, 2017; Lahiri, Maloney, & Zawia, 2009). More specifically, the phenotypes or symptoms caused by adverse early-life conditions often require additional factors to fully emerge, such as environmental stressors/insults or incubation period, in contrast to epigenetic changes which accumulate and interact with external factors to gradually disrupt homeostasis over the life course.

Several mental disorders, such as anxiety, schizophrenia, and bipolar disorder, emerge during adolescence, with symptoms materializing following the onset of puberty and its associated changes in brain plasticity and function. The emergence of certain mental health disorders later in life suggests a complex interplay between genetic and environment factors to produce long-term symptoms. While genetics certainly play a role in the predisposition to these mental illnesses, epigenetic mechanisms likely mediate gene by environment interactions in these instances. This is supported by the fact that monozygotic twins often display discordant prevalence of mental disorders (Dempster et al., 2011; van Dongen, Slagboom, Draisma, Martin, & Boomsma, 2012). Given that the epigenome of twins becomes increasingly divergent with age, epigenetic mechanisms could play a key role in the etiology or emergence of their symptoms (Kaminsky et al., 2009; Kuratomi et al., 2008).

Although the etiology of these disorders remains mostly unknown and likely include multifactorial contributions from genetic variation and environmental influences, they have also been linked to alterations in epigenetic patterns and regulators. To date, multiple studies have linked both epigenetic factors and environmental factors to numerous disorders, including Alzheimer's disease, autism, fetal alcohol spectrum disorder, schizophrenia, Huntington's, bipolar disorder, and depression (Berko et al., 2014; De Souza et al., 2016; Guintivano, Aryee, & Kaminsky, 2013; Hannon et al., 2015; Horvath et al., 2016; Jaffe et al., 2015; Portales-Casamar et al., 2016). For instance, the promoter region of the *COMT* gene, which catalyzes dopamine breakdown in the brain, displays decreased DNA methylation levels in the frontal cortex of individuals with schizophrenia and bipolar disorder (Abdolmaleky et al., 2006). Furthermore, a polymorphism in the *COMT* produces a more active protein variant and has been directly linked to schizophrenia and bipolar disorder in adults, supporting a role for both epigenetic and genetic factors in the emergence of these disorders (Glatt, Faraone, & Tsuang, 2003; Shifman et al., 2004). In addition, altered DNA methylation patterns of the *Reelin* gene, crucial for neuronal migration and early brain organization, have also been identified in the postmortem brain of schizophrenics, further supporting a role for epigenetic dysregulation in the manifestation of mental disorders and abnormal brain development (Grayson et al., 2005). Finally, Parkinson's disorder (PD) is another classic example of a heritable disorder that emerges later in life in only some individuals. Previous studies demonstrating that intronic DNA methylation results in reduced expression of alpha-synuclein (*SNCA*), a major risk gene for PD, along with reports of miRNA deregulation in human PD brain samples (Jowaed, Schmitt, Kaut, & Wüllner, 2010; Miñones-Moyano et al., 2011; Mullin & Schapira, 2015). Taken together, these studies provide compelling evidence for the contribution of epigenetic variation to the development of neurobiological disorders, and suggest a potential role for epigenetic mechanisms in the design of future therapeutic strategies and/or diagnostic tools.

7.8.5 Aging

The epigenome does not remain stable over time, as nearly one-third of DNA methylation patterns are associated with the aging process (reviewed in Jones, Goodman, & Kobor, 2015). Indeed, DNA methylation levels within certain regions of the genome gradually decrease, particularly in regions of low CpG density and gene bodies (Heyn et al., 2012). Repetitive elements, which contain high

CpG densities and typically display high levels of DNA methylation, also lose DNA methylation with age. By contrast, DNA methylation levels within CpG islands and other high CpG density regions tend to increase with age (Christensen et al., 2009; Johansson, Enroth, & Gyllensten, 2013). Given that the majority of CpG sites within the genome are fully methylated, these changes ultimately lead to a global decrease in DNA methylation levels over time. Furthermore, the epigenome undergoes gradual drift over time, with interindividual epigenetic patterns becoming increasingly divergent over time. For instance, the concordance of epigenetic patterns in monozygotic twins, born with very similar epigenomes, decreases with age, which is likely due to both stochastic events and differing environmental conditions (Fraga et al., 2005; Jones et al., 2015; Teschendorff, West, & Beck, 2013). Although the increase of epigenetic variability over time may potentially contribute to certain mental health illnesses, such as depression, the contribution of epigenetic aging to altered health outcomes in adulthood or later life remains relatively unknown (Córdova-Palomera et al., 2015). However, given the importance of epigenetic patterns in modulating neural functions and maintaining brain homeostasis, it is likely that epigenetic aging contributes to the gradual development and ultimate decline of the brain in adult life (Farré et al., 2015).

7.9 CONCLUSION

The convergence of genetic and epigenetic variation forms a powerful and dynamic regulatory framework that works to sculpt developmental trajectories across a lifespan, particularly in the context of shifting environmental cues. Over the past few decades, scientific discoveries positioned at the nexus of these fields have considerably illuminated our understanding of developmental regulation. It is now understood that genetic variation, comprised of structural sequence variants or single base polymorphisms, represent the heritable factors that drive developmental course. By contrast, the epigenome represents a structural overlay of the genome, comprised of DNA modifications, chromatin compaction and noncoding RNA activity, which allow for enduring ontogenetic programming and modulating environmentally influenced reorganization. Importantly, epigenetic processes are now thought to serve as potential mediators of gene–environment interplay, permitting brain-specific plasticity during key developmental windows. Despite these important research advances, a number of key issues remain: how these factors cooperatively function during different developmental periods of susceptibility; if these marks may be used as biomarkers to indicate disease risk; and finally, the potential reversibility of epigenetic marks in developmental diseases.

REFERENCES

Abdolmaleky, H. M., Cheng, K., Faraone, S. V., Wilcox, M., Glatt, S. J., Gao, F., ... Thiagalingam, S. (2006). Hypomethylation of MB-COMT promoter is a major risk factor for schizophrenia and bipolar disorder. *Human Molecular Genetics*, *15*, 3132–3145.

Agis-Balboa, R. C., Arcos-Diaz, D., Wittnam, J., Govindarajan, N., Blom, K., Burkhardt, S., ... Fischer, A. (2011). A hippocampal insulin-growth factor 2 pathway regulates the extinction of fear memories. *EMBO Journal*, *30*, 4071–4083.

Alaghband, Y., Bredy, T. W., & Wood, M. A. (2016). The role of active DNA demethylation and Tet enzyme function in memory formation and cocaine action. *Neuroscience Letters*, *625*, 40–46.

Albert, I., Mavrich, T. N., Tomsho, L. P., Qi, J., Zanton, S. J., Schuster, S. C., & Pugh, B. F. (2007). Translational and rotational settings of H2A.Z nucleosomes across the Saccharomyces cerevisiae genome. *Nature*, *446*, 572–576.

Allis, C. D., & Jenuwein, T. (2016). The molecular hallmarks of epigenetic control. *Nature Reviews. Genetics*, *17*, 487–500.

Almstrup, K., Lindhardt Johansen, M., Busch, A. S., Hagen, C. P., Nielsen, J. E., Petersen, J. H., & Juul, A. (2016). Pubertal development in healthy children is mirrored by DNA methylation patterns in peripheral blood. *Science Reports*, *6*, 28657.

Andreu-Vieyra, C. V., Chen, R., Agno, J. E., Glaser, S., Anastassiadis, K., Stewart Francis, A., & Matzuk, M. M. (2010). MLL2 is required in oocytes for bulk histone 3 lysine 4 trimethylation and transcriptional silencing. *PLoS Biology*, *8*, 53–54.

Aoki, V. W., & Carrell, D. T. (2003). Human protamines and the developing spermatid: Their structure, function, expression and relationship with male infertility. *Asian Journal of Andrology*, *5*, 315–324.

Aoki, V. W., Moskovtsev, S. I., Willis, J., Liu, L., Mullen, J. B. M., & Carrell, D. T. (2005). DNA integrity is compromised in protamine-deficient human sperm. *Journal of Andrology*, *26*, 741–748.

Arpanahi, A., Brinkworth, M., Iles, D., Krawetz, S. A., Paradowska, A., Platts, A. E., ... Miller, D. (2009). Endonuclease-sensitive regions of human spermatozoal chromatin are highly enriched in promoter and CTCF binding sequences. *Genome Research*, *19*, 1338–1349.

Badcock, C. (2011). The imprinted brain: How genes set the balance between autism and psychosis. *Epigenomics*, *3*, 345–359.

Baillie, J. K., Barnett, M. W., Upton, K. R., Gerhardt, D. J., Richmond, T. A., De Sapio, F., ... Faulkner, G. J. (2011). Somatic retrotransposition alters the genetic landscape of the human brain. *Nature*, *479*, 534–537.

Bannister, A. J., & Kouzarides, T. (2011). Regulation of chromatin by histone modifications. *Cell Research*, *21*, 381–395.

Barker, D. J. (1990). The fetal and infant origins of adult disease. *British Medical Journal*, *301*, 1111.

Barker, D. J. (1995). Fetal origins of coronary heart disease. *British Medical Journal*, *311*, 171–174.

Barker, D. J. (2007). The origins of the developmental origins theory. *Journal of Internal Medicine*, *261*, 412–417.

Barlow, D. P., & Bartolomei, M. S. (2014). Genomic imprinting in mammals. *Cold Spring Harbor Perspectives in Biology*, 6.

Barry, G., Briggs, J., Vanichkina, D., Poth, E., Beveridge, N., Ratnu, V., ... Mattick, J. (2013). The long non-coding RNA Gomafu is acutely regulated in response to neuronal activation and involved in schizophrenia-associated alternative splicing. *Molecular Psychiatry*, *19*, 486–494.

Baubec, T., & Schuebeler, D. (2014). Genomic patterns and context specific interpretation of DNA methylation. *Current Opinion in Genetics & Development*, *25*, 85–92.

Berger, S. L., Kouzarides, T., Shiekhattar, R., & Shilatifard, A. (2009). An operational definition of epigenetics. *Genes & Development*, *23*, 781–783.

Berko, E. R., Suzuki, M., Beren, F., Lemetre, C., Alaimo, C. M., Calder, R. B., ... Greally, J. M. (2014). Mosaic epigenetic dysregulation of ectodermal cells in autism spectrum disorder. *PLoS Genetics*, *10*, e1004402.

Berletch, J. B., Yang, F., Xu, J., Carrel, L., & Disteche, C. M. (2011). Genes that escape from X inactivation. *Human Genetics*, *130*, 237–245.

Bernstein, B. E., Mikkelsen, T. S., Xie, X., Kamal, M., Huebert, D. J., Cuff, J., ... Lander, E. S. (2006). A bivalent chromatin structure marks key developmental genes in embryonic stem cells. *Cell, 125*, 315–326.

Bestor, T. H. (2000). The DNA methyltransferases of mammals. *Human Molecular Genetics, 9*, 2395–2402.

Bhutani, N., Burns, D. M., & Blau, H. M. (2011). DNA demethylation dynamics. *Cell, 146*, 866–872.

Bird, A. (2007). Perceptions of epigenetics. *Nature, 447*, 396–398.

Bock, C. (2009). Epigenetic biomarker development. *Epigenomics*, 99–110.

Boggs, B. A., Connors, B., Sobel, R. E., Chinault, A. C., & Allis, C. D. (1996). Reduced levels of histone H3 acetylation on the inactive X chromosome in human females. *Chromosoma, 105*, 303–309.

Boggs, B. A., Cheung, P., Heard, E., Spector, D. L., Chinault, A. C., & Allis, C. D. (2002). Differentially methylated forms of histone H3 show unique association patterns with inactive human X chromosomes. *Nature Genetics, 30*, 73–76.

Bohacek, J., Gapp, K., Saab, B. J., & Mansuy, I. M. (2012). Transgenerational epigenetic effects on brain functions. *Biological Psychiatry, 73*, 313–320.

Borgel, J., Guibert, S., Li, Y., Chiba, H., Schübeler, D., Sasaki, H., ... Weber, M. (2010). Targets and dynamics of promoter DNA methylation during early mouse development. *Nature Genetics, 42*, 1093–1100.

Borghol, N., Suderman, M., Mcardle, W., Racine, A., Hallett, M., Pembrey, M., ... Szyf, M. (2012). Associations with early-life socio-economic position in adult DNA methylation. *International Journal of Epidemiology, 41*, 62–74.

Borsani, G., Tonlorenzi, R., Simmler, M. C., Dandolo, L., Arnaud, D., Capra, V., ... Ballabio, A. (1991). Characterization of a murine gene expressed from the inactive X-chromosome. *Nature, 351*, 325–329.

Bouniol-Baly, C., Hamraoui, L., Guibert, J., Beaujean, N., Szöllösi, M. S., & Debey, P. (1999). Differential transcriptional activity associated with chromatin configuration in fully grown mouse germinal vesicle oocytes. *Biology of Reproduction, 60*, 580–587.

Bourc'his, D., & Bestor, T. H. (2004). Meiotic catastrophe and retrotransposon reactivation in male germ cells lacking Dnmt3L. *Nature, 431*, 96–99.

Bowman, G. D., & Poirier, M. G. (2015). Post-translational modifications of histones that influence nucleosome dynamics. *Chem Rev, 115*, 2274–2295.

Boyce, W. T. (2016). Differential susceptibility of the developing brain to contextual adversity and stress. *Neuropsychopharmacology, 41*, 142–162.

Boyce, W. T., & Kobor, M. S. (2015). Development and the epigenome: The "synapse" of gene–environment interplay. *Developmental Science, 18*, 1–23.

Brown, C. J., Ballabio, A., Rupert, J. L., Lafreniere, R. G., Grompe, M., Tonlorenzi, R., & Willard, H. F. (1991). A gene from the region of the human X inactivation centre is expressed exclusively from the inactive X chromosome. *Nature, 349*, 38–44.

Cairns, B. R. (2007). Chromatin remodeling: Insights and intrigue from single-molecule studies. *Nature Structural & Molecular Biology, 14*, 989–996.

Cao, F., Li, X., Hiew, S., Brady, H., Liu, Y., & Dou, Y. (2009). Dicer independent small RNAs associate with telomeric heterochromatin. *RNA, 15*, 1274–1281.

Carone, D. M., Longo, M. S., Ferreri, G. C., Hall, L., Harris, M., Shook, N., ... O'Neill, R. J. (2009). A new class of retroviral and satellite encoded small RNAs emanates from mammalian centromeres. *Chromosoma, 118*, 113–125.

Carrel, L., & Willard, H. F. (2005). X-inactivation profile reveals extensive variability in X-linked gene expression in females. *Nature, 434*, 400–404.

Carrell, D. T., Emery, B. R., & Hammoud, S. (2007). Altered protamine expression and diminished spermatogenesis: What is the link? *Human Reproduction Update, 13*, 313–327.

Carrell, D. T., & Hammoud, S. S. (2009). The human sperm epigenome and its potential role in embryonic development. *Molecular Human Reproduction, 16*, 37–47.

Carroll, J., & Marangos, P. (2013). *The DNA damage response in mammalian oocytes. Frontiers in Genetics* (4, p. 117).

Caspi, A., Hariri, A. R., Holmes, A., Uher, R., & Moffitt, T. E. (2010). Genetic sensitivity to the environment: The case of the serotonin transporter gene and its implications for studying complex diseases and traits. *American Journal of Psychiatry, 167*, 509–527.

Caspi, A., Sugden, K., Moffitt, T. E., Taylor, A., Craig, I. W., Harrington, H., ... Poulton, R. (2003). Influence of life stress on depression: Moderation by a polymorphism in the 5-HTT gene. *Science (80-), 301*, 386–389.

Cattanach, B. M., & Kirk, M. (1985). Differential activity of maternally and paternally derived chromosome regions in mice. *Nature, 315*, 496–498.

Chadwick, B. P., & Willard, H. F. (2001). Histone H2A variants and the inactive X chromosome: Identification of a second macroH2A variant. *Human Molecular Genetics, 10*, 1101–1113.

Champagne, F. A., Curley, J. P., Swaney, W. T., Hasen, N. S., & Keverne, E. B. (2009). Paternal influence on female behavior: The role of Peg3 in exploration, olfaction, and neuroendocrine regulation of maternal behavior of female mice. *Behavioral Neuroscience, 123*, 469–480.

Chen, D. Y., Stern, S. A., Garcia-Osta, A., Saunier-Rebori, B., Pollonini, G., Bambah-Mukku, D., ... Alberini, C. M. (2011). A critical role for IGF-II in memory consolidation and enhancement. *Nature, 469*, 491–497.

Chen, L., Pan, H., Tuan, T. A., Teh, A. L., MacIsaac, J. L., Mah, S. M., ... Group, G. S. (2015). Brain-derived neurotrophic factor (BDNF) Val66Met polymorphism influences the association of the methylome with maternal anxiety and neonatal brain volumes. *Development and Psychopathology, 27*, 137–150.

Chodroff, R. A., Goodstadt, L., Sirey, T. M., Oliver, P. L., Davies, K. E., Green, E. D., ... Ponting, C. P. (2010). Long noncoding RNA genes: Conservation of sequence and brain expression among diverse amniotes. *Genome Biology, 11*, R72.

Chotalia, M., Smallwood, S. A., Ruf, N., Dawson, C., Lucifero, D., Frontera, M., ... Kelsey, G. (2009). Transcription is required for establishment of germline methylation marks at imprinted genes. *Genes & Development, 23*, 105–117.

Christensen, B. C., Houseman, E. A., Marsit, C. J., Zheng, S., Wrensch, M. R., Wiemels, J. L., ... Kelsey, K. T. (2009). Aging and environmental exposures alter tissue-specific DNA methylation dependent upon CPG island context. *PLoS Genetics*, 5.

Churikov, D., Siino, J., Svetlova, M., Zhang, K., Gineitis, A., Morton Bradbury, E., & Zalensky, A. (2004). Novel human testis-specific histone H2B encoded by the interrupted gene on the X chromosome. *Genomics, 84*, 745–756.

Clarke, H. J., & Vieux, K. F. (2015). Epigenetic inheritance through the female germ-line: The known, the unknown, and the possible. *Seminars in Cell & Developmental Biology, 43*, 106–116.

Cordaux, R., & Batzer, M. A. (2009). The impact of retrotransposons on human genome evolution. *Nature Reviews Genetics, 10*, 691–703.

Córdova-Palomera, A., Fatjó-Vilas, M., Gastó, C., Navarro, V., Krebs, M.-O., & Fañanás, L. (2015). Genome-wide methylation study on depression: Differential methylation and variable methylation in monozygotic twins. *Translational Psychiatry, 5*, e557, March.

Corzett, M., Mazrimas, J., & Balhorn, R. (2002). Protamine 1: Protamine 2 stoichiometry in the sperm of eutherian mammals. *Molecular Reproduction and Development, 61*, 519–527.

Cotton, A. M., Chen, C. Y., Lam, L. L., Wasserman, W. W., Kobor, M. S., & Brown, C. J. (2014). Spread of X-chromosome inactivation into autosomal sequences: Role for DNA elements, chromatin features and chromosomal domains. *Human Molecular Genetics, 23*, 1211–1223.

Cotton, A. M., Price, E. M., Jones, M. J., Balaton, B. P., Kobor, M. S., & Brown, C. J. (2015). Landscape of DNA methylation on the X chromosome reflects CpG density, functional chromatin state and X-chromosome inactivation. *Human Molecular Genetics, 24*, 1528–1539.

Coufal, N. G., Garcia-Perez, J. L., Peng, G. E., Yeo, G. W., Mu, Y., Lovci, M. T., ... Gage, F. H. (2009). L1 retrotransposition in human neural progenitor cells. *Nature, 460*, 1127–1131.

Crespi, B. (2008). Genomic imprinting in the development and evolution of psychotic spectrum conditions. *Biological Reviews*, 441–493.

Csankovszki, G., Panning, B., Bates, B., Pehrson, J. R., & Jaenisch, R. (1999). Conditional deletion of Xist disrupts histone macroH2A localization but not maintenance of X inactivation. *Nature Genetics, 22*, 323–324.

Cukier, H. N., Rabionet, R., Konidari, I., Rayner-Evans, M. Y., Baltos, M. L., Wright, H. H., ... Gilbert, J. R. (2010). Novel variants identified in methyl-CpG-binding domain genes in autistic individuals. *Neurogenetics*, *11*, 291–303.

Davies, S. M. (1994). Developmental regulation of genomic imprinting of the IGF2 gene in human liver. *Cancer Research*, *54*, 2560–2562.

De La Fuente, R., Viveiros, M. M., Burns, K. H., Adashi, E. Y., Matzuk, M. M., & Eppig, J. J. (2004). Major chromatin remodeling in the germinal vesicle (GV) of mammalian oocytes is dispensable for global transcriptional silencing but required for centromeric heterochromatin function. *Developmental Biology*, *275*, 447–458.

de Mateo, S., Gazquez, C., Guimera, M., Balasch, J., Meistrich, M. L., Ballesca, J. L., & Oliva, R. (2009). Protamine 2 precursors (Pre-P2), protamine 1 to protamine 2 ratio (P1/P2), and assisted reproduction outcome. *Fertility and Sterility*, *91*, 715–722.

De Souza, R. A. G., Islam, S. A., McEwen, L. M., Mathelier, A., Hill, A., Mah, S. M., ... Leavitt, B. R. (2016). DNA methylation profiling in human Huntington's disease brain. *Human Molecular Genetics*.

Deal, R. B., & Henikoff, S. (2010). Capturing the dynamic epigenome. *Genome Biology*, 11.

Dean, W. (2013). Epigenetic regulation of oocyte function and developmental potential. In G. Coticchio, F. D. Albertini, & L. De Santis (Eds.), *Oogenesis* (pp. 151–167). London: Springer.

Delgado-Morales, R., & Esteller, M. (2017). Opening up the DNA methylome of dementia. *Molecular Psychiatry*.

Dempster, E. L., Pidsley, R., Schalkwyk, L. C., Owens, S., Georgiades, A., Kane, F., ... Mill, J. (2011). Disease-associated epigenetic changes in monozygotic twins discordant for schizophrenia and bipolar disorder. *Human Molecular Genetics*, *20*, 4786–4796.

Dias, B. G., & Ressler, K. J. (2014). Parental olfactory experience influences behavior and neural structure in subsequent generations. *Nature Neuroscience*, *17*, 89–96.

Domcke, S., Bardet, A. F., Ginno, P. A., Hartl, D., Burger, L., & Schuebeler, D. (2015). Competition between DNA methylation and transcription factors determines binding of NRF1. *Nature*, *528*, 575–579.

Dominissini, D., Moshitch-Moshkovitz, S., Schwartz, S., Salmon-Divon, M., Ungar, L., Osenberg, S., ... Rechavi, G. (2013). Topology of the human and mouse m6A RNA methylomes revealed by m6A-seq. *Nature*, *485*, 201–206.

Donnelly, S. R., Hawkins, T. E., & Moss, S. E. (1999). A conserved nuclear element with a role in mammalian gene regulation. *Human Molecular Genetics*, *8*, 1723–1728.

Drake, N. M., Devito, L. M., Cleland, T. A., & Soloway, P. D. (2011). Imprinted Rasgrf1 expression in neonatal mice affects olfactory learning and memory. *Genes, Brain and Behavior*, *10*, 392–403.

Edgar, R., Tan, P. P. C., Portales-Casamar, E., & Pavlidis, P. (2014). Meta-analysis of human methylomes reveals stably methylated sequences surrounding CpG islands associated with high gene expression. *Epigenetics Chromatin*, 7.

Edgar, R. D., Jones, M. J., Meaney, M. J., Turecki, G., & Kobor, M. S. (2017). 'BECon: A tool for interpreting DNA methylation findings from blood in the context of brain. *bioRxiv*.

Ehrlich, M., Sanchez, C., Shao, C., Nishiyama, R., Kehrl, J., Kuick, R., ... Hanash, S. M. (2009). ICF, an immunodeficiency syndrome: DNA methyltransferase 3B involvement, chromosome anomalies, and gene dysregulation. *Autoimmunity*, *41*, 253–271.

Ender, C., Krek, A., Friedländer, M. R., Beitzinger, M., Weinmann, L., Chen, W., ... Meister, G. (2008). A human snoRNA with microRNA-like functions. *Molecular Cell*, *32*, 519–528.

Essex, M. J., Thomas Boyce, W., Hertzman, C., Lam, L. L., Armstrong, J. M., Neumann, S. M. A., & Kobor, M. S. (2013). Epigenetic vestiges of early developmental adversity: Childhood stress exposure and DNA methylation in adolescence. *Child Development*, *84*, 58–75.

Esumi, S., Kakazu, N., Taguchi, Y., Hirayama, T., Sasaki, A., Hirabayashi, T., ... Yagi, T. (2005). Monoallelic yet combinatorial expression of variable exons of the protocadherin-alpha gene cluster in single neurons. *Nature Genetics*, *37*, 171–176.

Fan, G., Beard, C., Chen, R. Z., Csankovszki, G., Sun, Y., Siniaia, M., ... Jaenisch, R. (2001). DNA hypomethylation perturbs the function and survival of CNS neurons in postnatal animals. *Journal of Neuroscience*, *21*, 788–797.

Farré, P., Jones, M. J., Meaney, M. J., Emberly, E., Turecki, G., & Kobor, M. S. (2015). Concordant and discordant DNA methylation signatures of aging in human blood and brain. *Epigenetics Chromatin*, *8*, 19.

Fatica, A., & Bozzoni, I. (2014). Long non-coding RNAs: New players in cell differentiation and development. *Nature Reviews Genetics*, *15*, 7–21.

Faunes, F., Sanchez, N., Moreno, M., Olivares, G. H., Lee-Liu, D., Almonacid, L., ... Larrain, J. (2011). Expression of transposable elements in neural tissues during Xenopus development. *PLoS ONE*, 6.

Feil, R., & Fraga, M. F. (2012). Epigenetics and the environment: Emerging patterns and implications. *Nature Reviews Genetics*, *13*, 97–109.

Fiore, R., Khudayberdiev, S., Saba, R., & Schratt, G. (2011). MicroRNA function in the nervous system. *Progress in Molecular Biology and Translational Science*, *102*, 47–100.

Fraga, M. F., Ballestar, E., Paz, M. F., Ropero, S., Setien, F., Ballestart, M. L., ... Esteller, M. (2005). Epigenetic differences arise during the lifetime of monozygotic twins. *Proceedings of the National Academy of Sciences of the United States of America*, *102*, 10604–10609.

Frank, M., Ebert, M., Shan, W., Phillips, G. R., Arndt, K., Colman, D. R., & Kemler, R. (2005). Differential expression of individual gamma-protocadherins during mouse brain development. *Molecular and Cellular Neuroscience*, *29*, 603–616.

Frazer, K. A., Murray, S. S., Schork, N. J., & Topol, E. J. (2009). Human genetic variation and its contribution to complex traits. *Nature Reviews Genetics*, *10*, 241–251.

Gamazon, E. R., Badner, J. A., Cheng, L., Zhang, C., Zhang, D., Cox, N. J., ... Berrettini, W. H., et al. (2013). Enrichment of cis-regulatory gene expression SNPs and methylation quantitative trait loci among bipolar disorder susceptibility variants. *Molecular Psychiatry*, *18*, 340–346.

Garcia-Peiro, A., Martinez-Heredia, J., Oliver-Bonet, M., Abad, C., Amengual, M., Navarro, J., ... Benet, J. (2011). Protamine 1 to protamine 2 ratio correlates with dynamic aspects of DNA fragmentation in human sperm. *Fertility and Sterility*, *95*, 105–109.

Gardiner-Garden, M., & Frommer, M. (1987). CpG islands in vertebrate genomes. *Journal of Molecular Biology*, *196*, 261–282.

Gatewood, J. M., Cook, G. R., Balhorn, R., Schmid, C. W., & Bradbury, E. M. (1990). Isolation of four core histones from human sperm chromatin representing a minor subset of somatic histones. *Journal of Biological Chemistry*, *265*, 20662–20666.

Ge, Z.-J., Liang, X.-W., Guo, L., Liang, Q.-X., Luo, S.-M., Wang, Y.-P., ... Sun, Q.-Y. (2013). Maternal diabetes causes alterations of DNA methylation statuses of some imprinted genes in murine oocytes. *Biology of Reproduction*, *88*, 117.

Gendrel, A.-V., & Heard, E. (2014). Noncoding RNAs and epigenetic mechanisms during X-chromosome inactivation. *Annual Review of Cell and Developmental Biology*, *30*, 561–580.

Gibbs, J. R., van der Brug, M. P., Hernandez, D. G., Traynor, B. J., Nalls, M. A., Lai, S.-L., ... Singleton, A. B. (2010). Abundant quantitative trait loci exist for DNA methylation and gene expression in human brain. *PLoS Genetics*, *6*, 29.

Glatt, S. J., Faraone, S. V., & Tsuang, M. T. (2003). Association between a functional catechol *O*-methyltransferase gene polymorphism and schizophrenia: Meta-analysis of case-control and family-based studies. *American Journal of Psychiatry*, *160*, 469–476.

Godfrey, K., & Robinson, S. (1998). Maternal nutrition, placental growth and fetal programming. *Proceedings of the Nutrition Society*, *57*, 105–111.

Godfrey, K. M., Lillycrop, K. A., Burdge, G. C., Gluckman, P. D., & Hanson, M. A. (2007). Epigenetic mechanisms and the mismatch concept of the developmental origins of health and disease. *Pediatric Research*, *61*(5 Part 2 Suppl), 5R–10R.

Goll, M. G., Kirpekar, F., Maggert, K. A., Yoder, J. A., Hsieh, C.-L., Zhang, X., ... Bestor, T. H. (2006). Methylation of tRNAAsp by the DNA methyltransferase homolog Dnmt2. *Science (80–)*, *311*, 395–398.

Gonzales, M. L., & LaSalle, J. M. (2010). The role of MeCP2 in brain development and neurodevelopmental disorders. *Current Psychiatry Reports*, 127–134.
Gowher, H., Liebert, K., Hermann, A., Xu, G., & Jeltsch, A. (2005). Mechanism of stimulation of catalytic activity of Dnmt3A and Dnmt3B DNA-(cytosine-C5)-methyltransferases by Dnmt3L. *Journal of Biological Chemistry, 280*, 13341–13348.
Grayson, D. R., Jia, X., Chen, Y., Sharma, R. P., Mitchell, C. P., Guidotti, A., & Costa, E. (2005). Reelin promoter hypermethylation in schizophrenia. *Proceedings of the National Academy of Sciences of the United States of America, 102*, 9341–9346.
Gregg, C., Zhang, J., Weissbourd, B., Luo, S., Schroth, G. P., Haig, D., & Dulac, C. (2010). High-resolution analysis of parent-of-origin allelic expression in the mouse brain. Supporting online material. *Science, 329*, 643–648.
Grewal, S. I. S., & Jia, S. (2007). Heterochromatin revisited. *Nature Reviews Genetics, 8*, 35–46.
Grunstein, M. (1997). Histone acetylation in chromatin structure and transcription. *Nature, 389*, 349–352.
Gu, T.-P., Guo, F., Yang, H., Wu, H.-P., Xu, G.-F., Liu, W., ... Xu, G.-L. (2011). The role of Tet3 DNA dioxygenase in epigenetic reprogramming by oocytes. *Nature, 477*, 606–610.
Guibert, S., & Weber, M. (2013). Functions of DNA methylation and hydroxymethylation in mammalian development. *Current Topics Developmental Biology, 104*, 47–83.
Guintivano, J., Aryee, M. J., & Kaminsky, Z. A. (2013). A cell epigenotype specific model for the correction of brain cellular heterogeneity bias and its application to age, brain region and major depression. *Epigenetics, 8*, 290–302.
Guo, F., Li, X., Liang, D., Li, T., Zhu, P., Guo, H., ... Xu, G. L. (2014). Active and passive demethylation of male and female pronuclear DNA in the mammalian zygote. *Cell Stem Cell, 15*, 447–458.
Guo, H., Zhu, P., Yan, L., Li, R., Hu, B., Lian, Y., ... Qiao, J. (2014). The DNA methylation landscape of human early embryos. *Nature, 511*, 606–610.
Guo, J. U., Su, Y., Shin, J. H., Shin, J., Li, H., Xie, B., ... Song, H. (2014). Distribution, recognition and regulation of non-CpG methylation in the adult mammalian brain. *Nature Neuroscience, 17*, 215–222.
Gustincich, S., Sandelin, A., Plessy, C., Katayama, S., Simone, R., Lazarevic, D., ... Carninci, P. (2006). The complexity of the mammalian transcriptome. *Journal of Physiology, 575*, 321–332.
Gutierrez-Arcelus, M., Lappalainen, T., Montgomery, S. B., Buil, A., Ongen, H., Yurovsky, A., ... Dermitzakis, E. T. (2013). Passive and active DNA methylation and the interplay with genetic variation in gene regulation. *Elife, 2*, e00523.
Guttman, M., Amit, I., Garber, M., French, C., Lin, M. F., Feldser, D., ... Lander, E. S. (2009). Chromatin signature reveals over a thousand highly conserved large non-coding RNAs in mammals. *Nature, 458*, 223–227.
Haberstick, B. C., Lessem, J. M., Hewitt, J. K., Smolen, A., Hopfer, C. J., Halpern, C. T., ... Mullan Harris, K. (2014). MAOA genotype, childhood maltreatment, and their interaction in the etiology of adult antisocial behaviors. *Biological Psychiatry, 75*, 25–30.
Hahn, M. A., Qiu, R., Wu, X., Li, A. X., Zhang, H., Wang, J., ... Lu, Q. (2013). Dynamics of 5-hydroxymethylcytosine and chromatin marks in mammalian neurogenesis. *Cell Reports, 3*, 291–300.
Hamatani, T. (2012). Human spermatozoal RNAs. *Fertility and Sterility*, 275–281.
Hammoud, S. S., Nix, D. A., Zhang, H., Purwar, J., Carrell, D. T., & Cairns, B. R. (2009). Distinctive chromatin in human sperm packages genes for embryo development. *Nature, 460*, 473–478.
Hannon, E., Spiers, H., Viana, J., Pidsley, R., Burrage, J., Murphy, T. M., ... Mill, J. (2015). Methylation QTLs in the developing brain and their enrichment in schizophrenia risk loci. *Nature Neuroscience, 19*, 48–54.
Hanson, M. A., & Gluckman, P. D. (2008). Developmental origins of health and disease: New insights. *Basic Clinical Pharmacology Toxicology, 102*, 90–93.
Hanson, M. A., Low, F. M., & Gluckman, P. D. (2011). Epigenetic epidemiology: The rebirth of soft inheritance. *Annals of Nutrition and Metabolism, 58*(Suppl. 2), 8–15.

Hashimoto, H., Liu, Y., Upadhyay, A. K., Chang, Y., Howerton, S. B., Vertino, P. M., ... Cheng, X. (2012). Recognition and potential mechanisms for replication and erasure of cytosine hydroxymethylation. *Nucleic Acids Research*, *40*, 4841–4849.

Hayakawa, K., Ohgane, J., Tanaka, S., Yagi, S., & Shiota, K. (2012). Oocyte-specific linker histone H1foo is an epigenomic modulator that decondenses chromatin and impairs pluripotency. *Epigenetics*, *7*, 1029–1036.

He, Y. F., Li, B. Z., Li, Z., Liu, P., Wang, Y., Tang, Q., & Ding, J. (2011). Tet-mediated formation of 5-carboxylcytosine and its excision by TDG in mammalian DNA. *Science (80–)*.

Heard, E., Rougeulle, C., Arnaud, D., Avner, P., Allis, C. D., & Spector, D. L. (2001). Methylation of histone H3 at Lys-9 Is an early mark on the X chromosome during X inactivation. *Cell*, *107*, 727–738.

Heijmans, B. T., Tobi, E. W., Stein, A. D., Putter, H., Blauw, G. J., Susser, E. S., ... Lumey, L. H. (2008). Persistent epigenetic differences associated with prenatal exposure to famine in humans. *Proceedings of the National Academy of Sciences of the United States of America*, *105*, 17046–17049.

Hellemans, K. G. C., Sliwowska, J. H., Verma, P., & Weinberg, J. (2010). Prenatal alcohol exposure: Fetal programming and later life vulnerability to stress, depression and anxiety disorders. *Neuroscience & Biobehavioral Reviews*, *34*, 791–807.

Henikoff, S., & Greally, J. M. (2016). Epigenetics, cellular memory and gene regulation. *Current Biology*, *26*, R644–R648.

Heyn, H., Li, N., Ferreira, H. J., Moran, S., Pisano, D. G., Gomez, A., ... Esteller, M. (2012). Distinct DNA methylomes of newborns and centenarians. *Proceedings of the National Academy of Sciences of the United States of America*, *109*, 10522–10527.

Hirschhorn, J. N., & Daly, M. J. (2005). Genome-wide association studies for common diseases and complex traits. *Nature Reviews Genetics*, *6*, 95–108.

Horvath, S., Langfelder, P., Kwak, S., Aaronson, J., Rosinski, J., Vogt, T. F., ... Yang, X. W. (2016). Huntington' s disease accelerates epigenetic aging of human brain and disrupts DNA methylation levels. *Aging*, *8*, 1496–1523.

Howell, C. Y., Bestor, T. H., Ding, F., Latham, K. E., Mertineit, C., Trasler, J. M., & Chaillet, J. R. (2001). Genomic imprinting disrupted by a maternal effect mutation in the Dnmt1 gene. *Cell*, *104*, 829–838.

Hung, T., Wang, Y., Lin, M. F., Koegel, A. K., Kotake, Y., Grant, G. D., ... Chang, H. Y. (2011). Extensive and coordinated transcription of noncoding RNAs within cell-cycle promoters. *Nature Genetics*, *43*, 621–629.

Illingworth, R. S., & Bird, A. P. (2009). CpG islands—'A rough guide'. *FEBS Letter*, *583*, 1713–1720.

Imamura, T., Uesaka, M., & Nakashima, K. (2014). Epigenetic setting and reprogramming for neural cell fate determination and differentiation. *Philosophical Transactions of the Royal Society of London, Series B: Biological Science*, *369*, 20130511.

Irizarry, R. A., Ladd-Acosta, C., Carvalho, B., Wu, H., Brandenburg, S. A., Jeddeloh, J. A., ... Feinberg, A. P. (2008). Comprehensive high-throughput arrays for relative methylation (CHARM). *Genome Research*, *18*, 780–790.

Ito, S., D'Alessio, A. C., Taranova, O. V., Hong, K., Sowers, L. C., & Zhang, Y. (2010). Role of Tet proteins in 5mC to 5hmC conversion, ES-cell self-renewal and inner cell mass specification. *Nature*, *466*, 1129–1133.

Jaffe, A. E., Gao, Y., Deep-Soboslay, A., Tao, R., Hyde, T. M., Weinberger, D. R., & Kleinman, J. E. (2015). Mapping DNA methylation across development, genotype and schizophrenia in the human frontal cortex. *Nature Neuroscience*, *19*, 40–47.

Jeltsch, A., Nellen, W., & Lyko, F. (2006). Two substrates are better than one: Dual specificities for Dnmt2 methyltransferases. *Trends in Biochemical Sciences*, *31*, 306–308.

Jeppesen, P., & Turner, B. M. (1993). The inactive X chromosome in female mammals is distinguished by a lack of histone H4 acetylation, a cytogenetic marker for gene expression. *Cell*, *74*, 281–289.

Jodar, M., Selvaraju, S., Sendler, E., Diamond, M. P., & Krawetz, S. A. (2013). The presence, role and clinical use of spermatozoal RNAs. *Human Reproduction Update*, *19*, 604–624.

Johansson, A., Enroth, S., & Gyllensten, U. (2013). Continuous aging of the human DNA methylome throughout the human lifespan. *PLoS ONE*, *8*, e67378.

Jones, M. J., Fejes, A. P., & Kobor, M. S. (2013). DNA methylation, genotype and gene expression: Who is driving and who is along for the ride? *Genome Biology*, *14*, 126.

Jones, M. J., Goodman, S. J., & Kobor, M. S. (2015). DNA methylation and healthy human aging. *Aging Cell*, 924–932.

Jones, P. A. (2012). Functions of DNA methylation: Islands, start sites, gene bodies and beyond. *Nature Reviews Genetics*, 484–492.

Jones, P. A., & Baylin, S. B. (2007). The epigenomics of cancer. *Cell*, *128*, 683–692.

Jones, P. A., & Takai, D. (2001). The role of DNA methylation in mammalian epigenetics. *Science*, *293*, 1068–1070.

Joti, Y., Hikima, T., Nishino, Y., Kamada, F., Hihara, S., Takata, H., ... Maeshima, K. (2014). Chromosomes without a 30-nm chromatin fiber. *Nucleus*, *3*, 404–410.

Joubert, B. R., Håberg, S. E., Nilsen, R. M., Wang, X., Vollset, S. E., Murphy, S. K., ... London, S. J. (2012). 450K epigenome-wide scan identifies differential DNA methylation in newborns related to maternal smoking during pregnancy. *Environmental Health Perspectives*, *120*, 1425–1431.

Jowaed, A., Schmitt, I., Kaut, O., & Wüllner, U. (2010). Methylation regulates alpha-synuclein expression and is decreased in parkinson's disease patients' brains. *Journal of Neuroscience*, *30*, 6355 LP-6359.

Kageyama, S. I., Liu, H., Kaneko, N., Ooga, M., Nagata, M., & Aoki, F. (2007). Alterations in epigenetic modifications during oocyte growth in mice. *Reproduction*, *133*, 85–94.

Kaminsky, Z. A., Tang, T., Wang, S., Ptak, C., Oh, G. H. T., & Wong, A. H. C. (2009). DNA methylation profiles in monozygotic and dizygotic twins. *Nature Genetics*, *41*, 240–245.

Kang, M. K., & Han, S. J. (2011). Post-transcriptional and post-translational regulation during mouse oocyte maturation. *BMB Reports*, 147–157.

Kapranov, P., Laurent, G., St, Raz, T., Ozsolak, F., Reynolds, C. P., Sorensen, P. H., ... Triche, T. J. (2010). The majority of total nuclear-encoded non-ribosomal RNA in a human cell is "dark matter" un-annotated RNA. *BMC Biology*, *8*, 149.

Kapranov, P., Ozsolak, F., Kim, S. W., Foissac, S., Lipson, D., Hart, C., ... Milos, P. M. (2010). New class of gene-termini-associated human RNAs suggests a novel RNA copying mechanism. *Nature*, *466*, 642–646.

Kernohan, K. D., & Bérubé, N. G. (2010). Genetic and epigenetic dysregulation of imprinted genes in the brain. *Epigenomics*, *2*, 743–763.

Khalil, A. M., Guttman, M., Huarte, M., Garber, M., Raj, A., Rivea Morales, D., ... Rinn, J. L. (2009). Many human large intergenic noncoding RNAs associate with chromatin-modifying complexes and affect gene expression. *Proceedings of the National Academy of Sciences of the United States of America*, *106*, 11667–11672.

Kim-Cohen, J., Caspi, A., Taylor, A., & Williams, B. (2006). MAOA, maltreatment, and gene–environment interaction predicting children's mental health: New evidence and a meta-analysis. *Molecular Psychiatry*, *11*, 903–913.

Kinde, B., Gabel, H. W., Gilbert, C. S., Griffith, E. C., & Greenberg, M. E. (2015). Reading the unique DNA methylation landscape of the brain: Non-CpG methylation, hydroxymethylation, and MeCP2. *Proceedings of the National Academy of Sciences*, *112*, 6800–6806.

Kishore, S., & Stamm, S. (2006). The snoRNA HBII-52 regulates alternative splicing of the serotonin receptor 2C. *Science*, *311*, 230–232.

Klengel, T., Mehta, D., Anacker, C., Rex-Haffner, M., Pruessner, J. C., Pariante, C. M., ... Binder, E. B. (2012). Allele-specific FKBP5 DNA demethylation mediates gene–childhood trauma interactions. *Nature Neuroscience*, *16*, 33–41.

Kobayashi, H., Sakurai, T., Imai, M., Takahashi, N., Fukuda, A., Yayoi, O., ... Kono, T. (2012). Contribution of intragenic DNA methylation in mouse gametic DNA methylomes to establish Oocyte-specific heritable marks. *PLoS Genetics*, 8.

Kohli, R. M., & Zhang, Y. (2013). TET enzymes, TDG and the dynamics of DNA demethylation. *Nature*, *502*, 472–479.

Kornberg, R. D. (1974). Chromatin structure: A repeating unit of histones and DNA. *Science (80−)*, *184*, 868−871.

Kouzarides, T. (2007). Chromatin modifications and their function. *Cell*, *128*, 693−705.

Koziol, M. J., Bradshaw, C. R., Allen, G. E., Costa, A. S. H., Frezza, C., & Gurdon, J. B. (2015). Identification of methylated deoxyadenosines in vertebrates reveals diversity in DNA modifications. *Nature Structural & Molecular Biology*, *23*, 24−30.

Kramer, J. M., & van Bokhoven, H. (2009). Genetic and epigenetic defects in mental retardation. *International Journal of Biochemistry and Cell Biology*, 96−107.

Kriaucionis, S., & Heintz, N. (2009). The nuclear DNA base 5-hydroxymethylcytosine is present in Purkinje neurons and the brain. *Science (80−)*, *324*, 929−930.

Kubota, T., Miyake, K., Hariya, N., & Mochizuki, K. (2015). Understanding the epigenetics of neurodevelopmental disorders and DOHaD. *Journal of Developmental Origins of Health and Disease*, *6*, 1−9.

Kuratomi, G., Iwamoto, K., Bundo, M., Kusumi, I., Kato, N., Iwata, N., ... Kato, T. (2008). Aberrant DNA methylation associated with bipolar disorder identified from discordant monozygotic twins. *Molecular Psychiatry*, *13*, 429−441.

Kurian, J. R., & Terasawa, E. (2013). *Epigenetic control of gonadotropin releasing hormone neurons. Frontiers in Endocrinology* (4, p. 61).

Lahiri, D. K., Maloney, B., & Zawia, N. H. (2009). The LEARn model: An epigenetic explanation for idiopathic neurobiological diseases. *Molecular Psychiatry*, *14*, 992−1003.

Lam, L. L., Emberly, E., Fraser, H. B., Neumann, S. M., Chen, E., Miller, G. E., & Kobor, M. S. (2012). Factors underlying variable DNA methylation in a human community cohort. *Proceedings of the National Academy of Sciences*, *109*(Supplement_2), 17253−17260.

Landerer, E., Villegas, J., Burzio, V. A., Oliveira, L., Villota, C., Lopez, C., ... Burzio, L. O. (2011). Nuclear localization of the mitochondrial ncRNAs in normal and cancer cells. *Cellular Oncology*, *34*, 297−305.

Lane, N., Dean, W., Erhardt, S., Hajkova, P., Surani, A., Walter, J., & Reik, W. (2003). Resistance of IAPs to methylation reprogramming may provide a mechanism for epigenetic inheritance in the mouse. *Genesis*, *35*, 88−93.

LaSalle, J. M., Powell, W. T., & Yasui, D. H. (2013). Epigenetic layers and players underlying neurodevelopment. *Trends Neuroscience*, *36*, 460−470.

Latham, K. E., Solter, D., & Schultz, R. M. (1991). Activation of a two-cell stage-specific gene following transfer of heterologous nuclei into enucleated mouse embryos. *Molecular Reproduction and Development*, *30*, 182−186.

Laurent, L., Wong, E., Li, G., Huynh, T., Tsirigos, A., Ong, C. T., ... Wei, C. L. (2010). Dynamic changes in the human methylome during differentiation. *Genome Research*, *20*, 320−331.

Lefebvre, L., Viville, S., Barton, S. C., Ishino, F., Keverne, E. B., & Surani, M. A. (1998). Abnormal maternal behaviour and growth retardation associated with loss of the imprinted gene Mest. *Nature Genetics*, *20*, 163−169.

Li, G., & Reinberg, D. (2011). Chromatin higher-order structures and gene regulation. *Current Opinion in Genetics & Development*, *21*, 175−186.

Li, L., Keverne, E. B., Aparicio, S. A., Ishino, F., Barton, S. C., & Surani, M. A. (1999). Regulation of maternal behavior and offspring growth by paternally expressed Peg3. *Science*, *284*, 330−333, April.

Li, L., Lu, X., & Dean, J. (2013). The maternal to zygotic transition in mammals. *Molecular Aspects of Medicine*, 919−938.

Liebers, R., Rassoulzadegan, M., & Lyko, F. (2014). Epigenetic regulation by heritable RNA. *PLoS Genetics ics*.

Lilja, T., Heldring, N., & Hermanson, O. (2012). Like a rolling histone: Epigenetic regulation of neural stem cells and brain development by factors controlling histone acetylation and methylation. *Biochimica et Biophysica Acta—General Subjects*, *1830*, 2354−2360.

Lin, C.-J., Conti, M., & Ramalho-Santos, M. (2013). Histone variant H3.3 maintains a decondensed chromatin state essential for mouse preimplantation development. *Development*, *140*, 3624−3634.

Lister, R., Mukamel, E. A., Nery, J. R., Urich, M., Puddifoot, C. A., Johnson, N. D., ... Ecker, J. R. (2013). Global epigenomic reconfiguration during mammalian brain development. *Science, 341*, 1237905.

Lister, R., Pelizzola, M., Dowen, R. H., Hawkins, R. D., Hon, G., Tonti-Filippini, J., ... Ecker, J. R. (2009). Human DNA methylomes at base resolution show widespread epigenomic differences. *Nature, 462*, 315–322.

Lister, R., Pelizzola, M., Kida, Y. S., Hawkins, R. D., Nery, J. R., Hon, G., ... Ecker, J. R. (2012). Hotspots of aberrant epigenomic reprogramming in human induced pluripotent stem cells. *Nature, 470*, 68–73.

Liu, N., Dai, Q., Zheng, G., He, C., Parisien, M., & Pan, T. (2015). *N*(6)-methyladenosine-dependent RNA structural switches regulate RNA-protein interactions. *Nature, 518*, 560–564.

Lock, L. F., Takagi, N., & Martin, G. R. (1987). Methylation of the Hprt gene on the inactive X occurs after chromosome inactivation. *Cell, 48*, 39–46.

Lomniczi, A., Loche, A., Castellano, J. M., Ronnekleiv, O. K., Bosch, M., Kaidar, G., ... Ojeda, S. R. (2013). Epigenetic control of female puberty. *Nature Neuroscience, 16*, 281–289.

Luger, K., Mäder, A. W., Richmond, R. K., & Sargent, D. F. (1997). Crystal structure of the nucleosome core particle at 2.8 Å resolution. *Nature, 389*, 251–260.

Lung, B., Zemann, A., Madej, M. J., Schuelke, M., Techritz, S., Ruf, S., ... Huttenhofer, A. (2006). Identification of small non-coding RNAs from mitochondria and chloroplasts. *Nucleic Acids Research, 34*, 3842–3852.

Lussier, A. A., Weinberg, J., & Kobor, M. S. (2017). Epigenetics studies of fetal alcohol spectrum disorder: where are we now? *Epigenomics, 9*, 291–311.

Lv, X., Jiang, H., Liu, Y., Lei, X., & Jiao, J. (2014). MicroRNA-15b promotes neurogenesis and inhibits neural progenitor proliferation by directly repressing TET3 during early neocortical development. *EMBO Reports, 15*, 1305–1314.

Lyon, M. F. (1961). Gene action in the X-chromosome of the mouse (*Mus musculus* L.). *Nature, 190*, 372–373.

Lyon, M. F. (1962). Sex chromatin and gene action in the mammalian X-chromosome. *American Journal of Human Genetics, 14*, 135–148.

Ma, D. K., Jang, M.-H., Guo, J. U., Kitabatake, Y., Chang, M., Pow-Anpongkul, N., ... Song, H. (2009). Neuronal activity-induced Gadd45b promotes epigenetic DNA demethylation and adult neurogenesis. *Science (80–), 323*, 1074–1077.

Ma, J.-Y., Li, M., Luo, Y.-B., Song, S., Tian, D., Yang, J., ... Sun, Q.-Y. (2013). Maternal factors required for oocyte developmental competence in mice: Transcriptome analysis of non-surrounded nucleolus (NSN) and surrounded nucleolus (SN) oocytes. *Cell Cycle, 12*, 1928–1938.

Ma, Y., Kanakousaki, K., & Buttitta, L. (2015). How the cell cycle impacts chromatin architecture and influences cell fate. *Bioinforma Computer Biology, 6*, 19.

Mak, W., Nesterova, T. B., de Napoles, M., Appanah, R., Yamanaka, S., Otte, A. P., & Brockdorff, N. (2004). Reactivation of the paternal X chromosome in early mouse embryos. *Science, 303*, 666–669.

Manolio, T. A., Collins, F. S., Cox, N. J., Goldstein, D. B., Hindorff, L. A., Hunter, D. J., ... Visscher, P. M. (2009). Finding the missing heritability of complex diseases. *Nature, 461*, 747–753.

Marcho, C., Cui, W., & Mager, J. (2015). Epigenetic dynamics during preimplantation development. *Reproduction, 150*, R109–R120.

Marmorstein, R., & Trievel, R. C. (2009). Histone modifying enzymes: Structures, mechanisms, and specificities. *BBA Gene Regulatory Mechanisms, 1789*, 58–68.

Martí, E., Pantano, L., Bañez-Coronel, M., Llorens, F., Miñones-Moyano, E., Porta, S., ... Estivill, X. (2010). A myriad of miRNA variants in control and Huntington's disease brain regions detected by massively parallel sequencing. *Nucleic Acids Research, 38*, 7219–7235.

Marzi, S. J., Meaburn, E. L., Dempster, E. L., Lunnon, K., Paya-Cano, J. L., Smith, R. G., ... Mill, J. (2016). Tissue-specific patterns of allelically-skewed DNA methylation. *Epigenetics, 11*, 24–35.

Massah, S., Beischlag, T. V., & Prefontaine, G. G. (2015). Epigenetic events regulating monoallelic gene expression. *Critical Reviews in Biochemistry and Molecular Biology, 50*, 337–358.

Maunakea, A. K., Chepelev, I., Cui, K., & Zhao, K. (2013). Intragenic DNA methylation modulates alternative splicing by recruiting MeCP2 to promote exon recognition. *Cell Research*, *23*, 1256–1269.

Maunakea, A. K., Nagarajan, R. P., Bilenky, M., Ballinger, T. J., D'Souza, C., Fouse, S. D., ... Costello, J. F. (2010). Conserved role of intragenic DNA methylation in regulating alternative promoters. *Nature*, *466*, 253–257.

McEwen, B. S., & Wingfield, J. C. (2003). The concept of allostasis in biology and biomedicine. *Hormones and Behavior*, 2–15.

McGowan, P. O., Sasaki, A., D'Alessio, A. C., Dymov, S., Labonté, B., Szyf, M., ... Meaney, M. J. (2009). Epigenetic regulation of the glucocorticoid receptor in human brain associates with childhood abuse. *Nature Neuroscience*, *12*, 342–348.

McKittrick, E., Gaften, P. R., Ahmad, K., & Henikoff, S. (2004). Histone H3.3 is enriched in covalent modifications associated with active chromatin. *Proceedings of the National Academy of Sciences*, *101*, 1525–1530.

Meaney, M. J. (2010). Epigenetics and the biological definition of gene X environment interactions. *Child Development*, 41–79.

Mercer, T. R., Qureshi, I. A., Gokhan, S., Dinger, M. E., Li, G., Mattick, J. S., & Mehler, M. F. (2010). Long noncoding RNAs in neuronal-glial fate specification and oligodendrocyte lineage maturation. *BMC Neuroscience*, *11*, 14.

Mercer, T. R., Wilhelm, D., Dinger, M. E., Soldà, G., Korbie, D. J., Glazov, E. A., ... Mattick, J. S. (2011). Expression of distinct RNAs from 3′ untranslated regions. *Nucleic Acids Research*, *39*, 2393–2403.

Meyer, K. D., & Jaffrey, S. R. (2016). Expanding the diversity of DNA base modifications with N6-methyldeoxyadenosine. *Genome Biology*, 1–4.

Meyer, K. D., Saletore, Y., Zumbo, P., Elemento, O., Mason, C. E., & Jaffrey, S. R. (2012). Comprehensive Analysis of mRNA Methylation Reveals Enrichment in 3′ UTRs and near Stop Codons. *Cell*, *149*, 1635–1646.

Mikkelsen, T. S., Ku, M., Jaffe, D. B., Issac, B., Lieberman, E., Giannoukos, G., ... Bernstein, B. E. (2007). Genome-wide maps of chromatin state in pluripotent and lineage-committed cells. *Nature*, *448*, 553–560.

Millar, J. K., Wilson-Annan, J. C., Anderson, S., Christie, S., Taylor, M. S., Semple, C. A., ... Porteous, D. J. (2000). Disruption of two novel genes by a translocation co-segregating with schizophrenia. *Human Molecular Genetics*, *9*, 1415–1423.

Miñones-Moyano, E., Porta, S., Escaramís, G., Rabionet, R., Iraola, S., Kagerbauer, B., ... Martí, E. (2011). MicroRNA profiling of Parkinson's disease brains identifies early downregulation of miR-34b/c which modulate mitochondrial function. *Human Molecular Genetics*, *20*, 3067–3078.

Moore, T., & Haig, D. (1991). Genomic imprinting in mammalian development: A parental tug-of-war. *Trends Genetics*, 7, 45–49.

Morgan, C. P., & Bale, T. L. (2011). Early prenatal stress epigenetically programs dysmasculinization in second-generation offspring via the paternal lineage. *Journal of Neuroscience*, *31*, 11748–11755.

Morgan, H. D., Sutherland, H. G., Martin, D. I., & Whitelaw, E. (1999). Epigenetic inheritance at the agouti locus in the mouse. *Nature Genetics*, *23*, 314–318.

Morrison, K. E., Rodgers, A. B., Morgan, C. P., & Bale, T. L. (2014). Epigenetic mechanisms in pubertal brain maturation. *Neuroscience*, *264*, 17–24.

Mullin, S., & Schapira, A. (2015). The genetics of Parkinson's disease. *British Medical Bulletin*, *114*, 39–52.

Muotri, A. R., Zhao, C., Marchetto, M. C. N., & Gage, F. H. (2009). Environmental influence on L1 retrotransposons in the adult hippocampus. *Hippocampus*, *19*, 1002–1007.

Nabel, C. S., Jia, H., Ye, Y., Shen, L., Goldschmidt, H. L., Stivers, J. T., ... Kohli, R. M. (2012). AID/APOBEC deaminases disfavor modified cytosines implicated in DNA demethylation. *Nature Chemical Biology*, *8*, 751–758.

Nakamura, T., Arai, Y., Umehara, H., Masuhara, M., Kimura, T., Taniguchi, H., . . . Nakano, T. (2007). PGC7/Stella protects against DNA demethylation in early embryogenesis. *Nature Cell Biology, 9*, 64–71.

Nakamura, T., Liu, Y.-J., Nakashima, H., Umehara, H., Inoue, K., Matoba, S., . . . Nakano, T. (2012). PGC7 binds histone H3K9me2 to protect against conversion of 5mC to 5hmC in early embryos. *Nature, 486*, 415–419.

Narlikar, G. J., Sundaramoorthy, R., & Owen-Hughes, T. (2013). Mechanisms and functions of ATP-dependent chromatin-remodeling enzymes. *Cell, 154*, 490–503.

Ng, S.-F., Lin, R. C. Y., Laybutt, D. R., Barres, R., Owens, J. A., & Morris, M. J. (2010). Chronic high-fat diet in fathers programs β-cell dysfunction in female rat offspring. *Nature, 467*, 963–966.

Oakley, A. E., Clifton, D. K., & Steiner, R. A. (2009). Kisspeptin signaling in the brain. *Endocrine Reviews*, 713–743.

Obata, Y., & Kono, T. (2002). Maternal primary imprinting is established at a specific time for each gene throughout oocyte growth. *Journal of Biological Chemistry, 277*, 5285–5289.

Ohhata, T., & Wutz, A. (2013). Reactivation of the inactive X chromosome in development and reprogramming. *Cellular and Molecular Life Sciences, 70*, 2443–2461.

Okamoto, I., Otte, A. P., Allis, C. D., Reinberg, D., & Heard, E. (2004). Epigenetic dynamics of imprinted X inactivation during early mouse development. *Science (80–), 303*, 644–649.

Okamura, K., Chung, W. J., Ruby, J. G., Guo, H., Bartel, D. P., & Lai, E. C. (2008). The drosophila hairpin RNA pathway generates endogenous short interfering RNAs. *Nature, 453*, 803–806.

Okano, M., Xie, S., & Li, E. (1998). Cloning and characterization of a family of novel mammalian DNA (cytosine-5) methyltransferases. *Nature Genetics, 19*, 219–220.

Oktem, O., & Oktay, K. (2008). The ovary: Anatomy and function throughout human life. *Annals of the New York Academy of Sciences, 1127*, 1–9.

Olins, A. L., & Olins, D. E. (1974). Spheroid chromatin units (ν bodies). *Science (80–), 183*, 330–332.

Oliva, R. (2006). Protamines and male infertility. *Human Reproduction Update*, 417–435.

Ooi, S. K. T., & Bestor, T. H. (2008). The colorful history of active DNA demethylation. *Cell, 133*, 1145–1148.

Orom, U. A., Derrien, T., Beringer, M., Gumireddy, K., Gardini, A., Bussotti, G., . . . Shiekhattar, R. (2010). Long noncoding RNAs with enhancer-like function in human cells. *Cell, 143*, 46–58.

Pan, G., Tian, S., Nie, J., Yang, C., Ruotti, V., Wei, H., . . . Thomson, J. A. (2007). Whole-genome analysis of histone H3 lysine 4 and lysine 27 methylation in human embryonic stem cells. *Cell Stem Cell, 1*, 299–312.

Pastori, C., Peschansky, V. J., Barbouth, D., Mehta, A., Silva, J. P., & Wahlestedt, C. (2014). Comprehensive analysis of the transcriptional landscape of the human FMR1 gene reveals two new long noncoding RNAs differentially expressed in fragile X syndrome and fragile X-associated tremor/ataxia syndrome. *Human Genetics, 133*, 59–67.

Peeters, S. B., Cotton, A. M., & Brown, C. J. (2014). Variable escape from X-chromosome inactivation: Identifying factors that tip the scales towards expression. *BioEssays, 36*, 746–756.

Penny, G. D., Kay, G. F., Sheardown, S. A., Rastan, S., & Brockdorff, N. (1996). Requirement for Xist in X chromosome inactivation. *Nature*, 131–137.

Petazzi, P., Sandoval, J., Szczesna, K., Jorge, O. C., Roa, L., Sayols, S., . . . Esteller, M. (2013). Dysregulation of the long non-coding RNA transcriptome in a Rett syndrome mouse model. *RNA Biology, 10*, 1197–1203.

Peters, A. H. F. M., Mermoud, J. E., O'Carroll, D., Pagani, M., Schweizer, D., Brockdorff, N., & Jenuwein, T. (2002). Histone H3 lysine 9 methylation is an epigenetic imprint of facultative heterochromatin. *Nature Genetics, 30*, 77–80.

Pfeifer, G. P., Kadam, S., & Jin, S.-G. (2013). 5-Hydroxymethylcytosine and its potential roles in development and cancer. *Epigenetics Chromatin, 6*, 10.

Piña, B., & Suau, P. (1987). Changes in histones H2A and H3 variant composition in differentiating and mature rat brain cortical neurons. *Developmental Biology, 123*, 51–58.

Plath, K., Fang, J., Mlynarczyk-Evans, S. K., Cao, R., Worringer, K. A., Wang, H., . . . Zhang, Y. (2003). Role of histone H3 lysine 27 methylation in X inactivation. *Science, 300*, 131–135, April.

Pollard, K. S., Salama, S. R., Lambert, N., Lambot, M.-A., Coppens, S., Pedersen, J. S., ... Haussler, D. (2006). An RNA gene expressed during cortical development evolved rapidly in humans. *Nature, 443*, 167–172.

Popp, C., Dean, W., Feng, S., Cokus, S. J., Andrews, S., Pellegrini, M., ... Reik, W. (2010). Genome-wide erasure of DNA methylation in mouse primordial germ cells is affected by AID deficiency. *Nature, 463*, 1101–1105.

Portales-Casamar, E., Lussier, A. A., Jones, M. J., MacIsaac, J. L., Edgar, R. D., Mah, S. M., ... Kobor, M. S. (2016). DNA methylation signature of human fetal alcohol spectrum disorder. *Epigenetics Chromatin, 9*, 81–101.

Powell, N. D., Sloan, E. K., Bailey, M. T., Arevalo, J. M. G., Miller, G. E., Chen, E., ... Cole, S. W. (2013). Social stress up-regulates inflammatory gene expression in the leukocyte transcriptome via β-adrenergic induction of myelopoiesis. *Proceedings of the National Academy of Sciences of the United States of America, 110*, 16574–16579.

Ramsahoye, B. H., Biniszkiewicz, D., Lyko, F., Clark, V., Bird, A. P., & Jaenisch, R. (2000). Non-CpG methylation is prevalent in embryonic stem cells and may be mediated by DNA methyltransferase 3a. *Proceedings of the National Academy of Sciences, 97*, 5237–5242.

Rando, O. J. (2012). Combinatorial complexity in chromatin structure and function: Revisiting the histone code. *Current Opinion in Genetics & Development, 22*, 148–155.

Rangasamy, D., Berven, L., Ridgway, P., & Tremethick, D. J. (2003). Pericentric heterochromatin becomes enriched with H2A.Z during early mammalian development. *EMBO Journal, 22*, 1599–1607.

Ranjan, A., Mizuguchi, G., FitzGerald, P. C., Wei, D., Wang, F., Huang, Y., ... Wu, C. (2013). Nucleosome-free region dominates histone acetylation in targeting SWR1 to promoters for H2A.Z replacement. *Cell, 154*, 1232–1245.

Rastan, S. (1983). Non-random X-chromosome inactivation in mouse X-autosome translocation embryos--location of the inactivation centre. *Journal of Embryology and Experimental Morphology, 78*, 1–22.

Rhee, I., Jair, K. W., Yen, R. W., Lengauer, C., Herman, J. G., Kinzler, K. W., ... Schuebel, K. E. (2000). CpG methylation is maintained in human cancer cells lacking DNMT1. *Nature, 404*, 1003–1007.

Rinn, J. L., Kertesz, M., Wang, J. K., Squazzo, S. L., Xu, X., Brugmann, S. A., ... Chang, H. Y. (2007). Functional demarcation of active and silent chromatin domains in human HOX loci by non-coding RNAs. *Cell, 129*, 1311–1323.

Robertson, K. D., Uzvolgyi, E., Liang, G., Talmadge, C., Sumegi, J., Gonzales, F. A., & Jones, P. A. (1999). The human DNA methyltransferases (DNMTs) 1, 3a and 3b: Coordinate mRNA expression in normal tissues and overexpression in tumors. *Nucleic Acids Research, 27*, 2291–2298.

Robinson, P. J. J., Fairall, L., Huynh, V. A. T., & Rhodes, D. (2006). EM measurements define the dimensions of the "30-nm" chromatin fiber: Evidence for a compact, interdigitated structure. *Proceedings of the National Academy of Sciences, 103*, 6506–6511.

Rodgers, A. B., Morgan, C. P., Leu, N. A., & Bale, T. L. (2015). Transgenerational epigenetic programming via sperm microRNA recapitulates effects of paternal stress. *Proceedings of the National Academy of Sciences, 112*, 13699–13704.

Romens, S. E., Mcdonald, J., Svaren, J., & Pollak, S. D. (2015). Associations between early life stress and gene methylation in children. *Child Development, 86*, 303–309.

Ropers, H. H. (2010). Genetics of early onset cognitive impairment. *Annual Review of Genomics and Human Genetics, 11*(X), 161–187.

Rozhdestvensky, T. S., Robeck, T., Galiveti, C. R., Raabe, C. A., Seeger, B., Wolters, A., ... Skryabin, B. V. (2016). Maternal transcription of non-protein coding RNAs from the PWS-critical region rescues growth retardation in mice. *Science Reports, 6*, 20398, October 2015.

Ruhl, D. D., Jin, J., Cai, Y., Swanson, S., Florens, L., Washburn, M. P., ... Chrivia, J. C. (2006). Purification of a human SRCAP complex that remodels chromatin by incorporating the histone variant H2A.Z into nucleosomes. *Biochemistry, 45*, 5671–5677.

Rutter, M. (2010). Gene–environment interplay. *Depress Anxiety, 27*, 1–4.

Sadakierska-Chudy, A., Kostrzewa, R. M., & Filip, M. (2014). A comprehensive view of the epigenetic landscape part I: DNA methylation, passive and active DNA demethylation pathways and histone variants. *Neurotoxicity Research*, *27*, 84–97.

Sado, T., Fenner, M. H., Tan, S. S., Tam, P., Shioda, T., & Li, E. (2000). X inactivation in the mouse embryo deficient for Dnmt1: Distinct effect of hypomethylation on imprinted and random X inactivation. *Developmental Biology*, *225*, 294–303.

Saitou, M., Barton, S. C., & Surani, M. A. (2002). A molecular programme for the specification of germ cell fate in mice. *Nature*, *418*, 293–300.

Sakashita, A., Kobayashi, H., Wakai, T., Sotomaru, Y., Hata, K., & Kono, T. (2014). Dynamics of genomic 5-hydroxymethylcytosine during mouse oocyte growth. *Genes Cells*, *19*, 629–636.

Sánchez, F., & Smitz, J. (2012). Molecular control of oogenesis. *Biochimica et Biophysica Acta—Molecular Basis of Disease*, 1896–1912.

Santiago, M., Antunes, C., Guedes, M., Sousa, N., & Marques, C. J. (2014). TET enzymes and DNA hydroxymethylation in neural development and function—How critical are they? *Genomics*, *104*, 334–340.

Santos, F., Hendrich, B., Reik, W., & Dean, W. (2002). Dynamic reprogramming of DNA methylation in the early mouse embryo. *Developmental Biology*, *241*, 172–182.

Sarmento, O. F., Digilio, L. C., Wang, Y., Perlin, J., Herr, J. C., Allis, C. D., & Coonrod, S. A. (2004). Dynamic alterations of specific histone modifications during early murine development. *Journal of Cell Science*, *117*(Pt 19), 4449–4459.

Sato, M., Kimura, T., Kurokawa, K., Fujita, Y., Abe, K., Masuhara, M., ... Nakano, T. (2002). Identification of PGC7, a new gene expressed specifically in preimplantation embryos and germ cells. *Mechanisms of Development*, *113*, 91–94.

Saxonov, S., Berg, P., & Brutlag, D. L. (2006). A genome-wide analysis of CpG dinucleotides in the human genome distinguishes two distinct classes of promoters. *Proceedings of the National Academy of Sciences of the United States of America*, *103*, 1412–1417.

Schaefer, M., Pollex, T., Hanna, K., Tuorto, F., Meusburger, M., Helm, M., & Lyko, F. (2010). RNA methylation by Dnmt2 protects transfer RNAs against stress-induced cleavage. *Genes & Development*, *24*, 1590–1595.

Schauer, T., Schwalie, P. C., Handley, A., Margulies, C. E., Flicek, P., & Ladurner, A. G. (2013). CAST-ChIP maps cell-type-specific chromatin states in the drosophila central nervous system. *Cell Reports*, *5*, 271–282.

Schuebeler, D. (2015). Function and information content of DNA methylation. *Nature*, *517*, 321–326.

Seisenberger, S., Andrews, S., Krueger, F., Arand, J., Walter, J., Santos, F., ... Reik, W. (2012). The dynamics of genome-wide DNA methylation reprogramming in mouse primordial germ cells. *Molecular Cell*, *48*, 849–862.

Shafik, A., Schumann, U., Evers, M., Sibbritt, T., & Preiss, T. (2016). The emerging epitranscriptomics of long noncoding RNAs. *Biochimica et Biophysica Acta—Gene Regulatory Mechanisms*, 59–70.

Sharpley, C. F., Palanisamy, S. K. A., Glyde, N. S., Dillingham, P. W., & Agnew, L. L. (2014). An update on the interaction between the serotonin transporter promoter variant (5-HTTLPR), stress and depression, plus an exploration of non-confirming findings. *Behavioural Brain Research*, *273*, 89–105.

Shen, T., Ji, F., & Jiao, J. (2015). Epigenetics: Major regulators of embryonic neurogenesis. *Science Bulletin*, *60*, 1734–1743.

Shifman, S., Bronstein, M., Sternfeld, M., Pisanté, A., Weizman, A., Reznik, I., ... Darvasi, A. (2004). COMT: A common susceptibility gene in bipolar disorder and schizophrenia. *American Journal of Medical Genetics*, *128B*, 61–64.

Shirane, K., Toh, H., Kobayashi, H., Miura, F., Chiba, H., Ito, T., ... Sasaki, H. (2013). Mouse oocyte methylomes at base resolution reveal genome-wide accumulation of non-CpG methylation and role of DNA methyltransferases. *PLoS Genetics*, 9.

Shogren-Knaak, M., Ishii, H., Sun, J. M., Pazin, M. J., Davie, J. R., & Peterson, C. L. (2006). Histone H4-K16 acetylation controls chromatin structure and protein interactions. *Science (80–)*, *311*, 844–847.

Shukla, S., Kavak, E., Gregory, M., Imashimizu, M., Shutinoski, B., Kashlev, M., . . . Oberdoerffer, S. (2011). CTCF-promoted RNA polymerase II pausing links DNA methylation to splicing. *Nature*, 74–79.

Silva, J., Mak, W., Zvetkova, I., Appanah, R., Nesterova, T. B., Webster, Z., . . . Brockdorff, N. : (2003). Establishment of histone H3 methylation on the inactive X chromosome requires transient recruitment of Eed-Enx1 polycomb group complexes. *Developmental Cell*, 481–495.

Simon, L., Castillo, J., Oliva, R., & Lewis, S. E. M. (2011). Relationships between human sperm protamines, DNA damage and assisted reproduction outcomes. *Reproductive BioMedicine Online*, *23*, 724–734.

Smallwood, S. A., Tomizawa, S.-I., Krueger, F., Ruf, N., Carli, N., Segonds-Pichon, A., . . . Kelsey, G. (2011). Dynamic CpG island methylation landscape in oocytes and preimplantation embryos. *Nature Genetics*, *43*, 811–814.

Smith, Z. D., Chan, M. M., Mikkelsen, T. S., Gu, H., Gnirke, A., Regev, A., & Meissner, A. (2012). A unique regulatory phase of DNA methylation in the early mammalian embryo. *Nature*, *484*, 339–344.

Smith, Z. D., & Meissner, A. (2013). DNA methylation: Roles in mammalian development. *Nature Reviews Genetics*, *14*, 204–220.

Smolle, M., & Workman, J. L. (2013). Transcription-associated histone modifications and cryptic transcription. *BBA Gene Regulatory Mechanisms*, *1829*, 84–97.

Soeno, Y., Taya, Y., Stasyk, T., Huber, L. A., Aoba, T., & Hüttenhofer, A. (2010). Identification of novel ribonucleo-protein complexes from the brain-specific snoRNA MBII-52. *RNA*, *16*, 1293–1300.

Soubry, A., Hoyo, C., Jirtle, R. L., & Murphy, S. K. (2014). A paternal environmental legacy: Evidence for epigenetic inheritance through the male germ line. *BioEssays*, *36*, 359–371.

Soubry, A., Murphy, S. K., Wang, F., Huang, Z., Vidal, A. C., Fuemmeler, B. F., . . . Hoyo, C. (2013). Newborns of obese parents have altered DNA methylation patterns at imprinted genes. *International Journal of Obesity (London)*, *39*, 650–657.

Soubry, A., Schildkraut, J. M., Murtha, A., Wang, F., Huang, Z., Bernal, A., . . . Hoyo, C. (2013). Paternal obesity is associated with IGF2 hypomethylation in newborns: Results from a Newborn Epigenetics Study (NEST) cohort. *BMC Medicine*, *11*, 29.

Spadaro, P. A., & Bredy, T. W. (2012). Emerging role of non-coding RNA in neural plasticity, cognitive function, and neuropsychiatric disorders. *Frontiers in Genetics*, *3*(July), 1–16.

Spadaro, P. A., Flavell, C. R., Widagdo, J., Ratnu, V. S., Troup, M., Ragan, C., . . . Bredy, T. W. (2015). Long noncoding RNA-directed epigenetic regulation of gene expression is associated with anxiety-like behavior in mice. *Biological Psychiatry*, *78*, 848–859.

Stouder, C., & Paoloni-Giacobino, A. (2010). Transgenerational effects of the endocrine disruptor vinclozolin on the methylation pattern of imprinted genes in the mouse sperm. *Reproduction*, *139*, 373–379.

Strahl, B. D., & Allis, C. D. (2000). The language of covalent histone modifications. *Nature*, *403*, 41–45.

Szenker, E., Ray-Gallet, D., & Almouzni, G. (2011). The double face of the histone variant H3.3. *Cell Research*, *21*, 421–434.

Taft, R., Hawkins, P., Mattick, J., & Morris, K. (2011). The relationship between transcription initiation RNAs and CCCTC-binding factor (CTCF) localization. *Epigenetics Chromatin*, *4*(August), 13.

Taft, R. J., Glazov, E. A., Cloonan, N., Simons, C., Stephen, S., Faulkner, G. J., . . . Mattick, J. S. (2009). Tiny RNAs associated with transcription start sites in animals. *Nature Genetics*, *41*, 572–578.

Taft, R. J., Simons, C., Nahkuri, S., Oey, H., Korbie, D. J., Mercer, T. R., . . . Mattick, J. S. (2010). Nuclear-localized tiny RNAs are associated with transcription initiation and splice sites in metazoans. *Nature Structural & Molecular Biology*, *17*, 1030–1034.

Tahiliani, M., Koh, K. P., Shen, Y., Pastor, W. A., Bandukwala, H., Brudno, Y., . . . Rao, A. (2009). Conversion of 5-methylcytosine to 5-hydroxymethylcytosine in mammalian DNA by MLL partner TET1. *Science (80–)*, *324*, 930–935.

Takagi, N., & Abe, K. (1990). Detrimental effects of two active X chromosomes on early mouse development. *Development*, *109*, 189–201.

Takagi, N., & Sasaki, M. (1975). Preferential inactivation of the paternally derived X chromosome in the extraembryonic membranes of the mouse. *Nature, 256*, 640–642.

Talbert, P. B., & Henikoff, S. (2010). Histone variants — ancient wrap artists of the epigenome. *Nature Reviews Molecular Cell Biology, 11*, 264–275.

Talbert, P. B., & Henikoff, S. (2006). Spreading of silent chromatin: Inaction at a distance. *Nature Reviews Genetics, 7*, 793–803.

Tan, J. H., Wang, H. L., Sun, X. S., Liu, Y., Sui, H. S., & Zhang, J. (2009). Chromatin configurations in the germinal vesicle of mammalian oocytes. *Molecular Human Reproduction, 15*, 1–9.

Tannan, N. B., Brahmachary, M., Garg, P., Borel, C., Alnefaie, R., Watson, C. T., ... Sharp, A. J. (2014). Dna methylation profiling in X;autosome translocations supports a role for L1 repeats in the spread of X chromosome inactivation. *Human Molecular Genetics, 23*, 1224–1236.

Tate, P. H., & Bird, A. P. (1993). Effects of DNA methylation on DNA-binding proteins and gene expression. *Current Opinion in Genetics & Development, 3*, 226–231.

Taylor, A., & Kim-Cohen, J. (2007). Meta-analysis of gene–environment interactions in developmental psychopathology. *Development and Psychopathology, 19*, 1029–1037.

Teh, A. L., Pan, H., Chen, L., Ong, M.-L., Dogra, S., Wong, J., ... Holbrook, J. D. (2014). The effect of genotype and in utero environment on interindividual variation in neonate DNA methylomes. *Genome Research, 24*, 1064–1074.

Teschendorff, A. E., West, J., & Beck, S. (2013). Age-associated epigenetic drift: Implications, and a case of epigenetic thrift? *Human Molecular Genetics, 22*, 7–15.

Tognini, P., Napoli, D., & Pizzorusso, T. (2015). Dynamic DNA methylation in the brain: A new epigenetic mark for experience-dependent plasticity. *Frontiers in Cellular Neuroscience, 9*, 611–671.

Tomizawa, S., Kobayashi, H., Watanabe, T., Andrews, S., Hata, K., Kelsey, G., & Sasaki, H. (2011). Dynamic stage-specific changes in imprinted differentially methylated regions during early mammalian development and prevalence of non-CpG methylation in oocytes. *Development, 138*, 811–820.

Torregrosa, N., Dominguez-Fandos, D., Camejo, M. I., Shirley, C. R., Meistrich, M. L., Ballesca, J. L., & Oliva, R. (2006). Protamine 2 precursors, protamine 1/protamine 2 ratio, DNA integrity and other sperm parameters in infertile patients. *Human Reproduction, 21*, 2084–2089.

Torres-Padilla, M. E., Bannister, A. J., Hurd, P. J., Kouzarides, T., & Zernicka-Goetz, M. (2006). Dynamic distribution of the replacement histone variant H3.3 in the mouse oocyte and preimplantation embryos. *International Journal of Developmental Biology, 50*, 455–461.

Tropberger, P., Pott, S., Keller, C., Kamieniarz-Gdula, K., Caron, M., Richter, F., ... Schneider, R. (2013). Regulation of transcription through acetylation of H3K122 on the lateral surface of the histone octamer. *Cell, 152*, 859–872.

Ulahannan, N., & Greally, J. M. (2015). Genome-wide assays that identify and quantify modified cytosines in human disease studies. *Epigenetics Chromatin*, 8.

van Dongen, J., Slagboom, P. E., Draisma, H. H. M., Martin, N. G., & Boomsma, D. I. (2012). The continuing value of twin studies in the omics era. *Nature Reviews Genetics, 13*, 640–653.

van Otterdijk, S. D., & Michels, K. B. (2016). Transgenerational epigenetic inheritance in mammals: How good is the evidence? *FASEB Journal, 30*, 1–9.

Venkatesh, S., Smolle, M., Li, H., Gogol, M. M., Saint, M., Kumar, S., ... Workman, J. L. (2013). Set2 methylation of histone H3 lysine 36 suppresses histone exchange on transcribed genes. *Nature, 489*, 452–455.

Venkatesh, S., & Workman, J. L. (2015). Histone exchange, chromatin structure and the regulation of transcription. *Nature Chemical Biology, 16*, 178–189.

Vertino, P. M., Yen, R. W., Gao, J., & Baylin, S. B. (1996). De novo methylation of CpG island sequences in human fibroblasts overexpressing DNA (cytosine-5-)-methyltransferase. *Molecular Cell Biology, 16*, 4555–4565.

Visscher, P. M., Hill, W. G., & Wray, N. R. (2008). Heritability in the genomics era—concepts and misconceptions. *Nature Chemical Biology, 9*, 255–266.

Visscher, P. M. (2008). Sizing up human height variation. *Nature Genetics, 40*, 489–490.

Vissers, L. E. L. M., Gilissen, C., & Veltman, J. A. (2015). Genetic studies in intellectual disability and related disorders. *Nature Reviews Genetics, 17*, 9–18.

Voigt, P., Tee, W. W., & Reinberg, D. (2013). A double take on bivalent promoters. *Genes & Development*, *27*, 1318–1338.

Waddington, C. H. (1968). Towards a theoretical biology. *Nature*, *218*, 525–527.

Wagner, K. D., Wagner, N., Ghanbarian, H., Grandjean, V., Gounon, P., Cuzin, F., & Rassoulzadegan, M. (2008). RNA induction and inheritance of epigenetic cardiac hypertrophy in the mouse. *Developmental Cell*, *14*, 962–969.

Walton, E., Hass, J., Liu, J., Roffman, J. L., Bernardoni, F., Roessner, V., ... Ehrlich, S. (2015). Correspondence of DNA methylation between blood and brain tissue and its application to schizophrenia research. *Schizophrenia Bulletin*, *42*, sbv074-.

Wan, L. Ben, & Bartolomei, M. S. (2008). Chapter 7 Regulation of imprinting in clusters: Noncoding RNAs versus insulators. *Advances in Genetics*, 207–223.

Wang, K. C., & Chang, H. Y. (2011). Molecular mechanisms of long noncoding RNAs. *Molecular Cell*, 904–914.

Wang, T., Pan, Q., Lin, L., Szulwach, K. E., Song, C. X., He, C., ... Li, X. (2012). Genome-wide DNA hydroxymethylation changes are associated with neurodevelopmental genes in the developing human cerebellum. *Human Molecular Genetics*, *21*, 5500–5510.

Weber, C. M., & Henikoff, S. (2014). Histone variants: Dynamic punctuation in transcription. *Genes & Development*, *28*, 672–682.

Weber, M., Davies, J. J., Wittig, D., Oakeley, E. J., Haase, M., Lam, W. L., & Schubeler, D. (2005). Chromosome-wide and promoter-specific analyses identify sites of differential DNA methylation in normal and transformed human cells. *Nature Genetics*, *37*, 853–862.

Weber, M., Hellmann, I., Stadler, M. B., Ramos, L., Pääbo, S., Rebhan, M., & Schübeler, D. (2007). Distribution, silencing potential and evolutionary impact of promoter DNA methylation in the human genome. *Nature Genetics*, *39*, 457–466.

Webster, K. E., O'Bryan, M. K., Fletcher, S., Crewther, P. E., Aapola, U., Craig, J., ... Scott, H. S. (2005). Meiotic and epigenetic defects in Dnmt3L-knockout mouse spermatogenesis. *Proceedings of the National Academy of Sciences*, *102*, 4068–4073.

Wei, Y.-P. P. Y., Yang, C.-R. R., Zhao, Z.-A. A., Hou, Y., Schatten, H., & Sun, Q.-Y. Y. (2014). Paternally induced transgenerational inheritance of susceptibility to diabetes in mammals. *Proceedings of the National Academy of Sciences of the United States of America*, *111*, 1873–1878.

Wickramasinghe, D., Ebert, K. M., & Albertini, D. F. (1991). Meiotic competence acquisition is associated with the appearance of M-phase characteristics in growing mouse oocytes. *Developmental Biology*, *143*, 162–172.

Widagdo, J., Zhao, Q., Kempen, M., Tan, M. C., Ratnu, V. S., Wei, W., ... Bredy, T. W. (2016). Experience-dependent accumulation of N6-methyladenosine in the prefrontal cortex is associated with memory processes in mice. *Journal of Neuroscience*, *36*, 6771–6777.

Williams, J. M., Beck, T. F., Pearson, D. M., Proud, M. B., Sau, W. C., & Scott, D. A. (2009). A 1q42 deletion involving DISC1, DISC2, and TSNAX in an autism spectrum disorder. *American Journal of Medical Genetics Part A*, *149*, 1758–1762.

Williams, S. K., Truong, D., & Tyler, J. K. (2008). Acetylation in the globular core of histone H3 on lysine-56 promotes chromatin disassembly during transcriptional activation. *Proceedings of the National Academy of Sciences*, *105*, 9000–9005.

Wolff, G. L., Kodell, R. L., Moore, S. R., & Cooney, C. A. (1998). Maternal epigenetics and methyl supplements affect agouti gene expression in Avy/a mice. *FASEB Journal*, *12*, 949–957.

Wong, M. M., Cox, L. K., & Chrivia, J. C. (2007). The chromatin remodeling protein, SRCAP, is critical for deposition of the histone variant H2A.Z at promoters. *Journal of Biological Chemistry*, *282*, 26132–26139.

Wossidlo, M., Nakamura, T., Lepikhov, K., Marques, C. J., Zakhartchenko, V., Boiani, M., ... Walter, J. (2011). 5-Hydroxymethylcytosine in the mammalian zygote is linked with epigenetic reprogramming. *Nature Communications*, *2*, 241.

Wu, H., & Zhang, Y. (2014). Reversing DNA methylation: Mechanisms, genomics, and biological functions. *Cell*, *156*, 45–68.

Wutz, A. (2011). Gene silencing in X-chromosome inactivation: Advances in understanding facultative heterochromatin formation. *Nature Reviews Genetics*, *12*, 542–553.

Yan, Z., Hu, H. Y., Jiang, X., Maierhofer, V., Neb, E., He, L., . . . Khaitovich, P. (2011). Widespread expression of piRNA-like molecules in somatic tissues. *Nucleic Acids Research*, *39*, 6596–6607.

Yen, K., Vinayachandran, V., & Pugh, B. F. (2013). SWR-C and INO80 chromatin remodelers recognize nucleosome-free regions near + 1 nucleosomes. *Cell*, *154*, 1246–1256.

Yokochi, T., & Robertson, K. D. (2002). Preferential methylation of unmethylated DNA by Mammalian de NovoDNA methyltransferase Dnmt3a. *Journal of Biological Chemistry*, *277*, 11735–11745.

Yu, L., Chibnik, L. B., Yang, J., McCabe, C., Xu, J., Schneider, J. A., . . . Bennett, D. A. (2016). Methylation profiles in peripheral blood CD4+ lymphocytes versus brain: The relation to Alzheimer's disease pathology. *Alzheimer's & Dementia: The Journal of the Alzheimer's Association*, *12*, 942–951.

Zannas, A. S., Provençal, N., & Binder, E. B. (2015). Epigenetics of posttraumatic stress disorder_ current evidence, challenges, and future directions. *BPS*, *78*, 327–335.

Zhang, D., Cheng, L., Badner, J. A., Chen, C., Chen, Q., Luo, W., . . . Liu, C. (2010). Genetic control of individual differences in gene-specific methylation in human brain. *The American Journal of Human Genetics*, 411–419.

Zhao, C., Sun, G., Li, S., & Shi, Y. (2009). A feedback regulatory loop involving microRNA-9 and nuclear receptor TLX in neural stem cell fate determination. *Nature Structural & Molecular Biology*, *16*, 365–371.

Ziats, M. N., & Rennert, O. M. (2013). Aberrant expression of long noncoding RNAs in Autistic brain. *Journal of Molecular Neuroscience*, *49*, 589–593.

Ziller, M. J., Müller, F., Liao, J., Zhang, Y., Gu, H., Bock, C., . . . Meissner, A. (2011). Genomic distribution and inter-sample variation of non-CpG methylation across human cell types. *PLoS Genetics*, 7, e1002389-15.

Zovkic, I. B., Paulukaitis, B. S., Day, J. J., Etikala, D. M., & Sweatt, J. D. (2015). Histone H2A.Z subunit exchange controls consolidation of recent and remote memory. *Nature*, *515*, 582–586.

Zuccotti, M., Garagna, S., Merico, V., Monti, M., & Redi, C. A. (2005). Chromatin organisation and nuclear architecture in growing mouse oocytes. *Molecular and Cellular Endocrinology*, *234*, 11–17.

Zuk, O., Hechter, E., Sunyaev, S. R., & Lander, E. S. (2012). The mystery of missing heritability: Genetic interactions create phantom heritability. *Proceedings of the National Academy of Sciences of the United States of America*, *109*, 1193–1198.

PART III

Behavior

CHAPTER 8

Visual Systems

Daphne Maurer[1,2] and Terri L. Lewis[1,2]
[1]McMaster University, Hamilton, ON, Canada
[2]The Hospital for Sick Children, Toronto, ON, Canada

The child's visual perception is limited by immaturities in the nervous system beginning in the retina and continuing through the primary visual cortex to extrastriate visual cortex and beyond (see Fig. 8.1). At any given age during development, there are differential effects of these immaturities on different aspects of visual perception. Because the elimination of these constraints does not progress on a single developmental trajectory, it is never the case that the child's visual perception overall is a fraction of that of an adult (e.g., never half as good overall). Most—but not all—of these developmental trajectories depend on visual experience, that is, visual input is critical for tuning the neural underpinnings of most aspects of visual perception. When the input is missing, even for a brief period of time, the refinement fails to occur.

For this chapter, we have chosen visual capabilities that illustrate these points. We will begin by describing retinal limitations and their effect on the development of visual acuity. We will continue by describing limitations in the primary visual cortex, illustrated by their impact on the control of eye movements despite sufficient function to mediate orientation perception. The next section concerns limitations in both the dorsal and ventral streams of the visual cortex, illustrated by perceptual capabilities that require integration of local signals processed in the primary visual cortex by higher level visual areas to yield a perception of global motion and global form, respectively. Throughout, we summarize the evidence on neural limitations and relate it to developmental changes in visual capabilities. We also consider how a period of visual deprivation from bilateral congenital cataracts affects subsequent development. The cataracts blocked all patterned visual input during infancy until they were removed surgically and the eyes fitted with compensatory optical correction, usually contact lenses. Despite treatment early in infancy, these patients had later deficits in many aspects of vision. These "sleeper effects" indicate that early visual input—at a time when the visual nervous system is very immature—sets up the neural architecture for later refinement. In its absence, those later improvements fail to occur. In the last section, we consider general principles that emerge about brain development and visual perception.

The Neurobiology of Brain and Behavioral Development
DOI: http://dx.doi.org/10.1016/B978-0-12-804036-2.00008-X

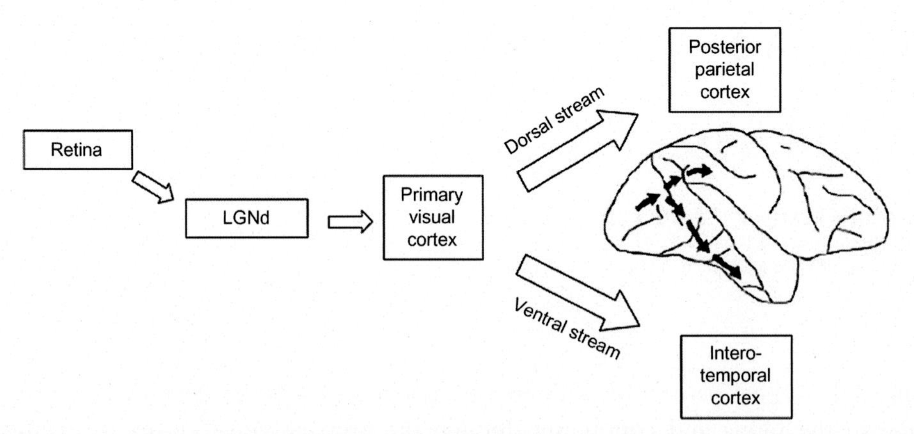

Figure 8.1 Simplified cartoon of the flow of visual information in the human adult brain from the retina to the dorsal lateral geniculate nucleus (LGNd) to the primary visual cortex and on to the extrastriate visual cortex where it divides into the dorsal visual stream projecting to the posterior parietal cortex and the ventral visual stream projecting to the infero-temporal cortex.

8.1 RETINAL LIMITATIONS

The retina of the human newborn is very immature, especially the central fovea, the portion of the retina that mediates the perception of fine detail. In the human adult, the central fovea forms a pit containing only densely packed cones. The outer segments are the same width as the inner segments (1.9 μm where 1 μm = 0.001 mm) resulting in very efficient transmission of light from the inner to the outer segment. The newborn, in contrast, has only the very beginnings of a foveal pit. Cones in the central retina are intermingled with, and covered by, outer ganglion cells (Yuodelis & Hendrickson, 1986). Moreover, cones in the central retina are much wider and shorter than those of the adult. Specifically, between birth and adulthood, the width of a foveal cone inner segment narrows on average from 6.5 to 1.9 μm, the length of the inner segment grows on average from 8.7 to 32 μm, and the length of outer segment grows on average from 3.1 to 45 μm (Yuodelis & Hendrickson, 1986). Moreover, in the central retina, the inner segments are more than five times wider than the outer segments at birth, but equivalent in width when mature. Together, these factors result in the newborn's central retina being 350 times less efficient than that of the adult at transmitting light (Banks & Bennett, 1988). The transition from the neonatal to the adult retina is gradual. Even at 45 months of age, immaturities in the central retina are evident: the outer segments of cones are only half as long as those of adults, the foveal pit still contains several layers of pale glia, and packing density is three-quarters of that in adults (Yuodelis & Hendrickson, 1986).

A very immature central retina at birth would be expected to coincide with very immature visual acuity. In fact, that is the case. Acuity in infants has been measured

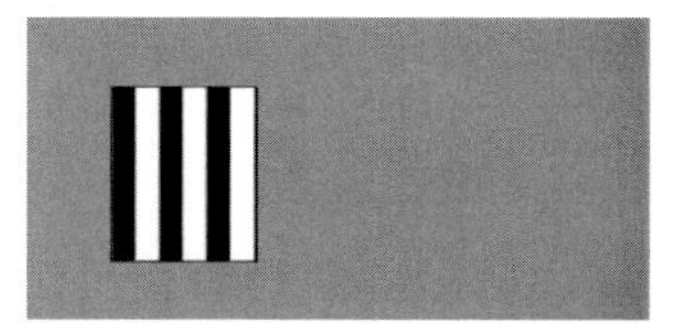

Figure 8.2 An acuity card with wide stripes typical of those used with the Acuity Card Procedure. Each card contains a patch of black and white stripes on a plain gray background of equal mean luminance. The stripes are placed to the right or left of a central peephole through which a tester watches the baby. The stripes become narrower and narrower on subsequent cards.

typically with the Acuity Card Procedure (McDonald et al., 1985; Teller, McDonald, Preston, Serbis, & Dobson, 1986). The infant is shown a card with wide black and white stripes on one side (to the right or left of center) against a plain gray background of equal mean luminance similar to that shown in Fig. 8.2. Because infants have a looking preference for something patterned over something gray (Fantz, Ordy, & Udelf, 1962), they will look to the side with the stripes if they can see them. Once a tester decides that the baby can see the stripes, because the baby looks at them or smiles, or gives any consistent indication, subsequent cards are shown with stripes that get narrower and narrower, sometimes to the right and sometimes to the left of center. The measure of acuity is the finest stripes for which the infant shows a response that the tester can pick up reliably. This method has been used to show that acuity is about 40 times worse in the newborn infant than in the adult, with an eightfold improvement over the next 6 months, and gradual improvement thereafter, until acuity reaches adult levels after 4 years of age (Mayer et al., 1995). Studies of spatial contrast sensitivity indicate that acuity is fully adult-like around age 7 (Ellemberg, Lewis, Liu, & Maurer, 1999).

Through a series of calculations based on computational modeling, anatomical data, and various assumptions, Banks and Bennett (1986) concluded that retinal immaturities underlie much, but not all, of infants' immature acuity. Another way to test this hypothesis is to test grating acuity developmentally in humans who had a normal retina but a compromised higher visual system. We did just that by testing the grating acuity of children born with dense cataracts in both eyes. These patients were born with cataracts so dense that they had no pattern vision whatsoever until the cataracts were removed during the first year of life by surgically removing the natural lens of the eye and replacing it with a suitable optical correction, typically a contact lens. Anatomical and physiological studies of monkeys that had been comparably deprived (reviewed in Boothe, Dobson, & Teller, 1985), suggest that cataract-reversal patients have an entirely normal retina but a compromised visual cortex (see below for a detailed discussion of cortical abnormalities). If normal grating acuity depends only on a normal retina, cataract-reversal patients should develop normal grating acuity.

To find out, we tested the grating acuity of a group of bilaterally deprived infants within 10 minutes of their receiving their first contact lenses, after 1 hour of waking time, and 1 month later (Maurer, Lewis, Brent, & Levin, 1999). When infants first received their contact lenses after surgical removal of the cataractous lens, in other words, when they first could see patterns, grating acuity was no better than that of normal newborns despite the fact that the duration of deprivation, and hence the age of the infants at test, ranged from 1 week to 9 months. However, after only 1 hour of being awake, visual acuity improved significantly—as much as the improvement that occurs over the first *month* of life during normal development. To verify that the rapid improvement was caused by the hour of visual experience rather than a practice effect, we carried out a second experiment where we followed the same protocol for the immediate and 1-hour tests except that we patched one of the treated eyes so that, for each patient, one eye received the hour of visual experience while the other eye did not. The improvement in visual acuity was far greater in the experienced eye than in the patched eye, showing that the improvement was caused mainly by the onset of visual experience. Vision continued to improve at a faster than normal rate over the next month and, in a comparable group of cataract-reversal patients, acuity was entirely normal in 85% of patients by 12 months of age (Lewis, Maurer, & Brent, 1995). Thus, it appears that visual experience is necessary to trigger the onset of acuity development. But the visual system is experience-expectant, resulting in a rapid improvement in acuity once the appropriate experience occurs. Moreover, it seems that a normal retina is sufficient to mediate normal visual acuity up to 12 months of age. However, after 1 year of age, the improvement in acuity was slower than normal so that, by 2 years of age, the mean acuity was below normal, and by 3 years of age, 74% of treated eyes had acuity outside the normal range (Lewis et al., 1995). That deficit persists into adulthood (Ellemberg, Lewis, Maurer, Brar, & Brent, 2002). Thus, it is tempting to conclude that a normal retina alone is not sufficient to mediate normal grating acuity after 1 year of age, and the refinements in acuity that occur after 1 year of age depend also on cortical involvement. Indirect support comes from evidence that cataracts with onset after 1 year of age also lead to acuity deficits, with the critical period for damage from visual deprivation not ending until about 10 years of age, several years after adult-like acuity is achieved. That pattern suggests that visual input refines the cortical circuitry underlying acuity from infancy throughout the 7 years of normal development and then consolidates or crystalizes it for several years thereafter.

8.2 LIMITATIONS OF THE PRIMARY VISUAL CORTEX

The next way station where visual perception is limited during development is the primary visual cortex. The primary visual cortex is very immature at birth. Its volume increases fourfold between birth and 4 months of age and continues to increase gradually until 4 years of age, at which time it is five times that of the newborn. Volume then gradually decreases by about 20% to reach the adult size by 11 years of age (reviewed in Huttenlocher, 1990). Synaptic density also reaches adult values at about 11 years of age by doubling during the first year of life and then decreasing by half over the next 10 years (reviewed in Huttenlocher, 1990). Moreover, neurons of the lateral geniculate nucleus (LGN) that feed into the primary visual cortex are smaller than those of adults until 2 years of age (Brauer, Leuba, Garey, & Winkelmann, 1985; Hickey, 1977), and studies of infant monkeys suggest that they are also less sensitive (Blakemore & Vital-Durand, 1986; Blakemore, 1990; Movshon & Kiorpes, 1993). Thus, it appears that the entire geniculo-striate pathway is very immature, especially during the first year or two of life.

One consequence arising from the immaturity of the visual cortex (and/or its input to subcortical eye movement areas) is the asymmetry of optokinetic nystagmus (OKN). OKN is a series of reflexive eye movements elicited by a repetitive pattern, such as stripes, moving across the visual field. The eyes alternately follow the movement of a stripe and then quickly saccade back to pick up fixation on another stripe. When visually normal adult humans, monkeys, or cats view moving stripes binocularly, OKN is symmetrical: it can be elicited easily when stripes move leftward or rightward. The same is true with monocular viewing: OKN can be elicited easily whether stripes move from the temporal visual field toward the nasal visual field or when they move in the opposite direction (nasally to temporally) (e.g., Braun & Gault, 1969; Lewis, Maurer, Smith, & Haslip, 1992; Pasik & Pasik, 1964). However, cats and monkeys made dependent only on subcortical pathways because of lesions of the primary visual cortex show normal monocular OKN for stripes moving temporally to nasally but no OKN for stripes moving in the opposite direction (Wood, Spear, & Braun, 1973; Zee, Tusa, Herdman, Butler, & Gucer, 1987). OKN to stripes moving nasally to temporally depends on an intact projection from LGN to primary visual cortex and then down to subcortical areas known to be involved in the mediation of OKN, namely the dorsal terminal nucleus of the accessory optic tract and the nucleus of the optic tract in the pretectum (reviewed in Maurer & Lewis, 1993). Measurements of the development of symmetrical OKN tested monocularly provide a window on the functional integrity of this pathway.

OKN eye movements can be elicited in humans even at birth, both when the stripes move from left to right or from right to left, providing that the infant is looking with both eyes open and that the width of stripe is above threshold, that is, for the newborn, quite wide (Atkinson, 1979; Krementizer, Vaughan, Kurtzberg, & Dowling, 1979; van Hof-van Duin, 1978). However, with monocular viewing, OKN is asymmetrical: it can be elicited easily when stripes move from the temporal visual field toward the nasal visual field but not at all when stripes move in the opposite direction (nasally to temporally) (e.g., Atkinson, 1979; Lewis et al., 1992; van Hof-van Duin, 1978). Not until 3 to 6 months of age does monocular OKN become symmetrical for wide stripes (Atkinson & Braddick, 1981; Atkinson, 1979; Lewis et al., 1992; Naegele & Held, 1982, 1983; Roy, Lachapelle, & Lepore, 1989; van Hof-van Duin & Mohn, 1984, 1985, 1986). However, even at 24 months of age, a small asymmetry exists so that the narrowest stripes eliciting OKN nasally to temporally are a bit wider than those eliciting OKN temporally to nasally (Lewis, Maurer, Chung, Holmes-Shannon, & Van Schaik, 2000). These results suggest that the visual cortex or pathways through it to the pretectum are very immature during early infancy and still not fully mature by 24 months of age. The resulting asymmetrical OKN could be mediated entirely by subcortical pathways from retina to the pretectum that favor OKN to stripes moving temporally to nasally (Hoffmann, 1989; reviewed in Lewis et al., 2000). Alternatively, asymmetrical OKN could be mediated by a cortical pathway that is especially immature at eliciting OKN to temporalward motion (reviewed in Lewis et al., 2000).

Our successful quantification of OKN asymmetry in individual normal infants suggests that this may be a useful tool for evaluating the degree of cortical insult caused by infantile ocular disorders such as cataract. We tested the symmetry of monocular OKN in 51 patients who had been deprived of patterned visual input at some point during childhood because of dense cataracts in one or both eyes (Maurer, Lewis, & Brent, 1989). OKN was asymmetrical in virtually every eye tested of the 23 patients born with cataracts in one eye and of the six patients born with cataracts in both eyes, regardless of the duration of deprivation (ranging from 1.4 to 29 months), regardless of acuity (ranging from 20/20 to light perception), and in unilateral cases, regardless of the patching regime. To determine the sensitive period for developing symmetrical OKN, we also measured the symmetry of OKN in patients born with normal eyes who later were deprived of pattern vision because of a trauma to one eye causing a cataract ($n = 13$) or who developed dense cataracts in both eyes ($n = 9$). We found that OKN was asymmetrical if the cataract occurred before 18–30 months of age, but not if its onset was later. These results suggest that patients who are deprived of patterned visual input beginning any time during the first 1–2 years of life, that is, any time during the normal period of development, suffer from abnormalities in the primary visual cortex and/or in its projection to the pretectum.

Note however that the cortex is functional at least to some extent right from birth. Even newborns can discriminate oblique stripes oriented to the left from those oriented to the right (Atkinson, Hood, Wattam-Bell, Anker, & Tricklebank, 1988; Slater, Morison, & Somers, 1988), and this discrimination is based on orientation cues rather than on differences in local contrast between stimuli (Maurer & Martello, 1980). Studies of monkeys show that the primary visual cortex is the first structure in the geniculostriate pathway capable of mediating such a discrimination (e.g., Hubel & Wiesel, 1968). By 4 weeks (youngest age tested), infants show changes in visually elicited cortical brain waves (known as visually evoked potentials or VEPs) when the orientation of a striped stimulus changes (Braddick, Birtles, Wattam-Bell, & Atkinson, 2005). However, the refinement of orientation tuning is not complete until middle childhood because, even at 5 years of age, the minimum tilt that can be discriminated from vertical is 4–5 times larger than in adults (Lewis, Kingdon, Ellemberg, & Maurer, 2007).

8.3 LIMITATIONS BEYOND THE PRIMARY VISUAL CORTEX

After information is processed in the primary visual cortex, it proceeds to higher-order visual areas along two streams: a dorsal stream and a ventral stream. The dorsal stream is comprised of middle temporal area (MT/MST complex, also sometimes referred to as V5), V3a, and other inputs to the posterior parietal cortex. The ventral stream is comprised of V3v, V4, and other inputs into the inferior temporal cortex. Although there are many interactions between the two streams, they are broadly specialized for detecting "where" a stimulus is located (dorsal: direction of motion, location, integration with action) and "what" the stimulus is (ventral: the identity of objects and faces). What is common across the two streams is that at higher levels of the visual nervous system the size of receptive fields increases, most neurons receive inputs from both eyes, and, as a result, information is increasingly integrated across time and space. Thus, detailed information processed in small regions of space at the level of the primary visual cortex is integrated to support more global percepts. Here we will illustrate the role of brain development in higher order visual cortex in mediating improvements in the detection of global motion (dorsal stream) and global form (ventral stream).

8.4 DORSAL STREAM LIMITATIONS: EXAMPLE OF GLOBAL MOTION

Sensitivity to local motion depends on neurons in the primary visual cortex that are tuned to direction. These neurons have small receptive fields and each one responds only to information in a small receptive field, often connected to only one eye (Movshon, 1990). Sensitivity to the overall, or global, direction of motion requires additional processing in the dorsal stream involving especially the MT/MST complex (also referred to as V5) where there is a convergence of inputs across receptive fields and eyes onto neurons turned to direction and speed (Hess, Hutchinson, Ledgeway, & Mansouri, 2007; Maunsell & Van Essen, 1983; Newsome & Pare, 1988). Higher areas in the dorsal stream (V3, TOC, LO, V6) also respond selectively to coherent global motion (Biagi, Crespi, Tosetti, & Morrone, 2015). Studies of the monkey indicate that neurons in MT already show directional-selectivity similar to that seen in adults by 1 week of age (youngest tested) (Movshon, Rust, Kohn, Kiorpes, & Hawken, 2004).

In nonhuman primates, there is converging evidence for the earlier development of MT/MST and other parts of the dorsal stream than of the ventral stream. Thus, in the macaque, local glucose utilization reaches adult levels throughout the dorsal pathway by 3 months of age, at which point the levels throughout the ventral pathway are still quite immature (Distler, Bachevalier, Kennedy, Mishkin, & Ungerleider, 1996). Correspondingly, behavioral studies indicate earlier evidence of the perception of global motion than of global form: when tested at 10–11 weeks of age, monkeys can detect the global direction of both translational and rotational motion but most fail comparable tests for detecting the structure of global form even in the concentric patterns that are easier for adult monkeys to detect than are linear patterns (Kiorpes, Price, Hall-Haro, & Movshon, 2012). Even at 15 weeks, only half of the tested monkeys were able to detect structure in the concentric patterns. Thereafter, there is steady improvement in threshold sensitivity for both global motion and global form, with a greater improvement in sensitivity for motion than form.

Anatomical studies in marmoset monkeys, like those in the macaque, suggest especially early development of MT/MST. Thus, labeling of cortical cells for neurofilament on the day of birth reveals activity in only V1 and an area identified by anatomical landmarks as MT (Bourne & Rosa, 2006). Subsequently, levels increase at the same rate in V1 and MT (or even slightly faster in MT), with gradual emergence, sequentially, in V2, V3, V4, and inferior temporal cortex, that is, the ventral stream. A similar pattern emerged in a study of the calcium binding proteins calbindin and paravalbumin and the neurofilament specifically of pyramidal neurons (Mundinano, Kwan, & Bourne, 2015): even before birth, all three were evident in V1, MT/MST, and DM (a dorsal medial area) but nowhere in the ventral stream. Based on the patterns of postnatal development, the authors speculate that MT and DM drive development of the dorsal stream, whereas the ventral stream's development occurs hierarchically beginning with V1 and continuing, in sequence, V2, V3, V4, and the inferotemporal cortex.

The largely parallel development of V1 and MT/MST in the marmoset summarized in the previous paragraph leads to the obvious hypothesis that, from birth, neurons in MT/MST respond to input from V1 neurons. However, an anatomical tracer study suggests an alternative possibility (Warner, Kwan, & Bourne, 2012): during the first few weeks after birth, the predominant input to MT/MST is via the pulvinar, which itself receives direct input from the retina. With age, the pulvinar inputs recede and the V1 inputs increase in magnitude. These results raise the possibility that early behavioral sensitivity to global motion is mediated by a pathway that bypasses V1.

Human infants do not show any evidence of cortical directional selectivity in VEP signals until about 7 weeks of age, several weeks after orientation selectivity is evident

(Braddick et al., 2005). Initially the selective VEP responses are evident only for elements moving at about 6 deg/s; with age, the effective velocity range extends to slower and faster speeds. Those data are consistent with behavioral evidence that infants fail tests of motion selectivity before about 7 weeks of age. In these studies, infants were given a choice of looking at a form defined by dots moving in a direction opposite the dots forming the background, contrasted with a field of uniformly oscillating dots. Infants show no preference for the side with the motion-defined form until about 7–8 weeks of age (Wattam-Bell, 1996a,b), with the preference first evident for velocities of about 5–10 deg/s and slowly expanding to higher and lower velocities (reviewed in Braddick, Atkinson, & Wattam-Bell, 2003). However, even at 9–12 weeks, infants show no evidence of discriminating between opposite directions of uniform motion (Armstrong, Maurer, Ellemberg, & Lewis, 2011; Wattam-Bell, 1996a).

In contrast, even by 7 weeks of age, infants can integrate motion signals into a global percept. This is surprising since one would have thought that integrating motion signals into a global percept would develop well after the ability to process the direction of motion in a simple stimulus such as stripes. However, the underlying processing for the integration of motion signals may be different in infants and adults. For example, a recent study using functional Magnetic Resonance Imaging (fMRI) of 7-week-old infants looked for differential cortical activation for coherent versus randomly moving dots. There was selective activation in both infants and adults throughout the dorsal stream, including areas identified anatomically as MT and V3, LO (lateral occipital cortex), TOS (transverse occipital sulcus), and V6 (Biagi et al., 2015). However, the low correlation of MT activity with V1 activity in the fMRI data and in separate resting state data suggested a different pattern of organization in the infants that might involve input to MT that bypasses V1, as has been documented in the immature marmoset (see above). Reorganization of the neural underpinnings for the perception of global motion is also suggested by a VEP study of global motion versus global form at 4–5 months of age: infants' responses to the two types of stimuli were distinct (and stronger to global motion than global form—see below) but their topography was different from that in the adult group (Wattam-Bell et al., 2010). From the locations, the authors speculate that global motion might be mediated in MT for infants but in the higher, more medial dorsal areas (V3, V6) in adults.

Little is known about the development after infancy of the cortical networks for the perception of global motion perception, except that individual differences in sensitivity to global motion among children 5–12 years old are correlated positively with the surface area of the parietal lobe and negatively with that of the occipital lobe (Braddick et al., 2016b) and in a complex pattern with measures of signal transmission (fractional anisotropy) of the superior longitudinal fasciculus that connects the parietal lobe to more anterior areas involved in attention and decision making (Braddick et al., 2016a). Behaviorally, sensitivity improves throughout early childhood, such that older children can detect the global direction of movement with an increasing percentage of randomly moving dots (reviewed in Hadad, Schwartz, Maurer, & Lewis, 2015). When sensitivity is adult-like depends on whether the moving elements have a limited lifetime (appear for only a short time before being replaced by another moving element) to eliminate local motion cues, on the speed at which the elements move, and on their density. With some parameters, children perform as well as adults as early as 3 years of age (Parrish, Giaschi, Boden, & Dougherty, 2005); with others, they continue to improve until about 12 years of age (Hadad, Maurer, & Lewis, 2011).

Patients treated for bilateral congenital cataracts have only small deficits in the perception of local motion, especially at slow velocities (Ellemberg et al., 2005) and can perceive the direction of global motion when most of the elements move in the same direction. However, when coherence is reduced by having a larger percentage of the dots move in random directions, they have difficulty seeing the direction of motion and, even when tested as adults, their sensitivity is roughly five times worse than that of adults with normal eyes (Ellemberg et al., 2002; Hadad, Maurer, & Lewis, 2012). The deficits occur even when the speed is in the range where the patients have minimal deficits in perceiving local motion, even in patients with steady fixation, and even when deprivation ended during the first two months of life, that is, before the onset of sensitivity to global motion in the child with normal eyes. This is an example of a sleeper effect (Maurer, Mondloch, & Lewis, 2007): visual input during early infancy, before a capability is even manifest, is essential for the *later* development of normal function. Presumably the early input establishes, or maintains, the necessary neural architecture, likely involving MT/MST, for later refinement. One possibility is that early input plays a role in the reorganization of cortical inputs to MT/MST seen in the marmoset (Warner et al., 2012) from mainly inputs bypassing V1 (e.g., the pulvinar) to predominantly V1 inputs. That hypothesis is bolstered by evidence that long-term binocular deprivation in cats leads to functional reliance on pathways bypassing V1 (Zablocka & Zernicki, 1996; Zablocka, Zernicki, & Kosmal, 1976, 1980; Zernicki, Zablocka, & Kosmal, 1978).

This deficit in patients treated for bilateral congenital cataract is much larger than their deficit in sensitivity to global form (see below), a pattern supporting the dorsal vulnerability hypothesis described below. Consistent with extrastriate deficits, patients treated for bilateral congenital cataract, unlike controls, fail to show late components of the event related potential (ERP) that differentiate in controls between different kinds of global motion (random, inward radial, outward radial) (Segalowitz, Sternin, Lewis, Dywan, & Maurer, 2017).

Patients treated for bilateral cataracts with postnatal onset between 1 and 10 years of age have acuity deficits (see above) but entirely normal sensitivity to global motion (Ellemberg et al., 2002; see also Fine et al., 2003). That pattern indicates that there are different sensitive periods for damage from visual deprivation for different visual capabilities, as would be expected from the fact that each depends on a different cortical network. The short critical period for global motion also illustrates that a long developmental trajectory (e.g., global motion *and* acuity) does not necessarily indicate a long critical period for damage (short for global motion; long for acuity).

8.5 VENTRAL STREAM LIMITATIONS: EXAMPLE OF GLOBAL FORM

Sensitivity to global form depends on the integration of information about the orientation of local elements, likely transmitted by V1 neurons. As early as 3 weeks of age (youngest tested), VEP responses of infants respond to changes in the orientation of gratings (Braddick, 1993) and in the same age range, infants show behavioral evidence of discriminating orientation (see above). Throughout the first few months, infants are more likely to show VEP and behavioral responses to a change of orientation than to a change in the direction of moving elements (Armstrong et al., 2011; Braddick et al., 2005). However, the perception of global form requires the integration of information about the orientation of individual parts of a stimulus. Detection of global structure is mediated in the ventral stream, with sensitivity to the patterning of orientation information first evident in V3 and V4 and increased responsiveness in the LOC (lateral occipital complex) in the occipitotemporal cortex (Ostwald, Lam, Li, & Kourtzi, 2008). As summarized above, in non-human primate models, the development of these ventral structures lags behind the development of the dorsal pathway. Consistent with that evidence, at 4–5 months, human infants show reliable cortical VEP responses to changes in the direction of motion (see above), but less reliable, more diffuse responses to changes in global form: the cortical VEP of over 90% of infants showed a response to the reversal of motion direction but only half showed any differential VEP response to concentrically arranged elements versus a random arrangement (Wattam-Bell et al., 2010). Moreover, the infants' response to global form was more central and diffuse than that of adults (Wattam-Bell et al., 2010).

It is difficult to deduce the age at which infants first show behavioral evidence of detecting the structure of global form because their visual preferences and dishabituation when tested with contrasting forms can often be based on local, as opposed to global, properties of the stimulus. For example, newborns' preference for face-like arrangements of pattern elements can be largely, if not wholly, explained by low-level features like visible spatial frequencies in the upper half of the image that are congruent with the shape of the contour and contained in features that are dark on a light background (Cassia, Turati, & Simion, 2004; Cassia, Valenza, Simion, & Leo, 2008; Farroni et al., 2005; Mondloch et al., 1999). In addition, these early preferences may be mediated by the superior colliculus (Morton & Johnson, 1991), which may be more mature at birth than the primary visual cortex. Over the next few months, face detection improves so that by 3 months infants prefer face-like structures even when low-level features cannot explain the preference (Mondloch et al., 1999).

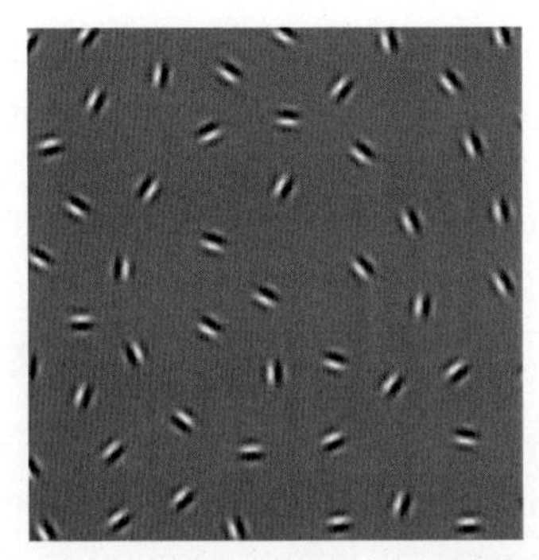

Figure 8.3 Gabor patches arranged as a circle within a background of randomly ordered Gabor patches (after Gerhardstein et al., 2004, *Vision Research*, *44*(26), 2982, Fig. 8.1).

This response to face structure is likely cortically mediated and depends on visual experience during the first three months of life: in infants whose visual experience was shortened by a period of visual deprivation caused by bilateral congenital cataracts, the preference is not evident at 3 months of age and only emerges after several months of "catch-up" visual experience (Mondloch et al., 2013).

As with faces, there is clear evidence of sensitivity to global structure in the arrangement of dots and lines at 3–4 months of age. For example, at 12 weeks, but not 10 weeks, infants dishabituate to a pattern of randomly oriented lines after habituation to a concentric arrangement (Curran, Braddick, Atkinson, & Wattam-Bell, 2000). At this age, infants can also discriminate a completely random pattern of Gabor patches from a stimulus with Gabor patches forming a circle that are arranged on a background of randomly ordered Gabors like that shown in Fig. 8.3 (Gerhardstein, Kovacs, Ditre, & Feher, 2004).

After infancy, sensitivity to global form increases gradually for many years, such that children become increasingly able to detect the form in higher levels of background noise. For example, Gunn et al. (2002) measured the highest level of background noise (randomly oriented line segments) that children can tolerate and still detect a concentric array of line segments. Children's thresholds were higher than those of adults until 6 years of age. Studies using Glass (Glass, 1969) patterns have documented a slightly longer developmental trajectory. Glass patterns use pairs of dots to form concentric or linear patterns within a background of randomly oriented dot pairs (see Fig. 8.4). Thresholds can be calculated by randomly orienting varying percentages of the signal dot pairs. In human adults, these patterns selectively activate area V4 in the ventral stream (Wilkinson et al., 2000). Children 9 years old, but not 6 years old, are as sensitive to the form in Glass patterns as are adults (Lewis et al., 2004). Similarly, between 5 and 13 years of age, children become able to find a form defined by oriented Gabors with increasingly greater background noise (Kovács, Kozma, Fehér, & Benedek, 1999). In addition to changes in V4, the long developmental trajectory may reflect, in part, the slow development of

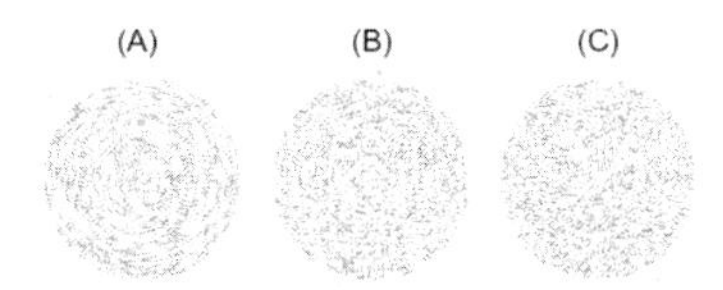

Figure 8.4 Concentric Glass Patterns. (A) 100% signal—every dot has a mate on the radial axis. (B) 50% signal—50% of the dots have a mate on the radial axis and the remaining dots are noise dots that are unpaired and distributed randomly. (C) Noise—there is 0% signal and every dot is distributed randomly.

long-range horizontal connections within V2, a relay station in the ventral stream between V1 and V4 (Burkhalter, Bernardo, & Charles, 1993; Taylor, Hipp, Moser, Dickerson, & Gerhardstein, 2014). Despite the long developmental trajectory for sensitivity to global form, the maturation of sensitivity to global motion takes even longer, at least under some testing conditions (Hadad et al., 2015).

Early binocular deprivation from congenital cataracts prevents the normal development of sensitivity to form in Glass patterns (Lewis et al., 2002). Such patients can distinguish concentric Glass patterns from random control patterns, but their threshold for tolerating randomized dot pairs within the patterns is 1.6 times higher than that of adults with normal vision. This contrasts with a threshold elevation of 4.9 for the perception of global motion (see above). Thus, early visual input is necessary to set up the neural architecture, likely involving V4, for the later integration of local elements into a global form structure, but there is more resiliency in the ventral system for processing global form than in the dorsal system for processing global motion. That conclusion is consistent with recent evidence that, unlike global motion, Glass patterns evoke a N170 (a negative ERP component about 170 msec after the onset of the stimulus) in adults treated for bilateral congenital cataracts. However, unlike controls, the amplitude of the signal is the same for random patterns and patterns in which the dot pairs define a circular pattern (Segalowitz et al., 2017). As with global motion, the effect of early visual deprivation may be to prevent the normal alteration of the balance of inputs to the relevant higher cortical areas (V4 in the case of global form) from extrageniculate pathways to pathways through V1. Studies of binocularly deprived cats provide further evidence that early visual deprivation prevents the normal alteration of the balance of inputs to extrastriate cortex: form perception in binocularly deprived cats, unlike that in visually normal cats, is resistant to lesions of V1 but is eliminated by lesions to the superior colliculus and pretectum (Zablocka et al., 1976, 1980).

8.6 DORSAL STREAM VULNERABILITY

Comparisons of sensitivity to global motion and global form in a variety of special populations have consistently reported that the deficit in global motion is much larger than the deficit in global form (reviewed in Braddick et al., 2016a). This has led to the "dorsal vulnerability" hypothesis (Braddick et al., 2003), which manifests as especially poor sensitivity to global motion, as well as poor visuo-motor control. In addition to patients treated for bilateral congenital cataract, especially poor sensitivity to global motion has been reported in patients who missed early binocular input because of misaligned eyes (Ho et al., 2005; Simmers, Ledgeway, & Hess, 2005) or who experienced unusual visual experience at birth because of extremely premature birth (e.g., Taylor, Jakobson, Maurer, & Lewis, 2009). In other conditions, the vulnerability is less clearly related to abnormal input, and may be better explained by a genetic abnormality that especially impacts the more slowly developing parts of the brain. The impact is likely greater on the dorsal stream than on the ventral stream because, although the dorsal stream appears functional earlier than the ventral stream (Distler et al., 1996; Kiorpes et al., 2012; Wattam-Bell et al., 2010), its sensitivity increases more during development, at least in monkeys (Kiorpes et al., 2012), and it is adult-like in humans at a later age (cf., Hadad et al., 2011 versus Lewis et al., 2004). The larger changes in development over a longer period as an underlying cause of a dorsal stream vulnerability is consistent with the Detroit Model (Levi, 2005): the last person hired is the first to be fired. Similarly, aspects of vision that are slower to mature might be the most likely to be affected by insult. This analysis might explain the especially poor sensitivity to global motion in children and adults with Williams Syndrome, autism spectrum disorder, developmental dyslexia, Fragile X, and hemiplegia (reviewed in Braddick et al., 2016a; Hadad et al., 2015). Alternatively, it may be the postnatal reorganization of inputs to MT, similar to that documented in the marmoset (Warner et al., 2012), that makes sensitivity to global motion especially vulnerable in humans.

8.7 CONCLUDING REMARKS

A common assumption in the literature is that cortical development is hierarchical. For vision, that means that the development of V1 is primary and that only as its neurons become more refined can input to the next relay stations in V2 and V3 begin to function. Those relay stations, in turn, it is assumed, must mature before there can be any functional input to higher cortical areas in the dorsal and ventral streams, namely MT/MST and V4, respectively. Our summary of brain and behavioral development of vision illustrates the fallacy of this assumption: MT/MST is surprisingly mature at an early age—as mature, or even more mature, than V1. This is evident in electrophysiology in the monkey and human, in neuroanatomy in the marmoset, and in behavioral sensitivity in the infant monkey and human. It is a reminder that there are many inputs to extrastriate cortex, some of which bypass V1, and hence may allow precocious development. It seems plausible that those inputs play a larger role in early development and after early binocular deprivation than they do in visually normal adults.

A similar caution against a hierarchical approach comes from a comparison of sensitivity to global versus biological motion in adults treated for bilateral congenital cataracts. In adults, a percept of a moving biological entity arises from the pattern of movement of the joints and occurs robustly even when the pattern is contained only in moving dots (Johansson, 1973). A critical structure for this decoding is the posterior region of the superior temporal sulcus, an area that receives its inputs from both the dorsal and ventral streams (Puce & Perrett, 2003). Since cataract-reversal patients have substantial deficits in global motion (dorsal stream) and in global form (ventral stream), from a hierarchical approach, one would expect that there also would be substantial deficits in sensitivity to biological motion. However, adults treated for bilateral congenital cataracts have completely normal sensitivity to biological motion, performing as accurately as controls even when a high percentage of noise dots are superimposed on the biological motion (Hadad et al., 2012). A similar caution comes from evidence that newborns are sensitive to biological motion (Simion, Regolin, & Bulf, 2008), months before they show the first evidence of sensitivity to global motion. This neonatal sensitivity, like the sensitivity after early binocular deprivation, may depend on alternative pathways than those that mediate the perception of biological motion in adults, such as those involving the cerebellum, premotor cortex, kinetic-occipital area, fusiform face area, ventral V3, and amygdala (reviewed in Hadad et al., 2012).

Studies of altered perception after early visual deprivation provide additional lessons. Here we reviewed evidence that deprivation of patterned visual input for as little as the first 2–3 months of life is sufficient to prevent the later development of normal acuity, OKN, and sensitivity to global form and global motion. Thus, in the child with normal eyes, visual experience in early infancy sets up the neural architecture for normal refinement, likely in the geniculostriate pathway and the extrastriate cortex to which it projects. In the absence of that input, there are *sleeper effects*: deficits in abilities that emerge only later, long after treatment. One reason suggested above is that early visual input may be critical for the reorganization of inputs to extrastriate cortex. Another—and not mutually exclusive possibility—is that early visual input is necessary to preserve the occipital cortex for visual responding and that in its absence, responding to other modalities is established, as in the congenitally blind (Ricciardi, Bonino, Pellegrini, & Pietrini, 2014). Support for this possibility comes from recent evidence that the cuneus (V3) responds to variations in auditory signals (moving footsteps and vowel sounds) in adults treated for bilateral congenital cataracts and in the congenitally blind but not in adults with normal eyes (Collignon et al., 2013, 2015). Similarly, in adults treated for bilateral congenital cataracts, adaptation to *auditory* motion elicits an unexpected *visual* motion aftereffect (Guerreiro, Putzar, & Röder, 2016). Processing of audio-visual signals is likewise reduced in ventral visual areas compared to responses to the visual signal alone (Guerreiro, Putzar, & Röder, 2015), perhaps as a developmental adaptation to noisy V1 signals. These unusual crossmodal remappings may occur to a lesser extent when the early visual deprivation is restricted to one eye, such that input to higher visual cortical areas through the nondeprived eye is able to better preserve the visual pathway for later refinement and/or allow more normal crossmodal reorganization to proceed. This possibility is suggested by evidence of *smaller* deficits in the deprived eye(s) after monocular than after binocular deprivation in sensitivity to global form (Lewis et al., 2002) and to global motion (Ellemberg, 2002).

Another lesson from patients treated for dense cataracts is that there are multiple sensitive periods during which visual deprivation can prevent later normal development (Lewis & Maurer, 2005). Here we illustrated three patterns: a short critical period for damage to global motion (~6 months) despite postnatal onset of sensitivity and a long developmental trajectory in infants with normal eyes; a longer critical period for symmetrical OKN (~30 months), roughly coincident with the period of normal development; and an extremely long critical period for acuity (~10 years), lasting past the age at which acuity is adult-like (~7 years). The lesson is that there are multiple critical periods during which visual experience is necessary for normal development and that one cannot infer their timing by knowing the normal developmental trajectory.

In this chapter we used "marker tasks" to probe specific aspects of developing brain function (e.g., symmetrical OKN for V1 influence on eye movements; global motion for MT; global form for V4). However, studies of the mature brain indicate that these visual capabilities are mediated by an interactive network involving feedforward and feedback connections that are further modulated by attention and previous experience of objects and events. Thus, limitations in childhood and deficits after visual deprivation could arise from immaturities anywhere in the network. Developmental improvements in visual perception could arise not from maturation in the "marker" area but from other parts of the network or even from top-down influences of attention and conceptual knowledge. Nevertheless, the brain—behavioral correlations we described here are supported by converging evidence of direct brain-behavior associations during development from animal models, anatomy, electrophysiology, neuroimaging, and behavior. A full understanding, however, will require a consideration of how those changes fit into a complex brain network.

REFERENCES

Armstrong, V., Maurer, D., Ellemberg, D., & Lewis, T. L. (2011). Sensitivity to first- and second-order drifting gratings in 3-month-old infants. *I-Perception*, *2*(5), 440—457. Available from http://dx.doi.org/10.1068/i0406.

Atkinson, J. (1979). Development of optokinetic nystagmus in the human infant and monkey infant: An analogue to development in kittens. In R. D. Freeman (Ed.), *Developmental neurobiology of Vision* (pp. 277—287). New York: Plenum.

Atkinson, J., & Braddick, O. (1981). Development of optokinetic nystagmus in infants: An indicator of cortical binocularity? In D. G. Fisher, R. A. Monty, & J. W. Senders (Eds.), *Eye movements: cognition and visual perception* (pp. 53–64). Hillsdale, NJ: Erlbaum.

Atkinson, J., Hood, B., Wattam-Bell, J., Anker, S., & Tricklebank, J. (1988). Development of orientation discrimination in infancy. *Perception, 17*, 587–595.

Banks, M. S., & Bennett, P. J. (1988). Optical and photoreceptor immaturities limit the spatial and chromatic vision of human neonates. *Journal of the Optical Society of America A, 5*, 2059–2079.

Biagi, L., Crespi, S. A., Tosetti, M., & Morrone, M. C. (2015). BOLD response selective to flow-motion in very young infants. *PLoS Biology, 13*(9), e1002260. Available from http://dx.doi.org/10.1371/journal.pbio.1002260.

Blakemore, C. (1990). *Vision: coding and efficiency. In C. Blakemore. Maturation of mechanisms for efficient spatial vision*. Cambridge: Cambridge University Press.

Blakemore, C., & Vital-Durand, F. (1986). Organization and postnatal development of the monkeys' lateral geniculate nucleus. *Journal of Physiology (London), 380*, 453–491.

Boothe, R. G., Dobson, V., & Teller, D. Y. (1985). Postnatal development of vision in human and non-human primates. *Annual Review of Neuroscience, 8*, 495–545.

Bourne, J. A., & Rosa, M. G. P. (2006). Hierarchical development of the primate visual cortex, as revealed by neurofilament immunoreactivity: Early maturation of the middle temporal area (MT). *Cerebral Cortex, 16*, 405–414. https://doi.org/10.1093/cercor/bhi119

Braddick, O. (1993). Orientation- and motion-selective mechanisms in infants. In K. Simons (Ed.), *Early visual development: Basic and clinical research*. New York: Oxford University Press.

Braddick, O., Atkinson, J., & Wattam-Bell, J. (2003). Normal and anomalous development of visual motion processing: Motion coherence and "dorsal-stream vulnerability." *Neuropsychologia, 41*(13), 1769–1784.

Braddick, O., Atkinson, J., Akshoomoff, N., Newman, E., Curley, L. B., Gonzalez, M. R., ... Jernigan, T. (2016a). Individual differences in children's global motion sensitivity correlate with tbss-based measures of the superior longitudinal fasciculus. *Vision Research*, Available from http://dx.doi.org/10.1016/j.visres.2016.09.013.

Braddick, O., Atkinson, J., Newman, E., Akshoomoff, N., Kuperman, J. M., Bartsch, H., ... Jernigan, T. L. (2016b). Global visual motion sensitivity: Associations with parietal area and children's mathematical cognition. *Journal of Cognitive Neuroscience, 28*(12), 1897–1908. Available from http://dx.doi.org/10.1162/jocn_a_01018.

Braddick, O., Birtles, D., Wattam-Bell, J., & Atkinson, J. (2005). Motion- and orientation-specific cortical responses in infancy. *Vision Research, 45*(25–26), 3169–3179. Available from http://dx.doi.org/10.1016/j.visres.2005.07.021.

Brauer, K., Leuba, G., Garey, L. J., & Winkelmann, E. (1985). Morphology of axons in the human lateral geniculate nucleus: A Golgi study in prenatal and postnatal material. *Brain Research, 359*, 21–33.

Braun, J. J., & Gault, F. P. (1969). Monocular and binocular control of horizontal optokinetic nystagmus in cats and rabbits. *Journal of Comparative and Physiological Psychology, 69*, 12–16.

Burkhalter, A., Bernardo, K. L., & Charles, V. (1993). Development of local circuits in human visual cortex. *Journal of Neuroscience, 13*(5), 1916–1931.

Cassia, V. M., Turati, C., & Simion, F. (2004). Can a nonspecific bias toward top-heavy patterns explain newborns' face preference?. *Psychological Science, 15*(6), 379–383. Available from http://dx.doi.org/10.1111/j.0956-7976.2004.00688.

Cassia, V. M., Valenza, E., Simion, F., & Leo, I. (2008). Congruency as a nonspecific perceptual property contributing to newborns' face preference. *Child Development, 79*(4), 807–820. Available from http://dx.doi.org/10.1111/j.1467-8624.2008.01160.

Collignon, O., Dormal, G., Albouy, G., Vandewalle, G., Voss, P., Phillips, C., & Lepore, F. (2013). Impact of blindness onset on the functional organization and the connectivity of the occipital cortex. *Brain: A Journal of Neurology, 136*(Pt 9), 2769–2783. Available from http://dx.doi.org/10.1093/brain/awt176.

Collignon, O., Dormal, G., de Heering, A., Lepore, F., Lewis, T. L., & Maurer, D. (2015). Long-lasting crossmodal cortical reorganization triggered by brief postnatal visual deprivation. *Current Biology, 25*(18), 2379–2383. Available from http://dx.doi.org/10.1016/j.cub.2015.07.036.

Curran, W., Braddick, O.J., Atkinson, J., & Wattam-Bell, J. (June, 2000). *Form coherence perception in infancy.* Paper presented at the international conference on infant studies, Brighton, England.

Distler, C., Bachevalier, J., Kennedy, C., Mishkin, M., & Ungerleider, L. G. (1996). Functional development of the corticocortical pathway for motion analysis in the macaque monkey: A 14c-2-deoxyglucose study. *Cerebral Cortex*, *6*(2), 184–195.

Ellemberg, D., Lewis, T. L., Defina, N., Maurer, D., Brent, H. P., Guillemot, J. P., & Lepore, F. (2005). Greater losses in sensitivity to second-order local motion than to first-order local motion after early visual deprivation in humans. *Vision Research*, *45*(22), 2877–2884. Available from http://dx.doi.org/10.1016/j.visres.2004.11.019.

Ellemberg, D., Lewis, T. L., Liu, C. H., & Maurer, D. (1999). Development of spatial and temporal vision during childhood. *Vision Research*, *39*(14), 2325–2333.

Ellemberg, D., Lewis, T. L., Maurer, D., Brar, S., & Brent, H. P. (2002). Better perception of global motion after monocular than after binocular deprivation. *Vision Research*, *42*(2), 169–179.

Fantz, R., Ordy, J., & Udelf, M. (1962). Maturation of pattern vision in infants during the first six months. *Journal of Comparative Physiological Psychology*, *55*, 907–917.

Farroni, T., Johnson, M. H., Menon, E., Zulian, L., Faraguna, D., & Csibra, G. (2005). Newborns' preference for face-relevant stimuli: Effects of contrast polarity. *Proceedings of the National Academy of Sciences of the United States of America*, *102*(47), 17245–17250. Available from http://dx.doi.org/10.1073/pnas.0502205102.

Fine, I., Wade, A. R., Brewer, A. A., May, M. G., Goodman, D. F., Boynton, G. M., ... MacLeod, D. I. (2003). Long-term deprivation affects visual perception and cortex. *Nature Neuroscience*, *6*(9), 915–916.

Gerhardstein, P., Kovacs, I., Ditre, J., & Feher, A. (2004). Detection of contour continuity and closure in three-month-olds. *Vision Research*, *44*(26), 2981–2988. Available from http://dx.doi.org/10.1016/j.visres.2004.06.023.

Glass, L. (1969). Moiré effect from random dots. *Nature*, *223*, 578–580.

Guerreiro, M. J., Putzar, L., & Röder, B. (2015). The effect of early visual deprivation on the neural bases of multisensory processing. *Brain: A Journal of Neurology*, Available from http://dx.doi.org/10.1093/brain/awv076.

Guerreiro, M. J., Putzar, L., & Röder, B. (2016). Persisting cross-modal changes in sight-recovery individuals modulate visual perception. *Current Biology*, *26*(22), 3096–3100. Available from http://dx.doi.org/10.1016/j.cub.2016.08.069.

Gunn, A., Cory, E., Atkinson, J., Braddick, O., Wattam-Bell, J., Guzzetta, A., & Cioni, G. (2002). Dorsal and ventral stream sensitivity in normal development and hemiplegia. *Neuroreport*, *13*(6), 843–847.

Hadad, B., Schwartz, S., Maurer, D., & Lewis, T. L. (2015). Motion perception: A review of developmental changes and the role of early visual experience. *Frontiers in Integrative Neuroscience*, *9*, 49. Available from http://dx.doi.org/10.3389/fnint.2015.00049.

Hadad, B. S., Maurer, D., & Lewis, T. L. (2011). Long trajectory for the development of sensitivity to global and biological motion. *Developmental Science*, *14*(6), 1330–1339. Available from http://dx.doi.org/10.1111/j.1467-7687.2011.01078.

Hadad, B. S., Maurer, D., & Lewis, T. L. (2012). Sparing of sensitivity to biological motion but not of global motion after early visual deprivation. *Developmental Science*, *15*(4), 474–481. Available from http://dx.doi.org/10.1111/j.1467-7687.2012.01145.

Hess, R. F., Hutchinson, C. V., Ledgeway, T., & Mansouri, B. (2007). Binocular influences on global motion processing in the human visual system. *Vision Research*, *47*(12), 1682–1692. Available from http://dx.doi.org/10.1016/j.visres.2007.02.005.

Hickey, T. L. (1977). Postnatal development of the human lateral geniculate nucleus: Relationship to a critical period for the visual system. *Science*, *198*, 836–838.

Ho, C. S., Giaschi, D. E., Boden, C., Dougherty, R., Cline, R., & Lyons, C. (2005). Deficient motion perception in the fellow eye of amblyopic children. *Vision Research*, *45*(12), 1615–1627. Available from http://dx.doi.org/10.1016/j.visres.2004.12.009.

Hoffmann, K.-P. (1989). Functional organization in the optokinetic system of mammals. In H. Rahman (Ed.), *Progress in zoology: neuronal plasticity and brain function* (pp. 261–271). New York: Fisher.

Hubel, D. H., & Wiesel, T. N. (1968). Receptive fields and functional architecture of monkey striate cortex. *Journal of Physiology, London, 195*, 215–243.
Huttenlocher, P. R. (1990). Morphometric study of human cerebral cortex development. *Neuropsychologia, 28*, 517–527.
Johansson, G. (1973). Visual perception of biological motion and a model for its analysis. *Perception & Psychophysics, 14*, 201–211.
Kiorpes, L., Price, T., Hall-Haro, C., & Movshon, J. A. (2012). Development of sensitivity to global form and motion in macaque monkeys (*Macaca nemestrina*). *Vision Research, 63*, 34–42. Available from http://dx.doi.org/10.1016/j.visres.2012.04.018.
Kovács, I., Kozma, P., Fehér, A., & Benedek, G. (1999). Late maturation of visual spatial integration in humans. *Proceedings of the National Academy of Sciences of the United States of America, 96*(21), 12204–12209.
Krementizer, J. P., Vaughan, H. G., Kurtzberg, D., & Dowling, K. (1979). Smooth-pursuit eye movements in the newborn infant. *Child Development, 50*, 442–448.
Levi, D. M. (2005). Perceptual learning in adults with amblyopia: A reevaluation of critical periods in human vision. *Developmental Psychobiology, 46*(3), 222–232. Available from http://dx.doi.org/10.1002/dev.20050.
Lewis, T. L., & Maurer, D. (2005). Multiple sensitive periods in human visual development: Evidence from visually deprived children. *Developmental Psychobiology, 46*(3), 163–183. Available from http://dx.doi.org/10.1002/dev.20055.
Lewis, T. L., Ellemberg, D., Maurer, D., Dirks, M., Wilkinson, F., & Wilson, H. R. (2004). A window on the normal development of sensitivity to global form in glass patterns. *Perception, 33*, 409–418. Available from http://dx.doi.org/10.1068/p5189.
Lewis, T. L., Ellemberg, D., Maurer, D., Wilkinson, F., Wilson, H. R., Dirks, M., & Brent, H. P. (2002). Sensitivity to global form in glass patterns after early visual deprivation in humans. *Vision Research, 42*(8), 939–948.
Lewis, T. L., Kingdon, A., Ellemberg, D., & Maurer, D. (2007). Orientation discrimination in 5-year-olds and adults tested with luminance-modulated and contrast-modulated gratings. *Journal of Vision, 7* (4). 9, 1–11. http://journalofvision.org/7/4/.
Lewis, T. L., Maurer, D., & Brent, H. P. (1995). Development of grating acuity in children treated for unilateral or bilateral congenital cataract. *Investigative Ophthalmology & Visual Science, 36*, 2080–2095.
Lewis, T. L., Maurer, D., Chung, J. Y. Y., Holmes-Shannon, R., & Van Schaik, C. S. (2000). The development of symmetrical OKN in infants: Quantification based on OKN acuity for nasalward versus temporalward motion. *Vision Research, 40*, 445–453.
Lewis, T. L., Maurer, D., Smith, R. J., & Haslip, J. K. (1992). The development of symmetrical optokinetic nystagmus during infancy. *Clinical Vision Sciences, 7*, 211–218.
Maunsell, J. H., & Van Essen, D. C. (1983). Functional properties of neurons in middle temporal visual area of the macaque monkey. II. Binocular interactions and sensitivity to binocular disparity. *Journal of Neurophysiology, 49*(5), 1148–1167.
Maurer, D., & Martello, M. (1980). The discrimination of orientation by young infants. *Vision Research, 20*, 201–204.
Maurer, D., & Lewis, T. L. (1993). Visual outcomes after infantile cataract. In K. Simons (Ed.), *Early visual development: Normal and abnormal. Committee on Vision, Commission on Behavioral and Social Sciences and Education, National Research Council* (pp. 454–484). Oxford: Oxford University Press.
Maurer, D., Lewis, T. L., & Brent, H. P. (1989). The effects of deprivation on human visual development: Studies of children treated for cataracts. In F. J. Morrison, C. E. Lord, & D. P. Keating (Eds.), *Applied developmental psychology, Vol. 3, Psychological development in infancy* (pp. 139–227). San Diego, CA: Academic Press.
Maurer, D., Lewis, T. L., Brent, H. P., & Levin, A. V. (1999). Rapid improvement in the acuity of infants after visual input. *Science, 286*, 108–110.
Maurer, D., Mondloch, C. J., & Lewis, T. L. (2007). Sleeper effects. *Developmental Science, 10*(1), 40–47. Available from http://dx.doi.org/10.1111/j.1467-7687.2007.00562.

Mayer, D. L., Beiser, A. S., Warner, A. F., Pratt, E. M., Raye, K. N., & Lang, J. M. (1995). Monocular acuity norms for the Teller acuity cards between ages one month and four years. *Investigative Ophthalmology & Visual Science, 36*, 671–685.

McDonald, M. A., Dobson, V., Serbis, S. L., Baitch, L., Varner, D., & Teller, D. Y. (1985). The acuity card procedure: A rapid test of infant acuity. *Investigative Ophthalmology & Visual Science, 26*, 1158–1162.

Mondloch, C. J., Lewis, T. L., Budreau, D. R., Maurer, D., Dannemiller, J. L., Stephens, B. R., & Kleiner-Gathercoal, K. A. (1999). Face perception during early infancy. *Psychological Science, 10*(5), 419–422.

Mondloch, C. J., Segalowitz, S. J., Lewis, T. L., Dywan, J., Le Grand, R., & Maurer, D. (2013). The effect of early visual deprivation on the development of face detection. *Developmental Science, 16*(5), 728–742. Available from http://dx.doi.org/10.1111/desc.12065.

Morton, J., & Johnson, M. H. (1991). CONSPEC and CONLERN: A two-process theory of infant face recognition. *Psychological Review, 98*, 164–181.

Movshon, J. A. (1990). *Visual processing of moving images: Images and understanding*. New York, NY: Cambridge University Press.

Movshon, J. A., & Kiorpes, L. (1993). Biological limits of visual development in primates. In K. Simons, Early visual development: normal and abnormal. *Commission on behavioural and social sciences and education. National Research Council* (pp. 296–304). Oxford: Oxford University Press.

Movshon, J. A., Rust, N. C., Kohn, A., Kiorpes, L., & Hawken, M. J. (2004). Receptive field properties of MT neurons in infant macaques. *Perception, 335*, 27.

Mundinano, I. C., Kwan, W. C., & Bourne, J. A. (2015). Mapping the mosaic sequence of primate visual cortical development. *Frontiers in Neuroanatomy, 9*, 132. Available from http://dx.doi.org/10.3389/fnana.2015.00132.

Naegele, J. R., & Held, R. (1982). The postnatal development of monocular optokinetic nystagmus in infants. *Vision Research, 22*, 341–346.

Naegele, J. R., & Held, R. (1983). Development of optokinetic nystagmus and effects of abnormal visual experience during in-fancy. In M. Jeannerod, & A. Hein (Eds.), *Spatially oriented behavior* (pp. 155–174). New York: Springer-Verlag.

Newsome, W., & Pare, E. (1988). A selective impairment of motion perception following lesions of the middle temporal visual area MT. *Journal of Neuroscience, 8*(6), 2201–2211.

Ostwald, D., Lam, J. M., Li, S., & Kourtzi, Z. (2008). Neural coding of global form in the human visual cortex. *Journal of Neurophysiology, 99*(5), 2456–2469. Available from http://dx.doi.org/10.1152/jn.01307.2007.

Parrish, E. E., Giaschi, D. E., Boden, C., & Dougherty, R. (2005). The maturation of form and motion perception in school age children. *Vision Research, 45*(7), 827–837. Available from http://dx.doi.org/10.1016/j.visres.2004.10.005.

Pasik, T., & Pasik, P. (1964). Optokinetic nystagmus: An unlearned response altered by section of chiasma and corpus callosum in monkeys. *Nature, 203*, 609–611.

Puce, A., & Perrett, D. (2003). Electrophysiology and brain imaging of biological motion. *Philosophical Transactions of the Royal Society of London: B Biological Sciences, 358*(1431), 435–445.

Ricciardi, E., Bonino, D., Pellegrini, S., & Pietrini, P. (2014). Mind the blind brain to understand the sighted one! Is there a supramodal cortical functional architecture? *Neuroscience and Biobehavioral Reviews, 41*, 64–77. Available from http://dx.doi.org/10.1016/j.neubiorev.2013.10.006.

Roy, M. S., Lachapelle, P., & Lepore, F. (1989). Maturation of the optokinetic nystagmus as a function of the speed of stimulation in fullterm and preterm infants. *Clinical Vision Sciences, 4*, 357–366.

Segalowitz, S. J., Sternin, A., Lewis, T. L., Dywan, J., & Maurer, D. (2017). Electrophysiological evidence of altered visual processing in adults who experienced visual deprivation during infancy. *Developmental Psychobiology*. Available from http://dx.doi.org/10.1002/dev.21502.

Simion, R., Regolin, L., & Bulf, H. (2008). A predisposition for biological motion in the newborn baby. *Proceedings of the National Academy of Sciences, USA, 195*, 809–813.

Simmers, A. J., Ledgeway, T., & Hess, R. F. (2005). The influences of visibility and anomalous integration processes on the perception of global spatial form versus motion in human amblyopia. *Vision Research, 45*(4), 449–460. Available from http://dx.doi.org/10.1016/j.visres.2004.08.026.

Slater, A., Morison, V., & Somers, M. (1988). Orientation discrimination and cortical function in the human newborn. *Perception, 17*, 597–602.

Taylor, G., Hipp, D., Moser, A., Dickerson, K., & Gerhardstein, P. (2014). The development of contour processing: Evidence from physiology and psychophysics. *Frontiers in Psychology, 5*, 719. Available from http://dx.doi.org/10.3389/fpsyg.2014.00719.

Taylor, N. M., Jakobson, L. S., Maurer, D., & Lewis, T. L. (2009). Differential vulnerability of global motion, global form, and biological motion processing in full-term and preterm children. *Neuropsychologia, 47*(13), 2766–2778. Available from http://dx.doi.org/10.1016/j.neuropsychologia.2009.06.001.

Teller, D. Y., McDonald, M. A., Preston, K., Serbis, S. L., & Dobson, V. (1986). Assessment of visual acuity in infants and children: The acuity card procedure. *Developmental Medicine and Child Neurology, 28*, 779–789.

van Hof-van Duin, J (1978). Direction preference of optokinetic responses in monocularly tested normal kittens and light-deprived cats. *Archives Italiennes de Biologie, 116*, 471–477.

van Hof-van Duin, J., & Mohn, G. (1984). Vision in the preterm infant. In H. F. R. Prechtl (Ed.), *Continuity of neural functions from prenatal to postnatal life* (pp. 93–115). Philadelphia: Lippincott.

van Hof-van Duin, J., & Mohn, G. (1985). The development of visual functions in preterm infants. *Ergebaisse der Experimentellen Medizin, 46*, 350–361.

van Hof-van Duin, J., & Mohn, G. (1986). Visual field measurements, optokinetic nystagmus and the visual threatening response: Normal and abnormal development. *Documenta Ophthalmologica Proceedings Series, 45*, 305–316.

Warner, C. E., Kwan, W. C., & Bourne, J. A. (2012). The early maturation of visual cortical area MT is dependent on input from the retinorecipient medial portion of the inferior pulvinar. *Journal of Neuroscience, 32*(48), 17073–17085. Available from http://dx.doi.org/10.1523/JNEUROSCI.3269-12.2012.

Wattam-Bell, J. (1996a). Visual motion processing in one-month-old infants: Habituation experiments. *Vision Research, 36*(11), 1679–1685.

Wattam-Bell, J. (1996b). Visual motion processing in one-month-old infants: Preferential looking experiments. *Vision Research, 36*(11), 1671–1677.

Wattam-Bell, J., Birtles, D., Nyström, P., von Hofsten, C., Rosander, K., Anker, S., ... Braddick, O. (2010). Reorganization of global form and motion processing during human visual development. *Current Biology, 20*(5), 411–415. Available from http://dx.doi.org/10.1016/j.cub.2009.12.020.

Wilkinson, F., James, T. W., Wilson, H. R., Gati, J., Menon, R., & Goodale, M. A. (2000). An fMRI study of the selective activation of human extrastriate forn vision areas by radical and concentric gratings. *Current Biology, 10*, 1455–1458.

Wood, C. C., Spear, P. D., & Braun, J. J. (1973). Direction-specific deficits in horizontal optokinetic nystagmus following removal of visual cortex in the cat. *Brain Research, 60*, 231–237.

Yuodelis, C., & Hendrickson, A. (1986). A qualitative and quantitative analysis of the human fovea during development. *Vision Research, 26*, 847–855.

Zablocka, T., & Zernicki, B. (1996). Discrimination learning of grating orientation in visually deprived cats and the role of the superior colliculi. *Behavioral Neuroscience, 110*(3), 621–625.

Zablocka, T., Zernicki, B., & Kosmal, A. (1976). Visual cortex role in object discrimination in cats deprived of pattern vision from birth. *Acta Neurobiology, 36*, 157–168.

Zablocka, T., Zernicki, B., & Kosmal, A. (1980). Loss of object discrimination after ablation of the superior colliculus-pretectum in binoculary deprived cats. *Behavioral Brain Research, 1*, 521–531.

Zee, D. S., Tusa, R. J., Herdman, S. J., Butler, P. H., & Gucer, G. (1987). Effects of occipital lobectomy upon eye movements in primate. *Journal of Neurophysiology, 58*, 883–907.

Zernicki, B., Zablocka, T., & Kosmal, H. (1978). An increased role of the superior colliculus-pretectum in cats deprived of pattern vision from birth. *Neuroscience Letters, Supplement, 1*, S397.

CHAPTER 9

The Development of the Motor System

Claudia L.R. Gonzalez[1] **and Lori-Ann R. Sacrey**[2]
[1]University of Lethbridge, Lethbridge, AB, Canada
[2]University of Alberta, Edmonton, AB, Canada

9.1 INTRODUCTION

Referring to motor development is like referring to the development of the entire nervous system. After all, the ultimate goal of brain function is to produce behavior (i.e., movement) within a perceptual world (Kolb & Whishaw, 2012). The relationship between motor and cognitive development has a long history, dating back to the 1930s with Gesell's maturation theory and the 1960s with Piaget's cognitive-developmental theory. Simply stated, a child's emerging motor skills provide ongoing opportunities to explore and understand their environment, allowing for the development and refinement of cognition. In order to understand the development of cognition, we must first have an understanding of motor development.

Research in humans and other species has shown that embryotic motor development follows a cephalocaudal (head to toes) and proximodistal (from torso to extremities) sequence that is later echoed at the behavioral level. Infants gain control of their head and torso before their hands and feet. Motor control thus begins with newborns learning to deal with gravity. Gaining head and torso stability are fundamental to the proper development of most other motor behaviors. Without postural control, interacting with the world would be impossible. In the big picture, the three major motor developmental achievements are walking, reaching and grasping, and talking. Together, these three functions define us as humans. This chapter discusses the trajectories of how humans learn to walk, to reach and grasp for objects, and how the emergence of these behaviors relate to the development of cognitive functions. The development of speech and language is described at length in Chapter 10, Language and Cognition.

9.2 THE DEVELOPMENT OF HUMAN WALKING

Walking, the ability to propel the body forward, is a highly coordinated, rhythmical, alternating movement of the legs. Although in most cases walking requires no thought, it involves all the joints of the lower limbs, vertebral column, and upper

The Neurobiology of Brain and Behavioral Development
DOI: http://dx.doi.org/10.1016/B978-0-12-804036-2.00009-1

limbs. Because of its rhythmicity, walking appears to be a stereotyped behavior, but it requires a great deal of flexibility and control. Walking has to adapt to a constantly changing environment. So, even though walking can be induced at lower levels of the nervous system (spinal cord), it is in fact a learned behavior that requires extensive training and constant monitoring.

9.2.1 Birth to 6-Months of Age

Spontaneous cyclic movements of the head and/or rump in utero have been detected in human embryos as early as 5 weeks postconception. These cyclic movements are replaced by more general, spontaneous, and complex movements, which appear at 7 weeks and remain throughout pregnancy (Felt et al., 2012). General movements are the most frequent type of fetal movement during the first trimester. They involve all body parts, are variable in speed and amplitude, and can be used as a tool for predicting the integrity of the developing nervous system (Prechtl, 2001). With respect to limb movements, single and double leg kicks as well as coordinated leg and arm movements increase in frequency and intensity during the second and third trimesters. Fetuses in the last trimester and newborns display locomotor-like behavior in the form of stepping. This behavior is elicited when the infant is in the upright position (supported under arms) and the feet soles make contact with the ground. Infant stepping is irregular, lacks postural control, and it is characterized by synchronous flexion-extension of the joints, uniform muscle activity, and the absence of adult-like heel strike (Dominici et al., 2011; *Science*). Nevertheless, a recent study has shown that during treadmill-induced locomotion 3-day-old infants are able to adapt their steps to graded velocities (Siekerman et al., 2015). This finding suggests that training interventions to promote independent locomotion in at-risk infants (e.g., when cerebral palsy is suspected) can begin from birth. Interestingly, this early stepping behavior is present during the first 2 months of life but then "disappears" only to re-appear as the foundation of intentional walking at around 6–9 months of age. The reason why infant stepping disappears at about 2 months is a mystery, but a recent study suggest that at this time there is relaxation of the lower limbs leading to a flexed posture that makes stepping difficult to elicit (Barbu-Roth et al., 2015). It has also been suggested that rearing practices that limit opportunities for practice during the early months are a contributing factor for the disappearance of the early stepping (Zelazo, 1983). Evidence supporting this view can be found in observations of rearing habits in different communities around the world and in carefully controlled experimental studies. Work by Geber and Dean (1957a,b), Konner (1973, 1977), Kilbride et al. (1970), and Super (1976) showed that infants in some parts of Africa are ahead of North American or European infants in motor development. This advantage was more notable for sitting, standing, and walking. In the case of walking, African infants could walk by 9–10 months compared to North American infants who begin walking around 12 months of age. Researchers that collected these observations indicate that part of this advantage arises from rearing practices. Soon after their infants are born, African mothers massage, bounce and shake their infant's legs in training for walking. In contrast to this practice, infants in some parts of China are left alone for hours with very little motor stimulation (so-called "sandbag rearing"), delaying walking until 15 months of age.

In a series of studies Zelazo et al. (1972, 1976) investigated if early stepping could be promoted and retained in normally developing infants. In their first study, 24 1-week-old babies were divided into two experimental and two control groups. The first experimental group received stepping and placing exercises for a total of 12 minutes every day for 7 weeks. To account for muscle activation and social engagement of the infant with the caregiver, a second experimental group received gross motor stimulation (i.e., passive exercises) for the same amount of time, but without stepping or placing behaviors. Infants in these two groups were tested once a week for the number of stepping and placing responses they could produce per minute. One control group received the same amount of testing as the two experimental groups. The other control group was only tested once, at the end of the experimental procedures, when babies were 8 weeks of age. The results were unambiguous; infants in the active exercise group stepped significantly more than infants in each of the other three groups. The increase in the number of steps was particularly steep during the last 2 weeks of testing. On week 8, infants in the control groups or in the passive exercise group stepped fewer than five times per minute, whereas infants in the active exercise group made more than 30 steps. This finding strongly suggests that, with practice, the early stepping behavior can be retained and augmented. A follow-up study (Zelazo, 1976) was designed to investigate if practice, regardless of the age at intervention, could increase early stepping. In other words, researchers were interested in finding out whether practice could elicit and promote early stepping in infants older than 8 weeks of age (the point where the disappearance of early stepping had been documented). Infants received active exercise for 6 weeks starting at different developmental times: 2 weeks (group 1), 6 weeks (group 2), or 10 weeks (group 3) of age. Group 1 served as a replication group for the previous study. Control groups were tested only once, at weeks 8, 12, and 16, which corresponded to the time where the intervention had ended for the experimental groups. Results show that infants subjected to active exercise increased their stepping behavior irrespective of when the training had begun. This result contradicts the standing belief that early stepping disappears after 2 months of age and instead suggests that, with practice, this behavior can be elicited at any developmental stage.

In addition to early stepping, infants also execute a variety of spontaneous alternate leg movements while lying on their backs (Thelen, 1979). This supine kicking occurs more often when the baby is excited or upset and shares numerous characteristics to upright stepping (e.g., similar temporal structure, same pattern, or joint angles) but kicking continues throughout the first 12 months of age. At 3 months of age, babies will purposefully kick to attain a goal. Studies have shown that babies increase the frequency of kicking so toys hanging from a mobile can move (Thelen & Fisher, 1983), they will switch their kicking to the leg that most reliably produce the toys to shake (Rovee-Collier, Morrongiello, Aron, & Kupersmidt, 1978), and they can even choose the degree of knee flexion required to make the mobile go (Angulo-Kinzler, Ulrich, & Thelen, 2002). This flexible and purposive behavior is the beginning of the prewalking stage where infants will do whatever it takes (rolling, pulling, hoisting, dragging) for them to transport their bodies from place to place.

9.2.2 Six to Twelve Months of Age

The prewalking stage (6–12 months) features a huge repertoire of "modes of propulsion," and none are the prerequisite of walking. Infants will use hands, knees, feet, buttocks, abdomen, and head in a variety of combinations to locomote (Adolph & Berger, 2005; Adolph & Robinson, 2013). They may begin by using their hands and bellies to drag themselves in a "commando" or "inchworm" style and progress into more advanced crawling strategies. These precursors of crawling have shown to increase the speed and size of their movements, turning them into more proficient crawlers (Adolph, Vereijken, & Denny, 1998). Cruising can take many shapes: iconic crawl (hands and knees), bear crawl (hands and feet), spider or crab crawl (hands and feet in supine position), bum shuffling (hands and buttocks), or knee walking. Infants will crawl for as little as 1 month and for as long as 8 months, but some babies do not crawl at all. This suggests that none of these modes of locomotion are requirements for walking. Furthermore, babies will experiment with one or all forms of crawling simultaneously or in succession. Adolph and Robinson (2013) argue against a universal developmental trajectory, pointing out that precursory forms of locomotion do not need to follow sequential stages; that the "typical" infant portraying developmental milestones and desirable scores in developmental scales is not a real child but the statistical average.

Nevertheless, the majority of infants will crawl for 2–4 months before venturing to walk. An intriguing question in the study of locomotion and its development is why do infants abandon crawling and start walking? Why are expert crawlers willing to give up an efficient mode of locomotion to stumble and fall as they learn to walk? In a study addressing this question, researchers compared experienced crawlers (12 months old), to novice walkers (12 months old), as they navigated freely in a playroom (Adolph et al., 2012). The number of falls and the number of steps were compared between the groups. Interestingly, the results showed that although novice walkers fell more than expert crawlers, walkers spent longer time in motion and thus traveled longer distances than crawlers did. So, when accounting for the total amount of distance traveled, the amount of falls were similar between the two groups. This may be one of the reasons infants abandon crawling: for the same amount of pain there is a bigger gain. Another study (Kretch, Franchak, & Adolph, 2014) placed a head-mounted eye-tracker in crawling and walking infants of the same age (13 month olds). They discovered that the visual experience of crawlers is limited to the ground and lower parts of walls. Walkers in contrast, enjoy a view of the entire surroundings, including, importantly, its inhabitants. As human interaction is central at this developmental stage, and walking provides them with this opportunity, this could be another good reason infants abandon crawling to start walking. Another consideration is the differences in the way caregivers interact with crawlers and walkers. A study has found that walkers receive more verbal and complex feedback than crawlers (Karasik, Tamis-LeMonda, & Adolph, 2014). This might be one of the reasons one can find correlations between the onset of walking and enhanced language abilities in infants (this topic is discussed in the last section of this chapter).

In sum, during the first 12 months of life, infants go from lying prone, to rolling over, to getting to hands and knees, to sitting, pulling to stand and eventually to walking.

9.3 DEVELOPMENT OF REACHING: THE FIRST 12 MONTHS OF LIFE

Evolutionarily, bipedalism afforded the specialization of the upper extremities to be freed for exploration and manipulation of objects. The reach-to-eat movement, in which a hand advances to grasp a target to bring to the mouth, is a natural act and is displayed in a number of forms by developing infants and adults. Newborn infants automatically close their hand around objects that are incidentally contacted (Butterworth & Harris, 1994; Twitchell, 1970) and will bring these objects to the mouth for oral exploration (Foroud & Whishaw, 2012; Rochat, 1989; Rochat & Senders, 1991; Whyte, McDonald, Baillargeon, & Newell, 1994). Targeted reaching emerges between 3 and 5 months of age (Thelen, Corbetta, & Spencer, 1996; von Hofsten et al., 1984), with 3-month-old infants clasping their hands at the midline in an attempt to reach an object (Atkinson, 1984; Bruner & Koslowski, 1972; Hopkins & Prechtl, 1984; Von Hofsten, 1991) and opening their mouths in preparation to grasp at the same time (Bruner & Koslowski, 1972; Butterworth & Hopkins, 1988; Foroud, 2008). Objects are grasped and brought to the mouth with bimanual movement by 4 months of age (Von Hofsten & Lindhagen, 1979), and with unimanual movements by 6 months of age (Sacrey, Karl, & Whishaw, 2012a; Von Hofsten, 1991). With increased experience, infants begin to produce reaching and grasping movements that approximate those of adults by 10–12 months of age (Karl &

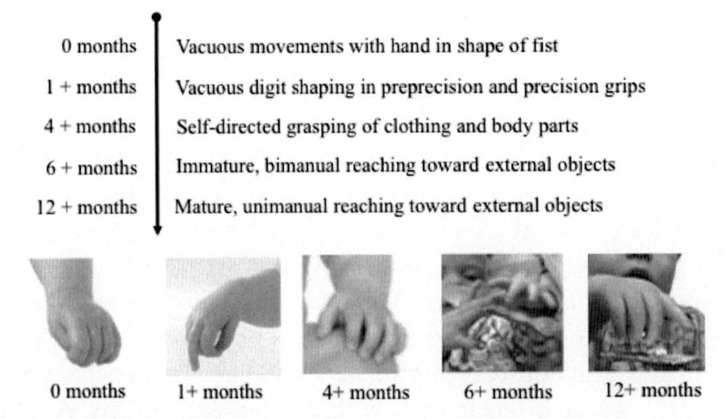

Figure 9.1 Milestones for goal-directed reaching during the first year of life.

Whishaw, 2014; Law, Lee, Hulse, & Tomassetti, 2011; Napier, 1956; Sacrey et al. 2012a; Touwen, 1976; von Hofsten & Fazel-Zandy, 1984; White, Castle, & Held, 1964). Important milestones for goal-directed reaching are illustrated in Fig. 9.1.

9.3.1 Birth to Six Months of Age

9.3.1.1 Reaching and grasping the self

From birth to 6 months of age, hand shaping and grasping postures undergo maturation in preparation for externally directed reaching, both when at rest (Sacrey & Whishaw, 2010) and while active (Thomas, Karl, & Whishaw, 2015; Wallace & Whishaw, 2003). When in a wakeful, resting state, the hand progresses through a series of developmental changes, from a fist resting beside the head to a collected, semiflexed state resting at the infants' side. This developmental progression was described from video-recordings of infants from birth to 6 months of age as they were alert but their hands were at rest. Four resting shapes were identified during this developmental period and occurred during each month; a fist, collection of the fingers (which are closed and semiflexed), intermediate position of the fingers (which are open and either semiflexed or extended), and a stationed position (fingers are extended and lay against a surface or the body). The prevalence of each shape changed depending on the age of the infant. Over the first month, infants shape their hand in a fist approximately 70% of the time. By 2 months, the fist is no longer the dominant resting shape, as the proportion of collected hand shapes rises and increases in incidence each month, reaching approximately 90% of resting shapes at 6 months of age. The stationed and intermediate shapes are also present in each of the 6 months, but they only comprise a small minority (combined 10%) of hand shapes during each month. The collected shape is the predominate resting posture of the hand by

6 months, suggesting that it may serve as an important milestone in the development of hand movements, including grasping (Sacrey & Whishaw, 2010).

Similarly, the hand undergoes a developmental maturation of pregrasps during a wakeful, active state, from a fist to preprecision grasps of the body (Wallace & Whishaw, 2003). This developmental progression was described from video-recordings of infants from birth to 5 months of age, as they were alert and actively moving their hands. Four pregrasping shapes were described from the video record; a fist, a preprecision grip (thumb pad meeting side of index or middle finger), precision grip (thumb pad meeting pad of index or middle finger), and self-directed grasps (closing of fingers on infants body, such as clothing). In the first month, the infant predominately holds a fist shape, but by the second month, infants make increasingly complex hand and finger movements, with preprecision and precision grip patterns appearing. At around 4 months, self-directed grasping appeared and included grasping the other hand or clothing on the body. Thus, spontaneous hand and finger movements show a developmental progression from vacuous grasping to self-directed grasping between birth and 5 months of age (Wallace & Whishaw, 2003).

The development of self-grasping in very young infants had also been examined by target location and hand position. That is, hand contacts were classified as rostral (head, trunk, arm, or other hand) or caudal (hips, leg, or feet), as well as by dorsum (back or side of the hand) or palmar (palm or fingertips) (Thomas et al., 2015). Video-recordings from infants between birth and 5 months of age suggest that there is a developmental progression from rostral to caudal contacts, as well as from dorsum to palmar contacts. In the first month, the greatest proportion of contacts was to the rostral area immediately surrounding the head and upper torso (in agreement with Sacrey & Whishaw, 2010). During the second month, the proportion of caudal contacts increase, and progress to contacts with the feet by 20 weeks (Thomas et al., 2015). Correspondingly, in the first month, contacts were mainly made using the back of the hand, with hand to body contacts increasingly made by the palm and fingertips by 12 weeks. The palmar contacts increased in complexity, with the first contacts featuring hooking one or two fingers into clothing, then clasping with the thumb or other fingers, and finally featuring whole hand grasps with manipulation by 5 months (Thomas et al., 2015).

9.3.1.2 Reaching and grasping distal targets

Directed reaching toward objects has been described for infants as young as 4 months of age (Karl & Whishaw, 2014). Interestingly, reach and grasp configurations in young infants (between 4 and 6 months) resemble reaching movements employed by adults when reaching with occluded vision. That is, 4- to 6-month-olds reach toward an object with an open hand and do not make attempts to preshape the hand prior to contact (Karl & Whishaw, 2014).

Interventions have been developed with the aim of manipulating the developmental trajectory of reaching in typical development. Williams and Corbetta (2016) examined the effect of contingent (toy activates following successful reach to touch) versus continuous (always active regardless of reaching attempts) manipulations to determine any impact on augmenting reaching behavior in 3-month-old infants. Overall, children in the contingent group show improvements in reaching compared to the continuous group. That is, although both groups showed gains in reaching between the first and last days of study, only the contingent group produced a greater number of hand-toy contacts compared to a control group of children who did not experience either the contingent or continuous conditions. Interestingly, infants in the contingent group also began to perform more visually guided target contacts relative to non-visually guided target contacts, suggesting that contingent reaching also augments the development of visually guided reaching (Williams & Corbetta, 2016).

9.3.1.3 Sensory guidance

Immature reaching movements appear to be partially mediated by vision. When identifying a target for reaching, the eye and head orient toward the object (Greenman, 1963; Kremenitzer, Vaughan, Kurtzberg, & Dowling, 1979; von Hofsten & Rosander, 1997), reaching movements begin with the head, as the mouth is open to grasp objects held in front (Foroud & Whishaw, 2012), followed by swiping at a visual target with a fisted or open hand (von Hofsten, 1982, 1984; White et al., 1964). In contrast, immature grasping movements, such as automatically closing the hand around an object that contacts the palm (Karl & Whishaw, 2014; Twitchell, 1970), spontaneously producing pincer and precision grips during hand babbling (Rönnqvist & von Hofsten, 1994; Wallace & Whishaw, 2003), and grasping one's own body or clothing (Thomas et al., 2015; Wallace & Whishaw, 2003) appear to be mediated by somatosensation.

9.3.1.4 Summary

The corticospinal tract develops extensively during the first year of life, resulting in increasing control of the hand and fingers (Armand, Olivier, Edgley, & Lemon, 1997; Berthier, Clifton, McCall, & Robin, 1999; Kuypers, 1981; Olivier, Edgley, Armand, & Lemon, 1997; White et al., 1964). It is therefore not surprising that the earliest hand shape, the fist, is associated with spasticity and neural immaturity and is a sign of spasticity of cerebral palsy if it prevails beyond 2 months of age (Perlstein & Barnett, 1952). With typical development, the hand continues to progress into preprecision grasping, including self-grasping of the clothing and the body (Wallace & Whishaw, 2003; Thomas et al., 2015). These movements allow infants to practice grasping and acquire body awareness in relation to a hand-centric schema. It is also important to note that these early grasping movements are largely guided by somatosensory feedback, as self-grasping occurs without visual feedback of the grasping hand (Thomas et al., 2015).

9.3.2 Six to Twelve Months of Age

Reaching and grasping undergoes dramatic development between 6 and 12 months of age. Over this time period, infants develop preparatory shaping of the hand, smooth trajectories of the reaching arm and hand, and accuracy in targeting both the object to be grasped, as well as placement of the grasped object into the mouth (Sacrey et al., 2012a,b).

9.3.2.1 Reach-to-grasp

Directed reaching in 6-month-old infants begins with a rotation of the shoulder to lift the hand from the lap. The fingers on the reaching hand are open and extended (not shaped) and the wrist does not rotate to match orientation of the object to be grasped. Arm trajectory toward the target is fragmented, and the hand does not preshape in preparation for grasping. The target is grasped using a whole hand grasp, in which the fingers are open and extended (or fan-like) and close around the target to capture it at contact (Karl & Whishaw, 2014; Sacrey, Karl, & Whishaw, 2012b). By 9 months, infants begin to flex their elbow to lift their hand, but continue to show little hand shaping or hand rotation in preparation for reaching toward the target. Aiming of the hand and end-point accuracy has improved, but preshaping of the fingers does not prepare the hand for grasping, but continues to capture the object using the whole hand at contact (Sacrey et al., 2012b). Reaching and grasping in 12-month-olds resembles adult reaching. Twelve-month-old infants lift their hand through flexion of the elbow, and the hand supinates and preshapes into a collected position at the midline of the body. The hand takes a direct path to the target, pronating over the target with fingers extended and open in preparation for grasping. The target is purchased using a precision grip, generally with the thumb and index finger (for a small food item), and the wrist extends to lift the grasped item from the substrate (Sacrey et al., 2012b). In comparison, when grasping larger objects that cannot be purchased using a pincer grasp, older infants utilize a capture strategy. With capture, the hand orients toward the target prior to contact, the digits open and extended until contact, at which time the digits close around the target (Karl & Whishaw, 2014). It is also important to note that in addition to using whole hand grasps, 6-month-old infants make a large number of errors when picking up targets compared to reaching at 9 and 12 months. Often, they first make contact with the target and then release and adjust hand placement before finalizing the grasp (Karl & Whishaw, 2014; Sacrey et al., 2012a).

9.3.2.2 *Oral exploration and eating*

Infants will generally manipulate a target once it has been grasped. These manipulations can include visually inspecting the target, haptically exploring the target with the other hand, placing the grasped object in the mouth for oral exploration, or in the case of food, placing the food item in the mouth for eating (Karl & Whishaw, 2014; Sacrey et al., 2012a,b). Mouthing of objects is a predominate form of object manipulation in infancy, particularly between 6 and 12 months of age. At 6 months of age, withdrawal of grasped objects, usually a toy, to the mouth is quite immature. Infants do not rotate their hand following grasp to place the object in a correct orientation for mouth placement, nor do they rotate their hand as the target nears the mouth to adjust during placement. Rotation of the wrist improves gradually over the next 6 months, when the movement resembles the adult substrate, with online corrections to adjust object orientation following grasp and during placement into the mouth (Sacrey et al., 2012a).

The accuracy at which the target is placed in the mouth also improves with age. At 6 months, infants are quite inaccurate when aiming the target toward the mouth, often first contacting the chin or the cheek before correcting and reorienting for proper placement. Aiming accuracy is improved by 9 months, as targets contact the lips prior to entry into the mouth. By 12 months of age, infants are able to place the target into the mouth without first contacting the lips, as seen with adult reaching-to-eat movements (Sacrey et al., 2012a).

9.3.2.3 *Kinematics*

The kinematic profile of the reach-to-oral exploration/eating movement has corroborated the behavioral description of the developmental profile of the movement. Comparisons of hand trajectories and velocity during a reaching task in 6-, 9-, and 12-month-old infants highlight the ongoing movement refinement during this time period. At 6 months of age, the movement of the hand is quite jerky and rife with an increased number of movement units (Berthier et al., 1999; Fallang, Saugstad, & Hadders-Algra, 2000; Mathew & Cook, 1990; Von Hofsten, 1991; Von Hofsten & Lee, 1982). By 9 months of age, there is an increase in movement smoothness (decrease in jerkiness) as well as a reduced number of movement units when completing the reaching movement. At 12 months of age, the reaching movement resembles that of the adults, being quite smooth and comprised of two movement units (reaching-to-grasp and grasp-to-mouth; Sacrey et al., 2012a).

Movement durations during reaching and grasping have also been examined and show a decrease with age. Using a task in which infants reach for a vertical rod, Karl et al. (2014) found younger infants (4 to 6 months) took significantly longer to reach out and grasp an object compared to older infants. Total movement time decreased as a function of age, so much so that movement durations of 24-month-old infants did not differ from adults who performed the same task (Karl & Whishaw, 2014). A similar finding was reported for infants who reached for small toys and food items. Movement times for reaching and grasping significantly decreased between 6 and 12 months of age, coinciding with improvements in movement smoothness (Sacrey et al., 2012b).

The behavioral descriptions of wrist rotation are also supported by kinematic analyses of the reaching wrist. That is, 6-month-old infants do not show any hand rotation until the grasped object (usually a toy) is being withdrawn toward the body for mouth placement (Sacrey et al., 2012b). Infants begin to demonstrate hand rotation at 9 months, particularly during grasp, when the hand pronates near the target prior to grasping (both for toy and small food item), and as the grasped item is withdrawn toward the mouth for placement (Karl & Whishaw, 2014; Sacrey et al., 2012b). By 12 months of age, wrist rotation in the infant resembles that of the adult, showing hand rotation following initial lift of the hand and prior to the grasp of a food object or toy (Karl & Whishaw, 2014; Sacrey et al., 2012b), and as the object is being withdrawn for mouth placement when reaching for small food items (Sacrey et al., 2012b). These developmental changes in wrist rotation echo earlier examinations of infant grasping (Lockman, Ashmead, & Bushnell, 1984; Morrongiello & Rocca, 1989; Schum, Jovanovic, & Schwarzer, 2011; von Hofsten & Fazel-Zandy, 1984; Wentworth, Benson, & Haith, 2000; Witherington, 2005).

9.3.2.4 Sensory guidance

Sensory guidance of reaching and grasping also undergoes dramatic maturation by 12 months of age. At 6 months of age, infants visually fixate on a target for a prolonged period of time prior to initiating a reach toward it and continue to fixate the target as the hand is brought toward it and as it is grasped. Interestingly, the infant continues to visually examine the object after it is grasped, including as it is being brought to the mouth for oral exploration (Sacrey et al., 2012a,b; see also, Bower, 1974; Bruner & Koslowski, 1972; Bushnell, 1985; McCarty, Clifton, Ashmead, Lee, & Goubet, 2001; Ruff, 1986). Although the objects are visually fixated, the information collected does not appear to affect reaching kinematics. Rather, reaching and grasping appear to be guided by haptic feedback from the target, in that the hand does not orient to match the target in preparation for grasping, but adjusts following contact with the object (Karl & Whishaw, 2014; Lockman et al. 1984; McCarty et al. 2001; Newell, Scully, McDonald, & Baillargeon, 1993; Sacrey et al., 2012a,b; Witherington, 2005). Furthermore, blocking visual feedback of the reaching hand prior to 9 months of age does not effect kinematic measurement of the reaching hand (e.g., there are no differences in movement durations, jerkiness, and units for reaches completed with and without visual information from the reaching hand;

Clifton, Muir, Ashmead, & Clarkson, 1993; McCall, Robin, Berthier, & Clifton, 1994; Robin, Berthier, & Clifton, 1996).

Following 9 months of age, the reach and grasp increasingly come under visual control, with the amount of time spent looking at the target both before movement initiation and following grasp decreasing. By 12 months of age, vision is tightly coupled to reaching and grasping, with infants looking at the target immediately before the reach begins and looking away from it just as it is grasped (Karl & Whishaw, 2014; Sacrey et al., 2012a,b). That reaching comes under visual control by 1 year of age is further supported by visual occlusion studies. Reaching kinematics of 1-year-old infants mirrors that of adults when performing the task without visual feedback of the reaching hand (by wearing a blindfold), resulting in slower reach durations and absence of preshaping of the hand in preparation for grasping (Berthier et al., 1999; Berthier & Carrico, 2010; Carrico & Berthier, 2008).

9.3.2.5 Handedness

One of the most salient characteristics of humans is the expression of population-level right-handedness. Ninety percent of the population is considered right-handed in that they prefer to use their right hands for a multitude of actions (e.g., writing, handling tools, grasping objects). The exact time during development when right-hand preference is established appears controversial in the literature. From birth to 6 months of age, grasping at self-targets, including ones' own clothing and other body parts, is equally distributed between the two hands (Thomas et al., 2015; Wallace & Whishaw, 2003). Distal reaching shows a similar pattern, with early reaching at 6 months showing a large proportion of bimanual grasps and equal distribution of unimanual grasping between the two hands. It is important to note that the high proportion of bimanual grasps at 6 months may be related to the size of the targets (e.g., toys) in relation to the size of the infants' hand. Over the next 5 months, infants increasingly progress to unimanual grasping, featuring increased use of the right hand by 12 months of age (Sacrey et al., 2012a; Sacrey et al., 2013). The lack of hand preference reported by some researchers may have to do with object exploration. For example, infants between 4 and 6 months of age who reached for wooden cylinders spent more time touching and passively exploring the haptic features of the cylinder with their left hands (Morange-Majoux, 2011). It is interesting that infants as young as 2 months of age are able to retain haptic information collected by the left hand compared to the right hand (Lhote & Streri, 1998), and that this left-hand preference for haptic information collection can result in an advantage for visually recognizing novel geometrical shapes that were previously manipulated by the left (but not the right) hand (Rose, 1984).

Other studies, however, have reported hand use asymmetries as early as the prenatal period, where fetuses show a right-hand preference for hand-to-mouth movements

(Hepper et al., 1998). What is more, a provoking new study showed that as early as 8 weeks postconception, fetuses display pronounced lateralized motor behavior of the arms (Ocklenburg et al., 2017). As at this developmental stage the cortex and the tracts that will connect it to the spinal cord (i.e., corticospinal tract and other descending pathways) are not formed, the results suggest that lateralized behaviors, such as handedness, may find their basis at the level of the spinal cord. Postnatally, further research has documented the expression of right-hand dominance in infants as young as 6 months (Claxton et al., 2003; Ferre et al., 2010; Hopkins & Rönnqvist, 2002; Michel et al., 2014; Morange-Majoux et al., 2000; Nelson et al., 2013; Rönnqvist & Domellöf, 2006). For example, Nelson and colleagues documented right-hand use in 6–14 month olds and discovered that 40% of the infants already exhibited a right-hand preference for reaching to grasp a toy. The lack of consensus on the time when right-handedness emerges is likely related to differences in methodology used by different researchers. For example, it has been shown that when the target is a food-object, children as young as one will demonstrate a robust right-hand preference for the reach-to-grasp action (Sacrey et al., 2013). In that study, hand preference in 3–5-year olds was recorded as the children reach-to-grasp either food items (Froot Loops) or nonfood items of comparable size and color (LEGO construction blocks) to the food. Consistent with other research (for a review see Scharoun & Bryden, 2014), the 4- and 5-year-olds displayed a right-hand preference for grasping the blocks but not the 3-year-old group. When the target was a food-object, however, 3-year-olds showed a greater than 80% right-hand preference for grasping. In fact, when younger groups were tested, this preference was observed in children as young as one year of age. This finding suggests that the expression right-handedness might follow different developmental trajectories depending on the goal of a task (e.g., eat, place, or play) and/or it may be tied to task parameters (e.g., object size, texture). The last consideration regarding the development of handedness (discussed in the last section of this chapter) is the evidence suggesting that the degree of right-hand preference is associated with earlier emergence of some cognitive abilities.

9.3.2.6 Summary

Reaching and grasping improves dramatically between 6 and 12 months of age. That is, 6-month-old infants perform the movement quite poorly, with the absence of digit shaping and wrist rotation in preparation for grasping. In addition, they show prolonged visual fixation on the target both before and following grasp but rely on haptic feedback from the target to complete the grasp. By 12 months of age, infants accurately reach toward the target and preshape the digits and orientate their wrist to grasp the target. The reach and grasp are also under visual control at this time, with a temporal coupling of visual fixation on the target occurring immediately prior to first hand movement and disengagement at grasp (Sacrey et al., 2012a,b). The gradual improvement in movement accuracy and sensory guidance with age is likely related to the maturation of sensory and motor control and also experience with the reaching during this time period (Bower, 1974; Martin, Choy, Pullman, & Meng, 2004; Martin, Friel, Salimi, & Chakrabarty, 2007; Rocha, Silva, & Tudella, 2006). Although great improvements in reaching are evident by 12 months of age, performance is not quite at adult levels, suggesting there are features of reaching and grasping that are not fully developed by this age (Heineman, Middelburg, & Hadders-Algra, 2010; Karl & Whishaw, 2014; Silva, Rocha, & Tudella, 2011).

9.4 INTERACTIONS BETWEEN THE DEVELOPMENT OF MOTOR AND COGNITIVE ABILITIES

Research has drawn many parallels between motor and cognitive development. Over 30 years ago, Zelazo (1983) suggested that, "the infant's cognitive development is not only synonymous with his or her sensorimotor development, but that it can be measured by his or her motor ability" (p.100). In fact, in the Bayley test, the widest used scale of infant development, cognitive and language skills are inferred through motor outputs. Fig. 9.2 illustrates the interactions between the development of motor and cognitive abilities.

There is compelling evidence that the emergence of independent walking and purposeful reach-to-grasp coincide with spurts in vocabulary. In two studies, Walle and Campos (2014) demonstrate the relationship between language development and the acquisition of walking. In the first study, 10–13-month-old infants were followed every 2 weeks for 3.5 months. Parents documented their infant's locomotor activity

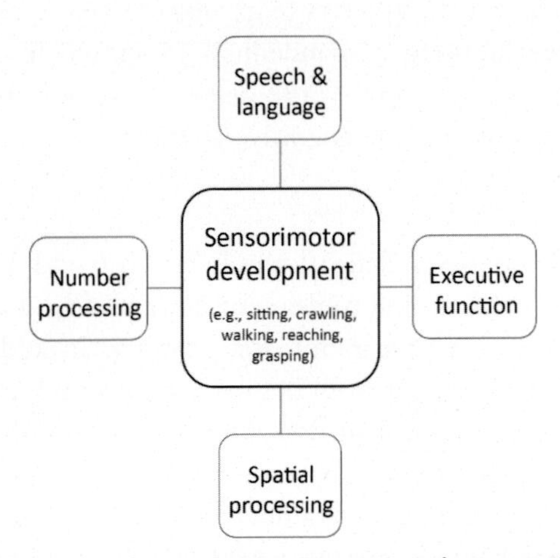

Figure 9.2 Schematic summarizing the section on: Interactions between the development of motor and cognitive abilities. Note how sensorimotor development sits at the heart of this interaction.

and linguistic development during this period. The results showed that walking experience (weeks of walking) was a significant predictor of language development and that the onset of walking was associated with an immediate increase in receptive language ability, independent of age. The second study demonstrated that walking infants had larger productive and receptive vocabularies than crawling infants regardless of the total locomotor experience. As walkers see the world differently from crawlers (Kretch et al., 2014), it has been argued that visual access to distal cues (including human bodies and faces) might be related to the increased language development at walking onset. A follow-up investigation asked whether the relationship between walking onset and enhanced language skill was true in other cultures/languages (He, Walle, & Campos, 2015). Researchers collected data from American and Chinese infants. Importantly, research has shown that Chinese infants start crawling and walking 1–2 months later than their American counterparts (Werner, 1972). As discussed earlier in the chapter, the later onset of these motor achievements appears to be related to rearing practices and not to genetics. The researchers therefore, were able to directly test if walking onset or developmental age determines the increase in vocabulary. Regardless of age and language (English or Mandarin), walking infants had greater receptive and productive vocabularies than same-aged crawling infants (He et al., 2015; Infancy). Taken together, these studies have shown that language development is related to infant walking acquisition.

There is also evidence that other cognitive abilities and not only language are related to motor development (Lifter & Bloom, 1989; Oudgenoeg-Paz, Leseman, & Volman, 2015). For example, Oudgenoeg-Paz et al. (2015) found that infants that learn to sit earlier display enhanced spatial memory, and that the earlier the onset of walking the better the infant's spatial processing. Less explored but unquestionably related is the relationship between numerical processing and motor development. It has been argued that finger counting represents the basis of numerical knowledge (for a review of this topic see Domahs, Kaufmann, & Fischer, 2012 Frontiers Research Topic: Handy numbers: Finger counting and numerical cognition) and finger gnosis is a better predictor of numerical performance in children than IQ (Noël, 2005).

Another motor achievement that has been associated with cognitive development is reaching and grasping. In a longitudinal study that follow infants from 9 to 26 months of age, Lifter and Bloom (1989) documented the infant's displacement of objects and their vocabulary extent. They found a close link between the infants' acting on objects and language progression. Although the authors did not report the incidence of right or left hand use during the manipulation of objects, other studies have shown a tight link between infants' hand preference and their cognitive development. Right-hand use correlates with analytic/receptive aspects of language at 13 months of age (Bates, O'Connell, Vaid, Sledge, & Oakes, 1986) and 14-month-old infants who use their right hands more often for pointing also demonstrated larger vocabulary (Esseily, Jacquet, & Fagard, 2011). Consistent with this research, our laboratory has demonstrated that when pointing to pictures, the strength of right-hand use positively correlates with a standard measure of language ability (using the Peabody Picture Vocabulary Test; PPVT-4) in 4–5-year-old children (Grandmont, Gibb, Li, & Gonzalez, 2017). With respect to grasping, consistency of right-hand use in infants is associated with advanced language skills in toddlers (Nelson, Campbell, & Michel, 2014). In this longitudinal study, infants that consistently showed a right-hand preference for grasping toys from 6 to 14 months of age demonstrated better language performance on the Bayley scales of infant and toddler development at 24 months of age. As such, we further investigated if hand preference for grasping would be a predictor of the maturity of the language production system. Studies have shown that the production of the "s" and "sh" sounds is relatively late in development given the high motor demand for articulation (Kent, 1992). Acquisition begins at 3 years of age and is mastered by age 7 (Sander, 1972). Using these sounds, we demonstrated that, in typically developing right-handed 4–5-year-olds, the greater the right-hand use for picking up small blocks, the greater the differentiation between the two sounds (Gonzalez et al., 2014a). This was a reciprocal relationship in that one predicts the other. Taken together, this evidence strongly links right-hand preference for pointing and grasping with enhanced reception and production of language.

Finally, a relationship between hand preference for grasping and executive function (EF) has also been established (Gonzalez et al., 2014b). EF is an umbrella term used to describe a variety of processes supported by the prefrontal cortex collectively known as higher cognitive functions. These include working memory, decision-making, planning and organization, and self-regulation. Importantly, it has been shown that EF is a better predictor of academic success than IQ (Diamond & Lee, 2011; Blair & Razza, 2007; Masten et al., 2012). Given the relationship demonstrated between language and right-hand preference for grasping (Esseily et al., 2011; Gonzalez et al., 2014a; Nelson et al., 2014) and the emerging evidence of the relationship between language (EF) in typically developing children (Kühn, Gleich, Lorenz, Lindenberger, & Gallinat, 2014; Miller & Marcovitch, 2015; Roello et al., 2014; Roello, Ferretti, Colonnello, & Levi, 2015), we wondered if the strength of right-hand preference would also predict EF. To achieve this end, right-handed typically developing children (5–6 years old) were assessed for their hand preference for grasping small blocks while their parent/caregiver completed the Behavioral Rating Inventory of EF (BRIEF; Gioia et al., 2000). The results showed that the stronger the right-hand preference, the better the child's reported EF (Gonzalez et al., 2014b). We have expanded this finding to include direct measures of EF by using a modified version of the Stroop (animal stroop) and of the Dimensional Change Card Sort (DCCS) test (Snap). Again, the greater the degree of right-hand preference, the better the performance on these EF tests (MacLean et al., in preparation).

In sum, many studies have documented the intertwined nature of motor and cognitive development (see Fig. 9.2). This evidence supports the embodied cognition theory in which the development of cognitive processes is rooted in the physical interactions of the body with the world.

9.5 CONCLUSION

Motor competency first emerges in the womb with reflexive movements and continues postnatally to comprise purposeful acts. Through these movements, infants learn about their environment, promoting parallel opportunities to develop and refine their cognitive ability. As such, there appears to be a bidirectional relationship between cognition and motor development, where advances in one skill promotes opportunities for advancement in the other.

REFERENCES

Adolph, K., & Berger, S. E. (2005). Physical and motor development. *Developmental science: An advanced textbook*, *5*, 223–281.

Adolph, K., & Robinson, S. R. (2013). The road to walking: What learning to walk tells us about development. In P. D. Zelazo (Ed.), *Oxford handbook of developmental psychology*. (1, pp. 403–443). NY: Oxford University Press.

Adolph, K., Cole, W. G., Komati, M., Garciaguirre, J. S., Badaly, D., Lingeman, J. M., ... Sotsky, R. B. (2012). How do you learn to walk? Thousands of steps and dozens of falls per day. *Psychological Science*, *23*(11), 1387–1394.

Adolph, K., Vereijken, B., & Denny, M. A. (1998). Learning to crawl. *Child Development*, *69*(5), 1299–1312.

Angulo-Kinzler, R. M., Ulrich, B., & Thelen, E. (2002). Three-month-old infants can select specific leg motor solutions. *Motor Control*, *6*(1), 52–68.

Armand, J., Olivier, E., Edgley, S. A., & Lemon, R. N. (1997). Postnatal development of corticospinal projections from motor cortex to cervical enlargement in the macaque monkey. *Journal of Neuroscience*, *17*, 251–266.

Atkinson, J. (1984). Human visual development over the first 6 months of life. A review and hypothesis. *Human Neurobiology*, *3*, 61–74.

Barbu-Roth, M., Anderson, D. I., Streeter, R., Combrouze, M., Park, J., Schultz, B., ... Provasi, J. (2015). Why does infant stepping disappear and can it be stimulated by optic flow? *Child Development*, *86*(2), 441–455.

Bates, E., O'Connell, B., Vaid, J., Sledge, P., & Oakes, L. (1986). Language and hand preference in early development. *Developmental Neuropsychology*, *2*(1), 1–15.

Berthier, N. E., & Carrico, R. L. (2010). Visual information and object size in infant reaching. *Infant Behavior and Development*, *33*, 555–566.

Berthier, N. E., Clifton, R. K., McCall, D. D., & Robin, D. J. (1999). Proximodistal structure of early reaching in human infants. *Experimental Brain Research*, *127*, 259–269.

Blair, C., & Razza, R. P. (2007). Relating effortful control, executive function, and false belief understanding to emerging math and literacy ability in kindergarten. *Child Development*, *78*(2), 647–663.

Bower, T. G. R. (1974). *Development in infancy*. San Francisco: Freeman.

Bruner, J. S., & Koslowski, B. (1972). Visually preadapted constituents of manipulatory action. *Perception*, *1*, 3–14.

Bushnell, E. W. (1985). The decline of visually guided reaching during infancy. *Infant Behavior & Development, 8*, 139–155.
Butterworth, G., & Harris, M. (1994). *Principles of developmental psychology*. Hove: Lawrence Erlbaum Associates.
Butterworth, G., & Hopkins, B. (1988). Hand-mouth coordination in the new-born baby. *British Journal of Developmental Psychology, 6*, 303–314.
Carrico, R., & Berthier, N. (2008). Vision and precision reaching in 15-month-old infants. *Infant Behavior and Development, 31*, 62–70.
Claxton, L. J., Keen, R., & McCarty, M. E. (2003). Evidence of motor planning in infant reaching behavior. *Psychological Science, 14*(4), 354–356.
Clifton, R. K., Muir, D. W., Ashmead, D. H., & Clarkson, M. G. (1993). Is visually guided reaching in early infancy a myth? *Child Development, 64*, 1099–1110.
Diamond, A., & Lee, K. (2011). Interventions shown to aid executive function development in children 4 to 12 years old. *Science, 333*(6045), 959–964.
Domahs, F., Kaufmann, L., & Fischer, M. H. (2012). Handy numbers: Finger counting and numerical cognition. *Frontiers (Psychology) Research Topics*, Available from http://dx.doi.org/10.3389/978-2-88919-059-1.
Dominici, N., Ivanenko, Y. P., Cappellini, G., d'Avella, A., Mondì, V., Cicchese, M., ... Giannini, C. (2011). Locomotor primitives in newborn babies and their development. *Science, 334*(6058), 997–999.
Esseily, R., Jacquet, A.-Y., & Fagard, J. (2011). Handedness for grasping objects and pointing and the development of language in 14-month-old infants. *Laterality: Asymmetries of Body, Brain and Cognition, 16*(5), 565–585.
Fallang, B., Saugstad, O. D., & Hadders-Algra, M. (2000). Goal directed reaching and postural control in supine position in healthy infants. *Behavioural Brain Research, 115*, 9–18.
Felt, R. H., Mulder, E. J., Lüchinger, A. B., Van Kan, C. M., Taverne, M. A., & de Vries, J. I. (2012). Spontaneous cyclic embryonic movements in humans and guinea pigs. *Developmental Neurobiology, 72*(8), 1133–1139.
Ferre, C. L., Babik, I., & Michel, G. F. (2010). Development of infant prehension handedness: A longitudinal analysis during the 6-to 14-month age period. *Infant Behavior and Development, 33*(4), 492–502.
Foroud, A. (2008). *Moving from stroke to development: A deconstruction of skilled reaching in humans. Doctor of Philosophy Thesis*. University of Lethbridge.
Foroud, A., & Whishaw, I. Q. (2012). The consummatory origins of visually guided reaching in human infants: a dynamic integration of whole-body and upper-limb movements. *Behavioural Brain Research, 231*(2), 343–355.
Geber, M., & Dean, R. (1957a). Gesell tests on African children. *Pediatrics, 20*, 1055–1065.
Geber, M., & Dean, R. (1957b). The state of development of newborn African children. *Lancet, 1*, 1216–1219.
Gioia, G. A., Isquith, P. K., Guy, S. C., & Kenworthy, L. (2000). Behavior rating inventory of executive function. *Child Neuropsychology, 6*(3), 235–238. Available from http://dx.doi.org/10.1076/chin.6.3.235.3152.
Gonzalez, C. L., Li, F., Mills, K. J., Rosen, N., & Gibb, R. L. (2014a). Speech in action: Degree of hand preference for grasping predicts speech articulation competence in children. *Frontiers in Psychology, 5*, 1267. Available from http://dx.doi.org/10.3389/fpsyg.2014.01267.
Gonzalez, C. L., Mills, K. J., Genee, I., Li, F., Piquette, N., Rosen, N., & Gibb, R. (2014b). Getting the right grasp on executive function. *Frontiers in Psychology, 5*, 285. Available from http://dx.doi.org/10.3389/fpsyg.2014.00285.
Grandmont, D., Gibb, R., Li, F., & Gonzalez, C. L. (2017). Pointing to the right meaning: Right hand use for pointing correlates with better receptive vocabulary in 3–6 year old children. *Abstract accepted in special issue Frontiers in Psychology*.
Greenman, G. W. (1963). Visual behavior of newborn infants. In A. J. Solnit, & S. A. Provence (Eds.), *Modern perspectives in child development*. New York: Hallmark.
He, M., Walle, E. A., & Campos, J. J. (2015). A cross-national investigation of the relationship between infant walking and language development. *Infancy, 20*(3), 283–305.

Heineman, K. R., Middelburg, K. J., & Hadders-Algra, M. (2010). Development of adaptive motor behaviour in typically developing infants. *Acta Paediatrics*, *99*(4), 618–624.

Hepper, P. G., Mccartney, G. R., & Shannon, E. A. (1998). Lateralised behaviour in first trimester human foetuses. *Neuropsychologia*, *36*, 531–534. Available from http://dx.doi:10.1016/S0028-3932(97)00156-5.

Hopkins, B., & Prechtl, H. F. R. (1984). A qualitative approach to the development of movements during early infancy. In H. F. R. Prechtl (Ed.), *Continuity of neural functions from prenatal to postnatal life. Clinics in Developmental Medicine* (Vol. 94, pp. 179–197). Oxford: Blackwell.

Hopkins, B., & Rönnqvist, L. (2002). Facilitating postural control: Effects on the reaching behavior of 6-month-old infants. *Developmental Psychobiology*, *40*(2), 168–182.

Karasik, L. B., Tamis-LeMonda, C. S., & Adolph, K. E. (2014). Crawling and walking infants elicit different verbal responses from mothers. *Developmental Science*, *17*(3), 388–395.

Karl, J. M., & Whishaw, I. Q. (2014). Haptic grasping configurations in early infancy reveal different developmental profiles for visual guidance of the Reach versus the Grasp. *Experimental Brain Research*, *232*, 3301–3316.

Kent, R. D. E. (1992). *Intelligibility in speech disorders: theory, measurement, and management* (Vol. 1), Amsterdam, Netherlands: John Benjamins Publishing.

Kilbride, J., Robbins, M., & Kilbride, P. (1970). The comparative motor development of Baganda, American white, and American black infants. *American Anthropologist*, *72*, 1422–1428.

Kolb, B., & Whishaw, I. Q. (2012). *An introduction to brain and behaviour* (4th ed). New York: Worth Publishers.

Konner, M. (1973). Newborn walking: Additional data. *Science*, *178*, 307.

Konner, M. (1977). Maternal care infant behavior and development among the Kalahari Deser San. In R. B. Lee, & I. DeVore (Eds.), *Kalahari hunter gatherers*. Cambridge, MA: Harvard University Press.

Kremenitzer, J. P., Vaughan, H. G., Kurtzberg, D., & Dowling, K. (1979). Smooth-pursuit eye movements in the newborn infant. *Child Development*, *50*, 442–448.

Kretch, K. S., Franchak, J. M., & Adolph, K. E. (2014). Crawling and walking infants see the world differently. *Child Development*, *85*(4), 1503–1518.

Kühn, S., Gleich, T., Lorenz, R. C., Lindenberger, U., & Gallinat, J. (2014). Playing Super Mario induces structural brain plasticity: gray matter changes resulting from training with a commercial video game. *Molecular Psychiatry*, *19*(2), 265–271.

Kuypers, H. G. J. M. (1981). Anatomy of the descending pathways. In J. M. Brookhart, & V. B. Mountcastle (Eds.), *Handbook of physiology. The nervous system, part II* (pp. 597–666). Bethesda: American Physiological Society.

Law, J., Lee, M., Hulse, M., & Tomassetti, A. (2011). The infant development timeline and its application to robot shaping. *Adapt Behavior*, *19*, 335–358.

Lhote, M., & Streri, A. (1998). Haptic memory and handedness in 2-month-old infants. *Laterality*, *3*, 173–192.

Lifter, K., & Bloom, L. (1989). Object knowledge and the emergence of language. *Infant Behavior and Development*, *12*(4), 395–423.

Lockman, J. L., Ashmead, D. H., & Bushnell, E. W. (1984). The development of anticipatory hand orientation during infancy. *Journal of Experimental Child Psychology*, *37*, 176–186.

Martin, J. H., Choy, M., Pullman, S., & Meng, Z. (2004). Corticospinal system development depends on motor experience. *The Journal of Neuroscience*, *24*(9), 2122–2132.

Martin, J. H., Friel, K. M., Salimi, I., & Chakrabarty, S. (2007). Activity- and use-dependent plasticity of the developing corticospinal system. *Neuroscience and Biobehavioral Reviews*, *31*, 1125–1135.

Masten, A. S., Herbers, J. E., Desjardins, C. D., Cutuli, J., McCormick, C. M., Sapienza, J. K., ... Zelazo, P. D. (2012). Executive function skills and school success in young children experiencing homelessness. *Educational Researcher*, *41*(9), 375–384.

Mathew, A., & Cook, M. (1990). The control of reaching movements by young infants. *Child Development*, *61*, 1238–1257.

McCall, D., Robin, D., Berthier, N. E., & Clifton, R. (1994). Vision and early reaching. *Infant Behavior and Development*, *17*, 521.

McCarty, M. E., Clifton, R. K., Ashmead, D. H., Lee, P., & Goubet, N. (2001). How infants use vision for grasping objects. *Child Development*, *72*, 973–987.

Michel, G. F., Babik, I., Sheu, C.-F., & Campbell, J. M. (2014). Latent classes in the developmental trajectories of infant handedness. *Developmental Psychology*, *50*(2), 349.

Miller, S. E., & Marcovitch, S. (2015). Examining executive function in the second year of life: coherence, stability, and relations to joint attention and language. *Developmental Psychology*, *51*(1), 101.

Morange-Majoux, F. (2011). Manual exploration of consistency (soft vs hard) and handedness in infants from 4 to 6 months old. *Laterality*, *16*, 292–312.

Morange-Majoux, F., Peze, A., & Bloch, H. (2000). Organisation of left and right hand movement in a prehension task: A longitudinal study from 20 to 32 weeks. Laterality: Asymmetries of Body. *Brain and Cognition*, *5*(4), 351–362.

Morrongiello, B. A., & Rocca, P. T. (1989). Visual feedback and anticipatory hand orientation during infants' reaching. *Perceptual Motor Skill*, *69*, 787–802.

Napier, K. M. (1956). The prehensile movements of the human hand. *The Journal of Bone and Joint Surgery*, *38B*, 902–913.

Nelson, E. L., Campbell, J. M., & Michel, G. F. (2013). Unimanual to bimanual: Tracking the development of handedness from 6 to 24 months. *Infant Behavior and Development*, *36*(2), 181–188.

Nelson, E. L., Campbell, J. M., & Michel, G. F. (2014). Early handedness in infancy predicts language ability in toddlers. *Developmental Psychology*, *50*(3), 809.

Newell, K. M., Scully, D. M., McDonald, P. V., & Baillargeon, R. (1993). Body scale and infant grip configurations. *Developmental Psychobiology*, *26*, 195–205.

Noël, M.-P. (2005). Finger gnosia: A predictor of numerical abilities in children? *Child Neuropsychology: A Journal on Normal and Abnormal Development in Childhood and Adolescence*, *11*(5), 413–430. Available from http://doi.org/10.1080/09297040590951550.

Ocklenburg, S., Schmitz, J., Moinfar, Z., Moser, D., Klose, R., Lor, S., ... Güntürkün, O. (2017). Epigenetic regulation of lateralized fetal spinal gene expression underlies hemispheric asymmetries. *Elife*, *6*, e22784. Available from http://dx.doi.org/10.7554/eLife.22784.

Olivier, E., Edgley, S. A., Armand, J., & Lemon, R. N. (1997). An electrophysiological study of the postnatal development of the corticospinal system in the macaque monkey. *Journal of Neuroscience*, *17*, 267–276.

Oudgenoeg-Paz, O., Leseman, P. P., & Volman, M. (2015). Exploration as a mediator of the relation between the attainment of motor milestones and the development of spatial cognition and spatial language. *Developmental Psychology*, *51*(9), 1241.

Perlstein, M. A., & Barnett, H. E. (1952). Nature and recognition of cerebral palsy in infancy. *Journal of the American Medical Association*, *148*, 1389–1397.

Prechtl, H. F. R. (2001). General movement assessment as a method of developmental neurology: new paradigms and their consequences. *Developmental Medicine & Child Neurology*, *43*, 836–842.

Robin, D., Berthier, N. E., & Clifton, R. (1996). Infants' predictive reaching for moving objects in the dark. *Developmental Psychology*, *32*, 824–835.

Rocha, N. A. C. F., Silva, F. P. S., & Tudella, E. (2006). The impact of object size and rigidity on infant reaching. *Infant Behavior & Development*, *29*, 251–261.

Rochat, P. (1989). Object manipulation and exploration in 2- to 5-month old infants. *Developmental Psychology*, *25*, 871–884.

Rochat, P., & Senders, S. J. (1991). Active touch in infancy: action systems in development. In M. J. Weiss, & P. R. Zelazo (Eds.), *Infant attention: biological constraints and the influence of experience*. Norwood: Ablex.

Roello, M., Ferretti, M. L., Colonnello, V., & Levi, G. (2014). When words lead to solutions: Executive function deficits in preschool children with specific language impairment. *Research in Developmental Disabilities*, *37C*, 216–222, Available from doi:S0891-4222(14)00500-9 [pii]10.1016/j.ridd.2014.11.017.

Roello, M., Ferretti, M. L., Colonnello, V., & Levi, G. (2015). When words lead to solutions: executive function deficits in preschool children with specific language impairment. *Research in Developmental Disabilities*, *37*, 216–222.

Rönnqvist, L., & Domellöf, E. (2006). Quantitative assessment of right and left reaching movements in infants: A longitudinal study from 6 to 36 months. *Developmental Psychobiology*, *48*(6), 444–459.

Rönnqvist, L., & von Hofsten, C. (1994). Neonatal finger and arm movements as determined by a social and an object context. *Early Development Parent, 3*, 81–94.

Rose, S. A. (1984). Developmental changes in hemispheric specialization for tactual processing in very young children: Evidence from cross-modal transfer. *Developmental Psychology, 20*, 568–574. Available from http://dx.doi.org/10.1037/0012-1649.20.4.568.

Rovee-Collier, C. K., Morrongiello, B. A., Aron, M., & Kupersmidt, J. (1978). Topographical response differentiation and reversal in 3-month-old infants. *Infant Behavior and Development, 1*, 323–333.

Ruff, H. A. (1986). Components of attention during infants' manipulative exploration. *Child Development, 57*, 105–114.

Sacrey, L. R., & Whishaw, I. Q. (2010). Development of collection precedes targeted reaching: Resting shapes of the hands and digits in 1-6-month-old human infants. *Behavior Brain Research, 214*(1), 125–129.

Sacrey, L. A., Karl, J. M., & Whishaw, I. Q. (2012a). Development of visual and somatosensory attention of the reach-to-eat movement in human infants aged 6 to 12. *Experimental Brain Research, 223*, 121–136.

Sacrey, L. R., Karl, J. M., & Whishaw, I. Q. (2012b). Development of rotational movements, hand shaping, and accuracy in advance and withdrawal for the reach-to-eat movement in human infants aged 6–12 months. *Infant Behavior and Development, 35*(5), 543–560.

Sacrey, L. A., Arnold, B., Whishaw, I. Q., & Gonzalez, C. L. (2013). Precocious hand use preference in reach-to-eat behavior versus manual construction in 1- to 5-year-old children. *Developmental Psychobiology, 55*(8), 902–911. Available from http://dx.doi.org/10.1002/dev.21083. PMID: 23129422.

Sander, E. K. (1972). When are speech sounds learned. *Journal of Speech and Hearing Disorders, 37*, 55–63.

Schum, N., Jovanovic, B., & Schwarzer, G. (2011). Ten- and twelve-monthholds' visual anticipation of orientation and size during grasping. *Journal of Experimental Child Psychology, 109*, 218–231.

Siekerman, K., Barbu-Roth, M., Anderson, D. I., Donnelly, A., Goffinet, F., & Teulier, C. (2015). Treadmill stimulation improves newborn stepping. *Developmental Psychobiology, 57*(2), 247–254.

Silva, F. P., Rocha, N. A., & Tudella, E. (2011). Can size and rigidity of objects influence infant's proximal and distal adjustments of reaching? *Revista Brasileira Fisioterapia, 15*(1), 37–44.

Super, C. M. (1976). Environmental effects on motor development: The case of 'African infant precocity'. *Developmental Medicine & Child Neurology, 18*(5), 561–567.

Thelen, E. (1979). Rhythmical stereotypies in normal human infants. *Animal Behaviour, 27*, 699–715.

Thelen, E., & Fisher, D. M. (1983). From spontaneous to instrumental behavior: Kinematic analysis of movement changes during very early learning. *Child Development*, 129–140.

Thelen, E., Corbetta, D., & Spencer, J. P. (1996). Development of reaching during the first year: role of movement speed. *Journal of Experimental Psychology and Human Perception and Performance, 22*, 1059–1076.

Thomas, B. T., Karl, J. M., & Whishaw, I. Q. (2015). Independent development of the Reach and the Grasp in spontaneous self-touching by human infants in the first 6 months. *Frontiers of Psychology*, Available from http://dx.doi.org/10.3389/fpsyc.2014.01526.

Touwen, B. (1976). *Neurological development in infancy. Clinics in developmental medicine, no 58*. London: Heinemann Medical Books.

Twitchell, T. E. (1970). Reflex mechanisms and the development of prehension. In K. Connolly (Ed.), *Mechanisms of motor skill development*. New York: Academic Press.

von Hofsten, C. (1982). Eye–hand coordination in the newborn. *Developmental Psychology, 18*, 450–461.

von Hofsten, C. (1984). Developmental changes in the organization of prereaching movements. *Developmental Psychology, 3*, 378–388.

von Hofsten, C. (1991). Structuring of early reaching movements: A longitudinal study. *Journal of Motor Behavior, 23*, 280–292.

von Hofsten, C., & Fazel-Zandy, S. (1984). Development of visually guided hand orientation in reaching. *Journal of Experimental Psychology, 38*, 208–219.

von Hofsten, C., & Lee, D. N. (1982). Dialogue on perception and action. *Human Movement Science, 1*, 125–138.

von Hofsten, C., & Lindhagen, K. (1979). Observation on the development of reaching for moving objects. *Journal of Experimental Child Psychology, 28*, 158–173.
von Hofsten, C., & Rosander, K. (1997). Development of smooth pursuit tracking in young infants. *Vision Research, 37*, 1799–1810.
Wallace, P. S., & Whishaw, I. Q. (2003). Independent digit movements and precision grip patterns in 1-5-month-old human infants: Hand babbling, including vacuous then self-directed hand and digit movements, precedes targeted reaching. *Neuropsychologia, 41*(14), 1912–1918.
Walle, E. A., & Campos, J. J. (2014). Infant language development is related to the acquisition of walking. *Developmental Psychology, 50*(2), 336.
Wentworth, N., Benson, J. B., & Haith, M. M. (2000). The development of infants' reaches for stationary and moving objects. *Child Development, 71*, 576–601.
Werner, E. E. (1972). Infants around the world: Cross-cultural studies of psychomotor development from birth to two years. *Journal of Cross-Cultural Psychology, 3*(2), 111–134.
White, B. L., Castle, P., & Held, R. (1964). Observations on the development of visually-directed reaching. *Child Development, 35*, 349–363.
Whyte, V. A., McDonald, P. V., Baillargeon, R., & Newell, K. M. (1994). Mouthing and grasping of objects by young infants. *Ecological Psychology, 6*, 205–218.
Williams, J. L., & Corbetta, D. (2016). Assessing the impact of movement consequences on the development of early reaching in infancy. *Frontiers in Psychology*, . Available from http://dx.doi.org/10.3389/fpsyc.2016.00587.
Witherington, D. C. (2005). The development of prospective grasping control between 5 and 7 months: a longitudinal study. *Infancy, 7*, 143–161.
Zelazo, P. R. (1976). From refelxive to instrumental behavior. In L. Lipsitt (Ed.), *Developmental psychobioloty: The significance of infancy.* Hillsdale, N.J.: Lawrence Erlbaum Associates, Inc.
Zelazo, P. R. (1983). The development of walking: new findings and old assumptions. *Journal of Motor Behavior, 15*(2), 99–137.
Zelazo, P. R., Zelazo, N. A., & Kolb, S. (1972). Newborn walking. *Science, 1972*(177), 1058–1059.

FURTHER READING

Adolph, K., Karasik, L., & Tamis-LeMonda, C. (2010). Handbook of cross-cultural development science. *Domains of development across cultures.*
Bril, B., & Sabatier, C. (1986). The cultural context of motor development: Postural manipulations in the daily life of Bambara babies (Mali). *International Journal of Behavioral Development, 9*(4), 439–453.
Forssberg, H. (1985). Ontogeny of human locomotor control I. Infant stepping, supported locomotion and transition to independent locomotion. *Experimental Brain Research, 57*(3), 480–493.
Forssberg, H. (1999). Neural control of human motor development. *Current Opinion in Neurobiology, 9*(6), 676–682.
Hopkins, B. (1976). Culturally determined patterns of handling the human infant. *Journal of Human Movement Studies, 2*, 1–27.
Kent, R. D. E. (1972). *Inteligibility in speech disorders: Theory, Measurement, and Management* (Vol. 1) Amsterdam: John Benjamins Publishing.
Kopp, C. B. (2011). Development in the early years: Socialization, motor development, and consciousness. *Annual Review of Psychology, 62*, 165–187.
Lacquaniti, F., Ivanenko, Y. P., & Zago, M. (2012). Development of human locomotion. *Current Opinion in Neurobiology, 22*(5), 822–828.
McLean, J., Gibb, R., & Gonzalez, C.L. (In preparation). Snap and Stroop: Two table-top tests of executive function correlate with the Behavioral Rating Inventory of Executive Function (BRIEF).
Rabain-Jamin, J., & Wornham, W. L. (1993). Practices and representations of child care and motor development among West Africans in Paris. *Infant and Child Development, 2*(2), 107–119.

CHAPTER 10

Neural Foundations of Cognition and Language

Lindsay C. Bowman[1,2,*,], Lara J. Pierce[1,2,*], Charles A. Nelson, III[1,2,3,4] and Janet F. Werker[4,5]**
[1]University of California, Davis, CA, United States
[2]Harvard Medical School, Boston, MA, United States
[3]Harvard University, Cambridge, MA, United States
[4]Canadian Institutes for Advanced Research, Toronto, ON, Canada
[5]University of British Columbia, Vancouver, BC, Canada

10.1 A PRECIS TO BRAIN DEVELOPMENT

Brain development is a protracted process that begins at conception and continues well into the second decade of life. We provide a brief outline of the prenatal stages of brain development, and highlight continued development in some cases long after birth. For additional detail, see Chapter 1 of this book, Overview of Brain Development.

Neural tissue is a derivative of the ectodermal (outer) layer of the embryo, and begins to form 12–18 days after conception. Over successive days the primitive neural plate folds over onto itself, forming a groove. By the end of the third week, this groove has become a tube, which then closes at the top and bottom ends. This process of *neurulation* provides the primitive scaffolding of brain development, and is complete by the end of the third week of gestation.

Once the neural tube has formed, progenitor cells that line the tube begin to give rise to the many classes of neurons and glia that will eventually swell to the tens of billions. This process of *neurogenesis* continues into the early postnatal years, when it finally comes to an end. Once cells begin to be formed, a subset of them begins to migrate radially and tangentially from their point of origin, eventually to create the cerebral cortex. This process of *cell migration* is largely complete by the 25th prenatal week, at which point there are 6 layers in all. Note that the cortex forms in an inside-out fashion, with the deepest layers forming first, followed by more superficial layers.

Once an immature neuron has migrated to its predetermined location (orchestrated by a combination of genetics and humoral signals), the cells mature, forming

* Cofirst authors.
** Work on this chapter was also conducted while author was affiliated with Boston Children's Hospital, Boston, MA, and Harvard Medical School, Boston, MA

The Neurobiology of Brain and Behavioral Development
DOI: http://dx.doi.org/10.1016/B978-0-12-804036-2.00010-8

cell bodies and processes such as dendrites and axons. This, in turn, enables neurons to begin the process of making connections with other neurons (*synaptogenesis*). The first synapses appear around the 23rd prenatal week, and continue well into the postnatal period. Importantly, we vastly overproduce synapses, followed by a period of pruning. In the visual system, for example, the peak of overproduction occurs in the first half year of life, followed by pruning to adult numbers, which occurs toward the end of the preschool period (e.g., Huttenlocher, 1994; Huttenlocher, de Courten, Garey, & Van der Loos, 1982). In contrast, the peak of overproduction of synapses in the prefrontal cortex does not occur until approximately 1 year of age, with pruning continuing not only through mid to late adolescence (Huttenlocher, 1979), but for some areas of prefrontal cortex, even into the third decade (Petanjik et al., 2011). The overproduction of synapses seems to be largely under genetic control, whereas the pruning process seems to be disproportionately dependent upon experience (Bourgeois, 2001).

The final phase of brain development is *myelination*, the process whereby some axons are coated with a fatty substance—myelin—that speeds the transmission of electrical impulses. Myelination begins in the late prenatal period and, depending on the area, continues well into postnatal life. As a rule, sensory regions myelinate long before regions concerned with higher cognitive functions; for example, sensory roots begin to myelinate in the third trimester of pregnancy, whereas regions of the prefrontal cortex continue to myelinate well into the second decade of life. (For an excellent overall review of brain development written for developmental scientists, see Nowakowski & Hayes, 2012; Stiles, 2008.)

10.2 CONNECTING BRAIN AND BEHAVIOR

10.2.1 The general value of developmental neuroscience

An understanding of the complex and lengthy process of brain development is critical to our understanding of the development of cognition and language. Of particular importance is an understanding of the timing at which each process unfolds and the varying influence of experience at different points in time. On the one hand, the brain in its earliest stages is not the same as the brain in its later stages, and thus how the brain supports cognition and language may also change across the lifespan. On the other, the development of certain abilities is well underway in infancy, and relatively early experiences may influence later outcomes in important ways. An understanding of how the brain supports cognition and language, and more importantly an understanding of whether and how the relation between brain and cognition changes with time and experience, offers a deeper understanding of cognition, language, and their respective developmental processes.

There are several contributions that the study of neuroscience makes to our understanding of developing cognition and language. Most simply, examining neural correlates of complex cognition and behavior offers an important complement to behavioral methods, as it can reveal underlying similarities where behavioral data show differences and vice versa, and it can highlight component processes that underlie complex reasoning. When neuroscience is used in the context of development it can be even more valuable. For example, identifying the neural correlates of components of cognition and language in infants and young children can reveal biological foundations of these domains. Such early neural correlates can be compared with those present in expert adults to investigate differences and similarities pointing to developmental continuity and change. Further, tracking these neural correlates longitudinally across development can potentially identify neural markers of biological change even prior to observable changes in behavior.

Finally, and most broadly, neuroscience adds a level of biological plausibility to our behavioral models of development (see Nelson, Moulson, & Richmond, 2006), and reveals the importance of considering the impact of experience on developing cognition. Classic arguments in development concern *experience-independent* models (which argue that certain language and cognitive abilities are essentially hard-wired at birth, and minimally affected by subsequent experience) versus *experience-dependent* or *experience–expectant* models (which emphasize the role of experience in shaping brain and behavior; see Greenough, Black, & Wallace 1987, for an early exposition of this view; Werker & Gervain, 2013, for a more detailed explication of this distinction). Viewing behavior through the lens of biology—via examination of its neural correlates—can shed light on these classic debates. Indeed, when adopting this biological lens, we might ask how likely it is that a set of neural circuits supporting a complex cognitive process is wired without benefit of experience. Such an experience-independent model would require an incredibly elaborate genetic network, the expression of which would require little experiential input, which seems unlikely. Are there any elements of cognition for which we think such a dedicated and resource-intensive process is plausible, without the shaping power of experience? Likewise, we could also ask whether there are limits to the extent that experience can shape developing cognition, language, and their neural foundations, and whether there are periods in development when the effect of experience is most or least influential. (Realistically, genes likely confer limits to the extent to which experience influences development but identifying what those limits are, and whether they can be shifted remains to be determined). Thus, a biological lens, as applied through a neuroscientific approach to the investigation of language and cognitive development, also sparks core questions in developmental psychology.

10.2.2 Developmental neuroscience methods

Developmental neuroscience research presents a special challenge given the application of complex and often time-consuming neuroscientific paradigms to pediatric populations. These populations are challenging given shorter attention spans, the increased need for noninvasive methods, and the increased need for flexibility during testing (e.g., to take breaks to rest, eat, change diapers). Three commonly used methods that navigate many of these challenges are: functional/structural magnetic resonance imaging (fMRI/MRI), functional near-infrared spectroscopy (fNIRS), and electroencephalography (EEG) (see Box 10.1).

The main strength of fMRI/MRI is its high spatial resolution: it yields a clear structural image of the brain (detailing white and gray matter), and it can localize (within limits) brain activity to specific regions with high resolution. Weaknesses of this method are that it is noisy, expensive, requires the participant to separate from their caregiver to lie inside the MRI scanner, and is very sensitive to motion artifact. Thus, infants typically need to be asleep or sedated during data collection to ensure best quality data. This, of course, limits the interrogation of brain function.

In contrast, fNIRS and EEG methods more easily allow testing of awake infants and children. Each of these methods offer a quiet testing environment in which the participant can be held by their caregiver, they are more motion tolerant, and considerably less expensive. While both methods have a lower spatial resolution compared to fMRI, fNIRS still offers reasonably precise localization given brain activity is measured in regions below and between given optode-pairs. In contrast, EEG measures electrical activity generated by large populations of neurons that spreads as it travels to sensors on the scalp surface, so its spatial resolution is low. However, temporal resolution for EEG is high given the speed at which electrical signals travel from source to sensor. Thus, EEG yields precise information on the timing of neural responses, which can elucidate components of complex processes, and potentially the order in which they occur.

The overall strength of each of these methods is that they yield a measurable response that can show variation across participants, critically even in very young participants for whom eliciting a behavioral response is difficult or at times impossible.

In the broadest sense, neuroscience helps further illuminate the picture of development. As discussed, it has many advantages that can apply to understanding development in general. More specifically for cognition and for language, as we outline next, it can shed light on open debates, it can reveal developments earlier in the lifespan than can be captured by behavioral methods alone, and it can deconstruct complex behaviors into component processes. We offer nuanced examples of the neural foundations of particular aspects of cognitive and language development next.

10.3 COGNITION

Broadly, *cognition* refers to a compilation of mental processes that support the acquisition of knowledge and understanding of the world. *Cognitive processes* can include how information is acquired, stored, evaluated, and used to guide behavior and further understanding. Some domains of cognitive processing are more complex, and include

BOX 10.1 Overview of Common Methods in Developmental Neuroscience

fMRI/MRI

- Uses magnetic principles to detect changes in oxygenated and deoxygenated blood flow in the brain (an indirect correlate of neuronal activity).
- Oxygenated and deoxygenated blood types have different magnetic properties. Thus, participants lie surrounded by large magnetic coils that generate magnetic fields, which detect changes in oxy- and deoxyhemoglobin concentrations.
- Critically, oxy- and deoxyhemoglobin concentrations change systematically as a function of neuronal activity. This neural-dependent bloodflow change is termed the *hemodynamic response function* (HRF), which peaks ~5 s after a neuronal event.
- The HRF is used to derive the blood-oxygen-level-dependent signal, which can then be compared across conditions and populations as an indirect correlate of neuronal activity.
- fMRI has a spatial resolution of ~1–3 mm (given that the magnetic field penetrates deep into the brain to detect blood flow changes very close to the site of neural activity).
- Temporal resolution is ~1–5 s (given the lag between neuronal events and corresponding HRF).
- For reviews of the fMRI/MRI method in developmental cognitive neuroscience see Casey, Giedd, & Thomas, 2000; Casey, Tottenham, Liston, & Durston, 2005; Lenroot & Giedd, 2006.

fNIRS

- Uses near-infrared light (~600–900 nm) to detect changes in oxy- and deoxyhemoglobin concentration as a function of neuronal activity.
- Near-infrared light is absorbed differently depending on the concentration of oxy- and deoxyhemoglobin in the blood. Thus, emitter optodes (that shine infrared light into the brain) and detector optodes (that detect light scattering back up to the scalp surface) are placed on a participant's scalp, and the temporal changes in absorption at each emitter-detector pair are used to calculate the HRF as an indirect correlate of neural activation.
- fNIRS can detect changes in hemodynamic signal (like fMRI), but can also measure responses of oxygenated and deoxygenated blood separately (unlike fMRI).
- Optimal measurements occur when emitter-detector pairs are placed 2–3 cm apart. At this distance, light penetrates maximally to ~3 cm below the scalp, allowing examination of hemodynamics on the cortical surface. The spatial resolution of fNIRS is dependent on the distance between emitter and detector pairs. A typical resolution is ~2–3 cm. Moreover, fNIRS provides reasonably precise localization of the brain activity occurring below and between paired emitters and detectors.

(*Continued*)

BOX 10.1 Overview of Common Methods in Developmental Neuroscience (Continued)

- Temporal resolution for this method is the same as for fMRI (~1–5 s) given it relies on changes in the HRF, although instrument sampling frequencies are much higher (10–100 Hz) and could theoretically determine more subtle changes in temporal response patterns.
- See Lloyd-Fox, Blasi, and Elwell (2010); Gervain, Mehler, Werker, Nelson, Csibra et al. (2011) for reviews of developmental fNIRS studies.

EEG

- Uses electrical sensors placed on a participant's scalp to detect the electrical potentials that are a direct consequence of the activity from underlying populations of neurons.
- Electrical activity is measured in waveform oscillations that have particular properties such as frequency, amplitude, and phase. Variations in each of these properties reveal information on underlying neural processes.
- Given the speed at which electrical activity travels from source to scalp, the temporal resolution of this method is high (on the order of milliseconds). Thus, EEG is often used to measure neural responses specific to a given event (e.g., neural response to the presentation of happy versus angry face). We can examine the amplitude, latency, and topographic distribution (i.e., the pattern of activity across sensors on the scalp) of event-related activity by examining *event-related potentials* (ERPs). (See Nelson & McCleery 2008 for a review of developmental ERP studies.)
- The EEG signal can also be broken down into activity in different frequency bands (e.g., delta, theta, alpha, beta, gamma, theta). We can examine frequency data in relation to specific events. We can also examine frequency data that are not to specific events, which can indicate functional connectivity and neural maturation.
- Spatial resolution of EEG data is associated with the number of sensors placed on the head. Thus it is on the order of centimeters and is somewhat ambiguous (given several scalp sensors could measure activity from one source). Source localization methods that statistically estimate the location of underlying sources can allow for more precise spatial resolution.
- See de Haan (2013) for more details on infant EEG and ERP.

component cognitions as part of a larger mechanism. The study of *cognitive development* can include examining the presence or absence of specific cognitions at different ages, as well as how cognitive processes change or stay the same over time, and how they set up future understanding.

There are many domains of cognitive processing that we could have selected for this chapter (e.g., symbolic representation, Callaghan & Corbit, 2015; spatial cognition, Olson & Bialystok, 2014; inductive reasoning, Fisher, Godwin, & Matlen, 2015), and many that have differing neural circuitries (e.g., attention, Langner & Eickhoff, 2013; number cognition, Ansari, 2008; executive functioning (see

Chapter 11: Toward an Understanding of the Neural Basis of Executive Function Development). Throughout the following subsections, we discuss the neural foundations of three cognitive processes in particular: Face-Processing, Memory, and Theory of Mind (ToM). Face-processing is an example of a simpler cognition that often becomes a component of more complex reasoning, and that includes the acquisition of information through perception. Memory is a cognitive process that consists of the storage of information and its recall once acquired. And theory of mind exemplifies a complex cognitive process that includes not only component cognitions such as those supporting the acquisition and storage of information, but also "higher level" reasoning and evaluation of information. Developments in each of these domains are manifest over infancy and childhood, and there is now a considerable amount of research on their neural foundations. Thus, together, these three examples help illustrate the incredible variability of cognition in terms of both function and underlying neural correlates, and in terms of development.

10.3.1 Face processing

The processing of faces, a key perceptual-cognitive domain, represents an early-emerging ability that plays a critical role in communication, before the onset of formal language; indeed, there are many aspects of face processing, including recognition of identity (Field, Cohen, Garcia, & Greenberg, 1984), gender (Leinbach & Fagot, 1993), race (Anzures, Quinn, Pascalis, Slater, & Lee, 2010), and emotion (Young-Browne, Rosenfeld, & Horowitz, 1977), that come on line in the first months of life yet have a protracted development. As faces are ubiquitous in the infants' world from the first moments of life and because of their role in nonverbal communication, considerable attention has been devoted to the study of face processing. For example, nearly 5 decades ago it was reported that newborns prefer face-like stimuli over nonface like stimuli (Fantz, 1963; Goren, Sarty, & Wu, 1975). This does not mean, of course, that there is an "innate" module for processing faces (which is what was claimed at the time); rather, it means that the brain, at birth, comes prepared to prioritize oval stimuli with openings for eyes and a mouth over other stimuli. Although many have argued that such preparedness reflects an innate mechanism, two factors must be considered. First, most studies of newborns focus on infants in the first days of life, which means that such infants have had experience viewing faces before being tested; thus, it is possible that the preference for faces over nonfaces could be attributed to a rapid learning mechanism. Second, unlike speech and language, infants are not exposed to faces prenatally. However, before birth they *are* capable of feeling their own face, both via proprioception and via hand-to-face activity (Quinn and Slater, 2003). Thus, we should not discount the possibility that infants are exposed to faces before birth and that the preference for faces that has been observed in the newborn period might be due to prenatal experience (although such evidence is currently lacking).

Be that as it may, shortly after birth infants attend preferentially to faces. This bias cascades into a sophisticated face processing "system" that evolves rapidly over the first 6–12 months of life. During this time infants develop the ability to discriminate faces from nonfaces, one face from another, one class of faces from another (e.g., male, female) and subtle differences in features of the face (such as facial emotion). Although face processing continues to develop through the school-age years, by the end of infancy there is in place a reasonably sophisticated face processing system (see Nelson, 2001; Pascalis & Kelly, 2009; Righi & Nelson, 2013 for additional reviews).

As with adults, the neural foundations of an infant face-processing system include the visual cortex and regions of the inferior temporal cortex, including the fusiform gyrus. Imaging studies in infants, using fMRI, PET, and event-related potentials (ERPs) have consistently shown activation in these regions for processing facial identity, and in adjacent regions for processing facial emotion. These regions are all online in the first 6 months of life, although they continue to mature over the next decade (de Haan, 2008; Johnson & De Haan, 2015; Righi & Nelson, 2013).

Overall, the ability to discriminate and recognize faces appears in the first weeks and months of life, and improves rapidly during the first few postnatal years. Adult levels of face processing (both identity and facial emotion) are likely not in place until around the time of adolescence, although long before that children are quite proficient in these skills.

10.3.2 Memory development

Memory refers to the critical cognitive processes of storing and recalling information, once acquired. Building on the ground breaking work of Fantz and Fagan (Fagan, 1970; Fantz, 1964), using novelty preferences, there have been countless demonstrations that infants show evidence of memory in the first year of life; for example, based on the principles of operant conditioning, infants as young as 3 months would discriminate a new mobile hanging over their crib from one 24 hours earlier; as they got older, the interval between initial exposure and retrieval lengthened from days to weeks to months (Rovee-Collier & Hartshorn, 1999; Rovee-Collier, 1999). Similarly, using a procedure known as elicited (Bauer & Hertsgaard, 1993) or deferred (Meltzoff, 1988, 1995) imitation, infants 12–18 months old have been shown to be able to remember a sequence of events (e.g., a marble dropped into a cup and then watch the cup being shaken) over several weeks. By the time children reach school-age, their memory begins to resemble that of adults, although on more difficult tasks adults still outperform children; the transition to adult levels of performance tends to occur by the age of 10 or so years. Changes from the preschool period to the preadolescent period have generally been attributed to the development of meta memory skills—that is, the ability to use strategies to remember things (Joyner & Kurtz-Costes, 1997; Kreutzer, Leonard, & Flavell, 1975).

The basic principles of encoding and retrieval (fundamentals of so-called *explicit* or *declarative* memory) are believed to be subserved by a distributed set of structures that reside in the medial temporal lobe, including the hippocampus and surrounding cortex (e.g., entorhinal and perirhinal cortices). Over the past 20 years, Nelson and colleagues (Bauer, 2006; Jabés & Nelson, 2015; Nelson, 1995; Newcombe, Lloyd, & Ratliff, 2007) have argued that much of this circuitry makes its appearance over the first year of life. However, delayed maturation of the hippocampus and later, the prefrontal cortex, make possible the more adult-like features of memory; the latter, in particular, appears to play an important role in the development of the strategies used to facilitate memory.

In some respects, the development of memory is paradoxical. On the one hand, infants in the first year of life are capable of impressive feats of memory, beginning with showing preferences for novel stimuli after just a few seconds of exposure to a (familiar) stimulus, to being able to recognize stimuli that had not been seen for weeks (Rovee-Collier, 1999). On the other, most adults do not remember anything before their second birthday, with the average age of first memories actually being closer to 4 years (Howe, Courage, & Edison, 2003). How do we reconcile this phenomenon known as *infantile amnesia* with the young infant's ability to remember stimuli over protracted periods of time? Although Freud posited that we repressed these early memories, a more nuanced neuroscience perspective would argue that long-term memories are stored in the neocortex, which is relatively immature until 4–5 years of age; and that *retrieval* strategies, which are subserved by the prefrontal cortex, are also immature until school age.

10.3.3 Theory of mind

ToM reflects our ability to make sense of and engage in human action and interaction. Specifically, ToM describes the ability to attribute mental states (such as beliefs, desires, intentions, and knowledge) to the self and others, and to understand that it is these internal mental states that motivate outward behavior. For example, we can understand the simple action of going to the refrigerator for milk as a behavior that is motivated by a *desire* for milk, and a *belief* that the milk is in the refrigerator. Even if in reality the milk was left in a grocery bag in the car, the belief (however false) that it is located in the refrigerator, and the desire to drink it, are ultimately what guide the behavior (Saxe, 2013; Wellman, 2002, 2011; Wimmer & Perner, 1983).

Though sometimes equated with achievement of understanding false beliefs, ToM is much broader than that, as demonstrated in the simple example above in which beliefs and desires interacted to motivate behavior (we might predict a different behavior if we knew someone thought there was milk in the refrigerator but did not desire it). The role of multiple mental states becomes even more apparent when we consider not just simple actions with objects, but interactions with others. Thus, most broadly, ToM describes the wide-ranging human understanding of agents' multiple interactive mental states and how each of these internal mental states ultimately shape human behavior (Wellman, 2014).

Behavioral development. From a behavioral perspective, infants and children demonstrate marked changes in their ability to reason about mental states. Over the first 2 years of life, infants' understanding advances from initial unsteady links between action and intention, to more solid conceptions of person-specific desires. For example, by 8 to 9 months, infants can identify the goals and intentions behind others' successful actions (e.g., reaching for and retrieving a ball; Woodward 1998). But by 10–12 months, infants can identify the intentions behind "failed" actions as well (e.g., reaching toward but never grasping a ball behind a barrier; Brandone & Wellman, 2009). At 12 months of age, infants understand that actions are based on desires (e.g., they show surprise when an actor chooses a toy that does not match her expressed desire) (Phillips, Wellman, & Spelke, 2002). But beyond 12 months, infants further recognize that desires are subjective and distinct; that they are specific to individuals and can sometimes contrast with the infants' own desires (e.g., Repacholi & Gopnik, 1997).

Some research also suggests that infants possess an implicit (nonverbal) understanding of beliefs. Investigations of implicit understanding typically measure infants' eye gaze (e.g., length of gaze, anticipatory saccades) as a way of uncovering infants' underlying expectations about the scenes they are viewing. In an initial, often-cited study, Onishi and Baillargeon (2005) found that 15-month-olds look longer (arguably indicating surprise) when an actor behaves in a way that is inconsistent with her prior belief (e.g., searches for a toy in location B when she originally placed the object in location A and did not see the object move). This finding has since been replicated with other eye-gaze measures (e.g., infants will shift their eye gaze to the location consistent with where an agent thinks her object is; Southgate, Senju, & Csibra, 2007; Surian, Caldi, & Sperber, 2007). For additional reviews, see Slaughter (2015) and Sodian (2011).

Later in development, children come to explicitly reason about internal mental states as person-specific representations of the world that are related to and constrained by a person's experiences. For example, children can predict that even though they themselves know the contents of a closed box, another person who has not seen inside the box would not know or would use the information pictured on the front of the box to form a belief. These conceptual developments are manifest roughly between 3 and 6 years old (e.g., Wellman & Liu, 2004; Wellman, Cross, & Watson,

2001). Children also begin to make connections between beliefs and past and future experiences (e.g., different past experiences result in different beliefs about future events; Lagattuta & Sayfan, 2013). Though less studied, older children and adults can continue to show changes and refinement in mental-state understanding (Apperly, 2013; Malle, 2004). These typical developments contrast with delays in ToM in individuals with autism and in deaf children born to nonsigning families (Peterson, Wellman, & Liu, 2005; Peterson, Wellman, & Slaughter, 2012).

Neural foundations. In adults, converging evidence demonstrates that mental-state reasoning (e.g., reasoning about one's own or others' beliefs, desires, intentions, and knowledge) consistently recruits a network of specific neural regions: The bilateral temporoparietal junction (TPJ) and medial prefrontal cortex (MPFC) are most consistently recruited (see Schurz, Radua, Aichhorn, Richlan, & Perner, 2014 for meta-analysis). In addition, the superior temporal gyrus/sulcus is often implicated more broadly, as well as cortical midline structures including the anterior cingulate cortex, posterior cingulate, and precuneus (see also Carrington & Bailey, 2009; Castelli, Happe, Frith, & Frith, 2000; Gallagher et al., 2000; Saxe & Kanwisher, 2003 for reviews).

Far fewer studies examine neural correlates of ToM early in development, though existing research suggests that children's explicit mental-state reasoning is associated with a similar neural network compared to adults (see Bowman & Wellman, 2014). For preschoolers, school-aged children, and adolescents—just as for adults—the MPFC and TPJ are most consistently associated with reasoning about mental states in general (Gweon, Dodell-feder, Bedny, & Saxe, 2012; Sabbagh, Bowman, Evraire, & Ito, 2009; Saxe, Whitfield-Gabrieli, Scholz, & Pelphrey, 2009). These same regions are recruited when children reason about specific mental states like beliefs (Sommer et al., 2010 ages 10–12 years). These regions are also activated when engaging in belief reasoning across multiple languages (Kobayashi, Glover, & Temple, 2008; ages 8–12 years), and when engaging in not only thoughts about others' beliefs, but thoughts about the self and one's own beliefs (Pfeifer et al., 2009; ages 11–14 years). Other neural regions implicated in the adult ToM neural network support aspects of ToM earlier in development as well. The STS is recruited when inferring intentions from action (Mosconi, Mack, McCarthy, & Pelphrey, 2005, Ohnishi, Moriguchi, Matsuda, Mori, Hirakata et al., 2004) and when thinking about the self (Pfeifer et al., 2009); and the precuneus is also recruited for general mental-state reasoning (Gweon et al., 2012; Saxe et al., 2009).

However, adult–child similarities in the neural systems supporting ToM are not without differences, and findings suggest emerging patterns of developmental change, particularly with respect to the role of the TPJ. For example, the TPJ appears to become increasingly selective for processing mental-states (beyond processing general social, nonmental characteristics) as children age from early to middle childhood

(Gweon et al., 2012; Saxe et al., 2009), and as their general mental-state reasoning advances in accuracy (Gweon et al., 2012). Moreover, as belief-reasoning becomes increasingly advanced and accurate, the TPJ perhaps plays a more prominent role in supporting belief-reasoning specifically, beyond supporting other mental states such as desires (Bowman, Kovelman, Hu, & Wellman, 2015; Bowman, Liu, Meltzoff, & Wellman, 2012; Liu, Meltzoff, & Wellman, 2009).

How do we characterize the relation between cognition and the brain, and what does it tell us about neural foundations for ToM? On the one hand, it might be tempting to conceptualize the implicated neural regions as specifically dedicated to supporting ToM, or even to supporting specific aspects of ToM like belief-reasoning. Existing research may point to such specificity. For example, there is evidence that functional maturation of the MPFC and TPJ is related to advancements in children's mental-state reasoning, even after controlling for cooccuring advancements in executive functioning, vocabulary development, and general aging (Sabbagh et al., 2009). Moreover, as noted above, the TPJ is recruited more strongly for belief-reasoning compared to desire-reasoning (Bowman et al., 2015). However, a more simplistic designation of function (in which a particular brain region is designated as "for mental state reasoning" or "for beliefs") may warrant caution, as the regions implicated in the ToM neural network are also implicated during other types of cognition. For example, TPJ is also recruited when reorienting attention, and updating or changing learned expectations and representations of the environment (Corbetta & Shulman, 2002; Mitchell, 2008). Thus, it is important to consider how knowledge of neural function can inform our understanding of how ToM develops (see Mahy, Moses, & Pfeifer, 2014 for similar argument). For example, if the TPJ is recruited when expectations or representations of the environment are updated *and* the TPJ also becomes increasingly recruited for mental-state and belief-reasoning as children's understanding advances, then it may be useful to investigate the process of "updating representations" in children's advancing ToM and in children's developing understanding of beliefs (see Sabbagh, Hopkins, Benson, & Flanagan, 2010 for similar discussions). Indeed, evidence for involvement of a particular structure or set of structures in any particular activity does not guarantee that the structure evolved only to support those activities, or that it uniquely does so.

The role of experience? More broadly, what accounts for relations between brain and cognition, and for any existing continuity or change in these relations across development? Some possibilities for local mechanisms of development include changes in strategic approaches to ToM reasoning, differences in performance accuracy, and ontogenetic increases and reorganization of gray matter. A broader mechanism, however, concerns the impact of experience more generally. There is behavioral evidence that different experiences shape development of ToM. For example, differences in the home and parenting environment (e.g., number of siblings, Perner, Ruffman, &

Leekam, 1994; exposure to parental discussions of mental states, Peterson & Slaughter, 2003) can delay or speed up advancements in explicit mental-state reasoning. In addition, the order in which different mental concepts are understood seems to vary based on the culture in which children are raised (Shahaeian, Peterson, Slaughter, & Wellman, 2011). It is therefore possible that these and other experiences may affect how ToM is supported in the brain. Yet, direct relations between experience and brain, and their association with ToM are little explored or understood. These kinds of investigations are key to uncovering whether and how experience can shape relations between brain and ToM.

10.4 LANGUAGE

Language allows us to express ourselves, to communicate, and to share experiences, while forming a foundation to acquire other complex abilities and knowledge. In most cases, language is universally developed, meaning that as long as children experience the expected environmental circumstances (i.e., are exposed to language in some modality from an early age) they will learn to speak and understand. In relatively recent history, research has begun to elucidate the role of the brain in language acquisition, development, and processing. This research has given us a greater understanding of how the various components of language (e.g., phonology, vocabulary, grammar) can be viewed as separate component processes, how these elements work together in an interconnected way, and how the interaction between brain maturation and experience can account for their developmental trajectory and ultimate outcomes. Throughout the following sections we will examine the neural foundations of three elements of language: phonology, morphysyntax, and vocabulary/lexicon.

10.4.1 Phonology

Phonology comprises the regularities of language at the level of form. In spoken language this includes *prosody*, or the rhythm/melody of a language; *phonemes*, the fundamental sounds of language that contrast meaning (i.e., in English the distinction between /b/ and /p/ allows for the different meanings of bat versus pat); and *phonotactics*, or the rules for combining speech sounds. While there are constraints on the types of sounds that can be used in a language, there are also differences across languages in the use of prosody, in the repertoire of phonemes and their distinctive phonetic features, and in acceptable phonotactics. Infants begin to become attuned to these properties of the native language early in development. That attunement, in turn, helps with word learning and with the acquisition of syntax (though note that vocabulary development also influences the formation of phonetic categories, illustrating the interconnected nature of language acquisition).

From the moment they are born, infants have perceptual biases that support learning about language. Behaviorally, infants only hours old will change their sucking pattern (i.e., while sucking on a pacifier) to preferentially listen to speech-like sounds over equally complex nonspeech sounds (Vouloumanos & Werker, 2007; see Price, 2012 for a review), and show different neural activation in response to speech over noncommunicative sounds (Shultz, Vouloumanos, Bennett, & Pelphrey, 2014). While this preference is broad enough to include monkey calls at birth, it rapidly sharpens to only human speech by 3 months of age (Vouloumanos, Hauser, & Werker, 2010), with the neural activation becoming more left hemisphere (LH) lateralized (Shultz et al., 2014), as in the adult (e.g., Binder, Rao, Hammeke, Frost, Bandettini et al., 1995; for a review see Price, 2012). Similarly, at birth, infants can distinguish languages from different rhythmical classes (Mehler et al., 1988), an ability, as discussed below, that helps bootstrap the acquisition of syntax. Learning is evident here, too, as even at birth neonates show a preference for the language heard in utero (Moon, Cooper, & Fifer, 1993) and show a different pattern of neural activation to the familiar over an unfamiliar language (May, Byers-Heinlein, Gervain, & Werker, 2011). Studies with bilingual exposed newborns reveal the interaction of perceptual constraints and learning: while infants exposed to two languages throughout gestation equally prefer both of their native languages, as evidence of learning they nonetheless still discriminate the two languages on the basis of rhythmical cues (Byers-Heinlein, Burns, & Werker, 2010). The equal preference for two native languages, along with a continued ability to discriminate them that is not erased by early exposure, could help bilingual infants to separately track the properties of each of their native languages in early perceptual learning.

By 4–5 months of age infants discriminate languages from within a rhythmic class (Nazzi & Ramus, 2003) even if both languages are familiar as in the case of bilingual infants (Bosch & Sebastian-Galles, 1997), and show greater neural activation to their native language (Minagawa-Kawai, Cristià, & Dupoux, 2011).

At the neural level, LH specialization for phonological processing, as is typically observed in adults (e.g., see Dehaene-Lambertz, 1997; Zatorre, Evans, Meyer, & Gjedde, 1992), has also been observed in neonates using both dichotic listening (e.g., Bertoncini et al., 1989) and ERP paradigms (e.g., Dehaene-Lambertz & Gliga, 2004). Moreover there is evidence from functional near-infrared spectroscopy (fNIRS) that preterm infants, at only 29 weeks gestational age, activate left inferior frontal brain regions during phonetic (ba/ga) discriminations (Mahmoudzadeh, Dehaene-Lambertz, Fournier, Kongolo, Goudjil et al., 2013). Biases such as these are thought to facilitate language acquisition by directing attention to the relevant stimuli that will form the basis of language, thus constraining the possibilities for what will later be acquired (Werker & Gervain, 2013). However, while there is evidence that certain biases are in place to support the acquisition of phonology, the fact that infants are able to learn any language they encounter demonstrates the essential role of experience in shaping this process. Experience has been found to influence the neural foundations of phonological acquisition through a process of perceptual attunement.

From the beginning of life infants possess a unique ability to distinguish speech sounds, including from unfamiliar languages (e.g., Aslin, Pisoni, Hennessy, & Perey, 1981; Cheour, Ceponiene, Lehtokoski, Luuk, Allik et al., 1998; Eimas, Siqueland, Jusczyk, & Vigorito, 1971; Werker & Tees, 1984; Werker, Gilbert, Humphrey, & Tees, 1981), a task that is difficult for adults (e.g., Pallier, Bosch, & Sebastián-Gallés, 1997; Werker et al., 1981). However, via a process of perceptual attunement that occurs during the first year of life, infants' become specialized at discriminating the sounds/sound categories of the language they hear around them. During this process, infants' ability to discriminate speech sounds that do not have functional relevance within their language environment declines (e.g., Rivera-Gaxiola, Silva-Pereyra, & Kuhl, 2005; Tsao, Liu, & Kuhl, 2006; Werker & Tees, 1984, 2005), while their ability to discriminate native-language speech sounds is maintained or improved (e.g., Kuhl et al., 2006; Narayan, Werker, & Beddor, 2010; Polka, Colantonio, & Sundara, 2001; for a review see Maurer & Werker, 2014). Infants show perceptual attunement for both vowels (by 6–8 months of age e.g., Kuhl, Williams, Lacerda, Stevens, & Lindblom, 1992; Polka & Werker, 1994) and consonants (by 8–10 months e.g., Kuhl, Tsao, & Liu, 2003; Werker & Tees, 1984), as well as other phonological elements such as lexical tone (e.g., Mattock, Molnar, Polka, & Burnham, 2008; Yeung, Chen, & Werker, 2013), clicks (Best, McRoberts, LaFleur, & Silver-Isenstadt, 1995), and components of sign language (Baker, Golinkoff, & Petitto, 2006; Palmer, Fais, Golinkoff, & Werker, 2012).

The majority of evidence suggests that perceptual attunement represents a reorganization of neural systems underlying phonological processing (e.g., Rivera-Gaxiola, Csibra, Johnson, & Karmiloff-Smith, 2000; Werker & Tees, 2005), and this process can be observed through neural changes, such as changes in ERP responses to native and nonnative language sounds. Many studies examining phonological processing in infants examine an ERP component known as the mismatch-negativity (MMN), which is typically elicited as an increased amplitude to deviant compared to standard stimuli during an oddball paradigm (for a review see Dehaene-Lambertz & Gliga, 2004). While the morphology of the MMN changes over the course of development, an MMN response has been observed even in newborns (e.g., Dehaene-Lambertz & Pena, 2001) and, when it is elicited during a phonetic discrimination task, it is thought to represent activation of a left temporal-parietal phoneme processing network (e.g., Dehaene-Lambertz, 1997; Näätänen, Lehtokoski, Lennest, Luuki, Alliki et al., 1997). Thus, it is a useful tool for assessing perceptual attunement and phonological processing across development. MMN studies have found that, by 3 months of age infants show a larger MMN response to acoustic changes that cross a phonemic boundary than those that remain within a phonemic category (Dehaene-Lambertz & Baillet, 1998), suggesting that infants have a dedicated system for processing phoneme categories even from that early age. However, there is also neural evidence for perceptual attunement, as both native and nonnative speech contrasts elicited different ERP responses in 7- but not 11-month-old English-learning infants. The older infants only showed sensitivity to native language contrasts (Rivera-Gaxiola et al., 2005). Similarly, while 6-month-old Finnish- and Estonian-learning infants showed similar MMN responses to an Estonian vowel contrast, Finnish-learning infants showed a smaller MMN to the nonnative Estonian vowel contrast by the time they were 12 months of age (Cheour et al., 1998). More recently, researchers have begun to look at oscillatory EEG activity (particularly in delta, theta, and gamma bands) as a measure of phonetic processing. The specialized processing seen in the adult brain for the native language (e.g., Poeppel, Idsardi, & Van Wassenhove, 2008), is also observed in infants, increasingly across the first year of life (Ortiz-Mantilla, Hämäläinen, Musacchia, & Benasich, 2013).

Importantly, there is evidence that perceptual attunement to native-language categories is beneficial to later language outcomes. For example 6-month-olds who were better at native-language phonetic discrimination, and 7.5 month olds who elicited brain responses indicating perceptual sensitivity to native, but not nonnative language phonemes, also produced and understood more words, used more complex sentences, and spoke in longer utterances when they were 13 to 30 months of age (Kuhl, Conboy, Coffey-Corina, Padden, Rivera-Gaxiola et al., 2008; Tsao, Liu, & Kuhl, 2004). Perceptual sensitivity to nonnative language discriminations, in contrast, predicted poorer language outcomes (Kuhl et al., 2008; Kuhl, Conboy, Padden, Nelson, & Pruitt, 2005).

Finally, there is evidence that the timing of exposure to language sounds influences the neural processing of those sounds. Specifically, there seems to be a window lasting from birth (or earlier) until roughly 10–12 months of age, during which phonological acquisition is facilitated, and after which it becomes much more difficult (for reviews see Werker & Hensch, 2015; Werker & Tees, 2005). Strong evidence that sensitive periods for phonological acquisition are active during the earliest stages of language acquisition comes from observing perceptual attunement to native language phonemes, as discussed previously. As infants maintain sensitivity to the phonetic contrasts they hear in their environment they become less adept at discriminating phonetic contrasts in other languages (for a review see Maurer & Werker, 2014). Importantly, while infants' phonetic categories can be modified up to about 10–12 months of age (Maye, Werker, & Gerken, 2002), beyond this point they become increasingly difficult to change (Yoshida, Pons, Maye, & Werker, 2010). Sensitive period timing may be at least partially maturationally determined. ERPs have demonstrated that premature infants continued to discriminate nonnative contrasts up to the same age as full-term infants, despite the fact that premature infants had had several more weeks postnatal exposure to language (Peña, Werker, & Dehaene-Lambertz, 2012; see also Peña, Pittaluga, & Mehler, 2010 for an example with rhythmical class). There is also evidence that sensitive period windows can be shifted via biological mechanisms. Prenatal exposure to serotonin reuptake inhibitors shifted the window for perceptual attunement earlier, such that infants both started and stopped discriminating nonnative phonetic contrasts earlier (i.e., by 6 months), while infants with mothers who were depressed during pregnancy continued to discriminate the same nonnative speech contrast at 10 months of age (Weikum, Oberlander, Hensch, & Werker, 2012).

10.4.2 Morphology and syntax

Syntax refers to the rules for ordering words. The languages of the world are primarily Subject—object—verb (SOV) in word order, or Subject—verb—object (SVO) with a scattering of other word orders. Morphemes are the smallest units of language that carry meaning on their own: content words such as nouns and verbs can be morphemes, but so can function words such as determiners and prepositions, as well as affixes (bound morphemes) such as "un" and "ing." Function words and affixes carry meaning, but only in relation to the content words (e.g., nouns, adjectives, verbs) with which they cooccur, i.e., they signal grammatical relations. Morphology refers to the rules for the use of morphemes, and hence bridges syntax, phonology, and lexical processes.

There is debate about the order of acquisition of syntax and morphology. If language production is used as the standard, there is no evidence of syntax until at least 2—3 word utterances are spoken, at 18—24 months of age. Even then, a "rich" interpretation is required to assume the use of word order rules and grammatical category interpretation. Evidence of morphological endings and such is not seen until 3—4 years of age. On the other hand, a growing body of research reveals that toddlers understand syntax from a much younger age than they can produce it (see Naigles, 1990).

Because syntax and morphology involve lexical knowledge, many theories of language acquisition assume that independent syntactic and morphological systems emerge only once a lexicon is in place (Pinker, 1984). Others, however, argue that the (syntactic) structure has to be in place to support the acquisition of lexical items, including their assignment to different grammatical categories (Landau & Gleitman, 1985; Yuan, Fisher, & Snedeker, 2012), and that a number of capabilities evident in infancy enable this to happen.

Neuroimaging studies reveal that a wide network, including syntactic, lexical and phonological processing circuits (Friederici, 2011) is activated in both adults and children when presented with spoken sentences. In adults, however, there are also independent networks for processing complex syntax at both the sentence (Makuuchi, Bahlmann, Anwander, & Friederici, 2009) and phrasal (Humphries, Binder, Medler, &

Liebenthal, 2006) levels. Although there are multiple studies that reveal that even children as young as age two engage similar brain systems as adults for processing distinct syntactic categories (Bernal, Dheaene-Lambertz, Millotte, & Christophe, 2010), unequivocal independence of syntactic versus semantic processing is not evident in children under about 10 years of age (Skeide, Brauer, & Friederici, 2015), however. Instead, some evidence suggests that semantic networks are activated first, and that syntactic analysis secondarily builds on semantic analysis (Brauer & Friederici, 2007; Wu, Vissiennon, Friederici, & Brauer, 2016). This kind of evidence, along with the long established fact that when they first begin to speak, infants and toddlers omit functional morphemes, using only content words (Bloom, 1975), is some of the evidence for the argument that semantics comes first, and that from that syntax must be built.

There is increasing evidence to refute this claim, however, and to indicate that the roots of morphology and syntax are in fact present in early infancy. One line of evidence comes from studies showing that infants can detect/learn/represent the kinds of structural relations necessary for syntax. When familiarized to a sequence of syllables that conform to an "ABB" grammar (e.g., bo-la-la, me-fo-fo) and then tested on a change to either a new sequence corresponding to that pattern or one that violates it (e.g., AAB), infants look longer to a checkerboard in a looking time task (Marcus, Vijayan, Rao, & Vishton, 1999) and show greater neural activation even at birth, for example as measured by NIRS (Gervain, Macagno, Cogoi, Peña, & Mehler, 2008) to a change to the new pattern. Moreover, the increased neural activation to the violation is most robust over fronto-temporal areas in the LH, corresponding to the neural areas involved in syntactic processing in adults.

While the world languages differ in their inventory of grammatical classes (e.g., English has prepositions and Japanese has postpositions), all have a fundamental distinction between content and functional elements. The second line of evidence for the foundations of syntax existing prior to the establishment of a lexicon comes from studies showing that at birth human neonates are able to use acoustic and phonological cues to perceptually categorize function versus content words. Function words are perceptually minimal: they are quieter than content words in natural speech, are shorter in duration, have a simpler syllable structure and often fewer syllables (Morgan, Shi, & Allopenna, 1996, Selkirk, 1996). When habituated in a high amplitude sucking task to a set of *content* items excised from natural speech (i.e., "baby, car, pretty") and then presented with new content versus new function words (i.e., "of, her"), neonates discriminate the change in grammatical category: they show a recovery in sucking rate only to the contrasting category (Shi, Werker, & Morgan, 1999). This is not based on lexical knowledge or even familiarity because this same pattern of results is evident when newborns are presented with words from a nonnative language.

A second line of evidence for syntax prior to, or along with, the development of semantics, comes from work showing that infants may use prosody to bootstrap the acquisition of syntax (Morgan & Demuth, 1996). In languages that have SVO word orders, functional elements tend to occur before content elements ("*from* Vancouver"), whereas SOV languages are the reverse ("Tokyo *kara*"). The acoustic characteristics of content words give them more prominence than function words, and phrasal prominence falls on content words. Correspondingly, SVO and SOV languages have different prosodic (rhythm: duration, pitch and amplitude) properties at the phrasal level, with initial prominence in SOV languages and final prominence in SVO languages. The prosodic bootstrapping hypothesis (Gleitman & Wanner, 1982; Morgan, 1986; Peters, 1983) predicts that infants, like adults (Langus, Marchetto, Bion, & Nespor, 2012), use prosody to pull out phrases from sentences (e.g., Hawthorne & Gerken, 2014; Soderstrom, Seidl, Kemler Nelson, & Jusczyk, 2003). In support, infants as young as 7 months can use prosody to parse the speech stream and increasingly use these cues in a language specific way (Bernard & Gervain, 2012), with bilingual infants able to use prosody to switch their parsing strategy fluidly from one language to another (Gervain & Werker, 2013). One recent fNIRS study indicates that even at birth infants are sensitive to the duration, pitch, and amplitude cues that are necessary for prosodic bootstrapping, with more sensitivity already by birth to those cues relevant to the native language, i.e., the language heard in utero (Gervain & Nazzi, personal communication).

10.4.3 Word learning (lexical development)

Lexical acquisition, or word learning, is a multifaceted process requiring several steps. Infants must use cues (e.g., phonology) in order to identify and discriminate words from a continuous stream of speech; they must establish representations for the forms of words and store those in memory; and, once word form representations are established, they must link those forms to real-world objects or concepts in a meaningful way (e.g., so they know that "ball" refers to the round object they want to roll) (e.g., Friedrich & Friederici, 2010, 2015). The following section will discuss these stages of lexical acquisition and what we know about the neural foundations that support them.

Neuroscientific methods for examining lexical acquisition. ERP components that are modulated by word familiarity, and that change over development, have been an effective tool for determining what infants know about words at different stages of lexical acquisition (e.g., Mills, Coffey-Corina, & Neville, 1993, 1997; Molfese, 1989, 1990). For example, the N200–500 is greater in amplitude to known than unknown words, is thought to reflect stored representations of word forms, and has been observed in children aged 6–20 months of age (e.g., Friedrich & Friederici, 2011; Mills et al., 1993). Word comprehension, or the linking of words to meaning, has been indexed by a negative component occurring between 600 and 1200 ms in 12–17 month olds (Conboy & Mills, 2006; Mills et al., 1997), and by the shorter latency, more "adult-like" N400 response in children 14 months and older (e.g., Friedrich & Friederici, 2008, 2010). While infants younger than 6 months do not appear to link words with meaning in the same way as older children and adults, they show a late negativity thought to reflect word-object associations that may be a necessary precursor to word learning (Friedrich & Friederici, 2015). Examining modulations in these and other components allows a precise understanding of the process and time course of lexical acquisition as a function of maturation and experience.

Word segmentation. By about 7 to 10 months infants begin producing language-like sounds in the form of canonical babbling (e.g., Oller, 1980), and soon after produce their first words. While word production begins slowly (i.e., 100 words produced by 12 months; Fenson, Dale, Reznick, Bates, Thal et al., 1994), word recognition appears much earlier. By 4.5–6 months infants segment familiar and frequent words from continuous speech (e.g., Bortfeld, Morgan, Golinkoff, & Rathbun, 2005), and by 7.5 months can do so for unknown words as well (e.g., Jusczyk & Aslin, 1995; Polka & Sundara, 2012). Infants use cues such as stress patterns, statistical regularities, and phonotactics to perform this difficult feat (e.g., Jusczyk, 1999; Kooijman, Hagoort, & Cutler, 2005; Saffran, Aslin, & Newport, 1996; Werker & Yeung, 2005), with strategy changes observed over time (e.g., Mannell & Friederici, 2013).

The neural foundations underlying word segmentation change over time. For example, the use of different cues for word segmentation (Mills et al., 2004; Mills et al., 1993, 1997), are marked by changes in infants' ERP responses. Early in development, ERP responses to word segmentation are typically positive and broadly distributed across electrodes. In contrast, more mature word segmentation strategies are reflected by ERP components that are negative in amplitude and are localized to LH electrode locations (Mannell & Friederici, 2013). The change in ERP response to "adult-like" LH localization is positively associated with vocabulary comprehension and word production (Kooijman, Junge, Johnson, Hagoort, & Cutler, 2013). This neural change may also result in children's vocabulary spurt, typically observed during the second year of life and signaled by a rapid increase in the number of words acquired per day. Supporting this, 20-month-old infants who had experienced a vocabulary spurt (and 13- and 20-month-olds with above average vocabularies) elicited adult-like ERPs localized to LH temporal and parietal locations. 13-month-olds who had not yet experienced a vocabulary spurt continued to elicit less-mature, broadly distributed ERP responses (Mills et al., 1993, 1997; see also Mills, Plunkett, Prat, & Schafer, 2005).

Word comprehension. Historically, it has been thought that infants must learn to segment word forms before they can map those forms to meaning. However, recent evidence has suggested that infants are aware of word meaning by 6 months of age, earlier than previously believed. This suggests that vocabulary and sound structure might actually be acquired in tandem (Bergelson & Swingley, 2012). Indeed, there is evidence that vocabulary size influences word comprehension because stored lexical items (words) are thought to help refine phonological categories. In other words, as vocabulary increases infants can use phonological detail in an increasingly mature way to discriminate new word forms and determine their meaning (e.g., Gaskell & Ellis, 2009; Swingley, 2009). Supporting this, although infants are able to discriminate minimal pairs of phonemes (e.g., "ba" and "da") at 14 months (Stager & Werker, 1997), they cannot map phonemically similar word forms (e.g., "bih" and "dih") to two novel objects until 17–20 months of age (Werker, Fennell, Corcoran, & Stager, 2002), around the time of the vocabulary spurt. However, at 14 months, they can map word forms to novel objects if the word forms are phonemically dissimilar (e.g., "lif" and "neem) (Stager & Werker, 1997). That changing abilities are linked to vocabulary is supported by the finding that 14-month-old children with above average vocabularies can precociously map phonetically similar words to novel objects (Mills et al., 1993; see Friedrich & Friederici, 2010, for a study of 12-month olds showing similar results; but see Swingley & Aslin, 2000, 2002 for evidence of earlier mapping of phonemically similar words). Once children have learned to map words to their referents, they face the task of understanding how words, or types of words, relate not only to those referents but to each other. Recent behavioral evidence suggests that by 2 years children can encode semantic relationships between novel words and visually similar novel objects (Wojcik & Saffran, 2013), suggesting an important avenue for future neural research.

Neuroscientific methods have demonstrated neural foundations that may facilitate infants' ability to attach new words to their referents. Mills et al. (2004) found that larger N200–400 amplitudes were elicited to known words (e.g., "bear") than to phonetically dissimilar nonsense words (e.g., "kobe") in 14-month-olds, indicating that they could discriminate those words. However it was not until 20 months that ERP responses differentiate words and phonetically similar nonsense words (e.g., "bear" and "gare").

Types of words. For adults, different types of words (e.g., verbs, nouns, function words) are represented in different brain areas or via separable processes (e.g., Caramazza & Hillis, 1991; Damasio & Tranel, 1993). Neuroscientific methods have also demonstrated that infants process different types of words in different ways that change over the course of development (for reviews see Neville & Bavelier, 1998; Neville & Mills, 1997). For example, semantically meaningful verbs and nouns (open class words; e.g., "car" "run") elicit different ERP patterns than function words (closed class; e.g., "the" "an") that carry grammatical information (Neville, Mills, & Lawson, 1992). However, while the ERP response to function words typically does not mature until late childhood or beyond, the ERP response to open class words is morphologically similar in children and adults (Neville & Weber-Fox, 1994). Supporting this, 20-month-old infants' ERP response to open versus closed class words does not differ. By 28–30 months ERP patterns can distinguish open and closed class words, but it is not until 3 years, when children begin to produce grammatical sentences, that an adult-like left asymmetry emerges to closed-class words. An exception to this occurs in children with larger than expected vocabularies whose ERP patterns already differentiate open and closed class items by 20 months. In contrast, late talkers continue to elicit similar ERP patterns to open and closed class words even at 28–30 months (Neville & Mills, 1997). This once again demonstrates the influence of stored vocabulary on subsequent lexical acquisition.

Timing of lexical acquisition. Of note, lexical acquisition appears to be less subject to timing effects than earlier acquired aspects of language (e.g., phonology). Older children and adults can typically acquire new words in a language at any age (e.g., Flege, 1999), and do so more rapidly than infants (see Muñoz, 2006). Delays of as long as 16 years (Weber-Fox & Neville, 2001) or more (Steinhauer, 2014) have been found to have little influence on the organization of brain systems underlying lexical processing, once sufficient proficiency has been obtained. Indeed, evidence suggests that highly proficient late learners of a language use the same neural substrates when processing lexical items as those who acquired a language from birth (for a review see Steinhauer, 2014). However, because links between phonology and lexical acquisition have been shown, it is nonetheless possible that experiences that occur during sensitive periods may have an indirect effect on later lexical acquisition. For example, Dietrich, Swingley, and Werker (2007) found that Dutch but not English-learning 18-month-olds used a Dutch phonological feature (vowel length) to map novel words to objects. In contrast, English but not Dutch learners used an English phonological feature (vowel quality, as in "tam" versus "tem"), to perform the same task. Thus, early formed phonological representations may form the basis for what is accepted as a lexical item later in development.

10.5 CONCLUSION

Throughout this chapter, we have summarized research that highlights advances in our knowledge and understanding of cognitive and language development. In particular, we have discussed how the tools of cognitive neuroscience have allowed us to probe the intersection between behavior and the brain, revealing neural foundations that support and promote the emergence and development of increasingly complex behaviors and abilities. Although a broad selection of topics was presented, several themes have emerged. Namely, evidence suggests that both timing and experience interact with biological processes in different ways to influence the development of cognition and language. There is also evidence for individual differences, both in biology and behavior, which might predict how different children develop in response to the experiences they encounter.

10.5.1 Timing, experience, and individual differences

Just as there are differences across systems in the extent to which they are already organized at birth, there are also differences across systems in the extent to which sensitivity to input is constrained by developmental level. The research to date indicates that those systems that are most organized in the young infant, speech and face processing, are also likely the most constrained by the timing of input (Maurer & Werker, 2014). Infants born with hearing loss, for example, are likely to have lifelong difficulties in the processing and the neural representations of speech sounds if augmentation (hearing aids) or cochlear implants are not given early (for a review, see Werker & Hensch, 2015). Similarly, infants born with cataracts that are not corrected by 3-months of age have life-long difficulties with certain (configural) aspects of face processing (Le Grand, Mondloch, Maurer, & Brent, 2004)—although importantly, most other aspects of face processing are spared and/or can show recovery. Yet, even these more constrained systems are influenced by early experience. For example, as reviewed in the chapter, the neural specialization for speech versus nonspeech, while robust (e.g., Peña, Bonatti, Nespor, & Mehler, 2002) shows slight but replicable differences in the organization of the neonate neural response to the language heard in utero versus an unfamiliar language (e.g., see May et al., 2011). Similarly, the precise neural signature reflecting attunement to native speech sound differences at the end of the first year of life is distinct in the infant exposed to two languages from birth to what it is in the infant exposed to only a single language (see Ramirez, Ramirez, Clarke, M., Taulu, S., & Kuhl, P.K., in press).

The impact of the timing of experience on other developmental systems, such as ToM, has been studied to a lesser degree. For example, while we presented research showing that children with more advanced explicit, verbal understanding of mental-states as measured behaviorally exhibit greater selectivity in certain brain regions that support ToM (Gweon et al., 2012), it is unknown whether similar neural specializations would exist earlier in development when mental state understanding is assessed with implicit, nonverbal measures. It is possible that these systems are more open at birth with neural specialization emerging across development, and thus the windows during which input is essential may be less constrained. Examination of the neural correlates of ToM in early infancy, and comparisons to those evident later in development are necessary to shed light on this possibility.

10.5.2 Thoughts for the future

Epigenetics. Individual differences at the level of the gene may be particularly influential in establishing the neural foundations for cognition and language. While it has long been observed that genetic differences affect how children grow and develop, a new area of research emerging is that of epigenetics, or the role of experience in determining which genes are expressed and how. While it was previously believed that genetic expression was more or less a stochastic process determined prior to birth, the study of epigenetic mechanisms has revealed that biology and the environment actually work together to dictate gene expression, in turn shaping the developing brain (Meaney, 2010; Murgatroyd & Spengler, 2011). Specifically, experiences cause neurons to produce gene regulatory proteins, which determine whether a particular gene will be turned "on" or "off." Such changes in gene expression can be temporary or long-lasting; however experiences that occur very early in life might have particularly enduring effects due to the rapid development of the brain during those early years. In effect, early changes to the epigenome have the capacity to actually modify *how* the brain develops. There is evidence from both animal and human studies that negative (e.g., maltreatment or neglect) and positive events (e.g., nurturing and enriching experiences), during the first years of life affect the development of brain architecture in a long-lasting/persistent way (Bredy, Humpartzoomian, Cain, & Meaney, 2003; Heim & Binder, 2012). In turn, the development of brain structure and function influences the acquisition and processing of complex cognition and language (Fox, Levitt, & Nelson, 2010). While epigenetic studies are in their early stages, particularly in the context of child development, they provide an important avenue for future research.

Continuity, change, and the "end state." Throughout this chapter, we have raised the issue of continuity and change in both neural systems and behaviors over the course of development. Understanding the neural foundations that are available to an infant just beginning to experience the world helps us to understand what they are capable of and how these capabilities change across time as they move toward a more "adult" state of development. However, the concept of developmental change also raises important questions. Namely, what are the "limits" of development and how do we know when a brain is truly "adult"? There is evidence that certain functions continue to develop long past the point at which neuroscience suggests that the brain has reached a mature state. We develop expertise, adopt new cognitive strategies, and adapt to changing circumstances. Thus, understanding that such an "end state" may be difficult to identify is important to consider in future research. Indeed, a more informative avenue of investigation may not be to understand an ultimate end state of maturation, but rather to understand how the brain functions at different stages throughout development and how stage-dependent brain processes interact with experiences and the environment.

REFERENCES

Ansari, D. (2008). Effects of development and enculturation on number representation in the brain. *Nature Reviews Neuroscience*, *9*(4), 278–291. Available from http://dx.doi.org/10.1038/nrn2334.

Anzures, G., Quinn, P. C., Pascalis, O., Slater, A. M., & Lee, K. (2010). Categorization, categorical perception, and asymmetry in infants' representation of face race. *Developmental Science, 13*(4), 553–564.

Apperly, I. (2013). Can theory of mind grow up? In S. Baron-Cohen, H. Tager-Flusburg, & M. Lombardo (Eds.), *Mindreading in adults, and its implications for the development and neuroscience of mindreading.* Understanding other minds: Perspectives from developmental social neuroscience (pp. 72–92). Oxford University Press.

Aslin, R. N., Pisoni, D. B., Hennessy, B. L., & Perey, A. J. (1981). Discrimination of voice onset time by human infants: New findings and implications for the effects of early experience. *Child development, 52*(4), 1135.

Baker, S. A., Golinkoff, R. M., & Petitto, L. A. (2006). New insights into old puzzles from infants' categorical discrimination of soundless phonetic units. *Language Learning and Development, 2*(3), 147–162.

Bauer, P. J. (2006). Constructing a past in infancy: A neuro-developmental account. *Trends in Cognitive Sciences, 10*(4), 175–181.

Bauer, P. J., & Hertsgaard, L. A. (1993). Increasing steps in recall of events: Factors facilitating immediate and long-term memory in 13.5-and 16.5-month-old children. *Child Development, 64*(4), 1204–1223.

Bergelson, E., & Swingley, D. (2012). At 6–9 months, human infants know the meanings of many common nouns. *Proceedings of the National Academy of Sciences, 109*(9), 3253–3258.

Bernal, S., Dheaene-Lambertz, G., Millotte, S., & Christophe, A. (2010). Two-year-olds computer syntactic structure on-line. *Developmental Science, 13*(1), 69–76.

Bernard, C., & Gervain, J. (2012). Prosodic cues to word order: what level of representation. *Frontiers in Psychology, 3*, 1–6.

Bertoncini, J., Morais, J., Bijeljac-Babic, R., McAdams, S., Peretz, I., & Mehler, J. (1989). Dichotic perception and laterality in neonates. *Brain and Language, 37*(4), 591–605.

Best, C. T., McRoberts, G. W., LaFleur, R., & Silver-Isenstadt, J. (1995). Divergent developmental patterns for infants' perception of two nonnative consonant contrasts. *Infant Behavior and Development, 18*(3), 339–350.

Binder, J. R., Rao, S. M., Hammeke, T. A., Frost, J. A., Bandettini, P. A., Jesmanowicz, A., & Hyde, J. S. (1995). Lateralized human brain language systems demonstrated by task subtraction functional magnetic resonance imaging. *Archives of Neurology, 52*(6), 593–601.

Bloom, L. (1975). *Language development.* Chicago: University of Chicago Press.

Bortfeld, H., Morgan, J. L., Golinkoff, R. M., & Rathbun, K. (2005). Mommy and me familiar names help launch babies into speech-stream segmentation. *Psychological Science, 16*(4), 298–304.

Bosch, L., & Sebastian-Galles, N. (1997). Native-language recognition abilities in 4-month-old infants from monolingual and bilingual environments. *Cognition, 65*, 33–69.

Bourgeois, J. P. (2001). Synaptogenesis in the neocortex of the newborn: The ultimate frontier for individuation. In C. A. Nelson, & M. Luciana (Eds.), *Handbook of developmental cognitive neuroscience* (pp. 23–34). Cambridge, MA: MIT Press.

Bowman, L. C., & Wellman, H. M. (2014). In O. N. Saracho (Ed.), *Neuroscience contributions to childhood theory-of-mind development.* Contemporary perspectives on research in theories of mind in early childhood education (pp. 195–224). Charlotte, NC: Information Age Publishing.

Bowman, L. C., Kovelman, I., Hu, X., & Wellman, H. M. (2015). Children's belief-and desire-reasoning in the temporoparietal junction: evidence for specialization from functional near-infrared spectroscopy. *Frontiers in Human Neuroscience, 9*, 560.

Bowman, L. C., Liu, D., Meltzoff, A. N., & Wellman, H. M. (2012). Neural correlates of belief- and desire-reasoning in 7- and 8-year-old children: an event-related potential study. *Developmental Science*, Available from http://dx.doi.org/10.1111/j.1467-7687.2012.01158.x.

Brandone, A. C., & Wellman, H. M. (2009). You can't always get what you want: Infants understand failed goal-directed actions. *Psychological Science, 20*, 85–91.

Brauer, J., & Friederici, A. D. (2007). Functional neural networks of semantic and syntactic processes in the developing brain. *Journal of Cognitive Neuroscience, 19*(10), 1609–1623.

Bredy, T. W., Humpartzoomian, R. A., Cain, D. P., & Meaney, M. J. (2003). Partial reversal of the effect of maternal care on cognitive function through environmental enrichment. *Neuroscience, 118*(2), 571–576.

Byers-Heinlein, K., Burns, T. C., & Werker, J. F. (2010). The roots of bilingualism in newborns. *Psychological Science, 21*(3), 343–348.

Callaghan, T., & Corbit, J. (2015). The Development of Symbolic Representation. In L. S. Liben, & U. Mueller (Eds.), *Handbook of Child psychology and developmental science*. New Jersey: John Wiley and Sons.
Caramazza, A., & Hillis, A. E. (1991). Lexical organization of nouns and verbs in the brain. *Nature*, *349*(6312), 788–790.
Carrington, S. J., & Bailey, A. J. (2009). Are there theory of mind regions in the brain? A review of the neuroimaging literature. *Human Brain Mapping*, *30*, 2313–2335.
Casey, B. J., Giedd, J. N., & Thomas, K. M. (2000). Structural and functional brain development and its relation to cognitive development. *Biological Psychology*, *54*(1), 241–257.
Casey, B. J., Tottenham, N., Liston, C., & Durston, S. (2005). Imaging the developing brain: what have we learned about cognitive development? *Trends in Cognitive Sciences*, *9*(3), 104–110.
Castelli, F., Happé, F., Frith, U., & Frith, C. (2000). Movement and mind: a functional imaging study of perception and interpretation of complex intentional movement patterns. *NeuroImage*, *12*(3), 314–325. Available from http://dx.doi.org/10.1006/nimg.2000.0612.
Cheour, M., Ceponiene, R., Lehtokoski, A., Luuk, A., Allik, J., Alho, K., & Näätänen, R. (1998). Development of language-specific phoneme representations in the infant brain. *Nature Neuroscience*, *1*(5), 351–353.
Conboy, B. T., & Mills, D. L. (2006). Two languages, one developing brain: Event-related potentials to words in bilingual toddlers. *Developmental Science*, *9*(1), F1–F12.
Corbetta, M., & Shulman, G. L. (2002). Control of goal-directed and stimulus-driven attention in the brain. *Nature Reviews Neuroscience*, *3*, 201–215. Available from http://dx.doi.org/10.1038/nrn755.
Damasio, A. R., & Tranel, D. (1993). Nouns and verbs are retrieved with differently distributed neural systems. *Proceedings of the National Academy of Sciences*, *90*(11), 4957–4960.
de Haan, M. (2008). *Neurocognitive mechanisms for the development of face processing. In Handbook of developmental cognitive neuroscience* (pp. 509–520). Cambridge: MIT Press.
de Haan, M. (2013). *Infant EEG and event-related potentials*. Hove: Psychology Press.
Dehaene-Lambertz, G. (1997). Electrophysiological correlates of categorical phoneme perception in adults. *NeuroReport*, *8*(4), 919–924.
Dehaene-Lambertz, G., & Baillet, S. (1998). A phonological representation in the infant brain. *NeuroReport*, *9*(8), 1885–1888.
Dehaene-Lambertz, G., & Gliga, T. (2004). Common neural basis for phoneme processing in infants and adults. *Journal of Cognitive Neuroscience*, *16*(8), 1375–1387.
Dehaene-Lambertz, G., & Pena, M. (2001). Electrophysiological evidence for automatic phonetic processing in neonates. *NeuroReport*, *12*(14), 3155–3158.
Dietrich, C., Swingley, D., & Werker, J. F. (2007). Native language governs interpretation of salient speech sound differences at 18 months. *Proceedings of the National Academy of Sciences*, *104*(41), 16027–16031.
Eimas, P. D., Siqueland, E. R., Jusczyk, P., & Vigorito, J. (1971). Speech perception in infants. *Science*, *171*(3968), 303–306.
Fagan, J. F. (1970). Memory in the infant. *Journal of Experimental Child Psychology*, *9*(2), 217–226.
Fantz, R. L. (1963). Pattern vision in newborn infants. *Science*, *140*, 296–297.
Fantz, R. L. (1964). Visual experience in infants: Decreased attention to familiar patterns relative to novel ones. *Science*, *146*(3644), 668–670.
Fenson, L., Dale, P. S., Reznick, J. S., Bates, E., Thal, D. J., Pethick, S. J., ... Stiles, J. (1994). Variability in early communicative development. *Monographs of the Society for Research in Child Development*, *59*, 1-185.
Field, T. M., Cohen, D., Garcia, R., & Greenberg, R. (1984). Mother-stranger face discrimination by the newborn. *Infant Behavior and Development*, *7*(1), 19–25.
Fisher, A. V., Godwin, K. E., & Matlen, B. J. (2015). Development of inductive generalization with familiar categories. *Psychonomic Bulletin & Review*, *22*(5), 1149–1173.
Flege, J. E. (1999). Age of learning and second language speech. In D. Birdsong (Ed.), *Second language acquisition and the critical period hypothesis* (pp. 101–131). Mahwah, NJ: Lawrence Erlbaum.
Fox, S. E., Levitt, P., & Nelson, C. A., III (2010). How the timing and quality of early experiences influence the development of brain architecture. *Child Development*, *81*(1), 28–40.

Friederici, A. D. (2011). The brain basis of language processing: from structure to function. *Physiological Reviews*, *91*(4), 1357–1392.

Friedrich, M., & Friederici, A. D. (2008). Neurophysiological correlates of online word learning in 14-month-old infants. *NeuroReport*, *19*(18), 1757–1761.

Friedrich, M., & Friederici, A. D. (2010). Maturing brain mechanisms and developing behavioral language skills. *Brain and language*, *114*(2), 66–71.

Friedrich, M., & Friederici, A. D. (2011). Word learning in 6-month-olds: Fast encoding–weak retention. *Journal of Cognitive Neuroscience*, *23*(11), 3228–3240.

Friedrich, M., & Friederici, A. D. (2015). The origins of word learning: Brain responses of 3-month-olds indicate their rapid association of objects and words. *Developmental Science*, Available from http://dx.doi.org/10.1111/desc.12357.

Gallagher, H. L., Happé, F., Brunswick, N., Fletcher, P. C., Frith, U., & Frith, C. D. (2000). Reading the mind in cartoons and stories: An fMRI study of "theory of mind" in verbal and nonverbal tasks. *Neuropsychologia*, *38*(1), 11–21, doi:10.1016/S0028-3932(99)00053-6.

Gaskell, M. G., & Ellis, A. W. (2009). Word learning and lexical development across the lifespan. *Philosophical Transactions of the Royal Society of London B: Biological Sciences*, *364*(1536), 3607–3615.

Gervain & Nazzi, personal communication.

Gervain, J., Mehler, J., Werker, J. F., Nelson, C. A., Csibra, F., Lloyd-Fox, S., Shulka, M., & Aslin, R. N. (2011). Near-infrared spectroscopy: A report from the McDonnell infant methodology consortium. *Developmental Cognitive Neuroscience*, *1*, 22–46.

Gervain, J., & Werker, J. F. (2013). Prosody cues word order in 7-month-old bilingual infants. *Nature Communications*, *4*, 1490.

Gervain, J., Macagno, F., Cogoi, S., Peña, M., & Mehler, J. (2008). The neonate brain detects speech structure. *Proceedings of the National Academy of Sciences*, *105*(37), 14222–14227.

Gleitman, L. R., & Wanner, E. (Eds.), (1982). *Language acquisition: The state of the art*. Cambridge, MA: Cambridge University Press.

Goren, C., Sarty, M., & Wu, P. (1975). Visual following and pattern discrimination of face-like stimuli by newborn infants. *Pediatrics*, *56*, 544–549.

Greenough, W. T., Black, J. E., & Wallace, C. S. (1987). Experience and brain development. *Child Development*, , 539–559.

Gweon, H., Dodell-feder, D., Bedny, M., & Saxe, R. (2012). Theory of mind performance in children correlates with functional specialization of a brain region for thinking about thoughts. *Child Development*, *83*, 1853–1868. Available from http://dx.doi.org/10.1111/j.1467-8624.2012.01829.x.

Hawthorne, K., & Gerken, L. (2014). From pauses to clauses: Prosody facilitates learning of syntactic constituency. *Cognition*, *133*(2), 420–428.

Heim, C., & Binder, E. B. (2012). Current research trends in early life stress and depression: Review of human studies on sensitive periods, gene–environment interactions, and epigenetics. *Experimental Neurology*, *233*(1), 102–111.

Howe, M. L., Courage, M. L., & Edison, S. C. (2003). When autobiographical memory begins. *Developmental Review*, *23*(4), 471–494.

Humphries, C., Binder, J. R., Medler, D. A., & Liebenthal, E. (2006). Syntactic and semantic modulation of neural activity during auditory sentence comprehension. *Journal of Cognitive Neuroscience*, *18*(4), 665–679.

Huttenlocher, P. R. (1979). Synaptic density in human frontal cortex—Developmental changes and effects of aging. *Brain Research*, *163*(2), 195–205.

Huttenlocher, P. R., de Courten, C., Garey, L. J., & Van der Loos, H. (1982). Synaptogenesis in human visual cortex—Evidence for synapse elimination during normal development. *Neuroscience Letters*, *33*(3), 247–252.

Huttenlocher, P. R. (1994). *Synaptogenesis, synapse elimination, and neural plasticity in human cerebral cortex*, *Threats to optimal development: Integrating biological, psychological, and social risk factors: The Minnesota Symposia on Child Psychology* (Vol. 27, p. 35). New Jersey: Lawrence Erlbaum Associates.

Jabés, A., & Nelson, C. A. (2015). 20 years after "The ontogeny of human memory: A cognitive neuroscience perspective". Where are we? *International Journal of Behavioral Development*, *39*(4), 293–303.

Johnson, M. H., & De Haan, M. (2015). *Developmental cognitive neuroscience: An introduction*. West. Sussex: John Wiley & Sons.

Joyner, M. H., & Kurtz-Costes, B. (1997). Metamemory development. The development of memory in childhood. In N. Cowan (Ed.), *The development of memory in childhood* (pp. 275–300). London: UCL Press.

Jusczyk, P. W. (1999). Narrowing the distance to language: One step at a time. *Journal of Communication Disorders, 32*(4), 207–222.

Jusczyk, P. W., & Aslin, R. N. (1995). Infants' detection of the sound patterns of words in fluent speech. *Cognitive Psychology, 29*(1), 1–23.

Kobayashi, C., Glover, G. H., & Temple, E. (2008). Switching language switches mind: Linguistic effects on developmental neural bases of "Theory of Mind". *Social Cognitive and Affective Neuroscience, 3*, 62–70.

Kooijman, V., Hagoort, P., & Cutler, A. (2005). Electrophysiological evidence for prelinguistic infants' word recognition in continuous speech. *Cognitive Brain Research, 24*(1), 109–116.

Kooijman, V., Junge, C., Johnson, E. K., Hagoort, P., & Cutler, A. (2013). Predictive brain signals of linguistic development. *Frontiers in Psychology, 4*, 1–13.

Kreutzer, M. A., Leonard, C., & Flavell, J. H. (1975). An interview study of children's knowledge about memory. *Monographs of the Society for Research in Child Development, 40*, 1–60.

Kuhl, P. K., Conboy, B. T., Coffey-Corina, S., Padden, D., Rivera-Gaxiola, M., & Nelson, T. (2008). Phonetic learning as a pathway to language: new data and native language magnet theory expanded (NLM-e). *Philosophical Transactions of the Royal Society of London B: Biological Sciences, 363*(1493), 979–1000.

Kuhl, P. K., Conboy, B. T., Padden, D., Nelson, T., & Pruitt, J. (2005). Early speech perception and later language development: implications for the "Critical Period". *Language Learning and Development, 1*(3-4), 237–264.

Kuhl, P. K., Stevens, E., Hayashi, A., Deguchi, T., Kiritani, S., & Iverson, P. (2006). Infants show a facilitation effect for native language phonetic perception between 6 and 12 months. *Developmental Science, 9*(2), F13–F21.

Kuhl, P. K., Tsao, F. M., & Liu, H. M. (2003). Foreign-language experience in infancy: Effects of short-term exposure and social interaction on phonetic learning. *Proceedings of the National Academy of Sciences, 100*(15), 9096–9101.

Kuhl, P. K., Williams, K. A., Lacerda, F., Stevens, K. N., & Lindblom, B. (1992). Linguistic experience alters phonetic perception in infants by 6 months of age. *Science, 255*(5044), 606–608.

Lagattuta, K. H., & Sayfan, L. (2013). Not all past events are equal: Biased attention and emerging heuristics in children's past-to-future forecasting. *Child Development, 84*, 2094–2111. Available from http://dx.doi.org/10.1111/cdev.12082.

Landau, B., & Gleitman, L. (1985). *Language and experience* (Vol. 174). Cambridge, MA: Harvard University Press.

Langner, R., & Eickhoff, S. B. (2013). Sustaining attention to simple tasks: A meta-analytic review of the neural mechanisms of vigilant attention. *Psychological Bulletin, 139*(4), 870.

Langus, A., Marchetto, E., Bion, R. A. H., & Nespor, M. (2012). Can prosody be used to discover hierarchical structure in continuous speech? *Journal of Memory and Language, 66*(1), 285–306.

Le Grand, R., Mondloch, C. J., Maurer, D., & Brent, H. P. (2004). Impairment in holistic face processing following early visual deprivation. *Psychological Science, 15*(1), 762–768.

Leinbach, M. D., & Fagot, B. I. (1993). Categorical habituation to male and female faces: Gender schematic processing in infancy. *Infant Behavior and Development, 16*(3), 317–332.

Lenroot, R. K., & Giedd, J. N. (2006). Brain development in children and adolescents: insights from anatomical magnetic resonance imaging. *Neuroscience & Biobehavioral Reviews, 30*(6), 718–729.

Liu, D., Meltzoff, A. N., & Wellman, H. M. (2009). Neural correlates of belief- and desire-reasoning. *Child Development, 80*, 1163–1171.

Lloyd-Fox, S., Blasi, A., & Elwell, C. E. (2010). Illuminating the developing brain: the past, present and future of functional near infrared spectroscopy. *Neuroscience & Biobehavioral Reviews, 34*(3), 269–284.

Mahmoudzadeh, M., Dehaene-Lambertz, G., Fournier, M., Kongolo, G., Goudjil, S., Dubois, J., & Wallois, F. (2013). Syllabic discrimination in premature human infants prior to complete formation of cortical layers. *Proceedings of the National Academy of Sciences, 110*(12), 4846–4851.

Mahy, C. E., Moses, L. J., & Pfeifer, J. H. (2014). How and where: Theory-of-mind in the brain. *Developmental Cognitive Neuroscience*, *9*, 68–81.

Makuuchi, M., Bahlmann, J., Anwander, A., & Friederici, A. D. (2009). Segregating the core computational faculty of human language from working memory. *Proceedings of the National Academy of Sciences*, *106*(20), 8362–8367.

Malle, B. F. (2004). *How the mind explains behavior: Folk explanations, meaning, and social interaction*. Cambridge, MA: MIT Press.

Männel, C., & Friederici, A. D. (2013). Accentuate or repeat? Brain signatures of developmental periods in infant word recognition. *Cortex*, *49*(10), 2788–2798.

Marcus, G. F., Vijayan, S., Rao, S. B., & Vishton, P. M. (1999). Rule learning by seven-month-old infants. *Science*, *283*(5398), 77–80.

Mattock, K., Molnar, M., Polka, L., & Burnham, D. (2008). The developmental course of lexical tone perception in the first year of life. *Cognition*, *106*(3), 1367–1381.

Maurer, D., & Werker, J. F. (2014). Perceptual narrowing during infancy: A comparison of language and faces. *Developmental Psychobiology*, *56*(2), 154–178.

May, L., Byers-Heinlein, K., Gervain, J., & Werker, J. F. (2011). Language and the newborn brain: does prenatal language experience shape the neonate neural response to speech? *Frontiers in Psychology*, *2*, 26–34.

Maye, J., Werker, J. F., & Gerken, L. (2002). Infant sensitivity to distributional information can affect phonetic discrimination. *Cognition*, *82*(3), B101–B111.

Meaney, M. J. (2010). Epigenetics and the biological definition of gene × environment interactions. *Child Development*, *81*(1), 41–79.

Mehler, J., Jusczyk, P., Lambertz, G., Halsted, N., Bertoncini, J., & Amiel-Tison, C. (1988). A precursor of language acquisition in young infants. *Cognition*, *29*(2), 143–178.

Meltzoff, A. N. (1995). What infant memory tells us about infantile amnesia: Long-term recall and deferred imitation. *Journal of Experimental Child Psychology*, *59*(3), 497–515.

Meltzoff, A. N. (1988). Infant imitation and memory: Nine-month-olds in immediate and deferred tests. *Child development*, *59*(1), 217.

Mills, D. L., Coffey-Corina, S., & Neville, H. J. (1997). Language comprehension and cerebral specialization from 13 to 20 months. *Developmental Neuropsychology*, *13*(3), 397–445.

Mills, D. L., Coffey-Corina, S. A., & Neville, H. J. (1993). Language acquisition and cerebral specialization in 20-month-old infants. *Journal of Cognitive Neuroscience*, *5*(3), 317–334.

Mills, D. L., Plunkett, K., Prat, C., & Schafer, G. (2005). Watching the infant brain learn words: Effects of vocabulary size and experience. *Cognitive Development*, *20*(1), 19–31.

Mills, D. L., Prat, C., Zangl, R., Stager, C. L., Neville, H. J., & Werker, J. F. (2004). Language experience and the organization of brain activity to phonetically similar words: ERP evidence from 14-and 20-month-olds. *Journal of Cognitive Neuroscience*, *16*(8), 1452–1464.

Minagawa-Kawai, Y., Cristià, A., & Dupoux, E. (2011). Cerebral lateralization and early speech acquisition: A developmental scenario. *Developmental Cognitive Neuroscience*, *1*(3), 217–232.

Mitchell, J. P. (2008). Activity in right temporo-parietal junction is not selective for theory-of-mind. *Cerebral Cortex*, *18*(2), 262–271.

Molfese, D. L. (1989). The use of auditory evoked responses recorded from newborn infants to predict later language skills. *Birth defects Original Article Series*, *25*(6), 47.

Molfese, D. L. (1990). Auditory evoked responses recorded from 16-month-old human infants to words they did and did not know. *Brain and Language*, *38*(3), 345–363.

Moon, C., Cooper, R. P., & Fifer, W. P. (1993). Two-day-olds prefer their native language. *Infant Behavior and Development*, *16*(4), 495–500.

Morgan, J. L. (1986). *From simple input to complex grammar*. Cambridge: The MIT Press.

Morgan, J. L., & Demuth, K. (1996). *Signal to syntax: An overview*. Signal to syntax: *Bootstrapping from speech to grammar in early acquisition*. New Jersey: Lawrence Erlbaum Associates.

Morgan, J. L., Shi, R., & Allopenna, P. (1996). Perceptual bases of rudimentary grammatical categories: Toward a broader conceptualization of bootstrapping. In J. L. Morgan, & K. Demuth (Eds.), *Signal to syntax: Bootstrapping from speech to grammar in early acquisition*. New Jersey: Lawrence Erlbaum Associates.

Mosconi, M. W., Mack, P. B., McCarthy, G., & Pelphrey, K. A. (2005). Taking an "intentional stance" on eye-gaze shifts: A functional neuroimaging study of social perception in children. *NeuroImage, 27*, 247–252.

Muñoz, C. (2006). The effects of age on foreign language learning: The BAF project. *Age and the Rate of Foreign Language Learning, 19*, 1–40.

Murgatroyd, C., & Spengler, D. (2011). Epigenetics of early child development. *Frontiers in Psychiatry, 2*, 1–15.

Näätänen, R., Lehtokoski, A., Lennest, M., Luuki, A., Alliki, J., Sinkkonen, J., & Alho, K. (1997). Language-specific phoneme representations revealed by electric and magnetic brain responses. *Nature, 385*, 432–434.

Naigles, L. (1990). Children use syntax to learn verb meanings. *Journal of Child Language, 17*(02), 357–374.

Narayan, C. R., Werker, J. F., & Beddor, P. S. (2010). The interaction between acoustic salience and language experience in developmental speech perception: Evidence from nasal place discrimination. *Developmental Science, 13*(3), 407–420.

Nazzi, T., & Ramus, F. (2003). Perception and acquisition of linguistic rhythm by infants. *Speech Communication, 41*(1), 233–243.

Nelson, C. A., & McCleery, J. P. (2008). The use of event-related potentials in the study of typical and atypical development. *Journal of the American Academy of Child and Adolescent Psychiatry, 47*(11), 1252–1261.

Nelson, C. A. (2001). The development and neural bases of face recognition. *Infant and Child Development, 10*(1-2), 3–18.

Nelson, C. A., Moulson, M. C., & Richmond, J. (2006). How does neuroscience inform the study of cognitive development? *Human Development, 49*(5), 260–272.

Nelson, C. A. (1995). The ontogeny of human memory: A cognitive neuroscience perspective. *Developmental Psychology, 31*, 723–738.

Neville, H. J., & Weber-Fox, C. M. (1994). Cerebral subsystems within language. In B. Albowitz, K. Albus, U. Kuhnt, H. C. Nothdurft, & P. Wahle (Eds.), *Structural and functional organization of the neocortex. A symposium in the memory of Otto D* (pp. 424–438). Creutzfeldt. New York: Springer-Verlag.

Neville, H. J., & Bavelier, D. (1998). Neural organization and plasticity of language. *Current Opinion in Neurobiology, 8*(2), 254–258.

Neville, H. J., & Mills, D. L. (1997). Epigenesis of language. *Mental Retardation and Developmental Disabilities Research Reviews, 3*(4), 282–292.

Neville, H. J., Mills, D. L., & Lawson, D. S. (1992). Fractionating language: Different neural subsystems with different sensitive periods. *Cerebral Cortex, 2*(3), 244–258.

Newcombe, N. S., Lloyd, M. E., & Ratliff, K. R. (2007). Development of episodic and autobiographical memory: A cognitive neuroscience perspective. *Advances in Child Development and Behavior, 35*, 37–85.

Nowakowski & Hayes (2012). Overview of Early Brain Development: Linking genetics to brain structure. In A. A. Benasich, & R. H. Fitch (Eds.), *Developmental dyslexia: Early precursors, neurobehavioral markers and biological substrates*. Baltimore, MD: Paul Brooks Publishers.

Ohnishi, T., Moriguchi, Y., Matsuda, H., Mori, T., Hirakata, M., Imabayashi, E., Hirao, K., et al. (2004). The neural network for the mirror system and mentalizing in normally developed children: An fMRI study. *NeuroReport, 15*, 1483–1487.

Oller, D. K. (1980). The emergence of the sounds of speech in infancy. In G. H. Yeni-Komshian, J. F. Kavanagh, & C. A. Ferguson (Eds.), *Child phonology* (Vol. 1). New York: Academic Press.

Olson, D. R., & Bialystok, E. (2014). *Spatial cognition: The structure and development of mental representations of spatial relations*. New York: Psychology Press.

Onishi, K. H., & Baillargeon, R. (2005). Do 15-month-old infants understand false beliefs?. *Science, 308*(5719), 255–258.

Ortiz-Mantilla, S., Hämäläinen, J. A., Musacchia, G., & Benasich, A. A. (2013). Enhancement of gamma oscillations indicates preferential processing of native over foreign phonetic contrasts in infants. *Journal of Neuroscience, 33*(48), 18746–18754.

Pallier, C., Bosch, L., & Sebastián-Gallés, N. (1997). A limit on behavioral plasticity in speech perception. *Cognition, 64*(3), B9–B17.

Palmer, S. B., Fais, L., Golinkoff, R. M., & Werker, J. F. (2012). Perceptual narrowing of linguistic sign occurs in the 1st year of life. *Child Development, 83*(2), 543–553.

Pascalis, O., & Kelly, D. J. (2009). The origins of face processing in humans: Phylogeny and ontogeny. *Perspectives on Psychological Science, 4*(2), 200–209.

Peña, M., Bonatti, L. L., Nespor, M., & Mehler, J. (2002). Signal-driven computations in speech processing. *Science, 298*(5593), 604–607.

Peña, M., Pittaluga, E., & Mehler, J. (2010). Language acquisition in premature and full-term infants. *Proceedings of the National Academy of Sciences, 107*(8), 3823–3828.

Peña, M., Werker, J. F., & Dehaene-Lambertz, G. (2012). Earlier speech exposure does not accelerate speech acquisition. *The Journal of Neuroscience, 32*(33), 11159–11163.

Perner, J., Ruffman, T., & Leekam, S. R. (1994). Theory of mind is contagious: You catch It from your sibs. *Child Development, 65*(4), 1228–1238.

Petanjik, Z., Judâs, M., Ŝimic, G., Rasin, M. R., Uylings, H. B., Rakic, P., & Kostovic, I. (2011). *Proceedings of the National Academy of Sciences, 108*(32), 13281–13286.

Peters, A. M. (1983). The units of language acquisition. *Cambridge Series of Monographs and Texts in Applied Psycholinguistics*. New York: Cambridge University Press.

Peterson, C. C., Wellman, H. M., & Liu, D. (2005). Steps in theory of mind development for children with autism and deafness. *Child Development, 76*, 502–517.

Peterson, C., & Slaughter, V. (2003). Opening windows into the mind: Mothers' preferences for mental state explanations and children's theory of mind. *Cognitive Development, 18*(3), 399–429.

Peterson, C. C., Wellman, H. M., & Slaughter, V. (2012). The mind behind the message: Advancing theory-of-mind scales for typically developing children, and those with deafness, autism, or Asperger syndrome. *Child Development, 83*(2), 469–485.

Pfeifer, J. H., Masten, C. L., Borofsky, L. A., Dapretto, M., Fuligni, A. J., & Lieberman, M. D. (2009). Neural correlates of direct and reflected self-appraisals in adolescents and adults: When social perspective-taking informs self-perception. *Child Development, 80*, 1016–1038.

Phillips, A. T., Wellman, H. M., & Spelke, E. S. (2002). Infants' ability to connect gaze and emotional expression to intentional action. *Cognition, 85*, 53.

Pinker, S. (1984). *Language learnability and language development*. Cambridge, MA: Harvard University.

Poeppel, D., Idsardi, W. J., & van Wassenhove, V. (2008). Speech perception at the interface of neurobiology and linguistics. *Philosophical Transactions of the Royal Society of London, Series B Biological Science, 363*, 1071–1086.

Polka, L., & Sundara, M. (2012). Word segmentation in monolingual infants acquiring Canadian English and Canadian French: native language, cross-dialect, and cross-language comparisons. *Infancy, 17*(2), 198–232.

Polka, L., & Werker, J. F. (1994). Developmental changes in perception of nonnative vowel contrasts. *Journal of Experimental Psychology: Human Perception and Performance, 20*(2), 421.

Polka, L., Colantonio, C., & Sundara, M. (2001). A cross-language comparison of/d/–/ð/perception: Evidence for a new developmental pattern. *The Journal of the Acoustical Society of America, 109*(5), 2190–2201.

Price, C. J. (2012). A review and synthesis of the first 20years of PET and fMRI studies of heard speech, spoken language and reading. *NeuroImage, 62*(2), 816–847.

Quinn, P. C., & Slater, A. (2003). Face perception at birth and beyond. In O. Pascalis, & A. Slater (Eds.), *The development of face processing in infancy and early childhood: Current perspectives*. Huntington, NY: Nova Science.

Ramirez, N.F., Ramirez, R.R., Clarke, M., Tayly, S., Kuhl, P.K. (in press). Speech discrimination in 11-month-old bilingual and monolingual infants: A magnetoencephalography study. *Developmental Science*. http://www.washington.edu/news/2016/04/04/bilingual-baby-brains-show-increased-activity-in-executive-function-regions/Wen-Jui

Repacholi, B. M., & Gopnik, A. (1997). Early reasoning about desires: Evidence from 14- and 18-month-olds. *Developmental Psychology, 33*, 12-12

Righi, G., & Nelson, C. A., III (2013). The neural architecture and developmental course of face processing. In J. Rubenstein, & P. Rakic (Eds.), *comprehensive developmental neuroscience: Neural circuit development and function in the healthy and diseased brain*. Oxford: Elsevier.

Rivera-Gaxiola, M., Csibra, G., Johnson, M. H., & Karmiloff-Smith, A. (2000). Electrophysiological correlates of cross-linguistic speech perception in native English speakers. *Behavioural Brain Research, 111*(1), 13–23.

Rivera-Gaxiola, M., Silva-Pereyra, J., & Kuhl, P. K. (2005). Brain potentials to native and non-native speech contrasts in 7-and 11-month-old American infants. *Developmental Science*, *8*(2), 162–172.

Rovee-Collier, C. (1999). The development of infant memory. *Current Directions in Psychological Science*, *8*(3), 80–85.

Rovee-Collier, C., & Hartshorn, K. (1999). Long-term maintenance of infant memory. *Contract*, *8854*, 8020.

Sabbagh, M. A., Bowman, L. C., Evraire, L. E., & Ito, J. M. B. (2009). Neurodevelopmental correlates of theory of mind in preschool children. *Child Development*, *80*, 1147–1162.

Sabbagh, M. A., Hopkins, S. F., Benson, J. E., & Flanagan, J. R. (2010). Conceptual change and preschoolers' theory of mind: Evidence from load–force adaptation. *Neural Networks*, *23*(8), 1043–1050.

Saffran, J. R., Aslin, R. N., & Newport, E. L. (1996). Statistical learning by 8-month-old infants. *Science*, *274*(5294), 1926–1928.

Saxe, R. (2013). The new puzzle of theory of mind development. In M. R. Banaji, & S. A. Gelman (Eds.), *Navigating the social world: What infants, children, and other species can teach us*. New York: Oxford University Press.

Saxe, R., & Kanwisher, N. (2003). People thinking about thinking people. The role of the temporo-parietal junction in "theory of mind." *NeuroImage*, *19*, 1835–1842.

Saxe, R. R., Whitfield-Gabrieli, S., Scholz, J., & Pelphrey, K. A. (2009). Brain regions for perceiving and reasoning about other people in school-aged children. *Child Development*, *80*, 1197–1209.

Schurz, M., Radua, J., Aichhorn, M., Richlan, F., & Perner, J. (2014). Fractionating theory of mind: A meta-analysis of functional brain imaging studies. *Neuroscience and Biobehavioral Reviews*, *42*, 9–34.

Selkirk, E. (1996). The prosodic structure of function words. In J. L. Morgan, & K. Demuth (Eds.), *The prosodic structure of function words. Signal to syntax: Bootstrapping from speech to grammar in early acquisition* (pp. 187–214). Mahwah: Erlbaum.

Shahaeian, A., Peterson, C. C., Slaughter, V., & Wellman, H. M. (2011). Culture and the sequence of steps in theory of mind development. *Developmental Psychology*, *47*(5), 1239.

Shi, R., Werker, J. F., & Morgan, J. L. (1999). Newborn infants' sensitivity to perceptual cues to lexical and grammatical words. *Cognition*, *72*(2), B11–B21.

Shultz, S., Vouloumanos, A., Bennett, R. H., & Pelphrey, K. (2014). Neural specialization for speech in the first months of life. *Developmental Science*, *17*(5), 766–774.

Skeide, M. A., Brauer, J., & Friederici, A. D. (2015). Brain functional and structural predictors of language performance. *Cerebral Cortex*, bhv042.

Slaughter, V. (2015). Theory of mind in infants and young children: A review. *Australian Psychologist*, *50*(3), 169–172.

Soderstrom, M., Seidl, A., Kemler Nelson, D. G., & Jusczyk, P. W. (2003). The Prosodic Bootstrapping of Phrases: Evidence from Prelinguistic Infants. *Journal of Memory and Language*, *49*(2), 249–267.

Sodian, B. (2011). Theory of mind in infancy. *Child Development Perspectives*, *5*(1), 39–43.

Sommer, M., Meinhardt, J., Eichenmüller, K., Sodian, B., Döhnel, K., & Hajak, G. (2010). Modulation of the cortical false belief network during development. *Brain Research*, *1354*, 123–131.

Southgate, V., Senju, A., & Csibra, G. (2007). Action anticipation through attribution of false belief by 2-year-olds. *Psychological Science*, *18*(7), 587–592.

Stager, C. L., & Werker, J. F. (1997). Infants listen for more phonetic detail in speech perception than in word-learning tasks. *Nature*, *388*(6640), 381–382.

Steinhauer, K. (2014). Event-related potentials (ERPs) in second language research: a brief introduction to the technique, a selected review, and an invitation to reconsider critical periods in L2. *Applied Linguistics*, *35*(4), 393–417.

Stiles, J. (2008). *The fundamentals of brain development: Integrating nature and nurture*. Cambridge, MA: Harvard University Press.

Surian, L., Caldi, S., & Sperber, D. (2007). Attribution of beliefs by 13-month-old infants. *Psychological Science*, *18*(7), 580–586.

Swingley, D. (2009). Contributions of infant word learning to language development. *Philosophical Transactions of the Royal Society of London B: Biological Sciences*, *364*(1536), 3617–3632.

Swingley, D., & Aslin, R. N. (2000). Spoken word recognition and lexical representation in very young children. *Cognition*, *76*(2), 147–166.

Swingley, D., & Aslin, R. N. (2002). Lexical neighborhoods and the word-form representations of 14-month-olds. *Psychological Science*, *13*(5), 480–484.

Tsao, F. M., Liu, H. M., & Kuhl, P. K. (2004). Speech perception in infancy predicts language development in the second year of life: A longitudinal study. *Child Development*, *75*(4), 1067–1084.

Tsao, F. M., Liu, H. M., & Kuhl, P. K. (2006). Perception of native and non-native affricate-fricative contrasts: Cross-language tests on adults and infants. *The Journal of the Acoustical Society of America*, *120*(4), 2285–2294.

Vouloumanos, A., & Werker, J. F. (2007). Listening to language at birth: Evidence for a bias for speech in neonates. *Developmental Science*, *10*(2), 159–164.

Vouloumanos, A., Hauser, M. D., Werker, J. F., & Martin, A. (2010). The tuning of human neonates' preference for speech. *Child Development*, *81*(2), 517–527.

Weber-Fox, C., & Neville, H. J. (2001). Sensitive periods differentiate processing of open-and closed-class words: An ERP study of bilinguals. *Journal of Speech, Language, and Hearing Research*, *44*(6), 1338–1353.

Weikum, W. M., Oberlander, T. F., Hensch, T. K., & Werker, J. F. (2012). Prenatal exposure to antidepressants and depressed maternal mood alter trajectory of infant speech perception. *Proceedings of the National Academy of Sciences*, *109*(Supplement 2), 17221–17227.

Wellman, H. M. (2002). Understanding the psychological world: developing a theory of mind. In U. Goswami (Ed.), *Blackwell handbook of childhood cognitive development* (pp. 167–187). Malden, MA: Blackwell Publishing.

Wellman, H. M. (2011). Developing a theory of mind. In U. Goswami (Ed.), *The Blackwell handbook of childhood cognitive development* (2nd ed., pp. 258–284). New York: Blackwell.

Wellman, H. M. (2014). *Making minds: How theory of mind develops*. Oxford University Press.

Wellman, H. M., & Liu, D. (2004). Scaling of theory-of-mind tasks. *Child Development*, *75*(2), 523–541.

Wellman, H. M., Cross, D., & Watson, J. (2001). Meta-analysis of theory of mind development: The truth about false belief. *Child Development*, *72*, 655–684.

Werker, J. F., & Gervain, G. (2013). Speech perception in infancy: A foundation for language acquisition. *The Oxford Handbook of Developmental Psychology*, *1*, 909–925.

Werker, J. F., & Hensch, T. K. (2015). Critical periods in speech perception: new directions. *Psychology*, *66*(1), 173.

Werker, J. F., & Tees, R. C. (1984). Cross-language speech perception: Evidence for perceptual reorganization during the first year of life. *Infant Behavior and Development*, 7(1), 49–63.

Werker, J. F., & Tees, R. C. (2005). Speech perception as a window for understanding plasticity and commitment in language systems of the brain. *Developmental Psychobiology*, *46*(3), 233–251.

Werker, J. F., & Yeung, H. H. (2005). Infant speech perception bootstraps word learning. *Trends in Cognitive Sciences*, *9*(11), 519–527.

Werker, J. F., Fennell, C. T., Corcoran, K. M., & Stager, C. L. (2002). Infants' ability to learn phonetically similar words: Effects of age and vocabulary size. *Infancy*, *3*(1), 1–30.

Werker, J. F., Gilbert, J. H., Humphrey, K., & Tees, R. C. (1981). Developmental aspects of cross-language speech perception. *Child Development*, *52*(1), 349–355.

Wimmer, H., & Perner, J. (1983). Beliefs about beliefs: Representation and constraining function of wrong beliefs in young children's understanding of deception. *Cognition*, *13*(1), 103–128.

Wojcik, E. H., & Saffran, J. R. (2013). The ontogeny of lexical networks toddlers encode the relationships among referents when learning novel words. *Psychological Science*, *24*(10), 1898–1905.

Woodward, A. (1998). Infants selectively encode the goal object of an actor' s reach. *Cognition*, *69*, 1–34.

Wu, C. Y., Vissiennon, K., Friederici, A. D., & Brauer, J. (2016). Preschoolers' brains rely on semantic cues prior to the mastery of syntax during sentence comprehension. *NeuroImage*, *126*, 256–266.

Yeung, H. H., Chen, K. H., & Werker, J. F. (2013). When does native language input affect phonetic perception? The precocious case of lexical tone. *Journal of Memory and Language*, *68*(2), 123–139.

Yoshida, K. A., Pons, F., Maye, J., & Werker, J. F. (2010). Distributional phonetic learning at 10 months of age. *Infancy, 15*(4), 420–433.

Young-Browne, G., Rosenfeld, H. M., & Horowitz, F. D. (1977). Infant discrimination of facial expressions. *Child Development*, 555–562.

Yuan, S., Fisher, C., & Snedeker, J. (2012). Counting the nouns: Simple structural cues to verb meaning. *Child Development, 83*(4), 1382–1399.

Zatorre, R. J., Evans, A. C., Meyer, E., & Gjedde, A. (1992). Lateralization of phonetic and pitch discrimination in speech processing. *Science, 256*, 846–849.

FURTHER READING

Männel, C., & Friederici, A. D. (2009). Pauses and intonational phrasing: ERP studies in 5-month-old German infants and adults. *Journal of Cognitive Neuroscience, 21*(10), 1988–2006.

CHAPTER 11

Toward an Understanding of the Neural Basis of Executive Function Development

Sammy Perone[1], Brandon Almy[2] and Philip D. Zelazo[2]
[1]Washington State University, Pullman, WA, United States
[2]University of Minnesota, Minneapolis, MN, United States

11.1 OVERVIEW

Executive function (EF) refers to a set of top-down neurocognitive processes that underlie the regulation of thought, emotion, and action (Blair & Raver, 2015; Carlson, Zelazo, & Faja, 2013; Müller & Kerns, 2015; Zelazo, 2015). Scientific interest in EF development has blossomed over the past two decades because individual differences in EF predict outcomes in a wide range of contexts (Carlson et al., 2013; Müller & Kerns, 2015). For example, individual differences in EF predict math and reading abilities during preschool and the early school years (Blair & Razza, 2007). Teachers often report that top-down, regulatory abilities such as sitting still in the classroom, paying attention, and following rules are the most important determinants of classroom success (McClelland et al., 2007). Individual differences in EF during early childhood also set the foundation for long-term developmental outcomes. For example, individual differences in self-control abilities during early childhood predict outcomes in health, wealth, and involvement in criminal activity nearly 30 years later (Moffitt et al., 2011). EF processes are sometimes described as skills because they depend on specific neural circuits and can be developed in a use-dependent fashion that ultimately enables children to behave adaptively (for a discussion, see Zelazo, 2015).

EF is most often studied as three neurocognitive processes—working memory (i.e., the ability to hold and manipulate information in mind), inhibitory control (i.e., the ability to act purposefully despite distraction), and set shifting or cognitive flexibility (i.e., the ability to switch between ways of thinking). Miyake et al. (2000) identified these three processes in a factor analysis based on adults' performance on a battery of cognitive tasks. These three neurocognitive processes work together for conceptualizing problems, making inferences, and maintaining goals in mind in order to guide behavior in a top-down fashion (Zelazo, 2015). It is notable that the

The Neurobiology of Brain and Behavioral Development
DOI: http://dx.doi.org/10.1016/B978-0-12-804036-2.00011-X

three-factor model may be a less appropriate representation of EF processes during childhood. For example, Wiebe et al. (2011) found that a factor analysis on a battery of cognitive tasks showed that EF is best modeled by one latent, domain-general factor for preschool-aged children as opposed to the three factors identified by Miyake et al. in adults. Differentiation into multiple factors might begin to emerge in middle childhood. Brydges, Fox, Reid, and Anderson (2014) found that by 10 years of age, children's performance on a range of measures of working memory, inhibitory control, and cognitive flexibility is best modeled as a working memory factor that is separate from a combined inhibition and cognitive flexibility factor. The late emergence of cognitive flexibility as a unique factor may be due to the requirement of certain levels of working memory and inhibition for successful cognitive flexibility (for a discussion, see Müller & Kerns, 2015). Recent evidence indicates that the differentiation of EF over development may have a more complex developmental trajectory. Howard, Okely, and Ellis (2015) showed that performance on working memory, inhibitory control, and cognitive flexibility tasks are unrelated during the early preschool years but highly related during later preschool years. This finding raises the possibility that EF is initially differentiated and then undergoes a period of integration.

There is also behavioral evidence that EF varies along a continuum from "Hot" to "Cool" (Peterson & Welsh, 2014; Zelazo & Müller, 2002; Zelazo, 2015). Cool EF are aspects of EF that are most commonly studied in decontextualized laboratory measures of working memory, inhibitory control, and cognitive flexibility. Hot EF refers to aspects of EF that emerge in motivationally significant contexts, such as when evaluating rewards or interacting with peers. Studies of hot EF place an emphasis on the interaction of cognition and emotion in guiding behavior. Hot EF is highly relevant in real-world contexts.

It has long been known that EF is dependent on the prefrontal cortex. This insight was initially made possible by case studies of wounded soldiers who sustained damage to the prefrontal cortex. These individuals lacked many stereotypical top-down cognitive functions, such as the ability to act in a purposeful, future-oriented fashion and the capacity to plan, anticipate, initiate behavior, and think flexibly (for a review, see Peterson & Welsh, 2014). John Harlow's case study of Phineas Gage is a famous example (Harlow, 1868). Phineas was a 19th-century railroad worker who suffered damage to the frontal region of his brain from an accident in which a tamping iron rod was blown through the front of his skull. Prior to the injury, Phineas was described as a hardworking, intelligent, and planful individual but, afterwards, he was described as lacking restraint and unable to formulate and execute plans. Phineas was described by acquaintances to be so radically different after the injury that he was someone else.

Our understanding of the neural basis of EF has grown immensely. Our early understanding was greatly informed by research that aimed to identify connections between localized brain regions and function, largely focused on the link between regions of prefrontal cortex and EF. During the past decade, research has begun to focus more on identifying the networks of regions distributed across the brain that underlie EF. This shift is attributable to technological and theoretical advances in neuroscience that enable researchers to characterize the brain regions that are connected and how information flows through networks (e.g., see Sporns, 2011). The present is an especially exciting time in our scientific understanding of the neural basis of EF during development. We are beginning to paint a picture of the codevelopment of EF and the brain. Nevertheless, there are still large gaps to fill. For example, we lack an understanding of the neurodevelopmental mechanisms that drive change in the neural architectures that underlie EF. Neurodevelopmental theories of EF are playing an important role in filling such gaps and inspiring new research.

This chapter is divided into four sections. In the first two sections, we review cool and hot EF, respectively, and what is known about their neural basis during development. In both sections, we focus on key studies that shed light on the brain regions and brain dynamics involved when performing EF tasks over development. We discuss open questions regarding the neurodevelopmental mechanisms driving change in brain and behavior. In the third section, we describe neurodevelopmental theories of EF that focus on brain dynamics in real time and over development. In the fourth section, we draw a series of conclusions about our current understanding of the neural basis of EF during development. We also identify a set of research priorities for the future.

11.2 COOL EXECUTIVE FUNCTION

Contemporary studies of EF are born of a historical emphasis on those cool aspects of EF involved in guiding future-oriented behavior, such as working memory, inhibitory control, and cognitive flexibility. We review research on these neurocognitive processes as they are often studied—in decontextualized, emotionally neutral laboratory tasks. We review key studies that provide insights into their neural basis, particularly in development, and discuss potential developmental mechanisms underlying change in these processes.

Research using the National Institutes of Health (NIH) Toolbox for Neurological and Behavioral Function-Cognition Battery (CB) has shown that cool EF develops markedly across childhood, into adolescence, and even into adulthood. The NIH Toolbox provides a set of brief, reliable, validated, and normed measures of all three aspects of cool EF, and they have helped characterize the development of cool EF across the lifespan, using the same measures from ages 3 to 85 years (Tulsky et al., 2013; Zelazo et al., 2013, 2014). Cool EF develops rapidly during the preschool years, from about 3 and 6 years of age. They continue to improve at a slower rate until early adolescence, when there is another period of rapid improvement, and then they again improve more gradually until reaching a peak in the early 1920s. It's all downhill from there.

11.2.1 Working memory

Working memory has been dubbed the heart of intelligent behavior (Necka, 1992). Working memory refers to the manipulation of information in mind in the service of behavior. Working memory has been distinguished from short-term memory, which refers more simply to the maintenance of information in mind over time. It is working memory that is considered to be an aspect of EF and to play a critical role in maintaining rules in mind, problem solving, planning, and so on (see Müller & Kerns, 2015). Working memory has been studied in a variety of task contexts (for a review, see Simmering & Perone, 2013). For example, working memory for digits or words has often been measured in span tasks in which participants are provided a list of items and asked to recall them in reverse order. Visuo-spatial working memory is often studied in variants of the Corsi block task in which participants observe an experimenter tap a number of spatially distinct locations and are asked to tap the locations in the reverse order. These tasks require participants go beyond maintaining information in mind over time and require them to manipulate that information respond correctly.

The dorsolateral prefrontal cortex and regions of parietal cortex are known to be involved in working memory processes. Improvements in working memory over development may correspond to the establishment of more efficient and reliable neural networks in these regions. Crone, Wendelken, Donohue, van Leijenhorst, and Bunge (2006) tested this possibility using functional magnetic resonance imaging (fMRI) by comparing performance in a maintenance (short-term memory) task and a manipulation (working memory) task in 8- to 12-year-old children, 13- to 17-year-old adolescents, and 18- to 25-year-old adults. Participants were presented with three objects (e.g., rooster, grapes, boat) and, across a 6-s delay, asked to repeat the items sequentially in the maintenance task or in the reverse order in the manipulation task. After the delay, participants were given a probe and asked to report the position that they were supposed to remember the item in. The behavioral results showed that participants performed more poorly on the manipulation task than the maintenance task, and the youngest age group performed particularly poorly on the manipulation task. All age groups more strongly recruited left ventrolateral prefrontal cortex during the manipulation task than the maintenance task. However, only adolescents and adults also more strongly recruited the right dorsolateral prefrontal cortex and superior parietal regions during the manipulation task than maintenance task. In addition, individual differences in neural activity were associated with accuracy in the manipulation task. Specifically, across age groups, greater activity in right dorsolateral prefrontal cortex and superior parietal cortex were associated with greater accuracy in the manipulation task. These findings suggest that the development of more effective working memory skills in childhood and adolescence involves an increased reliance on networks linking right dorsolateral prefrontal cortex and superior parietal regions.

The brain regions recruited during working memory tasks may depend on the type of working memory being probed. Scherf, Sweeny, and Luna (2006) used fMRI to assess the brain regions recruited during a memory-guided saccade task by 10- to 13-year-old children, 14- to 17-year-old adolescents, and adults greater than 18 years of age. Participants were required to remember the location of a target, and then after a delay execute an eye movement to that location. The behavioral results showed that with age participants' final saccade landed closer to the target. The fMRI results revealed both quantitative and qualitative differences in neural activity in the brain regions recruited over development. First, the quantitative differences were that with age participants exhibited an increased reliance on core working memory brain regions including left dorsolateral prefrontal cortex and parietal cortex. Second, the qualitative differences were that children also recruited different brain regions than adolescents and adults. Whereas children did not recruit premotor regions (involved in the voluntary control of saccades) that were components of the older participants' neural networks, they did recruit regions involved in encoding and visual object recognition, such as the fusiform. These results raise the possibility that children rely on ventral processes involved in object processing rather than dorsal processes involved in spatial processing to perform the memory-guided saccade task, which, in turn, might yield poorer performance.

A central hallmark of working memory is its limited capacity. With age, children's ability to maintain more items in mind increases, an ability that is associated with reading and mathematical abilities (see Gathercole, Alloway, Willis, & Adams, 2006). Klingberg, Forssberg, and Westerberg (2002) used fMRI to identify the brain regions involved in working memory capacity between 9 and 18 years of age. Participants performed a visuo-spatial working memory task in which they were presented with spans of three to five dots at a single location sequentially in the scanner. Spans were followed by a probe dot at one location, and participants were asked to report whether the probe appeared in one of the original positions. Critically, loads were intentionally kept low in the scanner to identify brain regions associated with working memory processes across age groups. Loads were increased outside the scanner to estimate capacity. To identify the brain regions associated with working memory processes, neural activity during the working memory task was compared to neural activity in a baseline task in which participants simply watched dots appear sequentially. The results showed that working-memory related neural activity in frontal (superior frontal sulcus, bilateral) and parietal (intraparietal sulcus, bilateral) regions, as well as left occipital cortex, increased with age. In addition, working-memory related activity in left superior frontal and left intraparietal areas was correlated with individual differences in capacity measured outside of the scanner. The results are consistent with the suggestion that the development of working memory capacity involves an increasing reliance on networks involving frontal and parietal regions. Edin et al. (2009) found that activity in dorsolateral and parietal cortex (intraparietal sulcus) were strongly correlated when working memory loads were high and that individuals who recruit dorsolateral cortex more strongly when working memory loads are high also exhibit better performance on high-load trials.

The vast majority of research on the neural basis of working memory development has been cross sectional, and limited to small groups of participants in middle childhood, adolescence, and adulthood. This is unfortunate because EF changes rapidly during the preschool years (see Carlson, 2005), and both in childhood and across development there are large group and individual differences in EF. Preschool-aged children's tendency to move and the loud scanning process are barriers to probing the neural basis of working memory using fMRI during this period. To overcome this barrier, Buss, Fox, Boas, and Spencer (2014) used a noninvasive imaging technology called functional near-infrared spectroscopy (fNIRS) to probe how frontal and parietal regions are linked to working memory processes and capacity limits in 3- and 4-year-old children. More specifically, they used the change detection task to estimate children's visual working memory capacity. Children were presented with a memory array of shapes in set sizes that varied in load from 1 to 3 and, following a delay, presented with a test array in which all the items remained the same or one item had changed. Children's task was to report whether the memory and test arrays were the same or different. The behavioral results showed that 4-year-old children performed more accurately across set sizes and had a higher estimated capacity than 3-year-old children. The fNIRS results showed that children exhibited a stronger neural response when the load was high, particularly in left parietal and left prefrontal regions. Four-year-old children exhibited a stronger response in parietal regions than did 3-year-old children. Interestingly, 3-year-old children showed greater left parietal and right frontal responses during the encoding phase of the memory array than during the later comparison phase, and this pattern of activity was associated with poorer performance in the change detection task. These findings suggest that engaging left parietal and left prefrontal regions, particularly during the comparison phase, play an important role in developmental improvements in performing tasks that estimate working memory capacity.

In summary, core-working memory areas including frontal and parietal regions are more strongly recruited over development and associated with improvements in children's performance across an array of working memory tasks. There is some evidence that children engage different regions while performing working memory tasks than adolescents and adults. This may be influenced by the details of the task children are participating in, or the type of working memory probed. The degree to which core working memory areas are recruited over development may arise from the formation of more efficient, reliable, and accessible neural networks involving specific frontal and parietal regions as well as the links among them. These networks are likely formed and strengthened through their use, and via neurodevelopmental processes such as pruning, myelination, and dendritic thickening (for a review, see Luna & Sweeny, 2004; Zelazo & Lee, 2010). For example, pruning may reduce redundant neural processes and enable neural systems to more efficiently process information. Computational approaches have shed some light on whether, and how, these neurodevelopmental processes might be a mechanism of change in the degree to which working memory areas are recruited. For example, Edin, Macoveanu, Olesen, Tegner, and Klingberg (2007; see also, Perone, Simmering, & Spencer, 2011; Simmering, 2016) used a neural network model that simulates visuo-spatial working memory processes in neural populations that represent frontal and parietal regions. Edin et al. converted the neural activity in the model into a blood oxygenation level-dependent (BOLD) signal measured in fMRI. They asked whether neuronal processes in the model related to pruning, myelination, and connectivity within and between regions could account for an increase in reliance on frontal and parietal regions when the model performed a visuo-spatial working memory task. The simulation results showed that an increase in the BOLD response in frontal and parietal regions emerged in the model from an increase in interregional connectivity together with local synaptic pruning processes that result in more precise coding of a stimulus. Simulation studies such as this help bridge the gap between neurodevelopmental processes and developmental changes in the brain regions observed to be recruited in functional imaging studies (see also Wijeakumar, Ambrose, Spencer, & Curt, 2017).

11.2.2 Inhibitory control

Inhibitory control refers to the suppression of attention and other responses to particular stimuli (e.g., distractors). Inhibitory control is often measured in tasks that require selecting a course of action in the presence of an alternative, often dominant, response. This form of inhibitory control is highly relevant in children's home and school environments when told, for example, "no" or "stop," in an effort to help them control their behavior in a manner appropriate for the situation. Response inhibition is often studied in variants of the Go—No—Go task which requires participants to respond to one stimulus but not another (e.g., respond to green but not blue), and in the Stroop task, which requires participants to respond to the ink color of a color word. This Stroop task has been adapted for young children in the Day-Night task in which children are asked to say "night" when shown a picture of a sun and "day" when shown a picture of a moon. Children exhibit steady improvement in the Day—Night task between 3 and 6 years of age (for a review, see Carlson, 2005).

Inhibitory control processes have been studied across many task contexts. There is evidence that the brain regions recruited depend on the specific inhibitory control task as well as age. For example, Rubia et al. (2006) used fMRI to examine developmental differences in brain regions recruited across three inhibitory control tasks. The first task was a response inhibition task, the Go—No—Go task, in which the go signal was arrows pointing left or right, and the no—go signal was an arrow pointing up. The second task was the Simon task, an interference selection task in which participants must report the direction of an arrow. Some trials are congruent, in which an arrow points in the same direction as the location it appears, and other trials are incongruent, in which an arrow points in the opposite direction as the location it appears. Participants are generally slower and less accurate on incongruent trials. The last task was an attention-shifting task that required participants to switch attention from the horizontal to vertical dimension across trials. Participants were 10- to 17-year-old adolescents and adults 20 to 43 years of age. The behavioral results showed that adolescents performed more poorly than adults across all tasks. The fMRI results showed that in the Go—No—Go task, adults more strongly recruited prefrontal regions including orbitofrontal, mesial frontal, and anterior cingulate, as well as the caudate. In the Simon task, adults more strongly recruited left dorsolateral prefrontal cortex, parietal cortex, and anterior cingulate. In the attention-switching task, adults more strongly recruited right inferior prefrontal cortex, left parietal cortex, and anterior cingulate than did adolescents.

The Rubia et al. (2006) study indicates that the integration of frontal regions with other brain regions increases over development and is associated with improved performance in inhibitory control tasks. There is evidence that connectivity within and between these regions influences top-down control. Hwang, Velanova, and Luna (2010) used fMRI with 8- to 12-year-old children, 13- to 17-year-old adolescents, and adults while performing two saccade tasks. One was an antisaccade task in which participants were instructed to direct an eye-movement to the location opposite a target, and the other was a prosaccade task in which they were instructed to simply look to the location of a target. The behavioral results showed that adults in the antisaccade task were more accurate than adolescents, and adolescents were more accurate than children, exhibiting greater inhibitory control with age. The fMRI results showed that connectivity *within* frontal and parietal regions decreased with age but long-range connections from frontal regions to many downstream sensorimotor and subcortical regions increased with age. Hwang et al. suggested that the development of inhibitory control emerges from increased top-down connectivity from frontal regions to parietal and subcortical regions, as well as decreased short-range connectivity.

Much like investigations into the neural basis of working memory, studies of the neural basis of inhibitory control have largely focused on middle childhood through adolescence and into adulthood. There are major changes in the organization of control networks that happen during this period of development. Fair et al. (2007) studied developmental change in the organization of two control networks in 6- to 9-year-old children, 10- to 15-year-old adolescents, and 21- to 31-year-old adults using resting state MRI. The two control networks they examined were the fronto-parietal network (includes regions such as dorsolateral prefrontal cortex, intraparietal sulcus, and midcingulate) and the cingulo-opercular network (includes regions such as dorsal anterior cingulate/medial superior frontal cortex, thalamus, and bilateral anterior insula/frontal operculum). Both networks are involved in implementing top-down control and processing bottom-up cues. The fronto-parietal network has been proposed to be involved in moment-to-moment (trial-to-trial) adjustments in control, whereas the cingulo-opercular network has been proposed to implement sustained control over the time scale of a task (Dosenbach et al., 2007). Fair et al. (2007) found that in adults the fronto-parietal and cingulo-opercular networks are distinct. During childhood, there is a bridge between the two networks and regions of the cingulo-opercular network are integrated with the fronto-parietal one. The organization of these networks in adolescents was intermediate. For adolescents, the two networks were largely distinct but regions of the cingulo-opercular network remain integrated with the fronto-parietal network. Over development there is a decrease in short-range connectivity, which effectively segregates the two networks, and an increase in long-range connectivity that integrates regions initially part of the fronto-parietal network with the cingulo-opercular network. The authors raise the possibility that one contribution to the decrease in short-range connectivity is synaptic pruning, whereas myelination may strengthen long-range connectivity. The degree to which these neurodevelopmental changes are experience-dependent remains unclear.

11.2.2.1 Cognitive flexibility

Cognitive flexibility refers to the ability to shift attention between task sets, attributes of a stimulus, responses, perspectives, or strategies (Miyake et al., 2000; Zelazo, 2015). The neural basis of cognitive flexibility has been studied during many periods of development, including early childhood. A canonical probe of cognitive flexibility during childhood is the dimensional change card sort (DCCS) task (Doebel & Zelazo, 2015; Zelazo, 2006). In the DCCS, children are presented with a pair of target cards depicting two-dimensional objects (e.g., red star, blue circle) and asked to sort a set of bivalent cards (e.g., red circle, blue star) by one rule (e.g., sort by shape) and then switch to sort by another rule (e.g., sort by color). Under these standard conditions of the DCCS task, 3- and 5-year-old children readily sort by the first rule but only 5-year-old children flexibly switch to sort by the second rule.

Performing the DCCS task in a flexible fashion requires children to resolve the conflict inherent in the bivalent stimuli. In order to resolve this conflict, they must first detect it. The detection of conflict, indexed by activation in anterior cingulate cortex, can initiate the recruitment of lateral prefrontal regions and permit performance adjustments (e.g., Botvinick, Braver, Barch, Carter, & Cohen, 2001; MacDonald, Cohen, & Carter, 2000; Ridderinkhof, van den Wildenberg, Wijnen, & Burle, 2004). Espinet, Anderson, and Zelazo (2012) tested the possibility that a neural index of conflict resolution in the DCCS might be seen in the downregulation of a neural signature of conflict detection in children who resolved the conflict versus those who did not. These researchers focused on the N2 component in the event related potential, which is a negative going wave initiated in anterior cingulate cortex and measured over frontal-midline sites about 200–400-ms poststimulus (300–500 ms in children) (e.g., Botvinick et al., 2001). Espinet et al. found that 3- to 5-year-old children who passed the DCCS exhibited smaller N2 amplitudes than did children who failed the DCCS. Source analysis showed that brain regions including the anterior cingulate cortex were a likely the origin of the signal. According to Espinet et al. (2012), children with better EF abilities detected the conflict on early trials, and this initiated reflection (i.e., the reprocessing of information, see below for a discussion) and the formulation and maintenance in working memory of a higher order rule (mediated by lateral prefrontal cortical networks) that effectively resolved the conflict inherent in the stimuli and downregulated anterior cingulate activation. Subsequent research found that children's performance in the DCCS could be improved through training over the course of a laboratory session, which, in turn, reduces the amplitude of the N2 (Espinet, Anderson, & Zelazo, 2013).

Waxer and Morton (2011) examined developmental change in the N2 in response to conflict in a variant of the DCCS that contained both conflict (bivalent) and no-conflict (univalent) stimuli. The focus was on the influence of prior conflict in resolving conflict in the moment. 9- to 11-year-old children, 14- to 15-year-old adolescents, and 18- to 25-year-old adults sorted bivalent stimuli (e.g., blue rabbit) on some trials and univalent stimuli (e.g., blue bar) on other trials. The key finding was that the N2 over frontal-midline sites was largest for bivalent trials when bivalent trials were preceded by univalent trials for adolescents and adults. For children, by contrast, the N2 was larger on bivalent trials than univalent trials but was not mediated by the prior trial. This finding suggests that children rely less on prior conflict to resolve conflict in the moment, and are more reactive than proactive.

Other research on the neural correlates of cognitive flexibility in children has focused on lateral prefrontal cortex. For example, Moriguchi and Hiraki (2009) found that young children who pass the DCCS show an increase in oxygenated hemoglobin in ventrolateral prefrontal cortex in response to the presentation of the stimuli (see also Moriguchi, Sakata, Ishibashi, & Ishikawa, 2015). Ezekiel, Bosma, and Morton (2013) tested whether the functional integration of lateral prefrontal cortex with other brain regions involved in cognitive control is associated with developmental change in cognitive flexibility. Ezekiel et al. used fMRI to observe neural activity in 12-year-old children and adults while performing a continuous variant of the DCCS that involved blocks of repeat trials (i.e., sort by the same rule on all trials) and mixed blocks of switch and repeat trials. Analyses were based on areas that showed greater activity in switch blocks than repeat blocks. The key finding was that there was greater coactivation of lateral prefrontal cortex and other regions involved in cognitive control (including anterior cingulate and inferior parietal cortex) in adults than in children. This indicates greater functional integration of lateral prefrontal cortex with other brain regions involved in cognitive control with age, which may contribute to developmental differences in cognitive flexibility.

Developmental change in the temporal dynamics of neuronal activity also contributes to developmental differences in cognitive flexibility. For example, Wendelken, Munakata, Baym, Souza, and Bunge (2012) used fMRI to probe how the temporal dynamics of neural activity are associated with rule switching. Their study aimed to separate the neural processes involved in two aspects of rule use, remembering the rules and switching between rules. Participants were children 8 to 13 years of age and adults 20 to 27 years of age. Participants were presented with stimuli that varied in color or direction (e.g., blue fish pointing left) and were given a cue (e.g., "color," "direction") indicating the relevant dimension. The behavioral results showed that participants were slower and less accurate on rule switching trials than rule representation (repeat) trials, and children were less accurate and slower than adults. The fMRI results showed that children and adults both recruited brain regions that included dorsolateral prefrontal cortex, left posterior parietal cortex, and left premotor cortex on both rule representation (repeat) and rule switching (switch) trials. These results suggest that the same brain regions underlie representing a rule and switching between rules. The time course of neural activity was slower for children than for adults, however. In particular, in children, activity at the onset of each trial appeared to be driven by the activity from the previous trial. This carryover in neural activity may slow children's responses and lead them to make errors on trials in which the rule has changed.

In summary, during early childhood, children show great improvement in resolving conflict, as measured, for example, in the DCCS. However, children continue to respond to conflict relatively well into middle childhood. Improvements in conflict resolution with age may reflect the integration of prefrontal regions with other brain regions involved in cognitive control processes, as well as a reduction in "sluggish" neuronal dynamics.

11.3 HOT EXECUTIVE FUNCTION

Hot EF refers to aspects of EF that are displayed in motivationally significant or emotionally salient situations, such as handling peer pressure, delaying a reward, or learning from gains and losses (Welsh & Peterson, 2014; Zelazo & Carlson, 2012; Zelazo

& Müller, 2002). These aspects of EF are common in real-world settings and noticeably absent in cool EF tasks, such as those described in the previous sections. Hot and cool EF, which typically work together in solving real-world problems, are both higher level forms of deliberate, effortful, top-down, self-regulatory processing that depend on similar brain regions, but they vary in the extent to which they require managing motivation and emotion, including the goal-directed management of basic approach and avoidance motivations (Zelazo & Carlson, 2012).

Hot EF is often studied in tasks that involve a reappraisal of whether to avoid or approach a stimulus (Zelazo, 2015). One hot EF task that is appropriate for young children is the delay of gratification task in which children must choose between a smaller reward (e.g., one marshmallow) available immediately and a larger but delayed reward (e.g., two marshmallows at a later point in time) (Mischel, Shoda, & Rodriguez, 1989; Prencipe & Zelazo, 2005). Delaying gratification involves avoiding a more salient immediate reward and approaching a less salient delayed reward. A number of studies have explored decision-making in gambling tasks with children as well, most notably variations of the Iowa Gambling Task, in which individuals must choose between decks of cards that offer both rewards and losses (e.g., Bechara, Damasio, Damasio, & Anderson, 1994; Crone & van der Molen, 2004; Kerr & Zelazo, 2004). In this task, the options that at first appear advantageous (higher rewards) are revealed gradually to be disadvantageous (higher rewards but even higher losses).

Behaviorally, the distinction between hot and cool EF can be observed in children's behavior by 3 years of age (e.g., Brock, Rimm-Kaufman, Nathanson, & Grimm, 2009; Carlson, Davis, & Leach, 2005; Hongwanishkul, Happaney, Lee, & Zelazo, 2005; Kim, Nordling, Yoon, Boldt, & Kochanska, 2013; Prencipe & Zelazo, 2005; Prencipe et al., 2011; Willoughby, Kupersmidt, Voegler-Lee, & Bryant, 2011), although Carlson and colleagues (e.g., Bernier, Carlson, & Whipple, 2010) have found that a principal components analysis (PCA) of 26-month-olds' performance on a battery of EF measures yielded two factors, corresponding to cool EF ("conflict" tasks, such as the DCCS) and hot EF ("delay" tasks, such as delay of gratification).

Poor hot EF in preschoolers is associated with inattentive-overactive problem behaviors, whereas good cool EF is positively related to academic outcomes, including math and reading (e.g., Brock et al., 2009; Kim et al., 2013; Willoughby et al., 2011). Deficits in hot EF also differentiate oppositional defiant disorder/conduct disorder (ODD/CD) from attention deficit hyperactivity disorder (ADHD) in adolescence (Hobson, Scott, & Rubia, 2011), and hot and cool EF may be differentially implicated in different forms of ADHD (e.g., Castellanos, Sonuga-Barke, Milham, & Tannock, 2006).

Hot EF may be especially important during adolescence. There is some evidence that the development of hot EF lags behind cool EF and shows marked change in adolescence (e.g., Crone & van der Molen, 2004; Crone, Bunge, Latenstein, & van der Molen, 2005; Prencipe et al., 2011). Age-related changes begin later and may continue longer in hot EF tasks, such as the Iowa Gambling Task, than in cool EF tasks, such as digit span. One challenge has been ensuring that hot and cool EF tasks are comparable in task difficulty, which makes determining the relative trajectories of acquiring hot and cool EF abilities difficult (for a discussion, see Welsh & Peterson, 2014).

van Duijvenvoorde, Jansen, Visser, and Huizenga (2010) examined adolescents' (13–15 years) performance on matched affective and cognitive decision-making tasks that both required a comparison of choices based on frequency of loss, amount of loss, and constant gain. Whereas in the cognitive task adolescents made adaptive choices, taking into consideration multiple dimensions, performance was poor in the affective task, where adolescents tended to focus only the frequency of loss. There was no relation between adolescents' performance on the two tasks.

A number of models focused on adolescent development emphasize the mismatch in developmental timing between earlier developing limbic processes (e.g., increasing reward seeking and risk taking) and later developing prefrontal networks (e.g., Casey, 2015; Ernst, 2014; Smith, Chein, & Steinberg, 2013; Somerville, Jones, & Casey, 2010), and the connectivity between them. In addition, however, psychosocial influences appear to play a prominent role in adolescent impulsivity. Studies have shown that the presence of same-aged peers increases the number of risky decisions compared to adolescents who make decisions alone (Chein, Albert, O'Brien, Uckert, & Steinberg, 2011; Gardner & Steinberg, 2005). It is hypothesized that the presence of peers engages a reward sensitive motivational state that biases adolescents to select a riskier choice over a safer alternative (Albert, Chein, & Steinberg, 2013).

Less is known about the neural basis of hot EF than cool EF, but there is good evidence that hot EF processes rely more heavily on ventromedial regions of the prefrontal cortex involved in emotion regulation and affective learning, including orbitofrontal cortex. Orbitofrontal cortex is part of a fronto-striatal circuit that has strong connections to the amygdala and other parts of the limbic system. Consequently, orbitofrontal cortex is anatomically well suited for the integration of affective and nonaffective information, and for the regulation of appetitive/motivated responses (e.g., Damasio, 1994; Rolls, 1999, 2004). Research with nonhuman animals, using object reversal learning and extinction paradigms (e.g., Butter, 1969; Butter, Mishkin, & Rosvold, 1963; Dias, Robbins, & Roberts, 1996; Iversen & Mishkin, 1970; Jones & Mishkin, 1972), together with research on human patients with acquired orbitofrontal damage (e.g., Fellows & Farah, 2003; Rolls, Hornak, Wade, & McGrath, 1994), have led to suggestions that whereas the amygdala is primarily involved in the initial learning of stimulus-reward associations (e.g., Killcross, Robbins, & Everitt, 1997; LeDoux, 1996), orbitofrontal cortex is heavily involved in the reversal and re-appraisal of those associations (e.g., Rolls, 1999, 2004).

Patients with damage to the ventromedial prefrontal region (perhaps especially on the right) exhibit impairments in decision-making in social and personal contexts, such as planning the day and choosing friends (for a review, see Bechara, 2004). In the Iowa Gambling Task, patients, unlike normal controls, do not learn to avoid decks yielding short-term gains and long-term loss (Bechara, Damasio, & Damasio, 2000) or utilize emotional signals to guide decisions in the task as measured by anticipatory skin conductance (Bechara, Tranel, Damasio, & Damasio, 1996).

There is evidence that individual differences in hot EF processes reflect individual differences in how ventromedial regions of prefrontal cortex are recruited in emotionally salient contexts. Casey et al. (2011) tested 4-year-old children in a delay of gratification task and then again when the children had reached their forties. They examined performance in cool and hot versions of the Go–No–Go task. The cool version used neutral male and female facial expressions as stimuli, whereas the hot version used fearful and happy facial expressions as stimuli. They examined performance in individuals they called low delayers and high delayers. Individuals who were less able to delay gratification as a child were classified as low delayers, and individuals who were better able to delay gratification as a child were classified as high delayers. One key finding was that low delayers tended to perform more poorly on the hot version than the cool version of the task, especially making errors on no–go trials to happy faces. These individual differences might reflect differences in brain areas recruited when attempting to reverse the motivational (approach-avoidance) significance of the stimuli (i.e., to avoid responding on the no–go trials, despite the tendency to respond), and this may be especially true to the extent that the stimuli are high in emotional significance (such as happy faces). Indeed, using fMRI Casey et al. (2011) found that low delay individuals less strongly recruited regions of the prefrontal cortex (inferior frontal gyrus) on no–go trials than high delay individuals and exhibited greater recruitment of reward circuitry (ventral striatal) than did high delay individuals on trials in which the happy face served as the no–go stimulus.

In summary, hot EF is common in real-world settings and may play an especially important role during adolescence when risk-taking is prevalent. Hot EF is often studied in tasks that require the reversal of the motivational significance of stimuli. These tasks typically recruit ventromedial regions of prefrontal cortex that are involved in emotion regulation and reward circuitry. There are numerous challenges that must be resolved going forward. For example, how do hot and cool aspects of EF interact, and does the nature of this interaction change over development? What makes a task hot (e.g., reversing the motivational significance of a stimulus, needing to downregulate emotions)? To what degree are cool processes recruited in hot contexts? Answering questions such as these is critical for advancing our understanding of hot and cool EF as well as their neural basis.

11.4 NEURODEVELOPMENTAL THEORIES OF EXECUTIVE FUNCTION

Our knowledge base of the neural basis of EF and its development is growing. A central challenge has been to map out how real-time brain dynamics control behavior in a top-down fashion as well as the neurodevelopmental mechanisms that create change in EF. Brain-based theories of EF are playing an important role on this front. For example, Bunge and Zelazo (2006) proposed that developmental change in children's ability to use increasingly complex rules is driven by the developmental course of the prefrontal cortex. During early childhood, children are first able to use a single rule (e.g., sort by shape), then switch between two rules (e.g., sort by shape, then sort by color), and eventually switch between sets of higher order rules (e.g., if there is a box, sort by shape; if no box, sort by color) (Zelazo, Müller, Frye, & Marcovitch, 2003). The different regions of the prefrontal cortex are associated with different aspects of rule representations. The orbitofrontal region is involved in differentiating between stimuli that are rewarding and those that are not, and reversing those associations (e.g., Wallis & Miller, 2003). Ventrolateral and dorsolateral regions are involved in guiding rule use for bivalent stimuli (Crone et al., 2006). The rostrolateral region is involved in representing higher order rules for switching between task sets or rules. Bunge and Zelazo (2006) proposed that developmental change in children's ability to use increasingly complex rules parallels the order with which these regions of the prefrontal cortex develop. Structural MRI studies have shown that gray matter (neuronal density and connectivity) increases first in orbitofrontal prefrontal cortex, next in ventrolateral prefrontal cortex, and finally in dorsolateral and rostrolateral prefrontal cortex (Gogtay et al., 2004; O'Donnell, Noseworthy, Levine, & Dennis, 2005).

Zelazo and colleagues (e.g., Cunningham & Zelazo, 2010; Zelazo, 2015) proposed a model of how prefrontal dynamics can guide behavior in a goal-directed fashion called the iterative reprocessing (IR) model. The IR model characterizes goal-directed behavior as a dynamic confluence of bottom-up and top-down processes. A key component of the IR model is reflection. At the neural level, reflection involves the use of neural circuits that allow for the active reprocessing of information, and for integration of hierarchically organized regions of prefrontal cortex. At the cognitive level, reflection occurs when one interrupts ongoing cognition and action to consider the context, which, in turn, enables one to guide future behavior by controlling attention in a selective fashion and using memory to guide goal-directed decision making.

The IR model posits that reflection happens in an iterative fashion. Consider a child situated in the DCCS task as an example. If the child is presented with a red rabbit, the child may initially attend to only one salient feature, such as its color. With reprocessing of the stimulus over time, the child may integrate additional information such as its shape, its category membership, its label, and so on. This, in turn, provides the child an opportunity to amplify attention to one or more of the attributes of the stimulus to guide behavior. Zelazo and colleagues (Cunningham & Zelazo, 2010; Zelazo, 2015) proposed that this IR engages increasingly anterior and lateral aspects of the prefrontal cortex involved in top-down control. Reflection enables a child to generate a sufficiently complex construal of the context to maintain a rule in mind (e.g., sort by color), attend selectively (e.g., amplify color information), and inhibit irrelevant information (e.g., ignore shape information).

Zelazo (2015) proposed that reflection develops in a use-dependent fashion, as children practice noticing when to adopt a more deliberate approach to problem solving, which initiates reflection (i.e., the reprocessing of information) and the formulation and maintenance in working memory of more complex representations. Espinet et al. (2013) showed that reflection training could help 3-year-old children think flexibly in the DCCS task. Specifically, 3-year-old children who failed the DCCS participated in another version (with different stimuli) of the task in which they were asked to pause before sorting cards and think about the rule (e.g., in the color game, blue stars go here). This reflection training improved children's performance in the DCCS as well as reduced the amplitude of the N2, which, as described previously, is characteristic of children who are able to resolve the conflict inherent in the bivalent stimuli in the DCCS task. Zelazo (2015) proposed that practice engaging the neural circuitry involved in reflective processes creates developmental change in EF and enables children to guide their behavior in a top-down fashion across a wide array of contexts.

The IR model provides insight into the real-time brain dynamics that might underlie EF and how those processes might change over development. Computational models have made a similar contribution by specifying how real-time brain dynamics drive behavior and the mechanisms by which those dynamics change over development. A recent example is a neural network of EF proposed by Buss and Spencer (2014) called a dynamic neural field model. The model consists of a frontal system that controls dimensional attention in an object-representation system. The object-representation system was inspired by the primate visual system that processes information through a dorsal processing pathway ("where") and a ventral processing pathway ("what") (Haxby et al., 1991). The object-representation system binds multiple features (e.g., blue, star) distributed over multiple dimensions (e.g., color, shape) into an object via their shared spatial location (e.g., blue star at the left location).

The model can be situated in the same DCCS procedure as children. Flexible rule use emerges from the dynamics between the frontal and object-representation systems. The frontal system uses the rule, for example, "sort by shape," to prime neural populations tuned to shape information in the object-representation system. This biases the model to sort by shape. Sorting leads the model to form strong memories for the first dimension, shape in this example. When the rule changes to "sort by color," the frontal system primes neural populations tuned to color information in the object-representation system. Critically, the strong memories for sorting by the first dimension lead the model to perseverate on the first dimension. The model fails the DCCS, much like 3-year-old children. Buss and Spencer (2014) proposed that developmental change in children's DCCS performance is attributable to an increase in the strength of the connectivity between the frontal and object-representation systems. Strong connectivity enables the frontal system to prime neural populations tuned to the postswitch dimension more strongly, overcoming the strong memories acquired by sorting during the preswitch phase of the task.

In summary, neurodevelopmental theories of EF are filling in gaps in our empirical understanding of the real-time brain dynamics underlying EF and the mechanisms by which these dynamics might change. For example, the IR model (e.g., Zelazo, 2015) posits that reprocessing of a stimulus over time engages the neural circuitry in the prefrontal cortex in a hierarchical fashion, yielding a more complex construal of the context and enabling top-down control of cognition, such as guiding attention selectively. This circuitry can be developed in a use-dependent way. Formal models bridge neuroscience with our psychological constructs by specifying how cognitive processes might be realized in the brain. For example, the frontal system in Buss and Spencer's (2014) model utilizes self-sustained neuronal activity associated with dimensional labels (e.g., "shape") to pass excitation via long-range connectivity to neural populations tuned to that dimension in the object-representation system. This implements a form of top-down selective attention.

11.5 CONCLUSIONS AND FUTURE DIRECTIONS

Our understanding of the neural basis of EF during development has grown immensely. The brain regions involved in a wide array of EF tasks are more strongly and selectively recruited with age, and the degree to which those regions are recruited is associated with performance (e.g., Crone et al., 2006). There is evidence that activity in prefrontal regions plays a key role in EF from a very early age, and also that maturation of prefrontal structure and function closely correspond to age-related changes in behavior. There are periods of relatively rapid structural and functional brain development that coincide with rapid increases in EF (e.g., Moriguchi & Hiraki, 2009). These include the preschool years, roughly 2–6 years, and between middle childhood (roughly 9–11 years) and adulthood. These networks appear to develop in a use-dependent fashion, which strengthens within-region networks as well as longer range connections between networks (e.g., fronto-parietal and fronto-limbic networks). Training strengthens the neural circuits that support EF (e.g., Espinet et al., 2013; Klingberg, 2010; Rueda, Rothbart, McCandliss, Saccomanno, & Posner, 2005), including networks within prefrontal cortex and parietal regions (intraparietal) and the functional and structural connectivity between these regions. These networks become more efficient as a function of various experience-dependent processes, including synaptic pruning, myelination, and dendritic thickening (Kolb, Mychasiuk, Muhammad, & Gibb, 2013).

We have identified three needs to further our understanding of the neural basis of EF development. The first need is a more complete understanding of the developmental trajectory of EF. It is widely accepted that EF consists of three neurocognitive processes in adults—working memory, inhibitory control, and cognitive flexibility (Miyake et al., 2000). It is unclear what the trajectory of this tripartite structure is in development. Some studies show that EF is best represented as a single factor during early childhood (Wiebe et al., 2011), perhaps differentiating into two factors in adolescence (Brydges et al., 2014). Yet other studies suggest that EF might undergo several transformations over a much shorter period of time, for example, EF may initially be fractionated during the early preschool years and integrate during the later preschool years (Howard et al., 2015). These mixed results may be due to differences in the EF tasks used across studies as well as the age groups studied. A more complete understanding of EF development can help guide neuroscientific investigations of the neural basis of EF development, and may have clinical implications for the optimization of EF interventions at different ages.

The second need is for an increase of research on brain development during early childhood. The preschool years are a period of time when neural plasticity is high and EF is undergoing rapid change (Carlson et al., 2013; Müller & Kerns, 2015). EF during this period sets the stage for later developmental outcomes during in middle childhood, such as academic success (e.g., Blair & Razza, 2007). EF during this period also predicts long-term developmental outcomes, such as socioeconomic status (Moffitt et al., 2011) and brain dynamics decades later (Casey et al., 2011). The majority of the research described in this chapter has focused on middle childhood, adolescence, and adults. This research has revealed a number of critical insights, including the reorganization of neural networks involved in top-down control processes (e.g., Fair et al., 2007). One barrier to studying the neural basis of EF during early childhood is using imaging methods, such as fMRI, because they involve a loud scanning process and are limited by the tendency of young children to move. Continued research using electroencephalography and fNRIS will help build a knowledge base of the neural basis of EF development during early childhood and beyond.

The last need is a greater emphasis on the neurodevelopmental mechanisms driving change in EF. Formal models may play an important role on this front. Models provide a platform for testing neurodevelopmental hypotheses. As described previously, Edin et al.'s (2007) simulation study tested whether pruning, myelination, or connectivity could yield adult-like recruitment of frontal and parietal regions in a visuo-spatial working memory task. Similarly, Buss and Spencer (2014) showed that developmental change in cognitive flexibility emerged from stronger connectivity between components of the model representing brain regions involved in top-down (frontal regions) and bottom-up (temporal and parietal regions) processes. Formal models also provide researchers a tool to explore the types of experience that might create developmental change in neural circuitry. For example, Perone, Molitor, Buss, Spencer, and Samuelson (2015) showed that in Buss and Spencer's model, experience with stimuli that varied along a feature dimension (e.g., color) influenced connectivity between the frontal and object-representation system, leading the 3-year-old version of the model to perform like the 4-year-old version of the model, a prediction confirmed empirically.

In conclusion, the present is an exciting time in our understanding of the neural basis of EF development. We have learned much about the brain regions recruited in hot and cool EF tasks. We are beginning to learn more about the networks that underlie EF and how these networks are changing over development. On the horizon, there are a number of challenges including a need to continue to map out the behavioral development of EF to guide investigations of their neural basis. Little is known about the mechanisms driving change in EF, but recent advances in network science and formal models have shed some light on these issues. Meeting these challenges will deepen our understanding of the neural basis of EF development.

REFERENCES

Albert, D., Chein, J., & Steinberg, L. (2013). The teenage brain: Peer influences on adolescent decision making. *Current Directions in Psychological Science, 22*(2), 114–120.

Blair, C., & Raver, C. C. (2015). School readiness and self-regulation: A developmental psychobiological approach. *Annual Review of Psychology, 66*, 711–731.

Blair, C., & Razza, R. P. (2007). Relating effortful control, executive function, and false belief understanding to emerging math and literacy ability in kindergarten. *Child Development, 78*(2), 647–663.

Bechara, A. (2004). The role of emotion in decision-making: Evidence from neurological patients with orbitofrontal damage. *Brain and Cognition, 55*(1), 30–40.

Bernier, A., Carlson, S. M., & Whipple, N. (2010). From external regulation to self-regulation: Early parenting precursors of young children's executive functioning. *Child Development, 81*(1), 326–339.

Bechara, A., Damasio, A. R., Damasio, H., & Anderson, S. W. (1994). Insensitivity to future consequences following damage to human prefrontal cortex. *Cognition, 50*(1), 7–15.

Bechara, A., Damasio, H., & Damasio, A. R. (2000). Emotion, decision making and the orbitofrontal cortex. *Cerebral Cortex, 10*(3), 295–307.

Bechara, A., Tranel, D., Damasio, H., & Damasio, A. R. (1996). Failure to respond autonomically to anticipated future outcomes following damage to prefrontal cortex. *Cerebral Cortex, 6*(2), 215–225.

Botvinick, M. M., Braver, T. S., Barch, D. M., Carter, C. S., & Cohen, J. D. (2001). Conflict monitoring and cognitive control. *Psychology Review, 108*(3), 624–652.

Brock, L. L., Rimm-Kaufman, S. E., Nathanson, L., & Grimm, K. J. (2009). The contributions of 'hot' and 'cool' executive function to children's academic achievement, learning-related behaviors, and engagement in kindergarten. *Early Childhood Research Quarterly, 24*(3), 337–349.

Brydges, C. R., Fox, A. M., Reid, C., & Anderson, M. (2014). The differentiation of executive functions in middle childhood: A longitudinal latent-variable analysis. *Intelligence, 47*, 34–43.

Bunge, S. A., & Zelazo, P. D. (2006). A brain-based account of the development of rule use in childhood. *Current Directions in Psychological Sciences, 15*(3), 118–121.

Buss, A. T., Fox, N., Boas, D. A., & Spencer, J. P. (2014). Probing the early development of visual working memory capacity with functional near-infrared spectroscopy. *NeuroImage, 85*, 314–325.

Buss, A. T., & Spencer, J. P. (2014). The emergent executive: A dynamic field theory of the development of executive function. *Monographs of the Society for Research in Child Development, 79*(2), 1–103.

Butter, C. M. (1969). Perseveration in extinction and in discrimination reversal tasks following selective frontal ablations in *Macaca mulatta*. *Physiology and Behavior, 4*(2), 163–171.

Butter, C. M., Mishkin, M., & Rosvold, H. E. (1963). Conditioning and extinction of a food-rewarded response after selective ablations of frontal cortex in rhesus monkeys. *Experimental Neurology, 7*(1), 65–75.

Carlson, S. M. (2005). Developmentally sensitive measures of executive function in preschool children. *Developmental Neuropsychology, 28*(2), 595–616.

Carlson, S. M., Davis, A. C., & Leach, J. G. (2005). Less is more executive function and symbolic representation in preschool children. *Psychological Science, 16*(8), 609–616.

Carlson, S. M., Zelazo, P. D., & Faja, S. (2013). Executive function. In P. D. Zelazo (Ed.), Oxford handbook of developmental psychology, Vol. 1: Body and mind (pp. 706–743). New York, NY: Oxford University Press.

Casey, B. J. (2015). Beyond simple models of self-control to circuit-based accounts of adolescent behavior. *Annual Review of Psychology, 66*, 295–319.

Casey, B. J., Sommerville, L. H., Gotlib, I. H., Ayduk, O., Franklin, N. T., Askren, M. K., & Shoda, Y. (2011). Behavioral and neural correlates of delay of gratification 40 years later. *Proceedings of the National Academy of Sciences of the United States of America, 108*(36), 14998–15003.

Castellanos, F. X., Sonuga-Barke, E. J., Milham, M. P., & Tannock, R. (2006). Characterizing cognition in ADHD: Beyond executive dysfunction. *Trends in Cognitive Sciences, 10*(3), 117–123.

Chein, J., Albert, D., O'Brien, L., Uckert, K., & Steinberg, L. (2011). Peers increase adolescent risk taking by enhancing activity in the brain's reward circuitry. *Developmental Science, 14*(2), F1–F10.

Crone, E. A., Bunge, S. A., Latenstein, H., & van der Molen, M. W. (2005). Characterization of children's decision making: Sensitivity to punishment frequency, not task complexity. *Child Neuropsychology, 11*(3), 245–263.

Crone, E. A., & van der Molen, M. W. (2004). Developmental changes in real life decision making: Performance on a gambling task previously shown to depend on the ventromedial prefrontal cortex. *Developmental Neuropsychology, 25*(3), 251–279.

Crone, E. A., Wendelken, C., Donohue, S., van Leijenhorst, L., & Bunge, S. A. (2006). Neurocognitive development of the ability to manipulate information in working memory. *Proceedings of the National Academy of Sciences of the United States of America, 103*(24), 9315–9320.

Cunningham, W. A., & Zelazo, P. D. (2010). The development of iterative reprocessing: Implications for affect and its regulation. In P. D. Zelazo, M. Chandler, & E. A. Crone (Eds.), *Developmental social cognitive neuroscience* (pp. 81–98). Mahwah, NJ: Lawrence Erlbaum Associates.

Damasio, A. (1994). *Descartes' Error: Emotion, Reason, and the Human Brain*. New York, NY: Putnam.

Dias, R., Robbins, T. W., & Roberts, A. C. (1996). Dissociation in prefrontal cortex of affective and attentional shifts. *Nature, 380*(6569), 69–72.

Doebel, S., & Zelazo, P. D. (2015). A meta-analysis of the Dimensional Change Card Sort: Implications for developmental theories and the measurement of executive function in children. *Developmental Review, 38*, 241–268.

Dosenbach, N. U., Fair, D. A., Miezin, F. M., Cohen, A. L., Wenger, K. K., Dosenbach, R. A., & Schlaggar, B. L. (2007). Distinct brain networks for adaptive and stable task control in humans. *Proceedings of the National Academy of Sciences of the United States of America, 104*(26), 11073–11078.

Edin, F., Klingberg, T., Johansson, P., McNab, F., Tegner, J., & Compte, A. (2009). Mechanism for top-down control of working memory capacity. *Proceedings of the National Academy of Sciences of the United States of America, 106*(16), 6802–6807.

Edin, F., Macoveanu, J., Olesen, P., Tegner, J., & Klingberg, T. (2007). Stronger synaptic connectivity as a mechanism behind development of working memory-related brain activity during childhood. *Journal of Cognitive Neuroscience, 19*(5), 750–760.

Ernst, M. (2014). The triadic model perspective for the study of adolescent motivated behavior. *Brain and Cognition, 89*, 104–111.

Espinet, S. D., Anderson, J. E., & Zelazo, P. D. (2012). N2 amplitude as a neural marker of executive function in young children: An ERP study of children who switch versus perseverate in the dimensional change card sort. *Developmental Cognitive Neuroscience, 2S*, S49–S58.

Espinet, S. D., Anderson, J. E., & Zelazo, P. D. (2013). Reflection training improves executive function in preschool-age children: Behavioral and neural effects. *Developmental Cognitive Neuroscience, 4*, 3–15.

Ezekiel, F., Bosma, R., & Morton, J. B. (2013). Dimensional change card sort performance associated with age-related differences in functional connectivity of lateral prefrontal cortex. *Developmental Cognitive Neuroscience, 5*, 40–50.

Fair, D. A., Dosenbach, N. U. F., Church, J. A., Cohen, A. L., Brahmbhatt, S., Miezin, F. M., & Schlaggar, B. L. (2007). Development of distinct control networks through segregation and integration. *Proceedings of the National Academy of Sciences of the United States of America, 104*(33), 13507–13512.

Fellows, L. K., & Farah, M. J. (2003). Ventromedial frontal cortex mediates affective shifting in humans: Evidence from a reversal learning paradigm. *Brain, 126*(8), 1830–1837.

Gardner, M., & Steinberg, L. (2005). Peer influence on risk taking, risk preference, and risky decision making in adolescence and adulthood: An experimental study. *Developmental Psychology, 41*(4), 625–635.

Gathercole, S. E., Alloway, T. P., Willis, C., & Adams, A. M. (2006). Working memory in children with reading disabilities. *Journal of Experimental Child Psychology, 93*(3), 265–281.

Gogtay, N., Giedd, J. N., Lusk, L., Hayashi, K. M., Greenstein, D., Vai-Tuzis, A. C., & Thompson, P. M. (2004). Dynamic mapping of human cortical development during childhood through early adulthood. *Proceedings of the National Academy of Sciences of the United States of America, 101*(21), 8174–8179.

Harlow, J. M. (1868). Recovery from the passage of an iron bar through the head. *Publications of the Massachusetts Medical Society, 2*(3), 327–346.

Haxby, J. V., Grady, C. L., Horwitz, B., Ungerleider, L. G., Mishkin, M., Carson, R. E., & Rapoport, S. I. (1991). Dissociation of object and spatial visual processing pathways in human extrastriate cortex. *Proceedings of the National Academy of Sciences of the United States of America, 88*(5), 1621–1625.

Hobson, C. W., Scott, S., & Rubia, K. (2011). Investigation of cool and hot executive function in ODD/CD independently of ADHD. *Journal of Child Psychology and Psychiatry, 52*(10), 1035–1043.

Hongwanishkul, D., Happaney, K. R., Lee, W. S., & Zelazo, P. D. (2005). Assessment of hot and cool executive function in young children: Age-related changes and individual differences. *Developmental Neuropsychology, 28*(2), 617–644.

Howard, S. J., Okely, A. D., & Ellis, Y. G. (2015). Evaluation of a differentiation model of preschoolers' executive functions. *Frontiers in Psychology, 6*, 1–7.

Hwang, K., Velanova, K., & Luna, B. (2010). Strengthening of top-down frontal cognitive control networks underlying the development of inhibitory control: An fMRI effective connectivity study. *Journal of Neuroscience, 30*(46), 15535–15545.

Iversen, S. D., & Mishkin, M. (1970). Perseverative interference in monkeys following selective lesions of the inferior prefrontal convexity. *Experimental Brain Research, 11*(4), 376–386.

Jones, B., & Mishkin, M. (1972). Limbic lesions and the problem of stimulus—Reinforcement associations. *Experimental Neurology, 36*(2), 362–377.

Kerr, A., & Zelazo, P. D. (2004). Development of "hot" executive function: The children's gambling task. *Brain and Cognition, 55*(1), 148–157.

Killcross, S., Robbins, T. W., & Everitt, B. J. (1997). Different types of fear-conditioned behaviour mediated by separate nuclei within amygdala. *Nature, 388*(6640), 377–380.

Kim, S., Nordling, J. K., Yoon, J. E., Boldt, L. J., & Kochanska, G. (2013). Effortful control in "hot" and "cool" tasks differentially predicts children's behavior problems and academic performance. *Journal of Abnormal Child Psychology, 41*(1), 43–56.

Klingberg, T. (2010). Training and plasticity of working memory. *Trends in Cognitive Sciences, 14*(7), 317–324.

Klingberg, T., Forssberg, H., & Westerberg, H. (2002). Increased brain activity in frontal and parietal cortex underlies the development of visuospatial working memory capacity during childhood. *Journal of Cognitive Neuroscience, 14*(1), 1–10.

Kolb, B., Mychasiuk, R., Muhammad, A., & Gibb, R. (2013). Brain plasticity in the developing brain. *Progress in Brain Research, 207*, 35–64.

LeDoux, J. (1996). Emotional networks and motor control: A fearful view. *Progress in Brain Research, 107*, 437–446.

Luna, B., & Sweeny, J. A. (2004). The emergence of collaborative brain function: fMRI studies of the development of response inhibition. *Annals of the New York Academy of Sciences, 1021*(1), 296–309.

McClelland, M. M., Cameron, C. E., Connor, C. M., Farris, C. L., Jewkes, A. M., & Morrison, F. J. (2007). Links between behavioral regulation and preschoolers' literacy, vocabulary, and math skills. *Developmental Psychology, 43*(4), 947–959.

MacDonald, A. W., Cohen, J. D., Stenger, V. A., & Carter, C. S. (2000). Dissociating the role of the dorsolateral prefrontal and anterior cingulate cortex in cognitive control. *Science, 288*(5472), 1835–1838.

Mischel, W., Shoda, Y., & Rodriguez, M. L. (1989). Delay of gratification in children. *Science, 244*(4907), 933–938.

Miyake, A., Friedman, N. P., Emerson, M. J., Witzki, A. H., Howerter, A., & Wager, T. (2000). The unity and diversity of executive functions and their contributions to complex "frontal lobe" tasks: A latent variable analysis. *Cognitive Psychology, 41*(1), 49–100.

Moffitt, T., Arseneault, L., Belsky, D., Dickson, N., Han-Cox, R. J., Harrington, H., & Caspi, A. (2011). A gradient of childhood self-control predicts health, wealth, and public safety. *Proceedings of the National Academy of Sciences of the United States of America, 108*(7), 2693–2698.

Moriguchi, Y., & Hiraki, K. (2009). Neural origin of cognitive shifting in young children. *Proceedings of the National Academy of Sciences of the United States of America, 106*(14), 6017–6021.

Moriguchi, Y., Sakata, Y., Ishibashi, M., & Ishikawa, Y. (2015). Teaching others rule-use improves executive function and prefrontal activations in young children. *Frontiers in Psychology, 6*, 1–9.

Müller, U., & Kerns, K. (2015). The development of executive function. In R. M. Lerner, L. S. Liben, & U. Mueller (Eds.), *Handbook of child psychology and developmental science* (pp. 1–53). Hoboken, NJ: Wiley.

Necka, E. (1992). Cognitive analysis of intelligence: The significance of working memory processes. *Perspectives on Individual Differences, 13*(9), 1031–1046.

O'Donnell, S., Noseworthy, M. D., Levine, B., & Dennis, M. (2005). Cortical thickness of the frontopolar area in typically developing children and adolescents. *NeuroImage, 24*(4), 948–954.

Perone, S., Molitor, S. J., Buss, A. T., Spencer, J. P., & Samuelson, L. K. (2015). Enhancing the executive functions of 3-year-olds in the dimensional change card sort task. *Child Development, 86*(3), 812–827.

Perone, S., Simmering, V. R., & Spencer, J. P. (2011). Stronger neural dynamics capture changes in infants' visual working memory capacity development. *Developmental Science, 14*(6), 1379–1392.

Peterson, E., & Welsh, M. C. (2014). The development of hot and cool executive functions in childhood and adolescence: Are we getting warmer? In S. Goldstein, & J. Naglieri (Eds.), *Executive functioning handbook* (pp. 45–65). New York, NY: Springer.

Prencipe, A., Kesek, A., Cohen, J., Lamm, C., Lewis, M. D., & Zelazo, P. D. (2011). Development of hot and cool executive function during the transition to adolescence. *Journal of Experimental Child Psychology, 108*(3), 621–637.

Prencipe, A., & Zelazo, P. D. (2005). Development of affective decision making for self and other evidence for the integration of first-and third-person perspectives. *Psychological Science, 16*(7), 501–505.

Ridderinkhof, K. R., van den Wildenberg, W. P., Wijnen, J., & Burle, B. (2004). Response inhibition in conflict tasks is revealed in delta plots. *Cognitive Neuroscience of Attention*, , 369–377.

Rolls, E. T. (1999). The functions of the orbitofrontal cortex. *Neurocase, 5*(4), 301–312.

Rolls, E. T. (2004). The functions of the orbitofrontal cortex. *Brain and Cognition, 55*(1), 11–29.

Rolls, E. T., Hornak, J., Wade, D., & McGrath, J. (1994). Emotion-related learning in patients with social and emotional changes associated with frontal lobe damage. *Journal of Neurology, Neurosurgery & Psychiatry, 57*(12), 1518–1524.

Rubia, K., Smith, A. B., Woolley, J., Nosarti, C., Heyman, I., Taylor, E., & Brammer, M. (2006). Progressive increase in frontostriatal brain activation from childhood to adulthood during event-related tasks of cognitive control. *Human Brain Mapping, 27*(12), 973–993.

Rueda, M. R., Rothbart, M. K., McCandliss, B. D., Saccomanno, L., & Posner, M. I. (2005). Training, maturation, and genetic influences on the development of executive attention. *Proceedings of the National Academy of Sciences of the United States of America, 102*(41), 14931–14936.

Scherf, K. S., Sweeney, J. A., & Luna, B. (2006). Brain basis of developmental change in visuospatial working memory. *Journal of Cognitive Neuroscience, 18*(7), 1045–1058.

Simmering, V. R. (2016). Working memory capacity in context: Modeling dynamic processes of behavior, memory, and development. *Monographs of the Society for Research Child Development, 81*, 1–152.
Simmering, V. R., & Perone, S. (2013). Working memory capacity as a dynamic process. *Frontiers in Psychology: Hypothesis and Theory, 3*, 1–26.
Smith, A. R., Chein, J., & Steinberg, L. (2013). Impact of socio-emotional context, brain development, and pubertal maturation on adolescent risk-taking. *Hormones and Behavior, 64*(2), 323–332.
Somerville, L. H., Jones, R. M., & Casey, B. J. (2010). A time of change: Behavioral and neural correlates of adolescent sensitivity to appetitive and aversive environmental cues. *Brain and Cognition, 72*(1), 124–133.
Sporns, O. (2011). The human connectome: A complex network. *Annals of the New York Academy of Sciences, 1224*(1), 109–125.
Tulsky, D. S., Carlozzi, N. E., Chevalier, N., Espy, K. A., Beaumont, J. L., & Mungas, D. (2013). V. NIH toolbox cognition battery (CB): Measuring working memory. *Monographs of the Society for Research in Child Development, 78*(4), 70–87.
van Duijvenvoorde, A. C., Jansen, B. R., Visser, I., & Huizenga, H. M. (2010). Affective and cognitive decision-making in adolescents. *Developmental Neuropsychology, 35*(5), 539–554.
Wallis, J. D., & Miller, E. K. (2003). Neuronal activity in primate dorsolateral and orbital prefrontal cortex during performance of a reward preference task. *European Journal of Neuroscience, 18*(7), 2069–2081.
Waxer, M., & Morton, J. B. (2011). The development of future-oriented control: An electrophysiological investigation. *NeuroImage, 56*(3), 1648–1654.
Wendelken, C., Munakata, Y., Baym, C., Souza, M., & Bunge, S. A. (2012). Flexible rule use: Common neural substrates in children and adults. *Developmental Cognitive Neuroscience, 2*(3), 329–339.
Welsh, M. C., & Peterson, E. (2014). Issues in the conceptualization and assessment of hot executive functions in childhood. *Journal of the International Neuropsychological Society, 20*(2), 1–5.
Wiebe, S. A., Sheffield, T., Mize Nelson, J., Clark, C. A. C., Chevalier, N., & Epsy, K. A. (2011). The structure of executive function in 3-year-olds. *Journal of Experimental Child Psychology, 108*(3), 436–452.
Wijeakumar, S., Ambrose, J. P., Spencer, J. P., & Curtu, R. (2017). Model-based functional neuroimaging using dynamic neural fields: An integrative cognitive neuroscience approach. *Journal of Mathematical Psychology, 76*, 212–235.
Willoughby, M., Kupersmidt, J., Voegler-Lee, M., & Bryant, D. (2011). Contributions of hot and cool self-regulation to preschool disruptive behavior and academic achievement. *Developmental Neuropsychology, 36*(2), 162–180.
Zelazo, P. D. (2006). The dimensional change card sort (DCCS): A method of assessing executive function in children. *Nature Protocols, 1*(1), 297–301.
Zelazo, P. D. (2015). Executive function: Reflection, iterative reprocessing, complexity, and the developing brain. *Developmental Review, 38*, 55–68.
Zelazo, P. D., Anderson, J. E., Richler, J., Wallner-Allen, K., Beaumont, J. L., & Weintraub, S. (2013). II. NIH toolbox cognition battery (CB): Measuring executive function and attention. *Monographs of the Society for Research in Child Development, 78*(4), 16–33.
Zelazo, P. D., Anderson, J. E., Richler, J., Wallner-Allen, K., Beaumont, J. L., Conway, K. P., & Weintraub, S. (2014). NIH Toolbox Cognition Battery (CB): Validation of executive function measures in adults. *Journal of the International Neuropsychological Society, 20*(6), 620–629.
Zelazo, P. D., & Carlson, S. M. (2012). Hot and cool executive function in childhood and adolescence: Development and plasticity. *Child Development Perspectives, 6*(4), 354–360.
Zelazo, P. D., & Lee, W. S. C. (2010). Brain development: An overview: Handbook of life-span development. In R. M. Lerner (Ed.), *Cognition, biology, and methods across the lifespan: Handbook of life-span development* (pp. 89–114). Hoboken, NJ: Wiley.
Zelazo, P. D., & Müller, U. (2002). Executive function in typical and atypical development. In U. Goswami (Ed.), *Handbook of childhood cognitive development* (pp. 445–469). Oxford, England: Blackwell Publishers.
Zelazo, P. D., Müller, U., Frye, D., & Marcovitch, S. (2003). The development of executive function in early childhood. *Monographs of the Society for Research in Child Development, 68*(3), 1–151.

FURTHER READING

Bunge, S. A., & Wright, S. B. (2007). Neurodevelopmental changes in working memory and cognitive control. *Current Opinion in Neurobiology, 17*(2), 243–250.

Huizenga, H. M., Crone, E. A., & Jansen, B. J. (2007). Decision-making in healthy children, adolescents and adults explained by the use of increasingly complex proportional reasoning rules. *Developmental Science, 10*(6), 814–825.

Wilk, H. A., & Morton, J. B. (2012). Developmental changes in patterns of brain activity associated with moment-to-moment adjustments in control. *NeuroImage, 63*(1), 475–484.

CHAPTER 12

Rough-and-Tumble Play and the Development of the Social Brain: What Do We Know, How Do We Know It, and What Do We Need to Know?

Sergio M. Pellis[1], Brett T. Himmler[2], Stephanie M. Himmler[1] and Vivien C. Pellis[1]
[1]University of Lethbridge, Lethbridge, AB, Canada
[2]University of Minnesota, Minneapolis, MN, United States

12.1 INTRODUCTION

Play fighting or rough-and-tumble play (RTP) is present in the juveniles of a wide range of mammals (Fagen, 1981) and involves the competition for some advantage over the partner, such as biting a particular body area (Aldis, 1975). Such competition involves the use of tactics of attack and defense that may resemble what is seen in serious fighting, but differs in important ways. Unlike serious fighting, RTP involves some restraint in the execution of the tactics of attack and defense, leading to some degree of reciprocity, so that play partners alternate in their roles as attackers and defenders (Pellis, Pellis, & Reinhart, 2010b). Moreover, RTP has some neural mechanisms that differ from those involved in serious fighting (Siviy & Panksepp, 2011; Trezza, Baarendse & Vanderschuren, 2010). Finally, for at least some species the experience of RTP during the juvenile period facilitates the development of social competence (Pellis, Pellis, & Himmler, 2014).

The literature on human RTP is sparser, but the findings are consistent with the view that, during childhood, RTP differs from serious fighting and functions to refine the development of social skills and in the postpubescent period it is used as a tool for social assessment and manipulation (for reviews, see Pellegrini, 2009; Pellis & Pellis, 2009). Of course, in humans, systematic experimentation is not possible, so the association between social skills and experience with RTP may simply be a noncausal correlation. Several decades of research on nonhuman animals, especially laboratory rats,

The Neurobiology of Brain and Behavioral Development
DOI: http://dx.doi.org/10.1016/B978-0-12-804036-2.00012-1

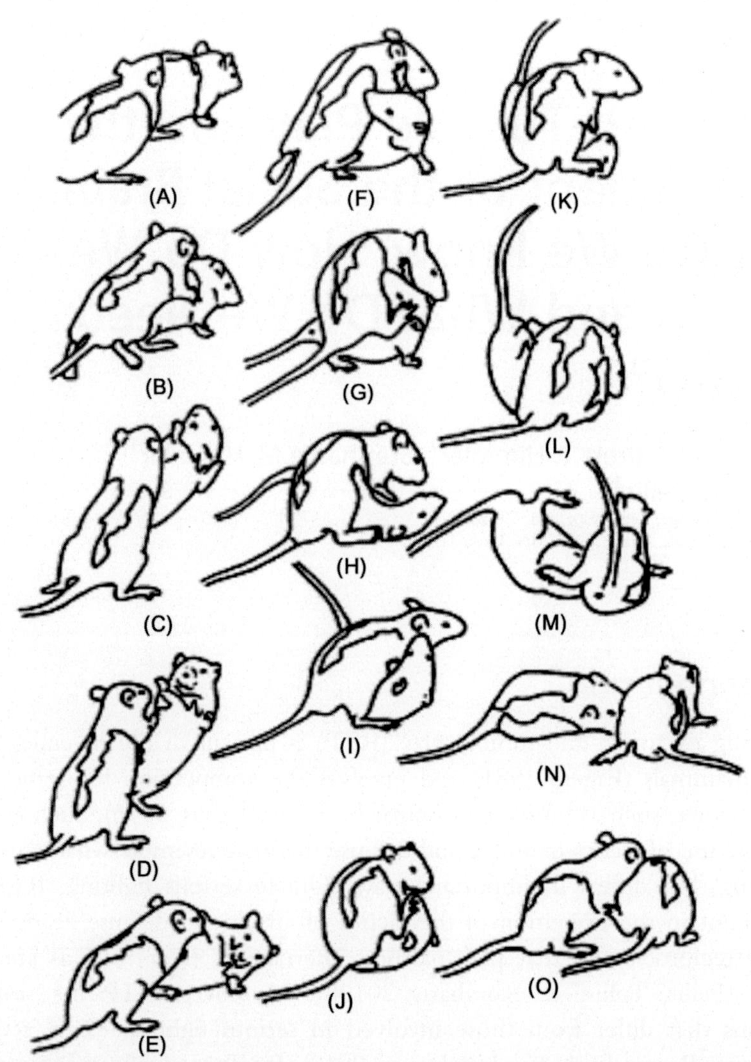

Figure 12.1 A sequence of play fighting is shown for a pair of juvenile rats. The rat on the left approaches another rat (A) and then, from the rear, pounces on it (B). However, before contact can be made, the defender rotates around its longitudinal axis (C) to face its attacker (D). By moving forward, the attacker pushes the defender onto its side (E). The defender then rolls over onto its back as its attacker continues to reach for its nape (F–H). Once in the supine position, the defender launches an attack to its partner's nape (I), but fails due to its partner's use of its hind foot (J and K). Eventually, the rat on top (L) is pushed off by the supine animal (M), which then regains its footing (N). The original defender then lunges towards its partner's nape (O). The whole sequence involves repeated attack and defense of the nape and frequent role reversals between the partners with regard to which one attacks and which one defends. *Reprinted from Pellis, S.M., & Pellis, V.C. (1987). Play-fighting differs from serious fighting in both target of attack and tactics of fighting in the laboratory rat* Rattus norvegicus. Aggressive Behavior, *13, 227–242 with permission.*

has provided overwhelming evidence showing that experience with RTP in the juvenile period improves social skills.

In rats, RTP involves competing for access to the nape of the neck, which is nuzzled with the snout if successfully contacted (Pellis & Pellis, 1987; Siviy & Panksepp, 1987), and is readily distinguishable from serious fighting which involves bites directed at the opponent's flanks and dorsum (Pellis & Pellis, 1987). To defend against nape contact, recipients adopt a variety of defensive tactics (Himmler, Pellis, & Pellis, 2013a). The defender can evade such contact, by swerving, jumping or running away, or can turn to face the attacker and actively block access to its nape. When turning to face the attacker, the defender can either remain standing or roll over to supine. Successful defense can then be coupled with counterattacks, as the original defender lunges to contact the original attacker's nape, which, if successful, can lead to a role reversal, whereby the original attacker assumes the defensive role (Fig. 12.1).

RTP in rats begins to emerge between 15 and 17 days postbirth but it does not reach its juvenile typical form and frequency until about 28–30 days (Pellis & Pellis, 1997). If the opportunity for engaging in RTP is prevented over the juvenile period, the resulting adults have reduced impulse control, over or underreact to fear-inducing situations, are less effective in solving cognitive tasks, and have impoverished social skills (for recent reviews, see Pellis et al., 2014; Vanderschuren & Trezza, 2014). That is, in the juvenile period, RTP has been implicated in improving a range of skills crucial for navigating complex social environments.

12.2 THE RTP OF RATS AND ITS LESSONS

A wide range of neural systems are involved in the production and regulation of RTP, especially subcortical systems involved in motor organization, motivation and reward (e.g., Siviy & Panksepp, 2011; Trezza, Baarendse, & Vanderschuren, 2010). In the absence of the cortex, RTP follows the same developmental trajectory and has all the typical complexity of attack and defense as in intact rats (Himmler, Himmler, Pellis, & Pellis, 2016a; Panksepp, Normansell, Cox, & Siviy, 1994; Pellis, Pellis, & Whishaw, 1992), suggesting that cortical mechanisms have a limited contribution to the motivation to play and in producing play (Pellis & Iwaniuk, 2004). However, at least in part, the effects of RTP on the development of emotional regulation, cognitive skills, and social skills is achieved by alterations of cortical function, especially those related to the executive functions of the prefrontal cortex (PFC) (Baarendse, Counotte, O'Donnell, & Vanderschuren, 2013; Bell, Pellis, & Kolb, 2010; Himmler, Pellis, & Kolb, 2013b).

The experience of play is not needed to develop species-typical behavioral repertoires (Martin & Caro, 1985), but it does appear to have a role in refining the ability to execute aspects of that repertoire in contextually adaptable ways, an ability that is partly mediated by the executive functions of the PFC (Vandershuren & Trezza, 2014).

For example, a rat will protect a piece of food being held in its mouth from another rat by executing a lateral dodging maneuver (Whishaw, 1988), and the ability to use this maneuver gradually matures between 3 and 4 weeks postbirth (Pellis, Field, & Whishaw, 1999). Being reared in social isolation, and so having no social experiences, from weaning to sexual maturity (approximately 25–60 days old) does not prevent rats from developing this maneuver. However, the rat's ability to coordinate its movements during the dodge with the movements of its partner is diminished (Pellis et al., 1999). Similarly, when adult rats that have been reared socially, and so have had peer–peer RTP, are given lesions of the medial PFC (mPFC), their ability to execute dodges to protect their food remains intact (Whishaw, 1988); however, their ability to coordinate their movements with those of the partner attempting to rob the food is diminished (Himmler et al., 2014). Given that the mPFC comprises part of the PFC, that is involved in executive functions (Euston, Gruber, & McNaughton, 2012), and the anatomical development of the mPFC is modified by being reared with peers with whom they can engage in RTP (Bell et al., 2010; Himmler et al., 2013b), these findings provide one potential mechanism by which play experience in the juvenile period can influence the development of executive function and so, social skills.

During RTP, animals seem to exaggerate the experience of loss of control and the unpredictability that has been hypothesized to be important for play to train subjects to deal with the unexpected (Špinka, Newberry, & Bekoff, 2001). To deal with unexpected events effectively, robust executive functions, such as good impulse control, are needed, and these are the very functions of the PFC that appear to be modified in rats by the experience of RTP (Baarendse et al., 2013). An example of such exaggerated loss of control has been documented in rats. When a defending rat rotates to supine (Pellis & Pellis, 1987), the attacker may stand over its partner, in what has been called a pin (Panksepp, 1981). Typically, the rat standing on top keeps its hind feet on the ground and uses its forepaws to restrain and block the countermoves of its supine partner (Fig. 12.2A). However, occasionally, the rat on top stands on its supine partner with all four of its feet (Fig. 12.2B) (Foroud & Pellis, 2003). From this unstable position, the on-top rat is less able to block its supine partner's counter attacks, greatly increasing the likelihood of a role reversal (Pellis, Pellis, & Foroud, 2005). Moreover, moving onto this unanchored position increases in frequency between 30 and 40 days postbirth (Foroud & Pellis, 2002), the age when RTP is most frequent (Thor & Holloway, 1984).

Even though we are persuaded that the general outline of the model of RTP and brain development that has emerged from studies of laboratory rats is correct (for detailed reviews see Pellis, Pellis, & Bell, 2010a, Pellis et al., 2014; Vandershuren & Trezza, 2014), there are some important gaps in that model. Based on our experiences with this model, we identify three main types of limitations and offer some suggestions on how to resolve them.

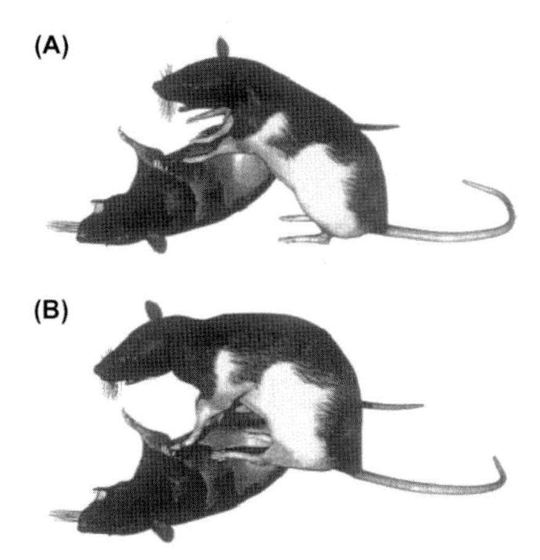

Figure 12.2 When engaged in play fighting, rats often adopt a posture in which one animal stands over its supine partner. However, the posture of the rat on top can take one of two forms: it can hold its partner with its forepaws while standing on the ground with its hind paws (A), or it can stand on its partner with all four of its paws. *Reprinted from Foroud, A., & Pellis, S.M. (2003). The development of "roughness" in the play fighting of rats: A Laban movement analysis perspective.* Developmental Psychobiology, *42, 35–43 with permission.*

12.3 LIMITATION 1: HOW DO WE KNOW THAT RTP IS THE CRITICAL SOCIAL EXPERIENCE?

The most common way to study the effects of social experience on development during the juvenile period has been to rear animals in social isolation and then compare these subjects with ones that had been reared with peers over the same age (Fone & Porkess, 2008). Even though being reared in isolation over the juvenile period deprives the animals of more than just RTP (Bekoff, 1976), there are compelling reasons to believe that the deficits that arise are at least, in part, due to the deprivation of RTP (for a more comprehensive review, see Pellis & Pellis, 2009).

First, it should be noted that social isolation during the juvenile period produces more pronounced deficits than comparable isolation in the postpubertal period (e.g., Amstislavskaya, Bulygina, Tikhonova, & Maslova, 2013; Seffer, Rippberger, Schwarting, & Wöhr, 2015; Varlinskaya & Spear, 2008), and second, rehousing rats socially as adults does not remediate all the effects of having been isolated in the juvenile period (e.g., Einon & Potegal, 1991; Lukkes, Mokin, Scholl, & Forster, 2009; Pellis et al., 1999). Moreover, the effects of rearing in isolation seem to be strongest when it encompasses the peak juvenile period (Arakawa, 2002, 2003). That is, the absence of social partners is most devastating during the juvenile period, especially in the period when RTP is most frequent.

Third, if juveniles reared in isolation are exposed daily to another juvenile for one hour per day they do not exhibit the cognitive and emotional deficits typical of rats reared in complete social isolation over the juvenile period (Einon & Morgan, 1977; Einon, Morgan, & Kibbler, 1978). Most critically, having a nonplayful peer placed in the home cage for that hour per day does not prevent the juveniles from developing the deficits typical of isolation rearing (Einon et al., 1978). Subsequent studies using variants of this paradigm have similarly shown that it is not just the presence of a social partner that is critical, but the presence of a playful one (Bean & Lee, 1991; Bell et al., 2010). That is, engaging in RTP with peers in the juvenile period provides some of the important experiences that are needed for normal development.

Nonetheless, the majority of our knowledge on the effects of RTP on development over the juvenile period is derived from the complete isolation paradigm, and it seems likely that the effects of complete isolation are more profound than those involving only the deprivation of RTP (Fone & Porkess, 2008; Hall, 1998; Robbins, Jones, & Wilkinson, 1996). Clearly, systematic comparisons are needed between rats reared in complete social isolation with those reared with nonplayful partners to delineate the facets of the isolation syndrome than can be attributed to the absence of RTP.

One avenue to explore is suggested by the work from several laboratories showing that, over the juvenile period, prolonged exposure to partners that provide atypical social experiences, including atypical RTP, can lead to the development of a variety of social deficits. For example, subjecting juvenile rats daily to novel social partners has a wide range of effects on their subsequent capacity to learn, regulate anxiety levels and ability to solve social problems (e.g., Cumming, Thompson, & McCormick, 2014; McCormick et al., 2013). Similarly, juvenile rats subjected to repeated social defeat are later shown to have various neural and behavioral changes that affect their ability to engage socially with other rats effectively (e.g., Burke, Renner, Forster, & Watt, 2010; Watt, Burke, Renner, & Forster, 2009). Finally, rearing a juvenile of a more playful strain with a partner from a less playful strain has been shown to result in decreased pain sensitivity for the rats of the more playful strain (Schneider et al., 2014; Schneider, Pätz, Spanagel, & Schneider, 2016), an alteration that could have implications for adult sexual and aggressive interactions.

Unlike the method used by Einon et al. (1978) in which the nonplayful peer was rendered nonplayful by the injection of a psychoactive drug, the one used by Schneider et al. (2016) capitalizes on the natural variation in playfulness across strains (see Section 12.2) as a means of reducing the experience of RTP over the juvenile period. Similarly, pairing juvenile females with adult females, which provides markedly less RTP experience (Pellis & Pellis, 1997), fails to produce the changes in the dendritic morphology of the mPFC present when juveniles are housed with peers (Bell et al., 2010; Himmler et al., 2013b).

Indeed, our preliminary data suggest that even when mismatched for age in the prepubertal period, there are changes in both the *quantity* and *quality* of the RTP

Table 12.1 The behavior of the partners of 32-day-old juveniles when playing in dyads

Play measure	Same age partner	Older partner	Matched pairs *t*-tests[a]
Attacks (#/10 min)	40.33 + 7.83	13.50 + 1.71	3.165
Defense (% of attacks)	89.83 + 3.06	75.80 + 6.94	ns
Evade (% of defense)	22.42 + 6.19	29.75 + 4.74	ns
Complete rotation (% of facing defense)	38.03 + 9.65	5.22 + 2.42	4.009

[a]Significance set at $P < 0.05$.

experienced by juveniles. Six quads of male rats of the Long-Evans strain were formed with two younger (26 days) and two older (40 days) pups in each. Following cohousing and habituation to the test arena, dyadic RTP was tested when the younger rats were 32 days, within the peak play period for juveniles (Panksepp, 1981; Pellis & Pellis, 1990; Thor & Holloway, 1984). The RTP from 10-min test sessions was assessed with regard to play experienced by the younger rats when paired with a same aged and different aged partner was compared. Using our scoring scheme (Himmler et al., 2013a), the frequency of launching nape attacks, the percentage of those attacks that elicited a defensive action, and the percentage use of evasive tactics and the tactic of rotating to supine were calculated (Table 12.1). There were no significant differences between same- and opposite-aged pairs in the likelihood of a nape attack leading to a defense or to the likelihood that the defense will be evasive. Juveniles with older partners, however, experienced significantly fewer attacks and when they attacked their older partner, there was a significant decrease in the percentage of attacks that were defended with a supine defense, and so wrestling. Clearly, if juveniles were reared with older cage mates, they would engage in less RTP and the experiences derived from that RTP would be significantly different.

By using a variety of pair mates with differing social capabilities, especially differing motivations to engage in RTP and different types of responses when they do so, as rearing partners for juvenile rats, the different contributions of RTP on neurobehavioral development may be selectively characterized. Such an approach would have the advantage of avoiding the more generalized and profound effects of isolation rearing.

12.4 LIMITATION 2: ARE ALL RATS THE SAME?

To expose juvenile rats to an atypical play partner, Schneider et al. (2014) housed Wistar rats with Fisher 344's. The latter strain initiates markedly less playful attacks than most commonly used laboratory strains and when playfully attacked defend themselves in a way that minimizes prolonged body contact (i.e., wrestling) (Siviy, Baliko, & Bowers, 1997; Siviy, Love, DeCicco, Giordano & Seifert, 2003). The reduced playfulness of Fisher 344 led to altered RTP experiences by the Wistar rats and this led to changes in their subsequent perception of pain (Schneider et al., 2016). That is, as noted above, manipulating the quality of the RTP experienced can be a useful tool by which to study the effects of RTP on development in the juvenile period. However, this example also illustrates something else—there are significant strain differences.

For a variety of reasons, usually practical ones, different laboratories tend to use different strains of rats for their research on RTP. However, unknown differences across strains in the pattern of RTP may limit our ability to compare studies from different laboratories readily. Moreover, laboratory rats have been derived from wild ancestors of the species *Rattus norvegicus* (Krinke, 2000), and they exhibit a syndrome of traits typical of domesticated mammals (Wilkins, Wrangham, & Fitch, 2014). This syndrome includes being less fearful of humans, tolerating greater social proximity, having reduced sensorimotor capabilities, and achieving sexual maturity at an earlier age (e.g., Blanchard, Flannelly, & Blanchard, 1986; Clark & Price, 1981; Garland, Gleeson, Aronovitz, Richardson, & Dohm, 1995; Trut, 1999)—all of which may influence the expression of RTP. Thus, caution must be exercised in generalizing to nondomesticated animals when a domesticated one is used as the model species.

There is mention of RTP in free-living, wild rats (Robitaille & Bovet, 1976), but details are insufficient for direct comparison with the play of laboratory rats. Therefore, using standardized housing, testing, and scoring procedures (Himmler et al., 2013a), we compared the RTP of captive wild rats with that of four strains of laboratory rats (Himmler et al., 2013c, 2014a). These four domesticated strains have different patterns of derivation from wild rats (Castle, 1947). Wistar rats were the first derived from wild ancestors, then a Wistar backcross with wild rats led to the development of the Long-Evans (LE) strain and additional selective breeding among Wistar rats led to the development of the Sprague-Dawley (SD) strain. Brown Norway rats are a more recent derivative of wild rats and so constitute an independent domestication event.

The frequency of initiating playful attacks, the probability of defending against such attacks and the relative use of different defensive tactics were scored. For defensive tactics, the probability of evasion relative to facing defense was scored and when the rats did engage in facing defense, the probability of using rotation to supine relative to standing defense was scored. Finally, as a proxy measure for reciprocity during RTP, the probability of a role reversal, so that the original attacker switches to becoming the defender, was scored.

The cross-strain comparisons show three important features of RTP when compared to that present in wild rats (Table 12.2). First, in all cases, domesticated strains initiate more playful attacks and are more likely to defend themselves. Second, for types of defense, there are major dissimilarities across the domesticated strains, with

Table 12.2 The components of the RTP of several strains of domesticated rats compared to that of wild rats[a]

Measures of RTP	Attack	Defense	Evasion	Supine defense	Role reversals
Wistar	++	+	+	+	Same
Sprague-Dawley	+++	++	+++	---	Same
Long-Evans	++	+	--	++++	Same
Brown Norway	++	+	--	Same	Same

[a]The plus sign indicates that component is more prevalent and the minus sign indicates less prevalence, with the number of signs indicating the strength of the difference; in contrast, the "same" means no significant difference. Based on Himmler, B.T., Stryjek, R., Modlinska, K., Derksen, S.M., Pisula, W., & Pellis, S.M. (2013c). How domestication modulates play behavior: A comparative analysis between wild rats and a laboratory strain of Rattus norvegicus. *Journal of Comparative Psychology*, 127, 453–464; Himmler, S.M., Modlinska, K., Stryjek, R., Himmler, B. T., Pisula, W., & Pellis, S.M. (2014a). Domestication and diversification: A comparative analysis of the play fighting of the Brown Norway, Sprague-Dawley, and Wistar strains of laboratory rats. *Journal of Comparative Psychology*, 128, 318–327; Himmler, S.M., Lewis, J.M., & Pellis, S.M. (2014b). The development of strain typical defensive patterns in the play fighting of laboratory rats. *International Journal of Comparative Psychology*, 27, 385–396; Himmler, B.T., Himmler, S.M., Stryjek, R., Modlińska, K., Pisula, W., & Pellis, S.M. (2015a). The development of juvenile-typical patterns of play fighting in juvenile rats does not depend on peer-peer play experience in the peri-weaning period. *International Journal of Comparative Psychology*, 28, 1–18.

some strains resembling wild rats more than other domesticated strains. Third, despite cross-strain variation in styles of defense, all rats have similar levels of role reversals. That domesticated rats are more playful than wild ones, as reflected in higher rates of initiating attacks, is consistent with what is seen in other domesticated mammals compared to their wild counterparts (Budiansky, 1999). In domesticated animals, the earlier onset of sexual maturity (Clark & Price, 1981) results in the retention of many juvenile characteristics, including greater playfulness (Burghardt, 1984; Coppinger & Coppinger, 2001).

It would be expected that the greater playfulness of domesticated rats should be accompanied by the greater adoption of defensive tactics that would facilitate prolonged bodily contact. This is not the case. In contrast to wild rats, LE rats diminish their use of evasion and increase their use of rotation to supine, allowing for more of their RTP to involve close quarter wrestling. In contrast, SD rats do the opposite, exaggerating the use of evasion and diminishing the use of supine defense. But it is not simply that changes in evasion and supine defense are complementary. In Wistar rats, evasion is increased slightly as is supine defense. Like LE rats, in Brown Norway rats, evasion is reduced, but unlike LE, supine defense remains at about the same level as seen in wild rats. These findings are consistent with earlier ones in two lines of rats, one selected for rapid kindling of the amygdala and the other for being resistant to kindling (Mohapel & McIntyre, 1998). Overall, the kindling prone rats retain more juvenile-like neural and behavioral features than the kindling resistant ones (Corcoran & Teskey, 2004). As expected, the kindling prone rats initiate more playful attacks, but unlike expected, both selected lines have low levels of supine defense. The lower level of supine defense in kindling prone rats is compensated for by a higher rate of standing defense, whereas in kindling resistant rats, the low rate of supine defense is associated with a higher rate of evasion (Reinhart, Pellis, & McIntyre, 2004).

That type of defense can vary independently of frequency of attack, and that despite differences in the latter, domestication shows a uniform increase in attack, suggest that attack and defense, especially type of defense, are controlled by different mechanisms (Pellis & Pellis, 2009). Cross-species comparisons in mouse-like rodents supports this conclusion as increases in the frequency of playful attack in different species does not necessarily lead to similar changes in tactics of defense (Pellis et al., 1989). The within species, cross-strain comparisons (Table 12.2) show that, relative to their wild progenitors, domesticated rats have increased their motivation to play (i.e., increased playful attacks), but that changes in the playful tactics of defense have diverged in many different ways. These strain differences in patterns of defense have repercussions on what we think may be providing the critical experiences during RTP that influence neurobehavioral development.

The relatively high frequency of the pin configuration present in some strains of domestic rats (Panksepp, 1981) has led to the view that it is the promoting of the bodily contact associated with pinning that provides the principle reward for RTP (Niesink & van Ree, 1989; Panksepp & Burgdorf, 1999). That some strains can sustain high levels of RTP even though they have relatively low rates of pinning (Himmler et al., 2014a,b) makes this unlikely. From a functional perspective, given that the facilitation of loss of control in RTP has been characterized in the pin configuration (Fig. 12.2), reduced pinning may diminish the ability of RTP to influence the development of the mPFC (Pellis et al., 2010a, 2014).

This may not be the case, however, as despite differences in the frequency of attack and styles of defense across strains, all have a comparable level of role reversals, with approximately 25%–30% of play fights involving the original defender successfully counterattacking and so forcing the original attacker to adopt the defensive role (Himmler et al., 2016a). This consistency suggests that the pin configuration is not essential to create the instabilities needed to facilitate successful counterattacks. That is, there may be many different ways in which rats can incorporate restraint in their attack and defense to produce similar levels of reciprocity. If so, the ability of RTP experience in the juvenile period to modify brain mechanisms should be comparable irrespective of strain differences in some aspects of their play. Indeed, play-induced, modified mPFC structure and function has been shown for more than one strain of domestic rats (Baarendse et al., 2013; Bell et al., 2010).

We used the same rearing and analytical paradigm as used previously (Bell et al., 2010; Himmler et al., 2013b) to test the effects of peer–peer play on the development of the mPFC in captive wild rats (Himmler, 2015). To our surprise, there were no differences in the neuronal architecture of the mPFC of rats reared with juveniles compared with rats reared with an adult female (Fig. 12.3). For LE rats, we have now

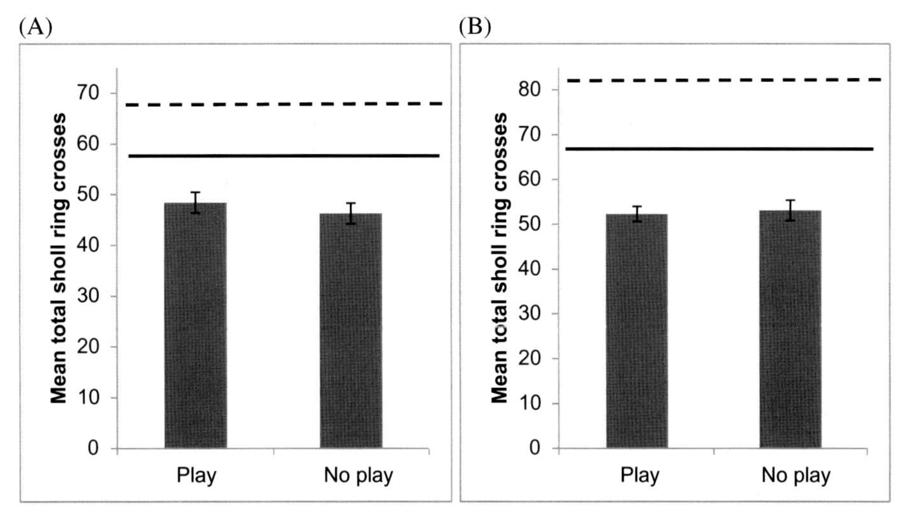

Figure 12.3 Mean dendritic length between groups for the apical (A) and basilar (B) dendritic fields are shown is show for captive wild rats that were reared with a peer or with an adult during the juvenile period. For comparison the average values for the effects of play experience by being reared with a peer (horizontal solid line) or the absence of play experience by being reared with an adult (horizontal dashed line) on the mPFC of LE rats from previous studies are shown. *From Himmler, B.T. (2015).* Exploring the brain–behaviour interface: The role of juvenile play experiences. *Unpublished Ph.D. thesis. Department of Neuroscience. University of Lethbridge: Lethbridge, AB, Canada.*

replicated the effects of juvenile RTP with peers on the pruning of the dendritic arbor of mPFC neurons four times (Bell et al., 2010; Himmler et al., 2013b; Himmler, Pellis, & Kolb, 2015b), so we are confident that this is a real effect. There are several possible explanations for the failure of the wild rats to show this play-induced change in the mPFC. First, the low frequency of wrestling in the RTP of wild rats (Himmler et al., 2013c) may be insufficient to provide the experiences necessary for the play-induced changes to occur. As noted above, however, the similar level of role reversals in wild compared to laboratory rats (Himmler et al., 2016a) seems inconsistent with this explanation. Second, adult wild rats may be more likely to provide the RTP experiences needed to achieve this effect, thus leading to comparable levels of pruning in both peer-reared and adult-reared animals. Our preliminary observations suggest that the playful interactions between adults and juveniles are fewer, mostly initiated by the juveniles and with few ever leading to prolonged wrestling (Himmler et al., work in progress). So, there are both qualitative and quantitative differences in the RTP experienced by rats reared with adults compared with ones reared with a peer, reducing the likelihood of this explanation. Third, the association between RTP and the development of the mPFC only emerges following domestication. That domestication may have a critical role is supported by the finding that in domesticated golden hamsters, the same association of juvenile peer play and pruning of the mPFC, as found in LE rats, is present (Burleson et al., 2016).

Across the four replications of the effects of juvenile mPFC on pruning the dendritic arbor of the neurons of the mPFC conducted in LE rats, the absolute differences between the peer-reared and adult-reared groups were fairly similar. Superimposing the values of the LE effects on the graphs showing the results from the rearing experiment in wild rats reveals that the complexity of the neurons in the wild rats is lower not only than the LE rats reared with an adult partner, but also lower than that of those that were reared with a peer (Fig. 12.3). That is, irrespective of type of partner, and so presumably of the amount of RTP experienced, the mPFC of wild rats is pruned to a greater extent than that of LE rats, including those reared with a peer. These comparisons support the possibility that the effect of RTP on the development of the mPFC is a by-product of domestication.

When left on their own, litters of rats are not fully weaned until about 34 days postbirth, much later than standard laboratory practice whereby rats are removed from their mothers around 23 days. That is, over the third to fourth week postbirth, weaning is a gradual process and even though the proportion of food, in the form of milk, provided by the mother slowly declines, members of the litter still remain near the mother and engage in social interactions with her, such as grooming and huddling (Cramer, Thiels, & Alberts, 1990; Thiels, Alberts, & Cramer, 1990). In the wild, close proximity with the mother, even when she no longer supplies the main source of food, likely provides some continued measure of protection from predators and information about foods available to eat. Sniffing the mouth of a conspecific is used as a clue about what is available and safe to eat (Galef, 1996). Once fully independent, however, young wild rats are likely to be at greater risk, and so it is unlikely that they can afford to wait until they are 60 days or so old before having a fully operational mPFC. The executive functions provided by the mPFC (Euston et al., 2012) are probably crucial for dealing with the vicissitudes of life at a younger age in wild rats than in cosseted domestic ones. In domesticated laboratory rats, there is more time to develop a fully functional mPFC. In this explanation, domestication would alter the relationship between peer—peer RTP and development of the mPFC by increasing the time window for such an experience-induced change.

The greater need for their executive functions to be fully formed as early as possible may also mean that, for wild rats, their mPFC may be more responsive to all social experiences, not just those arising from peer—peer RTP. That is, even though the adult-juvenile RTP experienced when the juveniles are reared only with an adult may not provide the same degree of exaggeration of the loss, and the regaining, of control as in domesticated rats (see Fig. 12.2), it may be sufficient. In this explanation, domestication would narrow the range of social experiences capable of pruning the mPFC.

We cannot draw any firm conclusions about why the association between juvenile RTP and the development of the mPFC is present in one strain of domesticated laboratory rat and absent in captive wild rats. Each of the possible hypotheses accounting for this difference needs to be tested empirically. However, what these discordant findings do suggest is that we need to be cautious in treating all rats as if they were the same. Wherever possible, experimental manipulations should be tested on more than one strain of rat (e.g., Pisula, Stryjek, & Nałęcz-Tolak, 2006; Stryjek, Modlinska, & Pisula, 2012), as different mechanisms may be more potent in their regulatory role in some strains compared to others.

12.5 LIMITATION 3: ARE ALL MEASUREMENTS OF PLAY EQUALLY INFORMATIVE?

Studying the brain–behavior relationship of RTP involves two steps. First, the brain mechanisms that produce and regulate RTP need to be characterized, and second, brain mechanisms and functional abilities that are influenced in their development by the experiences derived from RTP during the juvenile period need to be identified. To evaluate the consequences of having had RTP experience as juveniles, several social and nonsocial test paradigms have been used (e.g., Baarendse et al., 2013; da Silva, Ferreira, Carobrez, & Morato, 1996; Einon & Morgan, 1977; Einon et al., 1978; Himmler et al., 2014; Pellis et al., 2006; van den Berg et al., 1999) each tapping into different aspects of social and nonsocial skills present in adults. With regard to measuring RTP, and so identifying the roles of different brain mechanisms in regulating this behavior, there are many different scoring systems currently being used. Such diversity in scoring schemes often makes it difficult to know whether differences in the findings across laboratories are due to the actual experimental manipulation involved or in the divergent methods used to assess the behavioral consequences of those manipulations (Blake & McCoy, 2015).

One common approach is to score a variety of actions, such as chase, pounce, box and wrestle that are characteristic of RTP and then combine these into an index that gives an overall measure of the frequency of RTP (e.g., Birke & Sadler, 1983; Varlinskaya & Spear, 2009). More simply, one behavior, such as pinning, which is highly correlated with RTP in at least some rat strains (Panksepp, 1981), is used to assess the frequency of RTP (e.g., Aguilar, Caramés, & Espinet, 2009; Flynn, Ferguson, Delclos, & Newbold, 2000). Either in combination with pinning or on its own, another measure is to score how often animals initiate RTP, which may be scored as pouncing or making contact with the dorsal surface of the partner (e.g., Panksepp et al., 1994; Porrini et al., 2005). All these measures provide an estimate of how frequently rats engage in RTP, but they all have a major limitation. To varying degrees, they all involve creating measures that are a composite of attack and defense.

Consequently, it is not possible to determine whether a treatment-induced change in the measured frequency is due to a change in the actions of the attacker, the defender, or both. Yet knowing *how* a treatment alters the frequency of RTP can be crucial in identifying the mechanisms altered by different treatments.

As shown in Fig. 12.1, as one rat reaches to contact the nape of its partner, the recipient takes defensive action to block access, and then, as it blocks the continued attempts by its attacker to contact its nape, the defender can launch retaliatory lunges at its partner's nape, which can, in turn, be blocked by its partner. Finally, a successful counterattack can lead to a role reversal, as the original attacker goes on the defensive. Therefore, the movements performed in a competitive interaction for access to a particular body target are functionally executed as tactics of attack and defense (see Pellis & Pellis, 2015). Based on the organization of RTP in rats, which involves competition for contacting the nape, we devised a scheme for scoring both attack and defense (Himmler et al., 2013a). Because the measurements in this scheme can differentiate between attack and defense, it can detect treatment-induced changes either to attack or to defense, providing a basis for developing a standardized method for scoring RTP (Blake & McCoy, 2015).

An attack is scored when one rat directs its snout towards its partner's nape. We originally specified a maximum snout-to-nape distance of around 2 cm, as an important criterion for scoring a nape attack (e.g., Pellis et al., 1992), but this, we discovered, could underestimate the number of attacks. For example, while 2 cm represents the mean distance at which LE rats commence the execution of a defensive action, for wild rats, the distance is over 4 cm (Himmler et al., 2013c). Commencing a defensive action at a larger interanimal distance increases the chance that that action can be successfully executed before the attacker closes the gap (Pellis, Pellis, & McKenna, 1994). This reduces the chances that the attacker actually makes contact with the nape and increases the chances that the defender can launch a counterattack. Thus, only scoring attacks if they make actual contact or when the snout comes to within 2 cm of the nape could grossly underestimate the frequency of attacks (Himmler et al., 2013c). A more accurate way to assess whether an attack occurs is to examine videotaped sequences frame-by-frame and determine whether the movements by animal A compensate for the movements of animal B, so that A's snout remains directed at B's nape (Pellis & Bell, 2011). If so, then A's move towards B can be scored as a nape attack. Our method dissociates the actions of the attacker from that of the defender, making the scoring of the frequency of nape attacks less ambiguous than schemes requiring actual nape contact to be made in detecting treatment-induced changes in attack behavior.

Similarly, defensive actions need to take into account the movements of the attacker. Because of the continual influence of one partner on the other, moments in the sequence of action in which the movement of the defender is as free as possible from the influence of the movements of the attacker need to be identified so as to detect the preferred action by the defender (Pellis, 1989). For this reason, in our scheme, the first 2–3 video frames from the commencement of a defensive action are used to determine the type of tactic initiated (Himmler et al., 2013a). Once the attacker makes contact, the defense tactic initially commenced may not be successfully completed as the attacker may force other movements on the defender. For example, when attacked from the rear, one defensive tactic used to block access to one's nape is for the rat to pivot around on its hind feet so as to face the oncoming attacker. However, if the attacker pounces and lands on the defender before it succeeds in completing the pivot, that defense is negated and the defender then adopts some other defense, such as rolling over to supine, to extricate its nape from its partner's snout (Pellis et al., 1994). By scoring the defense tactic that is initiated first, even if not completed, what is measured is the preferred action by the defender rather than the action forced upon it by the continued actions of the attacker.

Pinning is highly correlated with RTP, at least in some strains, so that when other measures of RTP show an increase, so does pinning (Panksepp & Beatty, 1980; Panksepp, 1981), and this is the reason as to why pinning is such a commonly used method for scoring RTP (e.g., Aguilar et al., 2009; Flynn et al., 2000). However, pinning arises from the combined actions of both partners, so that a decrease in pinning could arise from a reduction in the frequency of launching playful attacks, or from a decrease in the likelihood that recipients of such attacks defend themselves, or in how they defend themselves. For example, juvenile rats that have been decorticated at birth show a 50% reduction in the frequency of pinning, but no such concomitant decrease in the frequency of nape attacks (Panksepp et al., 1994; Pellis et al., 1992). Using our scoring scheme, we found that decorticate rats were just as likely to defend themselves and do so by using facing defense as frequently as intact controls, but the facing defense that they were most likely to adopt was standing defense, not supine defense (Pellis et al., 1992). As supine defense is more likely to lead to a pin than is standing defense (Pellis & Pellis, 1987), the reduction in pinning was accounted for by a decortication-induced change in the preferred defensive tactic.

The downside to our scheme is that the period of training to master the system is longer than is the case for many other schemas. Also, given the necessity for many components to be determined by repeated observation of the same sequence at normal speed, slow motion and frame-by-frame, the scoring can be quite time consuming and so is not well-suited to studies requiring the testing of large numbers of animals. In this regard, scoring a more readily recognizable marker for RTP, like pinning, can be advantageous, despite its drawbacks as a composite measure of attack and defense. Our recent cross-strain comparisons suggest how a measure, such as pinning, can be used effectively, if due caution is given to limitations inherent in this measure.

The frequency of pinning can vary dramatically across different strains of rats: in wild rats, it is very rare, so that in a standard 10-minute test trial, only about 15 pins occur per pair, whereas in LE and SD rats, there are about 60 pins, with other strains (i.e., Wistar, Brown Norway) having intermediate values (Himmler et al., 2013b, 2014a). Comparing these values with that of the frequency of launching playful attacks, especially for the three most divergent strains, show discordance between the measures. On average, pairs of wild rats launch 50 attacks, LE about 85 and SD rats, 140. Even when corrected for differences in their probability of defending against a nape attack (ranging between 0.8 and 1.0), there are differences in the frequency with which attacks lead to pinning: over 80% for LE and less than 40% for wild and SD rats. Therefore, for wild and SD rats, scoring nape attacks and pins does not provide complementary measures of the frequency of RTP, as is the case for LE rats. However, using our schema to score the initial tactics of defense does not account for these strain differences either—as the initial defense, rotating to supine accounts for about 57% of pins in LE rats, 43% in wild rats and only 15% for SD rats.

Reassessing how pins arose across these divergent strains (wild, LE, and SD) reveals two separate mechanisms involved in contributing to these differences (Himmler et al., 2016b). Pins that met the criteria for being so labeled (Panksepp, 1981) were identified. Then, for each pin, the video was returned to the beginning of the attack and a frame-by-frame inspection was used to characterize the immediate cause of the pin. In wild rats, the majority of pins arose arise because the rat that ends on its back is pushed over, whereas for the majority for LE and SD rats, it is due to the defender of a nape attack rotating to supine (Fig. 12.4). In the case of the wild rats, scoring the initial defense tactics does not help in explaining the frequency of pinning because it is not the defensive tactic used that produces pins but rather, it is due to the attack behavior of the partner. In the case of the SD rats, even though they are less likely to use the rotation to supine tactic as their initial defensive tactic (Himmler et al., 2014a,b), when they do end up in the pin configuration, it is because, in the majority of cases, they actively rotate to supine (Fig. 12.4). Initially, SD rats attempt other tactics, before switching to rotating to supine later in the play sequence (Himmler et al., 2016b).

Given that the majority of pins in both LE and SD rats arise from the use of the rotation to supine tactic, a treatment-induced change in the frequency of pinning likely has the same causal explanation. A decrease in the frequency of pinning may result from a treatment-induced reduction in initiating RTP that would be detected by scoring nape attacks. If this measurement fails to reveal a reduced frequency of launching nape attacks, then the next likely explanation is that the defenders have reduced their reliance on the rotation to supine tactic. Such a reduction, however, would have to be scored at different phases of play fighting in the two strains. If using wild rats, or some other strain in which rotation to supine is not the main contributor to pinning, this interpretation for a treatment-induced change in pinning cannot be assumed.

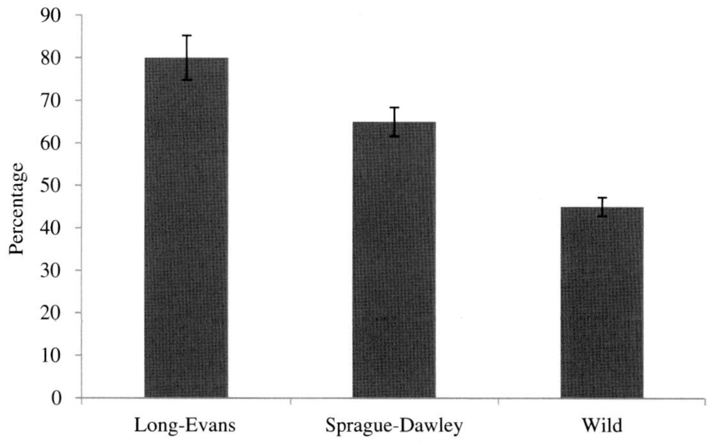

Figure 12.4 The mean and standard error are shown for the percentage of pins, among the three strains (Long-Evans (LE), Sprague-Dawley (SD), and wild), that arise from the defender actively rotating to the supine position. *From Himmler, S.M., Himmler, B.T., Modlinska, K., Stryjek, R., Pisula, W., & Pellis, S.M. (2016b). Pinning in the play fighting of rats: A comparative perspective with methodological recommendations.* International Journal of Comparative Psychology, *in press. Copyright held by authors.*

Based on the above considerations, we recommend that nape attacks be used instead of nape contacts, and that along with nape attacks, pins be scored. When there are concordant changes in these two measures—that is, an increase or decrease in the frequency of nape attacks is matched by a proportional increase or decrease in pins—sufficient information is available to reliably draw causal inferences as to how the treatment affects the frequency of RTP and the underlying attack and defense components. In such cases, scoring the full range of options for defense that is detailed in our scoring scheme (Himmler et al., 2013a) would provide little additional useful information to interpret the effects (Siviy et al., 1997, 2003). However, when nape attacks and pins are discordant—that is, an increase in the frequency of pins when nape attacks are decreased, or a decrease or unchanged frequency of pins in the face of a major increase in nape attacks—then additional information is needed. In terms of the level of effort involved, we suggest that the first step to resolve the discordance is to apply our scoring scheme, as this could reveal changes in defense preferences that may be sufficient to account for the frequency of pins (Himmler et al., 2013a). If this approach fails to produce a sufficient causal mechanism to account for the frequency of pins relative to nape attacks, then the details of how each pin arises needs to be determined (Himmler et al., 2016b).

Along with measures of attack and defense, we also recommend that role reversals (that is, how often a play fight involves the attacker becoming the defender) be scored (Himmler et al., 2016a). Role reversals represent successfully executed counterattacks that may involve actions by one or both partners that facilitate their successful execution (Pellis et al., 2005). However, scoring counterattacks (Pellis et al., 1989) and the movements that facilitate their success (Foroud & Pellis, 2003) can be labor intensive. Role reversals are easier to measure and establishing a base-line level for their frequency provides a means by which to assess whether there are treatment-induced changes to that frequency (Himmler et al., 2015a, 2016a). Scoring role reversals is important because it provides a measure of reciprocity in playful exchanges, and reciprocity can be a critical factor in ensuring that RTP remains playful (Pellis et al., 2010b). Moreover, changes in the frequency of role reversals can alert researchers to the possibility that some higher-level function, such as cognitive processes or communication, has been affected by an experimental treatment (Kisko, Himmler, Himmler, Euston, & Pellis, 2015).

A flexible scoring scheme that starts with scoring nape attacks, pins and role reversals, then proceeds to incorporate the scoring of initial defense tactics and tactics enacted later in the RTP sequence, would provide a method that can be adapted to the needs of each new study, but, at the same time, would provide a degree of standardization that is necessary for comparing results across studies (Blake & McCoy, 2015). Thus, our revised scheme advocates that one begins an analysis at a low level of resolution and then shift to finer levels of resolution as needed.

12.6 CONCLUSION

Experiencing RTP with peers during the juvenile period is an important developmental phase that shapes a range of skills, including emotional regulation, cognitive processing, and social competence. There are three levels of analysis needed to understand how play achieves these developmental effects: identify the critical experiences generated by RTP, identify the neural systems that are altered, and characterize how these neural changes mediate later adult skills. In this chapter, we have shown that although there are some strong clues as to the kinds of experiences generated by RTP that are important for shaping the development of these skills, there is much work left to do in fully characterizing what it is about these experiences that are able to modify brain development. As for the relevant neural mechanisms involved, the PFC, with its role in executive control, appears to be a major contributor in RTP-induced influences on the development of various skills (Pellis et al., 2014). The methodological adjustments to the study of RTP elaborated in the present chapter would facilitate identifying the key experiences derived from RTP. Also, as reviewed more fully elsewhere (Vanderschuren & Trezza, 2014), the nature of the neural changes involved and how these affect adult skills (e.g., Himmler et al., 2014; van Kerkhof et al., 2014) are also beginning to be characterized. Together, these disparate studies are beginning to reveal the causal mechanisms by which the experiences derived from RTP in the juvenile period influences the development of socially competent adults.

REFERENCES

Aguilar, R., Caramés, J. M., & Espinet, A. (2009). Effects of handling on playfulness by means of reversal of the desire to play in rats (*Rattus norvegicus*). *Journal of Comparative Psychology, 123*, 347–356.

Aldis, O. (1975). *Play fighting*. New York, NY: Academic Press.

Amstislavskaya, T. G., Bulygina, V. V., Tikhonova, M. A., & Maslova, L. N. (2013). Social isolation during peri-adolescence or adulthood: Effects on sexual motivation, testosterone and corticosterone response under conditions of sexual arousal in male rats. *Chinese Journal of Physiology, 56*, 36–43.

Arakawa, H. (2002). The effects of age and isolation period on two phases of behavioral response to foot-shock in isolation-reared rats. *Developmental Psychobiology, 41*, 15–24.

Arakawa, H. (2003). The effects of isolation rearing on open-field in male rats depends on developmental stages. *Developmental Psychobiology, 43*, 11–19.

Baarendse, P. J. J., Counotte, D. S., O'Donnell, P., & Vanderschuren, L. J. M. J. (2013). Early social experience is critical for the development of cognitive control and dopamine modulation of prefrontal cortex function. *Neuropsychopharmacology, 38*, 1485–1494.

Bean, G., & Lee, T. (1991). Social isolation and cohabitation with haloperidol-treated partners: Effect on density of striatal dopamine D2 receptors in the developing rat brain. *Psychiatry Research, 36*, 307–317.

Bekoff, M. (1976). The social deprivation paradigm: Who's being deprived of what? *Developmental Psychobiology, 9*, 499–500.

Bell, H. C., Pellis, S. M., & Kolb, B. (2010). Juvenile peer play experience and the development of the orbitofrontal and medial prefrontal cortices. *Behavioural Brain Research, 207*, 7–13.

Birke, L. I., & Sadler, D. (1983). Progestin-induced changes in play behaviour of the pre-pubertal rat. *Physiology Behavior, 30*, 341–347.

Blake, B. E., & McCoy, K. A. (2015). Hormonal programming of rat social play behavior: Standardized techniques will aid synthesis and translation to human health. *Neuroscience & Biobehavioral Reviews, 55*, 184–197.

Blanchard, R. J., Flannelly, K. J., & Blanchard, D. C. (1986). Defensive behaviors of laboratory and wild *Rattus norvegicus*. *Journal of Comparative Psychology, 100*, 101–107.

Budiansky, S. (1999). *The covenant of the wild*. New Haven, CT: Yale University Press.

Burghardt, G. M. (1984). On the origins of play. In P. K. Smith (Ed.), *Play in animals and man*. Oxford, UK: Blackwell.

Burke, A. R., Renner, K. J., Forster, G. L., & Watt, M. J. (2010). Adolescent social defeat alters neural, endocrine and behavioral responses to amphetamine in adult male rats. *Brain Research, 1352*, 147–156.

Burleson, C. A., Pedersen, R. W., Seddighi, S., DeBusk, L. E., Burghardt, G. M., & Cooper, M. A. (2016). Social play in juvenile hamsters alters dendritic morphology in the medial prefrontal cortex and attenuates effects of social stress in adulthood. *Behavioral Neuroscience, 130*, 437–447.

Castle, W. E. (1947). The domestication of the rat. *Proceedings of the National Academy of Sciences of the United States of America, 33*, 109–117.

Clark, B. R., & Price, E. O. (1981). Sexual maturation and fecundity of wild and domestic Norway rats (*Rattus norvegicus*). *Journal of Reproduction & Fertility, 63*, 215–220.

Coppinger, R., & Coppinger, L. (2001). *Dogs. A startling new understanding of canine origin, behavior & evolution*. New York: Scribner.

Corcoran, M. E., & Teskey, G. C. (2004). *Kindling: An inquiry into experimental epilepsy*. Oxford, UK: Oxford University Press.

Cramer, C. P., Thiels, E., & Alberts, J. R. (1990). Weaning in rats. I. Maternal behavior. *Developmental Psychobiology, 23*, 479–493.

Cumming, M. J., Thompson, M. A., & McCormick, C. M. (2014). Adolescent social instability stress increases aggression in a food competition task in adult male Long-Evans rats. *Developmental Psychobiology, 56*, 1575–1588.

da Silva, N. L., Ferreira, V. M. M., Carobrez, A. P., & Morato, G. S. (1996). Individual housing from rearing modifies the performance of young rats on the elevated plus-maze apparatus. *Physiology & Behavior, 60*, 1391–1396.

Einon, D., & Potegal, M. (1991). Enhanced defense in adult rats deprived of playfighting experience in juveniles. *Aggressive Behavior, 17*, 27–40.

Einon, D. F., & Morgan, M. J. (1977). A critical period for social isolation in the rat. *Developmental Psychobiology, 10*, 123–132.

Einon, D. F., Morgan, M. J., & Kibbler, C. C. (1978). Brief periods of socialization and later behavior in the rat. *Developmental Psychobiology, 11*, 213–225.

Euston, D. R., Gruber, A. J., & McNaughton, B. L. (2012). The role of medial prefrontal cortex in memory and decision making. *Neuron, 76*, 1057–1070.

Fagen, R. A. (1981). *Animal play behavior*. New York, NY: Oxford University Press.

Flynn, K. M., Ferguson, S. A., Delclos, K. B., & Newbold, R. R. (2000). Effects of genistein exposure on sexually dimorphic behaviors in rats. *Toxicological Science, 55*, 311–319.

Fone, K. C. F., & Porkess, M. V. (2008). Behavioural and neurochemical effects of post-weaning social isolation in rodents—relevance to developmental neuropsychiatric disorders. *Neuroscience & Biobehavioral Reviews, 32*, 1087–1102.

Foroud, A., & Pellis, S. M. (2002). The development of "anchoring" in the play fighting of rats: Evidence for an adaptive age-reversal in the juvenile phase. *International Journal of Comparative Psychology, 15*, 11–20.

Foroud, A., & Pellis, S. M. (2003). The development of "roughness" in the play fighting of rats: A Laban movement analysis perspective. *Developmental Psychobiology, 42*, 35–43.

Galef, B. G., Jr. (1996). Social enhancement of food preferences in Norway rats: A brief review. In C. M. Heyes, & B. G. Galef, Jr. (Eds.), *Social learning in animals: The roots of culture* (pp. 49–64). San Diego: Academic Press.

Garland, T., Jr., Gleeson, T. T., Aronovitz, B. A., Richardson, C. S., & Dohm, M. R. (1995). Maximal sprint speeds and muscle fiber composition of wild and laboratory house mice. *Physiology & Behavior, 58*, 869–876.

Hall, F. S. (1998). Social deprivation neonatal, adolescent, and adult rats has distinct neurochemical and behavioral consequences. *Critical Reviews in Neurobiology, 12*, 129–162.

Himmler, B.T. (2015). *Exploring the brain–behaviour interface: The role of juvenile play experiences*. Unpublished Ph. D thesis. Department of Neuroscience, University of Lethbridge: Lethbridge, AB, Canada.

Himmler, B. T., Bell, H. C., Horwood, L., Harker, A., Kolb, B., & Pellis, S. M. (2014). The role of the medial prefrontal cortex in regulating inter-animal coordination of movements. *Behavioral Neuroscience, 128*, 603–613.

Himmler, B. T., Himmler, S. M., Stryjek, R., Modlińska, K., Pisula, W., & Pellis, S. M. (2015a). The development of juvenile-typical patterns of play fighting in juvenile rats does not depend on peer-peer play experience in the peri-weaning period. *International Journal of Comparative Psychology, 28*, 1–18.

Himmler, B. T., Pellis, S. M., & Kolb, B. (2013b). Juvenile play experience primes neurons in the medial prefrontal cortex to be more responsive to later experiences. *Neuroscience Letters, 556*, 42–45.

Himmler, B.T., Pellis, S.M., & Kolb, B. (2015b). *The effects of post-pubertal aging on the juvenile-induced changes in the development of the prefrontal cortex*. Society for Neuroscience, 30.08/A40, Chicago, IL, October.

Himmler, B. T., Pellis, V. C., & Pellis, S. M. (2013a). Peering into the dynamics of social interactions: Measuring play fighting in rats. *Journal of Visualized Experiments, 71*, 1–8.

Himmler, B. T., Stryjek, R., Modlinska, K., Derksen, S. M., Pisula, W., & Pellis, S. M. (2013c). How domestication modulates play behavior: A comparative analysis between wild rats and a laboratory strain of *Rattus norvegicus*. *Journal of Comparative Psychology, 127*, 453–464.

Himmler, S. M., Himmler, B. T., Modlinska, K., Stryjek, R., Pisula, W., & Pellis, S. M. (2016b). Pinning in the play fighting of rats: A comparative perspective with methodological recommendations. *International Journal of Comparative Psychology, 153*, 1103–1137.

Himmler, S. M., Himmler, B. T., Pellis, V. C., & Pellis, S. M. (2016a). Play, variation in play and the development of socially competent rats. *Behaviour, 29*, 1–14.

Himmler, S. M., Lewis, J. M., & Pellis, S. M. (2014b). The development of strain typical defensive patterns in the play fighting of laboratory rats. *International Journal of Comparative Psychology, 27*, 385–396.

Himmler, S. M., Modlinska, K., Stryjek, R., Himmler, B. T., Pisula, W., & Pellis, S. M. (2014a). Domestication and diversification: A comparative analysis of the play fighting of the Brown Norway, Sprague-Dawley, and Wistar strains of laboratory rats. *Journal of Comparative Psychology, 128*, 318–327.

Kisko, T. M., Himmler, B. T., Himmler, S. M., Euston, D. R., & Pellis, S. M. (2015). Are 50-kHz calls used as play signals in the playful interactions of rats? II. Evidence from the effects of devocalization. *Behavioural Processes, 111*, 25–33.

Krinke, G. J. (2000). *The laboratory rat*. San Diego: Academic Press.

Lukkes, J. L., Mokin, M. V., Scholl, J. L., & Forster, G. L. (2009). Adult rats exposed to early-life social isolation exhibit increased anxiety and conditioned fear behavior, and altered hormonal stress responses. *Hormones & Behavior, 55*, 248–256.

Martin, P., & Caro, T. (1985). On the function of play and its role in behavioral development. *Advances in the Study of Animal Behavior, 15*, 59–103.

McCormick, C. M., Green, M. R., Cameron, N. M., Nixon, F., Levy, M. J., & Clark, R. A. (2013). Deficits in male sexual behavior in adulthood after social instability stress in adolescence in rats. *Hormones and Behavior, 63*, 5–12.

Mohapel, P., & McIntyre, D. C. (1998). Amygdala kindling-resistant (SLOW) or –prone (FAST) rat strains show differential fear responses. *Behavioral Neuroscience, 112*, 1402–1413.

Niesink, R. J. M., & van Ree, J. M. (1989). Involvement of opioid and dopaminergic systems in isolation-induced pining and social grooming of young rats. *Neuropharmacology, 28*, 411–418.

Panksepp, J. (1981). The ontogeny of play in rats. *Developmental Psychobiology, 14*, 327–332.

Panksepp, J., & Beatty, W. W. (1980). Social deprivation and play in rats. *Behavioral & Neural Biology, 30*, 197–206.

Panksepp, J., & Burgdorf, J. (1999). Laughing rats? Playful tickling arouses high frequency ultrasonic chirping in young rodents. In S. Hameroff, C. Chalmergs, & A. Kazniak (Eds.), *Toward a science of consciousness III* (pp. 231–244). Cambridge, MA: MIT Press.

Panksepp, J., Normansell, L., Cox, J. F., & Siviy, S. M. (1994). Effects of neonatal decortication on the social play of juvenile rats. *Physiology & Behavior, 56*, 429–443.

Pellegrini, A. D. (2009). *The role of play in human development*. New York: Oxford University press.

Pellis, S. M. (1989). Fighting: The problem of selecting appropriate behavior patterns. In R. J. Blanchard, P. F. Brain, D. C. Blanchard, & S. Parmigiani (Eds.), *Ethoexperimental approaches to the study of behavior* (pp. 361–374). Dordrecht, The Netherlands: Kluwer Academic Publishers.

Pellis, S. M., & Bell, H. C. (2011). Closing the circle between perceptions and behavior: A cybernetic view of behavior and its consequences for studying motivation and development. *Developmental Cognitive Neuroscience, 1*, 404–413.

Pellis, S. M., & Iwaniuk, A. N. (2004). Evolving a playful brain: A levels of control approach. *International Journal of Comparative Psychology, 17*, 90–116.

Pellis, S. M., & Pellis, V. C. (1987). Play-fighting differs from serious fighting in both target of attack and tactics of fighting in the laboratory rat *Rattus norvegicus*. *Aggressive Behavior, 13*, 227–242.

Pellis, S. M., & Pellis, V. C. (1990). Differential rates of attack, defense, and counterattack during the developmental decrease in play fighting in male and female rats. *Developmental Psychobiology, 23*, 215–231.

Pellis, S. M., & Pellis, V. C. (1997). The pre-juvenile onset of play fighting in rats (*Rattus norvegicus*). *Developmental Psychobiology, 31*, 193–205.

Pellis, S. M., & Pellis, V. C. (2009). *The playful brain. Venturing to the limits of neuroscience*. Oxford, UK: Oneworld Press.
Pellis, S. M., & Pellis, V. C. (2015). Are agonistic behavior patterns signals or combat tactics—or does it matter? Targets as organizing principles of fighting. *Physiology & Behavior, 146*, 73–78.
Pellis, S. M., Field, E. F., & Whishaw, I. Q. (1999). The development of a sex-differentiated defensive motor-pattern in rats: A possible role for juvenile experience. *Developmental Psychobiology, 35*, 156–164.
Pellis, S. M., Hastings, E., Shimizu, T., Kamitakahara, H., Komorowska, J., Forgie, M. L., & Kolb, B. (2006). The effects of orbital frontal cortex damage on the modulation of defensive responses by rats in playful and non-playful social contexts. *Behavioral Neuroscience, 120*, 72–84.
Pellis, S. M., Pellis, V. C., & Bell, H. C. (2010a). The function of play in the development of the social brain. *American Journal of Play, 2*, 278–296.
Pellis, S. M., Pellis, V. C., & Foroud, A. (2005). Play fighting: Aggression, affiliation and the development of nuanced social skills. In R. Tremblay, W. W. Hartup, & J. Archer (Eds.), *Developmental origins of aggression* (pp. 47–62). New York: Guilford Press.
Pellis, S. M., Pellis, V. C., & Himmler, B. T. (2014). How play makes for a more adaptable brain: A comparative and neural perspective. *American Journal of Play, 7*, 73–98.
Pellis, S. M., Pellis, V. C., & McKenna, M. M. (1994). A feminine dimension in the play fighting of rats (*Rattus norvegicus*) and its defeminization neonatally by androgens. *Journal of Comparative Psychology, 108*, 68–73.
Pellis, S. M., Pellis, V. C., & Reinhart, C. J. (2010b). The evolution of social play. In C. Worthman, P. Plotsky, D. Schechter, & C. Cummings (Eds.), *Formative Experiences: The interaction of caregiving, culture, and developmental psychobiology* (pp. 406–433). Cambridge, UK: Cambridge University Press.
Pellis, S. M., Pellis, V. C., & Whishaw, I. Q. (1992). The role of the cortex in play fighting by rats: Developmental and evolutionary implications. *Brain, Behavior & Evolution, 39*, 270–284.
Pisula, W., Stryjek, R., & Nałęcz-Tolak, A. (2006). Response to novelty of various types in laboratory rats. *Acta Neurobiologie Experimentalis, 66*, 235–243.
Porrini, S., Belloni, V., Seta, D. D., Farabollini, F., Giannelli, G., & Dessì-Fulgheri, F. (2005). Early exposure to a low dose of bisphenol A affects socio-sexual behavior of juvenile female rats. *Brain Research Bulletin, 65*, 261–266.
Reinhart, C. J., Pellis, S. M., & McIntyre, D. C. (2004). Development of play fighting in kindling-prone (FAST) and kindling-resistant (SLOW) rats: How does the retention of phenotypic juvenility affect the complexity of play?. *Developmental Psychobiology, 45*, 83–92.
Robbins, T. W., Jones, G. H., & Wilkinson, L. S. (1996). Behavioural and neurochemical effects of early social deprivation in the rat. *Journal of Psychopharmacology, 10*, 39–47.
Robitaille, J. A., & Bovet, J. (1976). Field observations on the social behaviour of the Norway rat, *Rattus norvegicus* (Berkenhout). *Biology of Behaviour, 1*, 289–308.
Schneider, P., Hannusch, C., Schmahl, C., Bohus, M., Spanagel, R., & Schneider, M. (2014). Adolescent peer-rejection alters pain perception and CB1 receptor expression in female rats. *European Neuropsychopharmacology, 24*, 290–301.
Schneider, P., Pätz, M., Spanagel, R., & Schneider, M. (2016). Adolescent social rejection alters pain processing in a CB1 receptor dependent manner. *European Neuropsychopharmacology, 26*, 1201–1212.
Seffer, D., Rippberger, H., Schwarting, R. K., & Wöhr, M. (2015). Pro-social 50-kHz ultrasonic communication in rats: Post-weaning but not post-adolescent social isolation leads to social impairments—phenotypic rescue by re-socialization. *Frontiers in Behavioral Neuroscience, 9*, 102. Available from http://dx.doi.org/10.3389/fnbeh.2015.00102.
Siviy, S. M., & Panksepp, J. (1987). Sensory modulation of juvenile play in rats. *Developmental Psychobiology, 20*, 39–55.
Siviy, S. M., & Panksepp, J. (2011). In search of the neurobiological substrates for social playfulness in mammalian brains. *Neuroscience & Biobehavioral Reviews, 35*, 1821–1830.
Siviy, S. M., Baliko, C. N., & Bowers, K. S. (1997). Rough-and-tumble play behavior in Fischer-344 and Buffalo rats: Effects of social isolation. *Physiology & Behavior, 61*, 597–602.

Siviy, S. M., Love, N. J., DeCicco, B. M., Giordano, S. B., & Seifert, T. L. (2003). The relative playfulness of juvenile Lewis and Fischer-344 rats. *Physiology & Behavior, 80*, 385–394.

Špinka, M., Newberry, R. C., & Bekoff, M. (2001). Mammalian play: Can training for the unexpected be fun? *Quarterly Review of Biology, 76*, 141–176.

Stryjek, R., Modlinska, K., & Pisula, W. (2012). Species specific behavioural patterns (digging and swimming) and reaction to novel objects in wild type, Wistar, Sprague-Dawley and brown Norway rats. *PLoS ONE*, 7, e40642.

Thiels, E., Alberts, J. R., & Cramer, C. P. (1990). Weaning in rats: II. Pup behavior patterns. *Developmental Psychobiology, 23*, 495–510.

Thor, D. H., & Holloway, W. R., Jr. (1984). Developmental analysis of social play behavior in juvenile rats. *Bulletin of the Psychonomic Society, 22*, 587–590.

Trezza, V., Baarendse, P. J. J., & Vanderschuren, L. J. M. J. (2010). The pleasures of play: Pharmacological insights into social reward mechanisms. *Trends in Pharmacological Science, 31*, 463–469.

Trut, L. M. (1999). Early canid domestication: The farm-fox experiment. *American Scientist, 8*, 160–168.

van den Berg, C. L., Hol, T., van Ree, J. M., Spruijt, B. M., Everts, H., & Koolhaas, J. M. (1999). Play is indispensable for an adequate development of coping with social challenges in the rat. *Developmental Psychobiology, 34*, 129–138.

van Kerkhof, L. W. M., Trezza, V., Mulder, T., Gao, P., Voorn, P., & Vanderschuren, L. J. M. J. (2014). Cellular activation in limbic brain systems during social play behavior in rats. *Brain Structure & Function, 219*, 1181–1211.

Vanderschuren, L. J. M. J., & Trezza, V. (2014). What the laboratory rat has taught us about social play behavior: Role in behavioral development and neural mechanisms. *Current Topics in Behavioral Neuroscience, 16*, 189–212.

Varlinskaya, E. I., & Spear, L. P. (2008). Social interactions in adolescent and adult Sprague–Dawley rats: Impact of social deprivation and test context familiarity. *Behavioural Brain Research, 188*, 398–405.

Varlinskaya, E. I., & Spear, L. P. (2009). Ethanol-induced social facilitation in adolescent rats: Role of endogenous activity at Mu opioid receptors. *Alcoholism: Clinical & Experimental Research, 33*, 991–1000.

Watt, M. J., Burke, A. R., Renner, K. J., & Forster, G. L. (2009). Adolescent male rats exposed to social defeat exhibit altered anxiety behavior and limbic monoamines as adults. *Behavioral Neuroscience, 123*, 564–576.

Whishaw, I. Q. (1988). Food wrenching and dodging: Use of an action pattern for the analysis of sensorimotor and social behavior in the rat. *Journal of Neuroscience Methods, 24*, 169–178.

Wilkins, A. S., Wrangham, R. W., & Fitch, W. T. (2014). The "domestication syndrome" in mammals: A unified explanation based on neural crest cell behavior and genetics. *Genetics, 197*, 795–808.

FURTHER READING

Moore, C. L. (1985). Development of mammalian sexual behavior. In E. S. Gollin (Ed.), *The comparative development of adaptive skills* (pp. 19–56). Hillsdale, NJ: Lawrence Erlbaum.

PART IV

Factors Influencing Development

CHAPTER 13

Brain Plasticity and Experience

Bryan Kolb
University of Lethbridge, Lethbridge, AB, Canada

13.1 INTRODUCTION

The general idea that experience modifies the brain is probably more than 100 years old and is a fundamental assumption of most of the chapters in this volume and even the topic of several popular trade books (e.g., Doidge, 2007). Darwin had noticed that the brains of domestic rabbits are smaller than those of wild rabbits and suggested that this resulted from reduced use of the "intellect, instincts, and senses." But knowing that brains change is quite different than knowing the mechanisms underlying the changing brain and how such changes relate to behavior. By its nature, most research on brain plasticity and behavior is correlational (for a thorough discussion, see Rutter & Pickles, 2016). For example, as a child learns language a vocabulary quickly develops and it is reasonable to conclude that changes in the brain are responsible for learning the vocabulary. But many questions remain. What changes in the brain and where in the brain do the changes occur? How did the changes happen? How did the brain changes alter behavior? Did the change in behavior alter the brain in return? These are not simple questions. The goal of this chapter is to identify some general principles of brain plasticity in development and outline answers (or at least hunches) to some of the questions just posed.

13.2 ASSUMPTIONS AND BIASES

As we consider the properties of the brain that make it plastic, we need to be mindful of several assumptions that underlie my thinking about brain plasticity.

13.2.1 Behavioral states, including mind states, correspond to brain states

Although this proposition is hardly novel and presumably appears self-evident to most neuroscientists, it remains a philosophical issue beginning at least by the time of Descartes. Until relatively recently, cognitive science and social psychology generally ignored this bias and saw brain activity as incidental to the understanding cognitive processes, but the emergence of cognitive neuroscience and social neuroscience as

The Neurobiology of Brain and Behavioral Development
DOI: http://dx.doi.org/10.1016/B978-0-12-804036-2.00013-3

disciplines has begun to shift thinking. Most of the new studies are using noninvasive imaging techniques such as functional Magnetic Resonance Imaging (fMRI), electroencephalography (EEG) in the form event-related potentials (ERPs), magnetoencephalography, and near infrared spectroscopy. The obvious conclusion from these types of studies is that the activity recorded with different techniques produces mind states, but this conclusion has proven to be controversial and Vul and Pashler (2012) go so far as to suggest that these "correlations in social neuroscience are voodoo." This assertion has led to strong disputations (e.g., Lieberman, Berkman, & Wagner, 2012). The arguments are complex, focus on the nature of noninvasive imaging data, and are continuing. Nonetheless, this debate does not impugn the general idea that brain states produce behavioral states.

13.2.2 The structural properties of the brain are important in understanding its function

If brain states are related to behavioral states, then it follows that because brain states are related to brain structure, then changes in the physical structure of the brain will be reflected in changes in its functioning. Thus, examining how brain morphology is altered by experience should be one way to study brain and behavior, and especially development of brain and behavior. One caution here, however, is that it is unlikely that every morphological change is meaningful, particularly because many changes in microstructure, such as the growth and loss of dendritic spines, is rapid and is likely often related to ongoing background activity in the brain, rather than a specific experience. Nonetheless, the details of cerebral morphology will place significant constraints on the computations of the brain and changes in the morphology may provide important clues for understanding changes in computations in development.

13.2.3 Plasticity is a property of the synapse

Although Ramon y Cajal (1928) first proposed that the efficiency of connections between neurons might be responsible for behavioral changes such as learning, Konorski (1948) was the first to propose a mechanism. Konorski proposed that combinations of sensory inputs could lead to two types of changes in neurons: (1) an invariant but transient change in the excitability of neurons and (2) an enduring plastic change in neurons. Stated differently, Konorski was proposing that when neurons are active they change. This idea is important because it implies that one could look at neurons and try to identify changes. But what part of the neuron would you look at? Hebb (1949) proposed that the logical place to look was at the synapse. He suggested that if neurons are coincidentally active the connection between them was strengthened. Hebb's idea was transformational because it specified the conditions under which plasticity would occur and it pointed to the possibility of changes in both the pre- and postsynaptic side of a connection. More importantly, however, Hebb proposed that during development there are changes in synaptic connectivity that allow the brain to be functionally plastic.

13.2.4 Behavioral plasticity results from the summation of plasticity of many neurons

This bias should be self-evident because it is unlikely that one, or even a few, of humans' 86 billion neurons will have much influence on function, but this needs to be stated explicitly. What about animals with simple nervous systems, such as *Caenorhabditis elegans*, which has 302 neurons and 56 glial cells? Changes in a single neuron could conceivably be very important. We might predict that such simple nervous systems might be less plastic than human brains because it might not be adaptive to have it changing every time one of the neurons was especially active.

13.2.5 Overall brain plasticity increases as the number of neurons increases

One implication of the *C. elegans* versus human plasticity comparison is that brains with smaller numbers of neurons might be less plastic than brains with very large numbers of neurons. But what about mammals? In a series of heroic studies, Suzanna Herculano-Houzel and her colleagues (e.g., 2011, 2012) have determined that although humans do not have the largest brains among mammals (or even close), human brains have the most neurons, and in fact orders of magnitude more neurons. As a general rule, primate brains contain more neurons than similarly sized rodent brain. For example, the human brain has about sevenfold more neurons than a hypothetical rodent with a human sized brain (86 billion versus 12 billion) (Herculano-Houzel, 2011). One consequence of having so many more neurons is an increase in cognitive reserve, which would facilitate increased plasticity.

13.2.6 Experience-dependent changes in the brain tend to be focal

Widespread changes are found from multidimensional experiences, such as living in complex versus isolated housing, but this is rarely the case. For example, training animals in specific tasks is associated with surprisingly focal changes in the brain (e.g., Comeau, McDonald, & Kolb, 2010; Greenough, Larson, & Withers, 1985; Greenough & Chang, 1988). Similarly, although psychoactive drugs may produce large behavioral changes and have widespread acute effects on neurons, the chronic plastic changes are surprisingly focal and largely confined to the prefrontal cortex and striatum (e.g., Robinson & Kolb, 2004).

13.2.7 Experience-dependent changes interact

It is convenient in laboratory studies to expose animals to singular experiences (e.g., stress, drugs, tactile stimulation) and then examine changes in the brain, but real life is very different. We are exposed sequential experiences over a lifetime. In particular, as the brain develops it is influenced by multiple experiences that may occur concurrently or sequentially. A reasonable question is how one early experience might alter the effect of other experiences, either in development or later in life. *Metaplasticity* refers to the idea that plastic changes in the brain at one point will in life modify how the brain responds to experience later in life (Abraham & Bear, 1996). There have not yet been extensive studies of metaplastic events in development but a few examples will illustrate the idea.

When adult rats are given daily doses of psychomotor stimulants for 2 weeks (nicotine, amphetamine, cocaine), there is a phenomenon called drug-induced behavioral sensitization. With each dose of the drug, there is an escalating effect on locomotor activity, and this effect persists after extended drug-free periods. It is correlated with changes in dendritic organization in the medial prefrontal cortex (mPFC), orbitofrontal cortex (OFC), and nucleus accumbens. Four months later the rats still show significant changes in mPFC, OFC, and nucleus accumbens dendritic morphology but no changes in parietal cortex. In contrast, when adult rats are placed in complex environments for 60–90 days they show few chronic changes in mPFC or OFC, but they do show significant increases in dendritic branching, length, and spine density in parietal cortex. But, if drug-treated rats are placed into the complex housing 2 weeks after the last drug treatment, they show no enrichment-related changes in parietal cortex (Hamilton & Kolb, 2005; Kolb, Gibb et al., 2003). This is an example of metaplasticity. The drug experience alters the brain's response to later experiences.

Similar studies have been done with developing animals. For example, Mychasiuk, Muhammad, and Kolb (2014) prenatally exposed rats to nicotine and then in adulthood placed the animals in complex environments versus standard group lab housing. As in the adult studies, prior exposure to the drug blocked some of the effects of complex housing in adulthood. Similarly, in a series of studies, Muhammad and Kolb (2011a, 2011b) gave developing rats gestational or postnatal tactile stimulation. When these animals were given amphetamine in adulthood they show a strongly attenuated behavioral sensitization as well as attenuated structural changes in mPFC and OFC. Thus, metaplastic effects can be both maladaptive (reduced plasticity to complex housing) or beneficial (reduced drug effects).

Metaplasticity is developmentally significant because it provides a mechanism for the development of individual differences in behavior, and in the extreme, can prime the brain to take a different developmental trajectory in response to later experiences that might otherwise be benign.

13.2.8 There are critical periods for some forms of plasticity

The developing brain is generally more responsive to experiences than the adult brain, which presumably enables individuals to adapt to a wide range of environments that cannot be predicted prior to birth. This brief postnatal epoch has been labeled the critical period (e.g., Wiesel & Hubel, 1965) (see Takesian & Hensch, 2013; for a more expanded discussion). The age of onset and the duration of the critical period vary by species and sensory modality. For example, there is a critical period for developing the whisker barrel fields in rats that begins at birth and ends around postnatal day 5 (p5) (Rice & van der Loos, 1977) but by contrast in the rat visual cortex it lasts about 2 weeks beginning in the third postnatal week (Fagiolini, Pizzorusso, Berardi, Domenici, & Maffei, 1994). The critical period in the auditory system is roughly from P9 to P28. For example, Zhang, Bao, and Merzenich (2002) exposed young rats to white noise from P9 to P28 and found severe disruptions in auditory processing in adults, but the same treatment after P30 did not have such effects. The physiological effects are mediated by cortical expression of certain NMDA and GABAa receptor subunits as well as brain-derived neurotrophic factor (BDNF). But the critical period for this effect turned out to be very brief indeed, namely P11–P13 (de Villers-Sidani, Chang, Bao, & Merzenich, 2007). Although the effects of restricted sensory inputs during the critical periods can be severe, some reversal of the effects is possible with intense sensory enrichment later in life (e.g., Zhu et al., 2014).

There are also critical periods for the response to brain perturbations, such as injury. For example, the rat brain is especially vulnerable to cortical injury in the first few postnatal days but shows dramatic functional recovery following injuries in the second postnatal week (e.g., Kolb, Mychasiuk, Muhammad, & Gibb, 2013) but not in the third week. These differences are associated with striking differences in the plastic changes seen throughout many cerebral regions, which appear to support the recovery (see discussion below in Section 13.5.9).

It can also be argued that adolescence comprises another critical period of brain plasticity relative to the juvenile and adult brain (e.g., Fuhrmann, Knoll, & Blakemore, 2015; Steinberg, 2016). The onset of enhanced plasticity in adolescence likely coincides with the release of gonadal hormones but the timing of the reduction in plasticity at the end of adolescence is not well studied and likely varies across the brain. A sensitive period in adolescence characterized by greatly increased plasticity is likely adaptive as there is considerable learning about the environment, and especially the social environment (Blakemore, 2008). But the enhanced plasticity in adolescence is not without potential negative consequences because the brain becomes vulnerable to a wide range of experiences such as stress, psychoactive drugs, brain trauma (e.g., concussion), and variations in the patterns of play that influence brain organization and function differently than in adults. It is no accident that many forms of mental illness become apparent in adolescence (e.g., Tottenham & Galvan, 2016).

13.3 TYPES OF BRAIN PLASTICITY

Three types of brain plasticity can be identified in the developing brain: experience-independent; experience-expectant, and experience-dependent.

13.3.1 Experience-independent plasticity

This is largely a developmental process (e.g., Shatz, 1992). The visual system provides two good examples. Consider the anatomical challenge of connecting the eyes to the rest of the nervous system. One critical characteristic of our visual experience is that it is spatial. That is, we have maps of where things are in visual space. In order to do this, the retinal neurons must project to the optic tectum and lateral geniculate nucleus (LGN) in such a way that the spatial organization of the receptors in the retina (the retinotopic map) is maintained. That is, receptors that have adjacent views of the visual world must send their information to adjacent regions of the tectum and thalamus. In early development axons arriving from the retina to the optic tectum are intermingled and do not respect the spatial organization of the retina's retinotopic map. Sperry (1963) first proposed that specific molecules in different cells in the retina and tectum give each cell a distinct identity. Presumably early in prenatal development each incoming axon seeks out a specific chemical that is similar to that in the retina (see Fig. 13.1). This leads the axons to right general location in the tectum.

The problem is that the axons are still not in the precise location required for an accurate visual spatial map. This requires neural activity. Thus, there are waves of spontaneous activity that travel across the retina. Because adjacent receptors in the retina tend to be activated at the same time, they tend to form synapses on the same neurons whereas receptors that are widely separated are not activated together. Through this simultaneous activity and with the passage of time, cells eventually line up correctly and form precise connections.

A similar process is used to organize the connections to the LGN of the thalamus but here there is an addition problem as there are inputs from the two eyes that must be coordinated so that each eye's retinotopic map are synchronous (see review by Penn & Shatz, 1999). Within the adult LGN the axons of the two eyes terminate in separate alternating layers such that each layer receives input from only one eye. It is not until the primary visual cortex that the inputs from the two eyes are integrated such that alternating 0.5-mm-wide columns (ocular dominance columns) of neurons are activated by one eye in such a way that the retinotopic map is preserved.

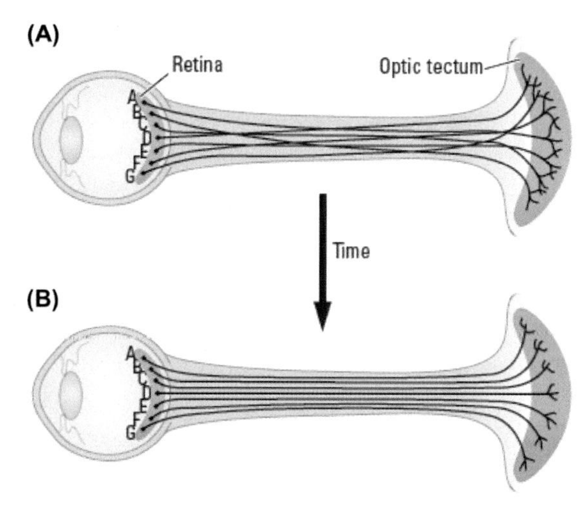

Figure 13.1 (A) Neurons A through G project from the retina to the tectum. Early in development the neurons use chemical affinity to grow to the approximately correct location in the tectum. The activity of the neurons tunes this organization as adjacent neurons in the retina are more likely to fire concurrently than neurons that are separated. (B) As a result, adjacent retinal neurons are more likely to establish permanent synapses on the same tectal neurons. *After* Kolb, B., Whishaw, I.Q., & Teskey, G.C. (2016). An Introduction to Brain and Behavior (5th edition). New York: Worth/ MacMillan *(Kolb et al., 2016).*

In the developing visual system neither the LGN layers nor the ocular dominance columns are present and like in the retina, the retinotopic map is crude. The emergence of eye-specific layers is related to the waves of spontaneous retinal activity. The waves in each eye are uncorrelated and last about 5 seconds, followed by a silent period of about 2 minutes. Then another wave occurs in another completely different and apparently random direction (Shatz, 1996). The wave of activity in each eye is uncorrelated whereas that in each eye is somewhat correlated, which would be a requirement for eye-specific layers to form.

The emergence of ocular dominance columns in primary visual cortex is also related to the prenatal correlated spontaneous activity of each eye such that at birth there are alternating columns but these are not precise and require postnatal experience. Weisel and Hubel (1963, 1965) demonstrated that closing one eye in a kitten during its early postnatal development disrupted the development of the ocular dominance columns such that the eye that was still open expanded its projections into the adjacent columns whereas the deprived eye lost its connections.

Spontaneously generated activity is not unique to the visual pathways but is present throughout the developing nervous system. It therefore is likely to be used throughout the brain to refine patterns of developing neural connections to form the precise adult patterns.

13.3.2 Experience-expectant plasticity

Experience-expectant plasticity largely occurs during development. A good example is the development of ocular dominance columns found in the primary visual cortex.

The emergence of ocular dominance columns in primary visual cortex is also related to the prenatal correlated spontaneous activity of each eye such that at birth there are alternating columns but these are not precise as there are still overlapping bands of afferent axons into layer IV of primary visual cortex. The refinement of the connections requires the neural activity derived from postnatal visual experience. If the input from one eye is degraded or blocked during this postnatal period, the development of the ocular dominance columns is altered such that the good eye fails to retract its projections into the adjacent columns whereas the degraded or deprived eye loses its connections. When the closed eye is eventually opened, its vision is compromised (Wiesel & Hubel, 1963). Thus, it is apparent that the visual system is expecting to receive experience-related input, which is needed to tune the system.

13.3.3 Experience-dependent plasticity

Experience-dependent plasticity refers to a process of changing existing neuronal ensembles. This involves at least two different processes, namely the selective regression or elimination of connections or the selective growth of new or altered connections. An experiment described by Greenough and Chang (1989) illustrates both processes. The somatosensory representation of the large facial whiskers in rodents is arranged in a series of rows of so-called cortical barrels, with each barrel representing a single whisker in each of the rows of whiskers. In adults, the neurons forming the barrels direct their dendrites into the center of the barrel. This arrangement could occur because the neurons just grew dendrites in the barrel hollow, indicating that something about the hollow attracted the dendrites, or because branches that grew toward the hollow tended to stabilize, whereas those growing in other directions were lost. On postnatal day 10 the dendrites were equally likely to orient in any direction but by day 30 virtually all of the dendrites oriented into the hollow. Thus, the development of highly ordered barrel fields involves both the selective growth of properly oriented dendrites as well as the selective loss of improperly oriented dendrites.

Experience-dependent plasticity can also be seen whenever animals learn problems (e.g., Comeau et al., 2010), intense environmental manipulations (e.g., Kolb, Gorny, Li, Samaha, & Robinson, 2003a; Kolb, Gibb, & Gorny, 2003b), electrical brain stimulation (Teskey, Monfils, Silasi, & Kolb, 2006), or are exposed to abnormal experiences such as psychoactive drugs (e.g., Robinson & Kolb, 2004) or injury (e.g., Kolb, 1995). These types of experiences both increase and decrease synapse numbers, often in the same animals, and in different brain regions. The critical point here is that experience-dependent plasticity reflects modification of an existing pattern of neuronal connections.

13.3.4 Plasticity in the adolescent brain

The place to start here is with a definition of adolescence. One common onset marker is the onset of puberty, which in humans would be around 12 years and in rats around 28–30 days (Spear, 2000). The more difficult question is when does it end? The end of obvious behavioral changes often has been used but when we consider changes in the brain, this becomes less certain. There are multiple potential markers including reduced neural pruning, myelination, gliogenesis (aside from oligodendrocytes), mature connectivity, and so on, but all of these change to different timetables and differently in different parts of the brain. I am persuaded by the general argument that adolescence is a period of heightened neural plasticity relative to the juvenile and adult brain (e.g., Fuhrmann et al., 2015; Knudsen, 2004; Steinberg, 2016).

But what is the nature of plasticity in adolescence? It seems likely that it is mostly experience-dependent plasticity but in principle there could be some form of experience-expectant plasticity, likely related to the production of gonadal hormones in puberty and changes in social behavior. Casey (2015) has suggested that there are three distinct forms of adolescent developmental changes. Changes in brain and behavior may be *adolescent nonspecific*, which reflect a continuation of processes already in progress, *adolescent emergent*, which increase rapidly up to adolescence and then are relatively constant for some period of time, or *adolescent specific*, which emerge for the adolescent period and then decline. The latter type would best reflect the contention that adolescence is a period of enhanced plasticity (for a longer discussion see, Kolb, 2017).

13.4 MEASURING BRAIN PLASTICITY

Brain plasticity can be inferred from data collected at several levels of analysis, ranging from behavior to changes in molecular structure of cell. There is no correct level of analysis but rather the level needs to be appropriate to the question asked. Understanding the emergence of language in development is unlikely to be helped by analyzing specific changes in protein production in some selected location in the brain because language learning involves vast areas of the brain. But understanding what controls neurogenesis in the hippocampus may be critical to understanding how new memories are formed.

13.4.1 Behavioral analyses

Three important sets of behavioral distinctions are relevant to understanding plastic changes in the developing brain. First, there is an important difference between moving around (e.g., getting exercise) and skill acquisition. Activity is important to developing bones and muscles as well as stimulating cardiovascular function including blood flow in the brain, and it may very well contribute to brain plasticity. But activity likely does not require *specific* plastic neural changes to support it. In contrast, learning a new skill, such as learning to play a musical instrument or to read, requires extensive practice, which in turn is likely key in changing the neural ensembles that will support it. This distinction is important because our inference that some plastic change has occurred in the brain to support a new behavior must come from a comparison of individuals who had the specific training and those who did not and some form of exercise regime is often used as a control to which those with specific training can be compared. A comparison to individuals who did nothing, such as laboratory animals sitting in small cages, may not be an appropriate comparison if simply doing something that requires moving around in space indeed does produce neural changes or perhaps prevents the loss of connections in unused neural ensembles.

Second, we must distinguish between voluntary behaviors and supporting reactions. Consider that the learning of most voluntary behaviors, such as making finger movements in learning to play an instrument, or advancing a limb to make grasping movements through a slot, require concomitant supporting movements. For example, when a child begins to play the violin it is not just about making appropriate finger movements in the left hand and arm and wrist movements in the right, but also the child must learn postures and movements that allow the violin-directed movements to occur. Similarly, when a quadraped such as a rat learns to lift a forelimb to reach and grasp a food object through a slot, the rest of the body must be positioned to balance itself with its other limbs and tail (e.g., Whishaw & Miklyaeva, 1996). Whereas it is widely accepted that making finger movements in musical training or forelimb movements in reaching for food are accompanied by changes in the morphology of neurons in the motor cortex (e.g., Elbert, Heim, & Rockstroh, 2001; Greenough et al., 1985), it is not known whether other supporting behaviors are skills in the same sense. This question could be answered empirically of course but recognizing this complication is important as we attempt to associate changes in neuronal morphology with behavioral changes.

Finally, when we consider plastic changes in the brain that follow early brain injury, we must be mindful that there is a distinction between recovery of function, sparing of function, and compensation. If young animals have brain injuries before behaviors such as language have emerged, the later appearance of these behaviors is not really recovery so much as it sparing of function. Changes in the brain, such as the shifting of language, or some language processes, from the left to the right hemisphere may be seen as sparing of language or a neural compensation in response to the injury. Similarly, if brain injury occurs later after a behavior such as language has emerged, the return of behavior could reflect either recovery in which spared neurons change morphology to support behavior or changes in more distal regions to compensate the loss of neurons normally engaged in the impaired behavior.

13.4.2 Functional organization

The idea that the brain had representations of maps of the sensory and motor world probably begins with Fritsch and Hitzig (1960 [1870]) who first showed that electrical stimulation of the frontal cortex of a dog could elicit movements on the opposite side of the dog's body and that stimulation of different regions led to movements of specific body parts. But perhaps the most influential paper was by Penfield and Boldery (1937) who provided evidence for separate motor and sensory maps in the human brain. Indeed, the summarization of their data as a homunculus became a standard figure in neurology textbooks. But it was not until more recently that it became clear that maps are not static but can change with experience. Pons et al. (1991) studied the somatosensory maps of monkeys in which the fibers entering the dorsal root ganglion in the arm region of the spinal cord were cut. Light touch of the skin of the affected arm no longer activated the somatosensory cortex but light touch of the face now activated the region that normally would respond to arm touch. Thus, the facial representation was now larger (see Fig. 13.2). Later studies by others showed that not only would peripheral injury perturb the maps but so would sensory experiences. For example, Elbert et al. (2001) showed that practicing playing a stringed musical instrument in childhood increased the representation of the fingers of the left hand and that the amount of change was proportional to the age that musical training began.

Although the original maps of Penfield and Boldery were made by using direct electrical stimulation of the cortex, maps are now identified using both direct and indirect (e.g., transcranial magnetic stimulation), fMRI, positron emission tomography (PET), ERP, among others. Most recently, Finn et al. (2015) demonstrated that although fMRI studies typically collapse data from many participants to identify functional organization maps, individual variability in functional connectivity is robust and can be used as a "brain fingerprint" that accurately identify individuals. The authors

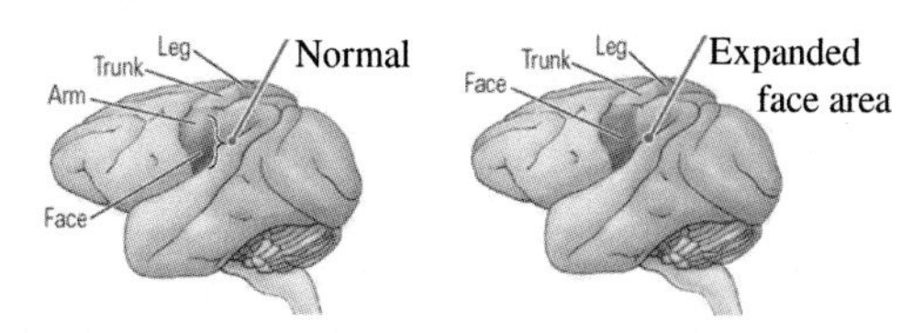

Figure 13.2 Schematic of the motor representations mapped by electrical recordings in a normal and denervated monkey. The face area is enlarged at the expense of the arm area in the denervated monkey. *After Kolb, B., & Whishaw, I. Q. (2015).* Fundamentals of human neuropsychology *(7th ed.). New York: Worth/MacMillan.*

conclude by suggesting that it is possible to follow phenotypic changes in the cerebral cortex that are related both to health and disease.

13.4.3 Brain structure

The first systematic studies on the effect of experience on brain structure began with the research group at Berkeley who demonstrated changes in brain weight, cortical thickness, acetylcholine levels, and dendritic structure in rats raised in complex environments versus standard lab caging (Bennett, Kiamond, Krech, & Rosenzweig, 1964; Diamond, Krech, & Rosenzweig, 1964). Later, beginning in the 1970s, Greenough and his colleagues initiated a multidisciplinary investigation of the effects of rearing animals in visually and/or motorically enriched environments. In a heroic series of experiments he and his colleagues measured neuron size, neuron density, size of axons, capillary vessels, glial size and numbers, dendritic structure, and more (e.g., Sirevaag & Greenough, 1987, 1988). The overall result was that there was a coordinated change in neuronal morphology as well as glial, vascular, and metabolic processes in response to different experiences. This means that the increase in synapses per neuron was accompanied by more astrocyte material, more blood capillaries, and a higher mitochondria volume (a measure of metabolic activity) in animals with enriched experience, and especially in development.

Greenough also asked whether these changes were specific to enriched housing but they are not. When rats learn specific tasks the learning stimulates area-dependent growth related to the specific training. Thus, for example, rats learning visual tasks show increased dendritic length in the visual cortex, whereas rats learning motor tasks show increased dendritic length in the motor cortex (see review by Greenough & Chang, 1988). Perhaps the most surprising feature of experience-dependent changes in the brain is that even gross measures such as brain weight and cortical thickness show robust effects from early experience. For example, several studies have shown that gestational or perinatal stress leads to smaller brains (measured by weight) and reduced cortical thickness, even though there is no difference in body weight (e.g., Muhammad & Kolb, 2011a). Given that these measures are so simple to collect, we recommend that at least brain weight be routinely collected at the end of behavioral studies.

One consistent finding in our lab has been that different measures of neuronal morphology change independently of each other and sometimes in opposite directions. Although there has been a tendency in the literature to view different neuronal changes (i.e., dendritic length, spine density) as surrogates for one other, they can and do vary independently and in opposite directions (e.g., Comeau et al., 2010; Kolb, Gibb, Pearce, & Tanguay, 2008). That is, sometimes dendritic length increases but spine density decreases and vice versa. Although it is not clear exactly what this means, the biophysics of the synapses would certainly be different. Finally, one surprising finding is that cells in different cortical layers, but in the same presumptive columns, can show very different responses to the same experiences (e.g., Teskey et al., 2006).

13.4.4 Synaptic structure

As noted by Greenough and Chang (1988), the most credible and dramatic form of plasticity in the CNS is that of changes in the number of synapses. The extent to which differences in experience, especially during development, can affect synapse number is massive, being on the order of thousands of synapses per neuron, and thus billions of synapses per brain. One important feature of the developing brain is the transient overproduction of synapses followed by a process of pruning that is guided by the experience of the animal, as reflected in the activity of its neurons (e.g., Purves & Lichtman, 1980). One example of how neural activity plays a role in neural perseveration or loss is in the development of the ocular dominance columns discussed earlier. In the absence of activity from a patched eye, there is a loss of synapses in that eye's column.

But the number of synapses can also increase in response to experiences in development such as when animals are trained on specific problems. For example, several studies have shown that training 1-day-old chicks on a passive-avoidance task results in a variety of morphological changes in a several regions of the chick brain 24 hours posttraining (e.g., Patel, Rose, & Stewart, 1988).

In addition to measuring the numbers of synapses many studies have shown that experience alters synaptic morphology. A tacit assumption of these studies is that greater sensory stimulation (or opportunity to store the effects of experience) will have, on average, stronger synapses. The aspects of synaptic structure that are most commonly associated with the strength of the synapse are: (1) the size of various synaptic structures; (2) the shape of postsynaptic spines; (3) the number, density, or location of vesicles in the presynaptic terminal; (4) the conformation of the contact or opposition zone; and (5) perforations in the postsynaptic density (PSD) (Fig. 13.3; Greenough & Chang, 1988). In general, the experience-dependent effects on these measures are much larger in young animals than in adults.

13.4.5 Physiological changes

Changes in cerebral organization, including synaptic structure, are presumably related to changes in the physiological activity of the brain. Noninvasive imaging methods (reviewed earlier) provide one measure of physiological changes but there are also more direct (and invasive) measures that show robust plastic modifications in the brain. As early as the 1950s a variety of investigators described short-term changes in

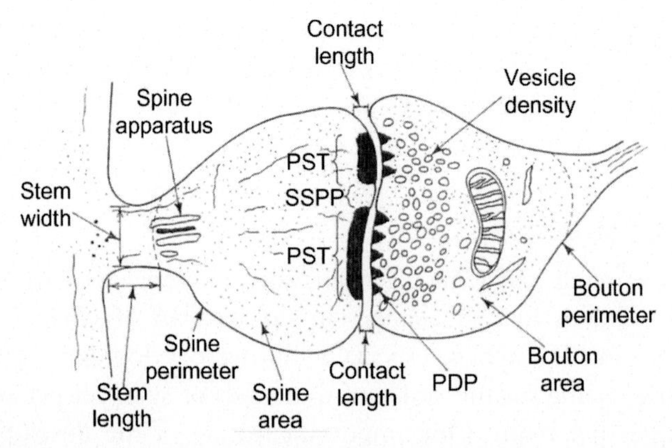

Figure 13.3 Measures of synapse characteristics. Abbreviations: *PST*, postsynaptic thickening; *SSPP*, subsynaptic plate perforation; *PDP*, presynaptic dense projection. *From Greenough, W. T., & Chang, F. F. (1989). Plasticity of synapse structure and pattern in the cerebral cortex. In A. Peters & E. G. Jones (Eds.),* Cerebral cortex *(Vol. 7, pp. 391–440). New York: Plenum Press.*

neuronal activity resulting from direct electrical stimulation, which are referred to as posttetanic potentiation (e.g., Green & Adey, 1956; Kandel & Spencer, 1968). But it was the landmark paper by Bliss and Lomo (1973) that provided a technique to identify long-lasting enhancement of synaptic transmission, which has come to be called long-term potentiation (LTP).

The demonstration of LTP involves measuring excitatory postsynaptic potentials (EPSPs) following a test pulse of brain stimulation and then again after an intense high-frequency (100 Hz) stimulation. The size of the EPSP is increased and this enhancement can last hours, days, or weeks, depending in part on the number of pulses of the high frequency stimulation. There are now literally hundreds of papers outlining the characteristics of LTP in many brain structures in species as diverse as *Aplysia* and humans (for an interesting review, see Teskey, 2016). The key message here is that LTP has been widely used as a model of learning and is sensitive to a wide range of experiences. For example, environmental enrichment enhances learning and LTP (e.g., Hullinger, O'Riordan, & Burger, 2015) and LTP varies with gonadal hormones and psychoactive drugs (e.g., Warren, Humphreys, Juraska, & Greenough, 1995). Although for technical reasons it is difficult to record LTP in very young animals, LTP can be used as a measure of the effects of early experiences on later brain activity and organization. Increased LTP is correlated with increased synaptogenesis in the hippocampus in adult rats (Greenough & Chang, 1988).

13.4.6 Mitotic activity

The traditional story regarding neurogenesis is that in both rodents and humans it is complete by birth, except in the cerebellum, olfactory bulb, and hippocampus. The neuronal population at birth is then reduced by cell death during development. Recently, however, it has become clear, at least in laboratory rats, that there is extensive postnatal neurogenesis in the neocortex, and although this has not yet been confirmed in primates, it has not been disconfirmed either (Bandeira, Lent, & Herculano-Houzel, 2009). The reason for the revised view is the development of a new method to cell counting called the isotropic fractionator, which allows precise determination of total numbers of neuronal and nonneuronal populations in the whole brain, or any specific structure (Herculano-Houzel & Lent, 2005). This method does not require injections of mitotic markers, which is difficult to do in young animals.

In a series of papers using rats, Herculano-Houzel and her colleagues have shown that there is a dormant period for about 3 days, followed by a several days in which the total number of cerebral cortical neurons doubles, then is reduced by 70% during the second postnatal week. A second round of postnatal neurogenesis occurs over the following two weeks followed by a final round of neurogenesis occurring between 1 and 2 mo of age, which is adolescence. This latter increase was observed across the brain and in the cerebral cortex represented about an 80% increase. By age 12 mo, there is a significant loss of neurons (~20%–30%) (Bandeira et al., 2009; Mortera & Herculano-Houzel, 2012).

The evidence of significant postnatal neurogenesis suggested that experience may have a previously unrecognized impact on postnatal neurons in brain development and the adult brain phenotype, and may be related to neurodevelopmental disorders. It is also important to recall from Greenough's enrichment studies that experience increases the number of astrocytes and the astrocyte/neuron ratio and during development experience therefore could influence both neurogenesis and gliogenesis.

13.4.7 Myelination

Although the main focus of the plasticity field over the past 100 years has been the synapse, there are data emerging suggesting that plasticity mechanisms operate beyond the synapse, or even gray matter (Fields, 2015). Technological advances in identify neural networks both functionally and connectively (termed "connectomics") have shown that conduction time through large networks is critical to the induction of neural plasticity. Until very recently it was assumed that once formed, myelin did not change except in disease states such as multiple sclerosis. Recent evidence has shown that experience, such as learning, can influence myelination. For example, Sampaio-Baptista et al. (2013) showed changes white matter tracts in somatosensory cortex in rats trained to reach through a slot to grasp food (see Fig. 13.4). Similar changes can be seen in humans too. Diffusion tensor MRI has shown that in humans trained for two hours on a car-racing video game there are changes in white matter in

Figure 13.4 Rats trained to reach through a slot to grasp a food pellet show an expansion of the white matter tracts in the external capsule.

hippocampus and parahippocampal gyrus (Hofstetter, Tavor, Tzur Moryosef, & Assaf, 2013) and more recently, the same group found micro changes in white matter in a language task (Hofstetter, Friedmann, & Assaf, 2016). In this study, the authors introduced new lexical items (flower names) to the lexicon of participants for about an hour. They found rapid changes in white matter tracts underlying the cortex, and the extent of change correlated with behavioral measures of the lexical learning rate.

A critical question is how do oligodendrocytes (i.e., myelin) "know" that neurons are learning or experiencing something? Although oligodendrocytes are not excitable, they do have ion channels and neurotransmitter receptors (see Fields 2015 for a review). So, activity in neurons could affect oligodendrocytes through transmitters, growth factors, other chemicals, or exosomes (see below). Exploration of white matter plasticity is just beginning and given the rapid growth of myelin during development it appears certain that white matter plasticity will influence brain development.

13.4.8 Molecular structure

As discussed in Chapter 1, Overview of Brain Development, development is a dynamic process of interplay between genes and experience. There is growing evidence that early life experiences induce epigenetic changes that account for variation in behavior, including cognition, response to stress, and sociality (see review by Kundakovic & Champagne, 2015). Measures of epigenetic changes in the brain thus provide another way to infer that experiences have induced plastic changes in brain organization and function. A series of studies by Harker and Gibb (e.g., Harker, Raza, Williamson, Kolb, & Gibb, 2015; Harker, Carroll, Raza, Kolb, & Gibb, 2016; Mychasiuk, Muhammad, Gibb, & Kolb, 2013) provide a nice example. They stressed male rats for 4 weeks prior to mating them with females and then studied the subsequent neurodevelopment, behavior, gene methylation, and neural morphology in the offspring. Preconception paternal stress changed all of these measures in the offspring leading to the hypothesis that stress in the preconception period changed gene expression in the paternal sperm, and this was transferred to the offspring and led to altered neurodevelopment and adult behavior. Although the authors did not do it, an obvious thing to do is to correlate the methylation changes with behavior as well as correlating the neuronal changes with behavioral measures. Unfortunately, it is not possible to correlate the epigenetic changes with neuronal changes in the same animals because processing the tissue for either measure precludes using the tissue for the other measure.

The power of experience to alter brain and behavior epigenetically can also be seen in studies relating tactile stimulation provided by parents to developing offspring. Weaver, Meaney, and Szyf (2006) showed that the amount licking and grooming of dams to their pups (i.e., high versus low amounts) was associated with epigenetic consequences, which have now been seen in multiple brain areas including hippocampus, prefrontal cortex, and amygdala (see Kundakovic & Champagne, 2015). Similarly, Seelke, Perkeybile, Grunewald, Bales, and Krubitzer (2016) measured the amount of contact between both parents and the infants in the biparental vole species *Microtus ochrogaster*. Offspring of parents with high levels of contact had a different pattern of connections within the somatosensory cortex as well as more distal connections with frontal and parietal cortex that those of offspring experiencing low levels of contact. Although the authors did not directly make epigenetic measures, they did speculate that like the pups in the Weaver et al. study there were epigenetic changes responsible for the individual differences in the effects of parental rearing styles associated with tactile contact.

Cells release a wide array of microvesicles into the extracellular space. Amongst those, *exosomes*, the cell-derived 30- and 100-nm nanovesicles released by cells and present in all biological fluids, have emerged as key regulators of cellular functioning. Exosomes are released by cells when vesicles housing them fuse with the plasma membrane, much as synaptic vesicles do at synapses, or they can be released directly from blood plasma. Exosomes have mostly been studied for their role in cancer, but appear to be vital for normal brain functions. Most cells in the CNS—neurons, astrocytes, oligodendrocytes and microglia—secrete exosomes. They participate in the development of the CNS, regulation of synaptic activity as well as regeneration following injury. The precise role of exosomes in development remains poorly understood but environmental enrichment stimulates exosome secretion promoting CNS myelination and may regulate inflammation (Pusic, Pusic, & Kraig, 2016) and will likely have a yet undetermined role in brain plasticity.

13.5 FACTORS INFLUENCING PLASTICITY IN DEVELOPMENT

Chapter 3, Overview of Factors Affecting Brain Development, reviewed how the brain is remarkably sensitive to a wide range of factors that influence brain and

Table 13.1 Factors inducing plastic changes during brain development

Factors
Age
Sensory and motor experience
Pre- and postnatal stress
Preconceptual parental experience
Parent–child relationships
Peer relations
Psychoactive drugs
Diet
Gonadal hormones
Gut bacteria
Perinatal brain injury

behavioral development. Here we return to the general topic but the focus will be on nature of the plastic changes in the brain that in associated with these factors (see Table 13.1).

13.5.1 Age

Although this may seem an obvious factor, it is important to emphasize that precise embryological age is critical in understanding brain plasticity. First, there is a tendency to think of birth as a constant point in development across species but it is not. Mammalian species are born at different points in their embryological development and thus exposure to stimulation may be in utero for monkeys but ex utero for rats or cats. For example, whereas monkeys are born quite mobile with their eyes and ears open, rodents or cats are relatively helpless because they are embryologically younger at birth. Thus, what appears to be a similar experience for different species may be very different because the brain is in a different developmental state.

Furthermore, there is natural tendency to think of an "infant" or "toddler" brain as though it was the same from day to day or week to week. But, it is not. The brain is changing rapidly. This was forcefully brought to my attention about 35 years ago in the midst of a series of studies on the effects of damage to the cortex of infant rats. I had been making prefrontal lesions on postnatal day 7 but owing to problems in timing and travel commitments, I performed surgery on two litters of rats on postnatal day 5. As described later (see Section 13.5.10), the difference was stunning. The same injury but spaced 2 days apart produced dramatically different outcomes as the day 5 operates were devastated whereas the day 7 operates showed dramatic recovery, and the compensatory changes following the injuries were very different. This likely was because the brains were so different at the time of surgery. In hindsight it was obvious but at the time a difference of 2 days seemed rather trivial. Clearly, it is not in a species that has an accelerated development and reaches weaning at about 21 days. I return to this point below but we should note that the time course of development is much faster in rats than in larger brained animals like cats and monkeys, who are faster than humans. Nonetheless, in all species the brain is changing rapidly, reflecting the rapid changes in the behavior of even human infants—something that is obvious to parents of infants.

In addition, even within a species such as humans, there is considerable variation in embryological development ranging from full term to several weeks premature. Thus, a human baby's brain is expecting stimulation ex utero at about 9 mo, but ex utero exposure occurring significantly earlier has a different effect. There is now considerable evidence that about 50% of very preterm children develop neurodevelopmental disabilities, including low IQ, cognitive impairments ranging from poor attention to impaired executive functions, as well as behavior and socialization problems (e.g., see review in Zhang et al., 2015). These behavioral abnormalities are associated with a wide range of morphological abnormalities including reductions in grey matter volume in frontal, parietal and temporal lobes, reduced white matter in the corpus callosum, reduced white matter in occipital lobe, reduced depth of some cerebral sulci, simpler cortical networks, and abnormalities in the cingulate cortex (e.g., Kwinto, Herman-Sucharska, Lesniak, & Klimek, 2015; Thompson, Chen, Beare, & Adamson, 2016; Zhang et al., 2015). Abnormalities in brain structure can even be seen at 38–44 weeks corrected gestational age as the preterm infants have smaller brains and less developed myelin in the posterior limb of the external capsule (Walsh, Doyle, Anderson, & Cheong, 2014).

Of course, premature infants are not just experiencing ex utero sensory stimulation sooner than the brain is prepared for but they are in incubators often attached to medical devices, which provide unusual somatosensory and auditory experiences, and no doubt are stressed. The key point, however, is that their brain is at a different embryological place at birth than those of full-term infants.

13.5.2 Sensory and motor experience

Historically, the two most common manipulations to examine effects of sensory and motor experience have been either placing animals in severely impoverished environments or so-called enriched environments (see Chapter 3: Overview of Factors Affecting Brain Development). As noted earlier, these experiences produce a wide range of plastic changes in the brain, and especially in sensorimotor regions. In our own studies on the effects of raising weanling rats for 60 days in complex environments we routinely see increases in brain weight (but not body weight) in the order of 7%–10%. The increased brain weight reflects increases in the number of glia and blood vessels, neuron soma size, dendritic elements, and synapses. It would be difficult to estimate the total number of increased synapses but it is probably on the order of 20% in the cortex, which is an extraordinary change.

It has only been more recently that investigators have given animals enhanced exposure to specific experiences such as visual, tactile, or auditory experiences. Consider a few examples.

Prusky, Silver, Tschetter, Alam, and Douglas (2008) used a novel form of visual stimulation in which rats were placed in a virtual optokinetic system in which vertical lines of differing spatial frequency moved past the animal. If the eyes are oriented toward the vertical grate it is impossible for animals to avoid tracking the lines, much like the effect of a train going by as we wait at crossing. The authors placed animals in the apparatus for 10 days following the day of eye opening (postnatal day 15 in rats). When acuity was tested at 25 days there was a 25% enhancement. However, if testing stopped, by day 72 the enhanced acuity was lost, but the enhanced acuity was maintained if the testing continued into adolescence, even though there was no additional visual benefit. Although the authors did not identify exactly what the plastic changes were, they did demonstrate that they were in the visual cortex and hypothesized that it was due to synaptic changes in layer II/III.

There is a long history of interest in the extent of licking and grooming of rat pups by their mothers with associated effects on offspring behavior and gene expression in hippocampus (e.g., McGowan et al., 2011). Parallel effects are observed in humans as well. For example, Schanberg and Field (1987) showed that when premature infants were tactilely stimulated there was an accelerated growth and earlier discharge from hospital. Although the precise mechanism is unknown in the preterm infant studies, it has been shown that preterm tactile stimulation accelerates EEG maturation, as well as increasing serum levels of insulin growth factor I and growth hormone (Field et al., 2014). Direct tactile stimulation of infant rats also has widespread effects on a variety of sensory, motor, and cognitive functions later in life (e.g., Kolb & Gibb, 2010).

13.5.3 Pre- and postnatal stress

There is a large literature showing that stress effects the brain throughout life, with significant effects on the structure and operation of the hippocampus, prefrontal cortex, and the amygdala (McEwen & Morrison, 2013; Tomason & Marusak, 2017). Functionally, chronic stress in adulthood has been associated with a multitude of cognitive, social, and physical symptoms that include deficits in emotional regulation, impaired motor function, impaired executive function including short-term memory, diminished self-regulatory behavior, and immunological impairment (e.g., Cohen, Janicki-Deverts, & Miller, 2007; McEwen, 2008; Metz, 2007; Segerstrom & Miller, 2004). It has been proposed that many of the cognitive symptoms, especially those related to cognitive functioning, are directly linked to dysfunctional connectivity between regions of the HPC, mPFC, and OFC (McEwen, 2007).

Less is known about the effect of stress during development but prenatal stress is now known to be a risk factor in the development of human disorders such as drug addiction, and various psychiatric conditions (e.g., Anda et al., 2006; van den Bergh & Marcoen, 2004). Indeed, evidence has accumulated to show that the epigenome, through which environment regulates gene expression, is responsible for long-lasting effects of gestational stress on behavior and brain (see review by Bock, Wainstock, Braun, & Segal, 2015). Experimental studies with laboratory animals have confirmed these findings, with the overall result being that perinatal stress in rodents produces a range of behavioral abnormalities including an elevated and prolonged stress response, impaired learning and memory, deficits in attention, altered exploratory behavior, altered social and play behavior, and increased preference for alcohol (Weinstock, 2008).

There are several studies looking at changes in the organization of brain regions from early stress in rodents, with the general finding being that both gestational and infant stress decreases overall brain, but not body, weight, and decreases spine density in both medial and orbital prefrontal regions, and these results are correlated with changes in gene expression in these regions (see review by Kolb, Whishaw, & Teskey, 2016).

It is only recently that studies of the brains of adults with early gestational or early childhood stressors have been conducted. For example, studies of adults with early institutionalization or abandonment show reduced prefrontal cortex volume, reduced prefrontal cortical thickness, and white matter disorganization (e.g., Eluvathingal et al., 2006; Hodel et al., 2015; McLaughlin et al., 2014), as well as reduced hippocampal and amygdala volumes (Hanson et al., 2010). In addition, prenatal stress is associated with reduced gray-matter volume in the left medial temporal cortex, and both amygdalae, but not hippocampus or prefrontal cortex (Favaro, Tenconi, Degortes, Manara, & Santonastaso, 2015). There are some differences in the details of the results of these studies, which are likely related to the type of stress, age of onset of stress, and the duration of stress.

In addition to changes in cerebral morphology, there is evidence that early life stress alters activation patterns to facial emotions in adult women with anxiety symptoms (Fonzo et al., 2016). The authors found greater amygdala engagement to both fear and anger that contrasts with decreased right dorsolateral prefrontal responses to those emotions. In addition, there was lower gray matter volumes in right prefrontal cortex.

Finally, it is worth noting that stressors do not have to be direct in order to have effects on the brains of offspring. Mychasiuk, Gibb, and Kolb (2011a, 2011d) generated a model of prenatal "bystander stress" by housing a pregnant mother with a female cage-mate who experienced daily chronic stress from E12 to E16. The authors then examined the effects on the offspring of the bystander stress on preweaning behavior, global methylation patterns, gene expression profiles, and cerebral morphology of the offspring. Similarly, in a parallel study pregnant mothers were housed with stressed males and behavior and cerebral morphology of the offspring was examined. The general finding was that all measures were affected in the offspring of the bystander stress dams but the effects were different, and in some cases qualitatively different, than the effects of direct gestational stress (see review by Kolb et al., 2017).

13.5.4 Preconception parental experience

A relatively new field of study is the examination of preconception parental experiences on brain and behavioral development in offspring. Because the parental environment is likely to be similar to that of their offspring, if adaptive traits gained by a parent could be transmitted to the offspring there might be increased survival. Somewhat surprisingly, paternal preconception experience may be more important, and certainly at least as important as maternal experience.

Paternal experience has been shown to alter the germline and mediate epigenetic changes through effects on spermatogenesis (Soubry, Hoyo, Jirtle, & Murphy, 2014). For example, paternal diet (high versus low calorie) is related to the risk of cardiovascular risk (with high calorie diet) and higher mortality (Bygren, Kaati, & Edvinsson, 2001). In fact, paternal obesity but not maternal obesity, appears to have a significant effect on body size and health of offspring. Similarly, paternal preconception alcohol consumption has been associated with various abnormalities in offspring (e.g., Passaro, Little, Savitz, & Noss, 1998), and a study in mice has shown that paternal alcohol exposure leaves an epigenetic imprint in offspring (Knezovich & Ramsay, 2012), as does paternal smoking, and exposure to various environmental toxins.

Soubry et al. (2014) hypothesize that there are four windows of susceptibility to paternal epigenetic influences. The first is paternal embryonic development. During embryonic development primordial stem cells undergo genome-wide epigenetic erasure as they migrate to form the genital ridge, which is the precursor of the gonads. However, some portions of the genome are reported to be resistant to DNA methylation erasure, providing a mechanism for transfer to later generations. The second and third are paternal prepuberty and spermatogenesis. Following epigenetic erasure DNA methylation is gradually reestablished postnatally, including the differentiation of spermatogonium, which are undifferentiated male germ cells, which are later related to the generation of sperm. Finally, the fourth window occurs around conception and in the zygote stage of the offspring. There is a selective retention of some histones in the sperm DNA. These histones are believed to be one mechanism for carrying epigenetic information to the next generation.

There are few studies of the effects of paternal experiences on behavior and brain development but a series of studies by Gibb and her colleagues is showing significant effects. For example, Mychasiuk et al. (2013) placed male rats in complex environments prior to mating males with females. The offspring showed a significant decrease in gene methylation in both the hippocampus and prefrontal cortex. These gene expression changes were virtually identical to those observed in the offspring of females who were housed in similar complex environments while pregnant. Parallel studies have shown that preconception paternal stress also alters gene expression. In these studies, male rats were stressed for 30 minutes, twice per day, for 27 consecutive days, and then mated with control female rats. This experience is associated with higher anxiety and skilled motor deficits in offspring as well as extensive differences in neuronal morphology in prefrontal cortex and hippocampus (Harker et al., 2015, 2016). Finally, a study by Hehar, Yu, Ma, and Mychasiuk (2016) showed that both paternal age and diet contribute to behavioral abilities and gene expression in offspring, and in addition, influenced recovery from concussion in offspring.

Less is known about the effects of preconception maternal effects on the brain of offspring. Bock et al. (2015a) gave females moderate stress for a week and then either mated them the following day or 15 days later. The brains of the offspring were collected in late adolescence (P65). Analysis of the dendritic complexity and spine density in AC, PL, and OFC found no effect in the animals whose dams were mated immediately after the stress whereas those mated 2 weeks later showed increased dendritic complexity and spine density in AC and PL but not in OFC. As in the males experiencing gestational stress the effects in males were larger in the left hemisphere. The authors speculated that this resulted from either a change in the maternal behavior or some type of epigenetic change in the dams or the offspring (see also Bock et al., 2015).

13.5.5 Parent–child relationships

In 1928, Watson published a book, *Psychological Care of Infant and Child*, which became very influential, selling over 100,000 copies in the first few months (Blum, 2002). A key premise in his book was that children should be treated as small adults and that too much mothering was dangerous; too much affection undermined the development of strong character and led to the development of weak and anxious adults. Although today it is easy to wonder why this perspective became so influential, it was widely accepted and admired. Even the US government supported the ideas and warned parents of the dangers of too much rocking and playing with their children (for an interesting review of this story see Blum, 2002).

It was against this background that Harry Harlow's studies of mother-infant bonds in rhesus macaques first appeared in the 1950s. Rather than finding that the infant-mother bond was largely a means for the infant to obtain food and drink, Harlow highlighted the importance of love (Harlow, 1959). Research over the past six decades since the early Harlow rearing studies has provided a remarkable demonstration that his rhesus monkey model of early childhood adversity (nursery-rearing) produces a wide array of behavioral, physiological, and neurobiological deficits that parallel those identified in human studies of early adversity (Bennett, 2010; Harlow & Harlow, 1965; Kaufman, Plotsky, Nemeroff, & Charney, 2000; Kraemer, 1992; Sackett, 1984; Suomi, 1997, 2002). Provencal et al. (2012) examined whether maternal- versus peer-rearing had differential effects on DNA methylation in early adulthood in the dorsolateral prefrontal cortex or T cells. The results revealed differential DNA methylation in both prefrontal cortex and T cells thus supporting the hypothesis that the response to early-life adversity is system-wide and genome-wide and persists to adulthood.

It is now clear that young mammals are dependent on their parents and must quickly learn to identify and prefer their caregivers. There is considerable variability in the patterns of parent–child relationships and the different patterns can initiate long-term developmental effects that last a lifetime (Myers, Brunelli, Squire, Shindledecker, & Hofer, 1989). Besides the monkey studies, there are two other lines of work showing the profound effects of variations in parent–child relationships. First, rodent studies have shown that the amount of maternal contact, including licking and grooming of infants, correlates with a range of somatic, behavioral, and neurological differences. For example, Meaney and his colleagues (e.g., Cameron et al., 2005) have been able to show that maternal–infant interactions in rats systematically modify the development of hypothalamic–adrenal stress response and a range of cognitive and emotional behaviors in adulthood. These changes are correlated with changes in gene expression that control hippocampal cell membrane testosterone receptors. Other studies have shown changes related to maternal–infant interactions in the mPFC, orbital frontal cortex, hypothalamus, and amygdala (Fenoglio, Chen, & Baram, 2006; Muhammad & Kolb, 2011).

Although most rodents are not biparental some species are, and variations in biparental–infant interactions also alter brain development. We noted earlier that the offspring of meadow vole parents with high levels of contact had a different pattern of connections within the somatosensory cortex as well as more distal connections with frontal and parietal cortex that those of offspring experiencing low levels of contact (Seelke et al., 2016). Similarly, Helmeke et al. (2009) showed that paternal deprivation in the biparental degu reduced dendritic development and spine density in OFC.

The second extensive line of work comes from studies of children who have been raised in institutions or abandoned. For example, when the Communist government in Romania fell, it was discovered that the state had sponsored large orphanages to house the children resulting from a misguided plan to increase productivity by increasing the number of workers through high birth rates. Many of these children were adopted out to loving families in western countries such as the United Kingdom, United States, Canada, and Australia with the expectation that the effects of the early deprived experiences could be reversed. Unfortunately, this was only true in children adopted out before 12–18 months. After this age, the children remained severely scarred and over 25 years later they still have significant cognitive and emotional deficits including an IQ drop of about 15 or more points, smaller brains, abnormal brain electrical activity, and a host of serious chronic cognitive and social deficits that do not appear to be easily reversed (e.g., Johnson et al., 2010; Lawler, Hostinar, Milner, & Gunnar, 2014; Rutter & O'Connor, 2004).

Hodel et al. (2015) studied the long-term neural correlates of early adverse rearing environments in a large sample of 12–14-year-old children who were adopted from institutional care at a mean age of 12 mo. The adopted children had reduced grey matter volumes and especially prefrontal volume, which was congruent with previous studies showing reduced prefrontal cortical thickness (e.g., McLaughlin et al., 2014). These effects were striking because the children were adopted quite young and had an average of 12 years in their adoptive families at the time of the study. The failure to detect a difference in structural brain development related to age at adoption given that longer duration of institutional care has been associated with larger physical, cognitive, and psychosocial deficits (e.g., Nelson, 2007; Rutter & O'Connor, 2004).

13.5.6 Peer relations

Peer relationships, and especially play, are a characteristic of all mammals and contribute to socioemotional temperament and cognitive functions through a form of problem solving. Harlow's studies on peer relationships in rhesus monkeys paralleled his studies of parent–infant interactions (e.g., Harlow & Harlow, 1965). In nonhuman primates affiliative behaviors, such as proximity, contact, grooming, and play, constitute a key component of social living (e.g., Weinstein & Capitanio, 2008). For example, young monkeys raised with their mothers, but without peers often do not socialize well and are fearful and/or inappropriately aggressive. The development of affiliative relationships (i.e., friendships) contributes to the stability of social groups. Although peer relationships must influence brain development, most studies have focused on the role of play behavior in brain development.

The prefrontal cortex plays a central role in play behavior and its development is strongly influenced by play. For example, the complexity of neurons in the mPFC of rats and hamsters is related to amount of play behavior through enhanced pruning (Bell, Pellis, & Kolb, 2010; Burleson et al., 2016) whereas the number of conspecifics interacted with during development, and not play per se, influences the pruning of the OFC (Bell et al., 2010). The chronic effects of play are not just related to social behavior as the manipulation of juvenile play behavior changes the brain's response to psychomotor stimulants (Himmler, Pellis, & Kolb, 2013). Furthermore, early experiences including gestational stress, tactile stimulation, and psychoactive drugs all alter play behavior and prefrontal cortex (e.g., Muhammad et al., 2011; Raza et al., 2015). Finally, perinatal injury to the prefrontal cortex compromises play behavior (e.g., Pellis et al., 2006).

13.5.7 Psychoactive drugs

Exposure to ethanol during gestation can have profound effects on neurobiological and behavioral development in children (e.g., Kodituwakku, 2007), leading to a range of disorders that can be grouped as Fetal Alcohol Spectrum Disorders that range from Fetal Alcohol Syndrome (FAS) characterized by severe morphological and neurological symptoms to less severe syndromes such as Partial FAS, Alcohol Related Neurodevelopmental Disorder, and Alcohol Related Birth Defects (see Hoyme et al., 2005). The incidence of FAS is relatively low (0.1% Abel, 1995) but because moderate ethanol consumption occurs in about 8.5% of pregnant women in the United States (Substance Abuse and Mental Health Services Administration, 2013), a much larger number of children are likely to affected than commonly believed (Centers for Disease Control and Prevention Report, 2007). But alcohol is only one of many psychoactive drugs commonly ingested by pregnant women. Estimates on the incidence of illicit drug use during pregnancy are about 5% and cigarette use is about 20%, but the actual incidence of infants exposed gestationally to psychoactive drugs may be higher because pregnant women tend to underreport substance use (Konijnenberg, 2015). And then, there are prescription drugs such as antidepressants, anxiolytics, and antipsychotics, as well as over the counter medications. It is difficult to get an exact number of women taking each type of medication, in part because of a high rate of women obtaining prescription drugs from friends, neighbors, and relatives (Mitchell et al., 2011). Nonetheless, Mitchell et al. report data showing that between 2006 and 2008 50% of women took four or more medications at any time during pregnancy. About 8% of the women reported taking antidepressants (mostly selective serotonin uptake inhibitors or SSRIs) during the first trimester.

Because nearly all psychoactive drugs enter the circulation of the developing fetus, maternal intake of these drugs can potentially impact the developmental brain events occurring at that time. The precise effects will vary with the drug, dose, timing during gestation, method of administration, and the duration of exposure (e.g., Thompson, Levitt, & Stanwood, 2009).

There is a very large literature on the effects of gestational exposure to psychoactive drugs on behavior, and especially nicotine, but this is beyond the scope of the current discussion. The most studied drugs are alcohol, nicotine, cocaine, amphetamine, and antidepressants. In general, exposure to each drug has effects on the neurochemical systems that they are designed to affect to alter behavior (see Thompson et al., 2009). To date there only a few MRI studies looking at human brains. For example, Lotfipour et al. (2009) assessed the likelihood of substance abuse in adolescents expose to maternal cigarette smoking ($n = 597$) and compared them to a matched cohort ($n = 314$) who were not. Among the exposed adolescents, the likelihood of drug experimentation correlated with the degree of thinning of the OFC. It thus appears that gestational exposure to cigarettes decreases OFC thickness and, in turn, increases the likelihood of drug use. Curiously, among the nonexposed participants drug experimentation increased OFC thickness, an effect moderated by a BDNF genotype (Val66Met). In a follow-up paper, Lotfipour et al. (2010) found that maternal smoking interacts with a polymorphism in the a6 nicotinic receptor gene in predicting adolescent smoking and substance use. Because the a6 receptor is highly expressed in the striatum, the authors evaluated differences in this structure, corrected for age, sex, and brain size, and found it larger in those with the polymorphism.

Exposure to psychoactive drugs in adult rats is associated with persistent changes in neuronal structure in mPFC, OFC, and nucleus accumbens (Robinson & Kolb, 2004). There is now considerable evidence that prenatal exposure of a wide range of psychoactive drugs including alcohol, nicotine, diazepam, and fluoxetine chronically alters both neuronal structure and cognitive and motor behaviors. For example, Kolb et al. (2008) showed that prenatal diazepam increased dendritic length in cortical pyramidal cells whereas prenatal fluoxetine had the opposite effect. Similarly, Hamilton et al. (2010) found decreases in spine density and dendritic length in OFC but an increase in dendritic length in mPFC in adults with prenatal alcohol exposure, and in later papers the same group showed reduced dendritic length and branching in nucleus accumbens of similar rats (Rice et al., 2012).

As might be expected from human MRI analyses of the effects of gestational exposure to smoking, animal studies have shown extensive changes in dendritic structure in prefrontal cortex at weaning and in adulthood (Muhammad et al., 2012; Mychasiuk, Harker, Ilnytskyy, & Gibb, 2013a). Mychasiuk et al. (2013b) also found that prenatal exposure to nicotine altered, and in some measures prevented, later effects of complex housing, suggesting that there was a change in brain plasticity in adult rats with prenatal nicotine exposure. This result parallels studies of adult exposure to nicotine, cocaine, or amphetamine, which all block later effects of complex housing (Hamilton & Kolb, 2005; Kolb et al., 2003a, 2003b). It would be interesting to determine if this is also true for prenatal exposure to these, or other, psychoactive drugs. This would have important implications for school performance in children with this type of exposure.

13.5.8 Diet

It has been recognized for at least 50 years that the long-term consequences of malnutrition during infancy is reduced brain size, thinner cerebral cortex, reduced myelination, diminished number of neurons, and reduced dendritic arborization and spine density (e.g., Dobbing, Hopewell, & Lynch, 1971; Winnick, 1969). Less is known, however, about the effect of optimal diets on brain and behavioral development.

Breast milk is thought to be an optimal source of nutrition for the human infant and several studies link breast milk to increased cognitive development, especially in preterm infants (e.g., Anderson, Johnstone, & Remley, 1999). Lucas and colleagues undertook a prospective study of the offspring of mothers who elected to breastfeed but varied in their success in expressing milk and therefore required supplementary feeding (e.g., Isaacs et al., 2010). The authors randomized infants to receive one of three supplements: a nutrient enriched preterm formula, a standard term formula used in the 1980s for feeding preterm infants, or banked breast milk provided by unrelated donors. MRI brain scans were obtained when the children were adolescents (mean age of 16 years). The studies have shown that there is a relationship between the dose of early breast milk intake and later IQ and whole brain volume in adolescence. Similarly, Kafouri et al. (2013) found that continuous exclusive breastfeeding is associated with thicker parietal cortex and improved cognitive performance in typically developing adolescents. The authors of both studies hypothesize that one or more constituents of mother's breast milk promote brain development.

Studies in laboratory animals do suggest that supplementation of milk with factors such as docosahexaenoic acid (DHA), lactoferrin, milk fat globule protein (MFGP), and prebiotics enhance brain development. Lactoferrin exerts antimicrobial effects and modulates the immune response, and piglets supplemented with lactoferrin show a variety changes in the hippocampus, including increased levels of BDNF (Chen et al., 2014). Mudd et al. (2016) fed piglets either a standard diet or one that was supplemented with dietary prebiotics, MFGP, and lactoferrin and found evidence of enhanced neural development and thus recommended supplementation of human infant formula with similar additives.

But other dietary factors, such as nutrients, influence brain development too. The role of nutrients has mostly been studied by examining the effects of nutrient deficiencies such as iron, copper, and choline (see review by Georgieff, 2007). Nutritional deficiencies can have either global or specific effects on brain development, depending upon the precise timing of the nutrient deficit. There has been more recent interest in examining the role of added nutrients in enhancing brain development, such as choline supplementation (e.g., Meck & Williams, 2003). Although individual supplements such as choline may provide some benefit, it seems more likely that supplementing with a combination of nutrients should work synergistically for an optimal outcome (e.g., Leung, Wiens & Kaplan, 2011; Rucklidge and Kaplan, 2013). Indeed, in the last 15 years many studies have demonstrated that nutrient supplementation for school-aged children can improve ADHD, bipolar disorder, and anxiety (e.g., Frazier, Fristad & Arnold, 2012; Harding, Judah, & Gant, 2003; Kaplan, Fisher, Crawford, Field, & Kolb, 2004).

One challenge for the nutrient supplements studies is the question of what the mechanism might be. Animal studies have shown that a combination of DHA and uridine increases the complexity of neurons in the hippocampus (Sakamoto, Cansev & Wurtman, 2007) and Halliwell (2011) showed that the supplement shown to be effective in children with ADHD (see above) increased dendritic complexity in cortex and enhanced recovery from perinatal cortical injury. It is unclear what mediated the synaptic changes in these studies, however. One possibility is that the behavior of the animals changed, which in turn changed the brain, whereas another is that there were epigenetic changes. We know of no evidence that postnatal dietary factors change gene expression but Dominguez-Salas et al. (2014) showed that maternal diet at conception significantly altered gene methylation in newborns. The researchers studied infants in rural Gambia that had been conceived in either the dry season or the rainy season. Gambians' diets are dramatically different during these two seasons and so was gene methylation in the infants' blood.

13.5.9 Gonadal hormones

The most obvious effect of gonadal hormones during development is the differentiation of the genitals into a female and male phenotype. But the brain has the same gonadal hormone receptors so it should not be surprising that there is significant sexual differentiation in the brain and thus behavior too. Although there appears to be some controversy over how big sex differences really are, there is considerable evidence in humans and laboratory animals for significant behavioral and anatomical differences (for a more detailed discussion of human data, see Kolb & Whishaw, 2015).

The issue here, however, is whether there are sex differences in brain plasticity in response to experiences during development. There are. As summarized in Chapter 10, Language and Cognition, gonadal hormones act to organize (differentiate) neural tissue in the direction of masculinization or feminization. In adulthood, gonadal hormones activate neural tissue that was organized early in development (e.g., Arnold, 2009). Studies in rats have shown that neurons in the mPFC have larger dendritic fields in males whereas the opposite is true in OFC, and the difference is hormone dependent. Neonatal gonadectomy abolishes the difference (Kolb & Stewart, 1991). Work by Juraska and colleagues (Drzewiekcki, Willing, & Juraska, 2016; Juraska & Willing, 2017; Willing & Juraska, 2015) has shown that the sex difference emerges during adolescence as females lose synapses and neurons in mPFC so it appears that the hormones are acting on tissue organized early in development. Males also lose synapses and neurons but it is not as closely tied to puberty onset.

Juraska (1984) initially showed that the effect of complex housing from weaning until P55 was larger on cortical neurons in males than females. Gibb, Gorny, and Kolb (2017) housed animals in complex environments from weaning until P120 showed a similar finding in both parietal and occipital cortex, and in fact there were only very small effects in the occipital cortex of females (Fig. 13.5). In contrast, Juraska (1990) found larger effects on neurons in the hippocampus in females than in males.

There are also a sex differences in the effect of developmental stress. Overall males are more affected by developmental stress than females. Both gestational stress and maternal separation have been shown to reduce adult brain weight and cortical thickness in males, but not females (Muhammad & Kolb, 2011a; Muhammad et al., 2011b) and a different pattern of dendritic changes in hippocampus (Bock, Murmu, Biala, Weinstock, & Braun, 2011). In a novel stress paradigm, which they called Bystander Stress, Mychasiuk et al. (2011a) showed that pairing pregnant females with other pregnant females that were given gestational stress altered prefrontal cortex and hippocampus of the offspring in a sexually dimorphic manner. First, female offspring showed an increase in neuronal cell numbers in mPFC, whereas males showed reductions. The authors also generated a gross estimation of synapse quantity by combining data collected for dendritic length, spine density, and neuronal number, finding an effect of the stress but the pattern was different in the two sexes. In addition, the authors found a sex difference in the effect of bystander stress on gene expression: there were significant changes in about 300 genes in each of prefrontal cortex and hippocampus but there was virtually no overlap (6 genes) in males and females (Mychasiuk et al., 2011a, 2011d).

Finally, there are sex differences in the effects of perinatal mPFC lesions in rats with males consistently showing better functional outcome and greater compensatory changes in perilesional cortical neurons (e.g., Kolb & Stewart, 1995).

13.5.10 Gut bacteria

The enteric nervous system, which functions largely independently to control digestion, interacts with gut bacteria, known collectively as the microbiome (see Chapter 3: Overview of Factors Affecting Brain Development for more details). Therapeutic modifications of the microbiome can be achieved by taking certain bacteria either orally or via rectal implant (Smythies & Smythies, 2014). Dinan, Stanton and Cryan (2013) proposed given that the microbiome interacts with the brain, the use of bacteria to alter the microbiome could be a novel class of psychotropic, which they called psychobiotics. Evidence is accumulating that psychobiotics can alter brain development. Diaz Heijtz et al. (2011) manipulated gut bacteria in newborn mice and

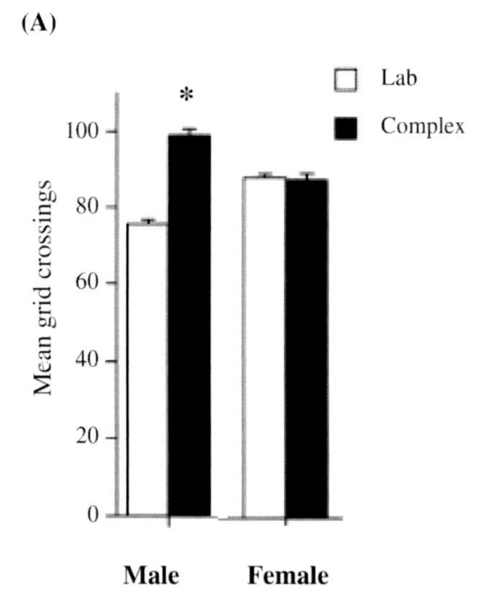

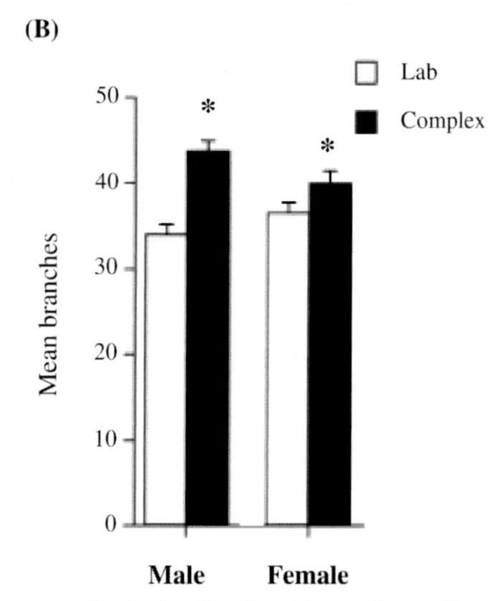

Figure 13.5 Summary of changes in the basilar dendrites of layer III pyramidal neurons in occipital cortex of rats placed in complex environments for 90 days. Whereas males show large changes in dendritic length and branching, females show no effect in length and only a small change in branching. (A) Basilar length, (B) basilar branches. *From Gibb, R., Gorny, G., and Kolb, B. (2017).*

found that gut bacteria influence motor and anxiety-like behaviors, which were associated with changes in the production of synaptic-related proteins in cortex and striatum. Similarly, Hoban et al. (2016) exposed germ free mice, which are devoid of any microbiota throughout maturation, until weaning and then exposed half of them to mice housed in a standard animal unit, thus allowing the germ-free mice to be exposed to environmental microbes, resulting in microbiota colonization of the gut. The germ-free animals had increased gene expression related to myelination, which was associated with hypermyelination of axons in the prefrontal cortex of the germ-free mice. Gut colonization of the germ-free mice reduced the gene expression and myelination in prefrontal cortex. (See a parallel study by the same authors on amygdala development, Stilling et al., 2015.) Although these latter studies did not measure anxiety, there is a growing body of evidence that germ-free animals are anxious and have impaired social behavior, as Diaz Heijtz et al. showed.

One final study by Callaghan, Cowan, and Richardson (2016) is provocative. These authors exposed neonatal male rats to maternal separation (a significant stressor) and then used two affective-learning paradigms in offspring of the stressed males. There was a paternal stress effect, which was reversed by giving the stressed animals, or their offspring, probiotics via drinking water for the dam. Given that paternal stress alters the gene expression and dendritic complexity in prefrontal cortex in paternally stressed rats (see above), it would be interesting to determine if the probiotics acted by changing these.

Taken together, these studies suggest that modifications of gut microbiota during development may contribute to a range of neurodevelopmental disorders and to their treatment.

13.6 BRAIN PLASTICITY AFTER EARLY BRAIN INJURY

In the late 19th century, it was widely believed that there was an advantage to having brain injury in the young brain, in part because children are very unlikely to suffer aphasia following perinatal brain injury. Studies using young monkeys in the 1930s and 1940s reinforced this view as Kennard (1942) reported that damage to the motor cortex in young monkeys had less severe outcomes than similar injuries in older animals and she proposed that there were changes in connectivity in the young injured brain that could support better outcomes. In his studies of children with frontal lobe injuries Hebb (1949) reached a different conclusion as his patients appeared to have more severe symptoms than adults with similar injuries. Hebb suggested that cognitive development is dependent on the early development of cognitive structures that are essential for later learning. Over the ensuing decades experimental support has gathered appearing to support both positions—at least under certain conditions (Kolb, 1995). Greenham et al. (Chapter 15) describe this difference in outcomes as "plasticity versus vulnerability." In their view, the plasticity argument holds that the young brain is thought to be more plastic because it is less rigidly specialized whereas the vulnerability view notes that the brain will be uniquely sensitive to insult, in part because ongoing developmental processes build on those that have come before.

But several factors will contribute to the plasticity versus vulnerability outcome. The outcomes depend upon the precise age at injury, the mechanisms of plasticity that are stimulated, the behavioral measurements used, the type of injury, the extent and location of injury, and metaplastic factors such as prenatal and preconception experiences. Our focus here will be on plastic changes in the brain and behavior in laboratory animals as Chapter 15 reviews the effects of brain injuries in children.

13.6.1 Age at injury

As noted earlier, age refers not to the actual postnatal age of animals but to their developmental age. In our studies of rats with cortical focal mPFC lesions (excisions) over the developmental lifespan we discovered that whereas damage in the first few days of life was associated with a miserable outcome and significant disruption of normal brain development, damage in the second week of life has a much better outcome, even though it is significantly smaller than adults with similar injury (e.g., Kolb, 1987). Damage to the juvenile brain allows better recovery than damage later in life but not as good as in the second week of life, even though the brain appears more normal, at least in size. I am only aware of one study of prenatal cortical lesions in rats on E18, which showed highly abnormal brain development yet remarkably normal behavior (Kolb, Cioe, & Muirhead, 1998). Hicks and D'Amato (1961) used ionizing radiation of fetal rat brains and also showed fairly good outcomes in spite of highly abnormal brains, at least if there was no hydrocephalus.

Injury in adolescence has largely been neglected in laboratory animal studies and the two studies I am aware of have contradictory results. Nemati and Kolb (2012) made bilateral medial prefrontal lesions on P35 and P55 and found that whereas the P35 rats were nearly normal at a variety of behavioral tests, the P55 rats were severely impaired. In contrast when Nemati and Kolb (2010) did a similar study in rats with unilateral motor cortex lesions, they found the reverse pattern, namely that the P35 animals were impaired and the P55 animals were not. Unfortunately, these studies are hard to compare because the mPFC lesions were bilateral and the motor cortex lesions were unilateral. Thus, we do not know if the rules are different for adolescent lesions in the two regions or if it is the unilateral versus bilateral difference that is important. My hunch is that it is the latter. We have shown in parallel studies of animals with hemidecortications that the role of age appears to be different (see 6.5 below).

In sum, if the developing rodent brain is injured during neurogenesis (in utero), the functional outcome is good whereas during neuronal migration and the beginning of synaptogenesis (P1–5), it is poor. Injury during peak synaptogenesis (P7-P25) allows much better outcomes. This pattern of outcomes is similar after lesions in other cortical regions including OFC, motor cortex, auditory cortex, posterior parietal cortex, posterior cingulate cortex, and visual cortex. The degree of plasticity and recovery is not as extensive after posterior parietal, visual, and auditory cortex lesions as it is after prefrontal or posterior cingulate lesions, however.

Studies of monkeys and cats with early lesions show a similar pattern of results when developmental stage is used as the time of injury. Cats are embryologically older than rats at birth, making their birthdate more similar to P7–10 rats. Cats with injuries shortly after birth have good recovery, much like P7–10 rats, whereas those with prenatal lesions (like P1–5 rats) have severe impairments (Villablanca, Hovda, Jackson, & Infante, 1993). In contrast, monkeys are more embryologically developed than cats (and humans) at birth so we would expect a somewhat different timeline. Although Kennard reported better outcomes with early motor cortex lesions, the bulk of the evidence from Goldman (see reviews by Goldman, 1974; Goldman, Isseroff, Schwrtz, & Bugbee, 1983) showed significant behavioral deficits with postnatal lesions whereas prenatal lesions showed significant recovery, much like newborn cats and P7–10 rats. Taken together, the data show that if the developing brain is injured during the original round of neurogenesis it can compensate, likely by prolonging neurogenesis, whereas damage during migration does not end well resulting in significant behavioral impairments.

One complication to the story arises from the recent findings by Herculano-Houzel and her colleagues discussed in Section 13.4.6. Recall that they showed periods of significant postnatal neurogenesis in rats, interspersed with periods of neuron die off. No studies have yet timed brain injuries with these postnatal periods of neurogenesis, nor have they looked for neurogenesis by labeling dividing cells postinjury. An exception would be our studies that have shown neurogenesis after mPFC and posterior cingulate lesions at P7-P10, which would coincide with the end of the first period of postnatal neurogenesis, but no studies have targeted specifically the later periods of neurogenesis or pruning. One problem is that it is not clear where the new postnatal neurons are coming from in Herculano-Houzel's studies. All of the prenatal neurons arise from the subventricular zone (SVZ) but this may not be true of the later neurogenesis—they could be the result of either neural precursor cells located in the cerebrum, or of astrocytes changing their phenotype to become neurons. The first two scenarios have some data support and although published evidence of mitosis of neurons is absent we have observed it in semithin sections so it is at least plausible.

13.6.2 Mechanisms of plasticity after early injury

The primary mechanisms of postinjury plasticity in the developing brain are the appearance of anomalous connections, failure of neuron pruning, dendritic growth and synaptogenesis, neurogenesis, and gliogenesis. Anomalous connections, which are connections that are not found in intact animals, have been shown to support functional recovery after motor cortex lesions and hemidecortications in rats and visual cortex lesions in cats (e.g., Kolb, Gibb, & van der Kooy, 1992; Payne & Lomber, 2001). But anomalous connections can also interfere with recovery when connections that normally would die off during development remain and appear to disrupt normal functions (Kolb, Gibb, & van der Kooy, 1994). Thus, anomalous connections could be new connections formed after an injury or the failure of existing connections, which would normally be pruned, to die off (see Kolb et al., 1994).

When cortical regions are damaged in adulthood the neurons supporting thalamocortical connections normally die but following early brain injury this sometimes does not occur. The dorsal medial-thalamus projects to the prefrontal cortex and in adults it degenerates after PFC lesions. In contrast, these neurons survive after prenatal PFC lesions in monkeys (Goldman & Galkin, 1978) or P10 PFC lesions in rats (Kolb & Nonneman, 1978). The remaining neurons appear to make connections with spared PFC regions, and presumably support recovery.

There is considerable evidence that functional recovery is correlated with dendritic growth and increased spine density (see Kolb, 1995; Kolb & Gibb, 2007; Nemati & Kolb, 2010). One example will illustrate. Kolb and Gibb (1993) made mPFC lesions on P1 or P10. When behavior was analyzed at P22–25 both groups had severe impairments relative to controls, whereas at P55–58, the P10 rats had recovered but the P1 rats had not. Analysis of dendritic complexity and spine density revealed that whereas both the lesion groups had stunted dendritic growth at P25, the P10 group showed increased spine density at P60 but the P1 group did not. The behavioral recovery in the P10 group was presumed to be at least partly supported by increased dendritic spines, and thus synapses.

Finally, under some circumstances there is spontaneous neurogenesis following midline lesions (i.e., olfactory bulb, mPFC, posterior cingulate cortex). For example, Kolb et al. (1998) showed that after P10 mPFC lesions that do not damage the SVZ there is a spontaneous filling in of the lesion cavity by neurons generated in the SVZ, and these neurons establish at least some connections with cortical and subcortical regions and the new neurons have electrical activity that is nearly normal (Driscoll, Monfils, Flynn, Teskey, & Kolb, 2007). Such animals show surprisingly good, but not complete, functional recovery. If the regenerated tissue is removed in adulthood the behavioral recovery is lost and if the regeneration is blocked by injections of the mitotic marker, bromodeoxyuridene injected on E12-E14, there is no neural generation and no functional recovery (Kolb, Pedersen, & Gibb, 2012). But it is likely not just neurons that are generated postinjury. Faiz, Acrain, Castellano, and Gonzalez (2005) showed that in normal brains there is extensive glial generation that peaks at P12, and which is enhanced after a P9 neurotoxic lesion.

13.6.3 Behavioral measurements

Not all behavioral measures show the same extent of plasticity after early injury. As a rule of thumb, there is better recovery of cognitive functions than motor functions, and rather poor recovery of socio-emotional functions, a result seen after lesions in both laboratory animals and children (see Fig. 13.6) (Kolb, 1995).

One important consideration is that it is important to do a broad battery of behavioral tests because symptoms after early brain injury can be quite unexpected. Rats with P1–P5 cortical lesions in motor or auditory cortex show spatial deficits not seen after similar injury in adults. Similarly, children with injuries in either cerebral hemisphere are impaired at spatial behaviors, a finding that would only be typical of adults with right hemisphere lesions (Akshoomoff, Feroleto, Doyle, & Stiles, 2002). Kolb and Whishaw (1989) report a case of an 18-year-old girl with a congenital posterior parietal cyst who showed no deficits on tests of parietal lobe function but rather she was impaired on tests of frontal lobe function. Had only tests of parietal lobe function been administered her impairments would have been missed.

13.6.4 Age at assessment

It is difficult to know when to assess the behavior children with brain injuries because the brain is still growing and behavioral plasticity may take a long time to stabilize. Goldman (1974) began studying the effects of frontal lobe injuries in infant monkeys when they were still young and she was impressed by the apparent recovery, much as Kennard had been. But she continued to study the monkeys as they developed and discovered that she had overestimated the amount of recovery. The normal monkeys continued to improve but the frontal-injured monkeys did not so that in a sense, they

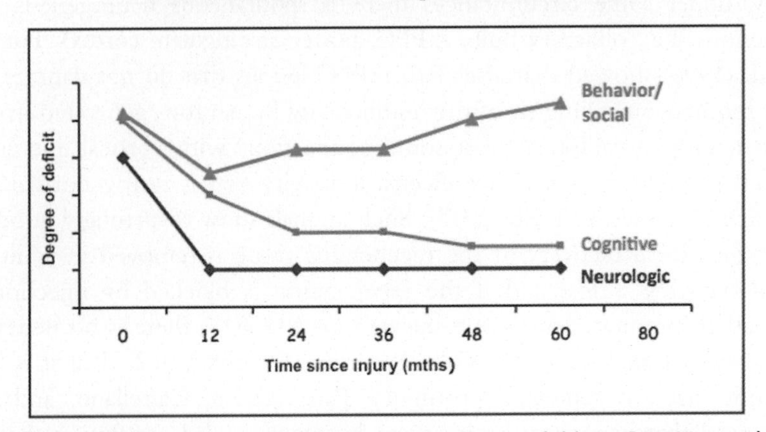

Figure 13.6 Differential recovery from traumatic brain injury in children. Cognitive and neurologic functions show much better recovery than social functions. *Graph courtesy of Vicki Anderson.*

"grew into their deficits" (Goldman et al., 1983). Similar observations have been seen in cognitive development of children with congenital brain injuries (Banich, Cohen-Levine, Kim, & Huttenlocher, 1990) and motor development in infants with hemiplegic cerebral palsy (Eyre, 2007).

13.6.5 Type of injury

Most of the laboratory animal studies have made fairly focal lesions by suction removal of tissue or by neurotoxic injections. The results of these studies can be compared to the effects of focal surgical removals of tissue in children but the surgical lesions rarely respect nice cortical boundaries as they do in the laboratory studies and are often quite large. In addition, childhood surgeries are being performed on neurologically abnormal brains so results may be different than the observations in lab animals.

The most common type of naturally occurring injury in developing children is traumatic head injuries (TBI), which are a very different in etiology than surgical injuries, especially because of the high probability of diffuse white matter damage (see Chapter 19). There are few lab animal studies of TBI in the developing brain but a model developed by Mychasiuk and her colleagues shows promise. They have shown, for example, that rats with TBI on P30 show little obvious brain damage but there are behavioral sequelae similar to those observed in children (Mychasiuk, Hehar, Ma, Kolb, & Esser, 2015a). In addition, they have shown that the relatively mild TBI in P30 rats has lasting changes in gene expression, dendritic morphology, and synaptic connectivity in the prefrontal cortex, apparently disrupting normal pruning processes at this point in development (Mychasiuk, Hehar, & Esser, 2015b). What is now needed is systematic study using this model and animals of different developmental ages to compare to children with TBIs.

13.6.6 Unilateral versus bilateral injury

As a general rule of thumb, bilateral injuries have more severe behavioral consequences than unilateral injuries. One reason is that in the case of unilateral injury the homologous region in the contralateral hemisphere is intact and can contribute to recovery. The most obvious example is seen in recovery from perinatal injuries to the speech zones of the left hemisphere. If the contralateral hemisphere is intact language can shift to the opposite hemisphere (e.g., Rasmussen & Milner, 1977) and although language is not normal, there is no aphasia and children do quite well. In contrast, if there is bilateral damage, including injury to a speech zone on the left but elsewhere on the right, language does not shift and the child is aphasic (Vargha-Kadem, & Watters, 1985).

Another reason for poorer recovery from bilateral injury is that the postinjury plasticity is different. Thus, for example, anomalous connections to the brainstem and spinal cord may be maintained from the normal hemisphere to support motor functions of the damaged hemisphere in unilateral injury. The critical window for optimal recovery may also be different after unilateral and bilateral injury as mentioned earlier (see Section 13.6.1). In fact, whereas rats with bilateral focal injuries show excellent recovery relative to earlier injuries, rats with hemidecortications on P10 fare worse, and have less extensive anatomical rewiring, than rats with earlier lesions (e.g., Kolb & Tomie, 1988; Kolb et al., 1992).

13.6.7 Metaplastic factors

As noted earlier, metaplasticity refers to the idea that plastic changes in the brain at one point will in life modify how the brain responds to later in life. The treatment of early brain injury provides a powerful example. We have seen that early brain injury alters brain development and this is dependent upon many factors. But a key question remains: how can we remediate the effects of early brain injury to improve behavior? There are two primary types of interventions that promise to have beneficial metaplastic effects on the injured developing brain: behavioral therapy, chemical therapy. We consider some examples.

13.6.7.1 Behavioral therapy

We have noted throughout the chapter that tactile stimulation has powerful effects on the developing brain. It also has an important role in remediating the effects of early brain injury. Kolb and Gibb (2010) gave rats with neonatal (e.g., P3) mPFC or posterior parietal injuries tactile stimulation for 15 minutes 3 × per day beginning the day after injury and continuing until weaning. When studied as adults the animals showed remarkable functional improvement on tests of both cognitive and motor behaviors, which was associated with increased brain weight and spine density in the tactilely stimulated rats. Although the details of the behavioral deficits and associated synaptic changes were somewhat different after mPFC and posterior parietal lesions, tactile stimulation improved behavior in both cases and was associated with lesion-specific effects on cortical morphology in adjacent cortex. Although the mechanism(s) of the action of the tactile stimulation is undoubtedly complex, Gibb, Kovalchuk and Kolb (2017) have shown that the tactile stimulation increases the production of Fibroblast Growth Factor-2 (FGF-2) in the skin and FGF-2 crosses the developing blood brain barrier to stimulate receptors in the cortex.

We have also noted that complex housing has large effects on cortical development and, like tactile stimulation, complex housing of rats with early brain injury stimulates significant functional improvement after neonatal cortical injury and this is associated with significant functional improvement on motor and cognitive behaviors, which is correlated with increased brain weight, cortical thickness, and dendritic length (e.g., Comeau, Gibb, Hastings, Cioe, & Kolb, 2008; Kolb & Elliott, 1987). Again, the mechanism is unknown but placing rats in complex environments also increases the production of FGF-2, as well as other growth factors (Kolb et al., 1998).

Not only does tactile stimulation stimulate recovery after neonatal mPFC lesions, gestational tactile stimulation enhances recovery from the same lesions, even though the therapy occurs before the brain injury (Gibb, 2004). Similarly, complex housing of the pregnant dam also has similar effects (Gibb, Gonzalez, & Kolb, 2014). These are remarkable outcomes and speak to the importance of gestational experiences on later brain plasticity.

13.6.7.2 Chemical therapy

In view of the correlation between recovery from early injury and the presence of increased FGF-2 it was logical to administer FGF-2 to see if that works too. It does. For example, Monfils et al. (2006) and Monfils, Driscoll, Vavrek, Kolb, and Fouad (2008) showed that subcutaneous FGF-2 administered to rats with P10 motor cortex lesions not only reversed all of the expected motor deficits but the lost cortex was actually regenerated and formed functional corticospinal connections. Similarly, administration of FGF-2 to rats with neonatal mPFC lesions stimulates functional recovery and increases cortical thickness (Comeau, Hastings, & Kolb, 2007; 2008). And, much like gestational tactile stimulation, gestational exposure to FGF-2 stimulates recovery and dendritic changes after neonatal lesions.

But there are negative metaplastic effects of gestation exposure to chemicals as well. Gestational exposure to fluoxetine, a commonly prescribed SSRI, blocks functional recovery and associated anatomical changes after P10 mPFC lesions (Kolb et al., 2008).

13.7 FUTURE DIRECTIONS

We have reviewed the types of brain plasticity, the ways to measure it, the potential mechanisms, and the factors that affect it in the normal and injured brain. Much of the work to date has been in laboratory animals but the explosion of work using non-invasive imaging in development promises to show us how the animal studies relate to the developing child. Understanding the mechanisms will remain largely an enterprise for laboratory animal studies, however, as such studies require more invasive procedures than is possible with human participants. One important challenge will be to understand how it will be possible to take advantage of metaplastic processes to remediate the effects of many negative early experiences, including brain injury, with the goal of normalizing behavior.

REFERENCES

Abraham, W. C., & Bear, M. F. (1996). Metaplasticity: The plasticity of synaptic plasticity. *Trends in Neuroscience, 19*, 126–130.

Akshoomoff, N., Feroleto, C., Doyle, R., & Stiles, J. (2002). The impact of early unilateral brain injury on perceptual organization and visual memory. *Neuropsychologia, 40*, 539–546.

Anda, R. F., Felitti, V. J., Bremmer, J. D., Walker, J. D., Whitfield, C., Perry, B. D., … Giles, W. H. (2006). The enduring effects of abuse and related adverse experiences in childhood. A convergence of evidence from neurobiology and epidemiology. *European Archives of Psychiatry and Clinical Neuroscience, 256*, 174–186.

Anderson, J. W., Johnstone, B. M., & Remley, D. T. (1999). Breast-feeding and cognitive development: a meta-analysis. *American Journal of Clinical Nutrition, 70*, 525–535.

Arnold, A. P. (2009). The organization-activational hypothesis as the foundation for a unified theory of sexual differentiation of all mammalian tissues. *Hormones and Behavior, 55*, 570–578.

Bandeira, F., Lent, R., & Herculano-Houzel, S. (2009). Changing numbers of neuronal and non-neuronal cells underlie postnatal brain growth in the rat. *Proceedings of the National Academy of Sciences of the United States of America, 106*, 14108–14113.

Banich, M., Cohen-Levine, S., Kim, H., & Huttenlocher, P. (1990). The effects of developmental factors on IQ in hemiplegic children. *Neuropsychologia, 28*, 35–47.

Bell, H., Pellis, S., & Kolb, B. (2010). Juvenile peer play experience and the development of the orbito-frontal and medial prefrontal cortices. *Behavioral Brain Research, 207*, 7–13.

Bennett, A. J. P. (2010). *Nonhuman Primate Research Contributions to Understanding Genetic and Environmental Influences on Phenotypic Outcomes Across Development.* New York: John Wiley & Sons, Inc.

Bennett, E. L., Kiamond, M. C., Krech, D., & Rosenzweig, M. R. (1964). Chemical and anatomical plasticity of the brain. *Science, 146*, 610–619.

Blakemore, S.-J. (2008). The social brain in adolescence. *Nature Reviews Neuroscience, 9*, 267–277.

Bliss, T., & Lomo, T. (1973). Long-lasting potentiation of synaptic transmission in the dentate area of the anaesthetized rabbit following stimulation of the perforant path. *Journal of Physiology, 232*, 331–356.

Blum, D. (2002). *Love at goon park.* Cambridge, MA: Perseus Books.

Bock, J., Murmu, M. S., Biala, Y., Weinstock, M., & Braun, K. (2011). Prenatal stress and neonatal handling induce sex-specific changes in dendritic complexity and dendritic spine density in hippocampal subregions of prepubertal rats. *Cellular and Molecular Neuroscience, 193*, 34–43.

Bock, J., Wainstock, T., Braun, K., & Segal, M. (2015). Stress in utero: prenatal programming of brain plasticity and cognition. *Biological Psychiatry, 78*, 315–326.

Burleson, C. A., Pedersen, R. W., Seddighi, S., DeBusk, L. E., Burghardt, G. M., & Cooper, M. A. (2016). Social play in juvenile hamsters alters dendritic morphology in the medial prefrontal cortex and attenuates effects of social stress in adulthood. *Behavioral Neuroscience, 130*, 437–447.

Bygren, L., Kaati, G., & Edvinsson, S. (2001). Longevity determined by paternal ancestors' nutrition during their slow growth period. *Acta Biotheoretica, 49*, 53–59.

Callaghan, B. L., Cowan, C. S. M., & Richardson, R. (2016). Treating generational stress: Effect of paternal stress on development of memory and extinction in offspring is reversed by probiotic treatment. *Psychological Science, 27*, 1171–1180.

Cameron, N., Champagne, F., Carine, P., Fish, E., Ozaki-Kurodoa, K., & Meaney, M. (2005). The programming of individual differences in defensive responses and reproductive strategies in a rat through variations in maternal care. *Neuroscience and Biobehavioral Reviews, 29*, 843–865.

Chen, U., Zheng, Z., Zhu, X., Shi, Y., Tian, D., Zhao, F., et al. (2014). Lactoferrin promotes early neurodevelopment and cognition in postnatal piglets by upregulating the BDNF signaling pathway and polysialylation. *Molecular Neurobiology, 52*, 256–269.

Cohen, S., Janicki-Deverts, D., & Miller, G. E. (2007). Psychological stress and disease. *Journal of the American Medical Association, 298*, 1685–1687.

Comeau, W., Gibb, R., Hastings, E., Cioe, J., & Kolb, B. (2008). Therapeutic effects of complex rearing or bFGF after perinatal frontal lesions. *Developmental Psychobiology, 50*, 134–146.

Comeau, W., Hastings, E., & Kolb, B. (2007). Pre- and postnatal FGF-2 both facilitate recovery and alter cortical morphology following early medial prefrontal cortical injury. *Behavioral Brain Research, 180*, 18–27.

Comeau, W., McDonald, R., & Kolb, B. (2010). Learning-induced structural changes in the prefrontal cortex. *Behavioural Brain Research, 214*, 91–101.

de Villers-Sidani, E., Chang, E. F., Bao, S., & Merzenich, M. M. (2007). Critical period window for spectral tuning defined in the primary auditory cortex (A1) in the rat. *Journal of Neuroscience, 27*, 180–189.

Diamond, M. C., Krech, D., & Rosenzweig, M. R. (1964). The effects of an enriched environment on the histology of the rat cerebral cortex. *Journal of Comparative Neurology, 123*, 111–120.

Diaz Heijtz, R., Wang, S., Anuar, F., Qian, Y., Bjorkholm, B., Samuelsson, A., . . . Pettersson, S. (2011). Normal gut microbiota modulates brain development and behavior. *Proceedings of the National Academy of Sciences of the United States of America, 108*, 3047–3052.

Dinan, T. G., Stanton, C., & Cryan, J. F. (2013). Psychobiotics: A novel class of psychotropic. *Biological Psychiatry, 74*, 720–726.

Dobbing, J., Hopewell, J., & Lynch, A. (1971). Vulnerability of developing brain. VII. Permanent deficit of neurons in cerebral and cerebellar cortex following early mild undernutrition. *Experimental Neurology, 32*, 439–447.

Doidge, N. (2007). *The brain that changes itself.* New York: Penguin.

Dominguez-Salas, P., Moore, S. E., Baker, M. S., Bergen, A. W., Cox, S. E., Dyer, R. A., . . . Laritsky, E., et al. (2014). *Maternal nutrition at conception modulates DNA methylation of human metastable epialleles, Nature Communications* (5, p. 3746). Available from http://dx.doi.org/10.1038/ncomms4746.

Driscoll, I., Monfils, H., Flynn, C., Teskey, G. C., & Kolb, B. (2007). Neurophysiological properties of cells filling the neonatal medial prefrontal cortex lesion cavity. *Brain Research, 1178*, 1209–1218.

Drzewiekcki, C. M., Willing, J., & Juraska, J. M. (2016). Synaptic number changes in the medial prefrontal cortex across adolescence in male and female rats: A role for pubertal onset. *Synapse, 70*, 361–368.

Elbert, T., Heim, S., & Rockstroh, B. (2001). Neural plasticity and development. In C. A. Nelson, & M. Luciana (Eds.), *Handbook of cognitive developmental cognitive neuroscience* (pp. 191–204). Cambridge, Mass: MIT Press.

Eluvathingal, T. J., Chugani, H. T., Behen, M. E., Juhasz, C., Muzik, O., Maqbool, M., & Makki, M. (2006). Abnormal brain connectivity in children after early severe socioemotional deprivation: a diffusion tensor imaging study. *Pediatrics, 117*, 2093–2100.

Eyre, J. A. (2007). Corticospinal tract development and its plasticity after perinatal injury. *Neuroscience and Biobehavioural Reviews, 31*, 1136–1149.

Fagiolini, M., Pizzorusso, T., Berardi, N., Domenici, L., & Maffei, L. (1994). Functional postnatal development of the rat primary visual cortex and the role of visual experience: dark rearing and monocular deprivation. *Vision Research, 34*, 709–720.

Faiz, M., Acrain, L., Castellano, B., & Gonzalez, B. (2005). Proliferation dynamics of germinative zone cells in the intact and excitotoxically lesioned postnatal rat brain. *BMC Neuroscience, 6*, 26. Available from http://dx.doi.org/10.1186/1471-2202-6-26.

Favaro, A., Tenconi, E., Degortes, D., Manara, R., & Santonastaso, P. (2015). Neural correlates of prenatal stress in young women. *Psychological Medicine, 45*, 2533–2543.

Fenoglio, K., Chen, Y., & Baram, T. (2006). Neuroplasticity of the hypothalamic–pituitary–adrenal axis early in life requires recurrent recruitment of stress-regulating brain regions. *Journal of Neuroscience, 26*, 2434–2442.

Fields, R. D. (2015). A new mechanism of nervous system plasticity: activity-dependent myelination. *Nature Reviews Neuroscience, 16*, 756–767.

Finn, E. S., Shen, X., Scheinost, D., Rosenberg, M. D., Huang, J., Chun, M. M., . . . Constable, R. T. (2015). Functional connectome fingerprinting: Identifying individuals using patterns of brain connectivity. *Nature Neuroscience, 18*, 1664–1671.

Fonzo, G. A., Ramsawh, H. J., Flagan, T. M., Simmons, A. N., Sullivan, S. G., Allard, C. B., ... Stein, M. B. (2016). Early life stress and the anxious brain: evidence for a neural mechanism linking childhood emotional maltreatment to anxiety in adulthood. *Psychological Medicine, 46*, 10378–11054.

Frazier, E. A., Fristad, M. A., & Arnold, L. E. (2012). Feasibility of a nutritional supplement as treatment for pediatric bipolar spectrum disorders. *Journal of Alternative and Complementary Medicine, 18*, 678–685.

Fritsch, G., & Hitzig, E. (1960). On the electrical excitability of the cerebrum. In G. von Bonin (Ed.), *The cerebral cortex* (pp. 73–96). Springfield, IL: Charles C. Thomas.

Fuhrmann, D., Knoll, L. J., & Blakemore, S.-J. (2015). Adolescence as a sensitive period of brain development. *Trends in Cognitive Sciences, 19*, 558–566.

Georgieff, M. K. (2007). Nutrition and the developing brain: Nutrient priorities and measurement. *The American Journal of Clinical Nutrition, 85S*, 614–620.

Gibb, R. (2004). Perinatal experience and recovery from brain injury. *Unpublished PhD Thesis*. University of Lethbridge.

Gibb, R., Gonzalez, C., & Kolb, B. (2014). Prenatal enrichment and recovery from perinatal cortical damage: effects of maternal complex housing. *Frontiers in Behavioral Neuroscience, 8*, 223. Available from http://dx.doi.org/10.3389/fnbeh.2014.00223.

Gibb R., Gorny G, and Kolb B. Behavioral and anatomical effects of complex rearing after day 7 neonatal frontal lesions vary with sex. In submission 2017.

Gibb, R., Kovalchuk, A., & Kolb, B. (2017). Tactile stimulation improves functional recovery after perinatal cortical injury by increasing the production of FGF-2 in the skin and brain. Manuscript in submission.

Goldman, P. S. (1974). An alternative to developmental plasticity: Heterology of CNS structures in infants and adults. In D. G. Stein, J. Rosen, & N. Butters (Eds.), *Plasticity and recovery of function in the central nervous system* (pp. 149–174). New York: Academic Press.

Goldman, P. S., & Galkin, T. W. (1978). Prenatal removal of frontal association neocortex in the fetal rhesus monkey: Anatomical and functional consequences. *Brain Research, 152*, 451–485.

Goldman, P. S., Isseroff, A., Schwrtz, M., & Bugbee, N. (1983). The neurobiology of cognitive development. In P. H. Mussen (Ed.), *Handbook of child psychology: biology and infancy development* (pp. 311–344). New York, NY: Wiley.

Green, J. D., & Adey, W. R. (1956). Electrophysiological studies of hippocampal connections and excitability. *Electroencephalography and Clinical Neurophysiology, 8*, 245–262.

Greenough, W. T., & Chang, F. F. (1989). Plasticity of synapse structure and pattern in the cerebral cortexIn A. Peters, & E. G. Jones (Eds.), *Cerebral cortex* (Volume 7, pp. 391–440). New york: Plenum Press.

Greenough, W. T., Larson, J. R., & Withers, G. S. (1985). Effects of unilateral and bilateral training in a reaching task on dendritic branching of neurons in the rat motor-sensory forelimb cortex. *Behavioral and Neural Biology, 44*, 301–314.

Halliwell, C. (2011). Treatment interventions following prenatal stress and neonatal cortical injury. *Unpublished PhD thesis*. Lethbridge, Alberta, Canada: University of Lethbridge.

Hamilton, D., & Kolb, B. (2005). Nicotine, experience, and brain plasticity. *Behavioral Neuroscience, 119*, 355–365.

Hamilton, D. A., Candelaria-Cook, F. T., Akers, K. G., Rice, J. P., Maes, L. I., Rosenberg, M., et al. (2010). Patterns of social-experience-related c-fos and Arc expression in the frontal cortices of rats exposed to saccharin or moderate levels of ethanol during prenatal brain development. *Behavioural Brain Research, 214*, 66–74.

Hanson, J. L., Adluru, N., Chung, M. K., Alexander, A. L., Davidson, R. J., & Pollak, S. D. (2010). Early neglect is associated with alterations in white matter integrity and cognitive functioning. *Child Development, 84*, 1566–1578.

Harding, K. L., Judah, R. D., & Gant, C. (2003). Outcome-based comparison of Ritalin versus food-supplement treated children with AD/HD. *Alternative Medicine Review, 8*, 319–330.

Harker, A., Carroll, C., Raza, S., Kolb, B., & Gibb, R. (2017). Preconception stress in rats alters brain and behaviour in offspring. In submission.

Harker, A., Raza, S., Williamson, K., Kolb, B., & Gibb, R. (2015). Preconception paternal stress in rats alters dendritic morphology and connectivity in the brain of developing male and females offspring. *Neuroscience*, *303*, 200–210.
Harlow, H., & Harlow, M. (1965). The affectional systems. In A. Schier, H. Harlow, & F. Stollnitz (Eds.), *Behaviour of nonhuman primates*. New York: Academic Press.
Harlow, H. F. (1959). Love in infant monkeys. *Scientific American*, *200*, 68–74.
Harlow, H. F., & Harlow, M. K. (1965). The effect of rearing conditions on behavior. *International Journal of Psychiatry*, *1*, 43–51.
Hebb, D. O. (1949). *The organization of behavior*. New York: McGraw-Hill.
Hehar, H., Yu, K., Ma, I., & Mychasiuk, R. (2016). Paternal age and diet: The contributions of a father's experience to susceptibility for post-concussion symptomology. *Neuroscience*, *332*, 61–75.
Helmeke, C., Seidel, K., Poeggel, G., Bredy, T. W., Abraham, A., & Braun, K. (2009). Paternal deprivation during infancy results in dendrite- and time-specific changes of dendritic development and spine formation in the orbitofrontal cortex of the biparental rodent *Octodon degus*. *Neuroscience*, *1632*, 790–798.
Herculano-Houzel, S., & Lent, R. (2005). Isotropic fractionator: A simple, rapid method for the quantification of cell and neuron numbers in the brain. *Journal of Neuroscience*, *25*, 2518–2521.
Herculano-Houzel, S. (2011). Not all brains are made the same: New views on brain scaling in evolution. *Brain, Behavior, and Evolution*, *78*, 22–36.
Herculano-Houzel, S. (2012). The remarkable, yet not extraordinary, human brain as a scaled-up primate brain and its associated cost. *Proceedings of the National Academy of Sciences of the United States of America*, *109*, 10661–10668.
Hicks, S., & D'Amato, C. (1961). How to design and build abnormal brains using radiation during development? In S. Fields, & M. Desmond (Eds.), *Disorders of the Developing Nervous System* (pp. 60–79). Springfield, IL: Thomas.
Himmler, B., Pellis, S. M., & Kolb, B. (2013). Juvenile play experience primes neurons in the medial prefrontal cortex to be more responsive to later experiences. *Neuroscience Letters*, *556*, 42–45.
Hoban, A. E., Stilling, R. M., Ryan, F. J., Shanahan, F., Claesson, M. J., Clarke, G., & Cryan, J. F. (2016). Regulation of prefrontal cortex myelination by the microbiota. *Translational Psychiatry*, *6*, e774. Available from http://dx.doi.org/10.1038/tp.2016.42.
Hodel, A. S., Hunt, R. H., Cowell, R. A., Van den Heuvel, S. E., Gunnar, M. R., & Thomas, K. M. (2015). Duration of early adversity and structural brain development in post-institutionalized adolescents. *NeuroImage*, *105*, 112–119.
Hofstetter, S., Tavor, I., Tzur Moryosef, S., & Assaf, Y. (2013). Short-term learning induces white matter plasticity in the fornix. *Journal of Neuroscience*, *33*, 12844–12850.
Hofstetter, S., Friedmann, N., & Assaf, Y. (2016). Rapid language-related plasticity: Microstructural changes in the cortex after a short session of new word learning. *Brain Structure and Function*, Available from http://dx.doi.org/10.1007/s00429-016-1273-2.
Hoyme, H., May, P., Kalberg, W., Kodituwakku, P., Gossage, J., Trujillo, P., et al. (2005). A practical clinical approach to diagnosis of fetal alcohol spectrum disorders: Clarification of the 1996 Institute of Medicine criteria. *Pediatrics*, *115*, 39–47.
Hullinger, R., O'Riordan, K., & Burger, C. (2015). Environmental enrichment improves learning and memory and long-term potentiation in young adult rats through a mechanism requiring mFluR5 signaling and sustained activation of p70s6k. *Neurobiology of Learning and Memory*, *125*, 126–134.
Isaacs, E. B., Fischl, B. R., Quinn, B. T., Chong, W. K., Gadian, D. G., & Lucas, A. (2010). Impact of breast milk on IQ, brain size and white matter development. *Pediatric Research*, *67*, 357–362.
Johnson, D. E., Guthrie, D., Smyke, A. T., Koga, S. F., Fox, N. A., Zeanah, C. H., & Nelson, C. A. (2010). Growth and associations between auxology, caregiving environment, and cognition in socially deprived Romanian children randomized to foster vs ongoing institutionalized care. *Archives of Pediatric Adolescent Medicine*, *164*, 507–516.
Juraska, J. M. (1984). Sex differences in dendritic response to differential experience in the rat visual cortex. *Brain Research*, *295*, 27–34.
Juraska, J. M. (1990). The structure of the rat cerebral cortex: Effects of gender and the environment. In B. Kolb, & R. C. Tees (Eds.), *The cerebral cortex of the rat* (pp. 484–505). Cambridge, MA: MIT Press.

Kafouri, S., Kramer, M., Leonard, G., Perron, M., Pike, B., Richer, L., Toro, R., Veillette, S., Pausova, Z., & Paus, T. (2013). Breastfeeding and brain structure in adolescence. *International Journal of Epidemiology, 42*, 150–159.

Kandel, E. R., & Spencer, W. A. (1968). Cellular neurophysiological approaches in the study of learning. *Physiological Review, 48*, 65–134.

Kaplan, B. J., Fisher, J. E., Crawford, S. G., Field, C. J., & Kolb, B. (2004). Improved mood and behavior during treatment with a mineral-vitamin supplement: an open-label case series of children. *Journal of Child and Adolescent Psychopharmacology, 14*, 115–122.

Kaufman, J., Plotsky, P. M., Nemeroff, C. B., & Charney, D. S. (2000). Effects of early adverse experiences on brain structure and function: clinical implications. *Biological Psychiatry, 48*, 778–790.

Kennard, M. (1942). Cortical reorganization of motor function. *Archives of Neurology, 48*, 227–240.

Knezovich, J. G., & Ramsay, M. (2012). The effect of preconception paternal alcohol exposure on epigenetic remodeling of the H19 and Rasgrfl imprinting control regions in mouse offspring. *Frontiers in Genetics*, Available from http://dx.doi.org/10.3389/fgene.2012.00010.

Knudsen, E. I. (2004). Sensitive periods in the development of the brain and behavior. *Journal of Cognitive Neuroscience, 16*, 1412–1425.

Kodituwakku, P. W. (2007). Defining the behavioral phenotype in children with fetal alcohol spectrum disorders: a review. *Neuroscience & Biobehavioral Reviews, 31*, 192–201.

Kolb, B., Gorny, G., Li, Y., Samaha, A. N., & Robinson, T. E. (2003a). Amphetamine or cocaine limits the ability of later experience to promote structural plasticity in the neocortex and nucleus accumbens. *Proceedings of the National Academy of Sciences of the United States of America, 100*, 10523–10528.

Kolb, B. (1987). Recovery from early cortical damage in rats. I. Differential behavioral and anatomical effects of frontal lesions at different ages of neural maturation. *Behavioural Brain Research, 25*, 205–220.

Kolb, B. (1995). *Brain plasticity and behavior.* Hillsdale, NJ: Lawrence Erlbaum.

Kolb, B. (2017). Brain plasticity in the adolescent brain. In A. A. Benasich & U. Ribary (Eds.), *Manifestations and mechanisms of dynamic brain coordination over development.* Strüngmann Forum Reports (vol. 25), J. Lupp, series editor. Cambridge, MA: MIT Press, in press.

Kolb, B., & Gibb, R. (2010). Tactile stimulation facilitates functional recovery and dendritic change after neonatal medial frontal or posterior parietal lesions in rats. *Behavioural Brain Research, 214*, 115–120.

Kolb, B., & Tomie, J. (1988). Recovery from early cortical damage in rats. IV. Effects of hemidecortication at 1, 5, or 10 days of age. *Behavioural Brain Research, 28*, 259–274.

Kolb, B., & Whishaw, I. Q. (1989). Plasticity in the neocortex: Mechanisms underlying recovery from early brain damage. *Progress in Neurobiology, 32*, 235–276.

Kolb, B., & Whishaw, I. Q. (2015). *Fundamentals of human neuropsychology* (7th ed.) New York: Worth/MacMillan.

Kolb, B., Cioe, J., & Muirhead, D. (1998). Cerebral morphology and functional sparing after prenatal frontal cortex lesions in rats. *Behavioral Brain Research, 91*, 143–155.

Kolb, B., & Elliott, W. (1987). Recovery from early cortical damage in rats. II. Effects of experience on anatomy and behavior following frontal lesions at 1 or 5 days of age. *Behavioral Brain Research, 26*, 47–56.

Kolb, B., Gibb, R., & Gorny, G. (2003b). Experience-dependent changes in dendritic arbor and spine density in neocortex vary with age and sex. *Neurobiology of Learning and Memory, 791*, 1–10.

Kolb, B., & Gibb, R. (1993). Possible anatomical basis of recovery of spatial learning after neonatal prefrontal lesion in rats. *Behavioral Neuroscience, 107*, 799–811.

Kolb, B., Gibb, R., & van der Kooy, D. (1992). Neonatal hemidecortication alters cortical and striatal structure and connectivity. *Journal of Comparative Neurology, 322*, 311–324.

Kolb, B., Gibb, R., & van der Kooy, D. (1994). Neonatal frontal cortical lesions in rats alter cortical structure and connectivity. *Brain Research., 645*, 85–97.

Kolb, B., Gibb, R., Pearce, S., & Tanguay, R. (2008). Prenatal exposure to prescription medications alters recovery following early brain injury in rats. *Society for Neuroscience Abstracts, 349*, 5.

Kolb, B., Mychasiuk, R., Muhammad, A., & Gibb, R. (2013). Brain plasticity in the developing brain. *Progress in Brain Research, 207*, 35–64.

Kolb, B., & Nonneman, A. J. (1978). Sparing of function in rats with early prefrontal cortex lesions. *Brain Research, 151*, 135–148.

Kolb, B., Pedersen, B., & Gibb, R. (2012). Embryonic pretreatment with bromodeoxyuridine blocks neurogenesis and functional recovery from perinatal frontal lesions. *Developmental Neuroscience, 34*, 228–239.

Kolb, B., & Stewart, J. (1991). Sex-related differences in dendritic branching of cells in the prefrontal cortex of rats. *Journal of Neuroendocrinology, 3*, 95–99.

Kolb, B., & Stewart, J. (1995). Changes in the neonatal gonadal hormonal environment prevent behavioral sparing and alter cortical morphogenesis after early frontal cortex lesions in male and female rats. *Behavioral Neuroscience, 109*, 285–294.

Kolb, B., Harker, A., Mychasiuk, R., & Gibb, R. (2017). Stress and prefrontal cortical plasticity in the developing brain. *Developmental Medicine & Child Neurology*, in press.

Kolb, B., Whishaw, I. Q., & Teskey, G. C. (2016). *Introduction to brain and behavior,* (5th ed.) New York: Worth/MacMillan.

Konijnenberg, C. (2015). Methodological issues in assessing the impact of prenatal drug exposure. *Substance Abuse: Research and Treatment, 9*(S2), 39–44.

Konorski, J. (1948). *Conditioned reflexes and neuron organization.* Cambridge: Cambridge University Press.

Kraemer, G. W. (1992). A psychobiological theory of attachment. *Behavioral and Brain Sciences, 15*, 493–541.

Kundakovic, M., & Champagne, F. A. (2015). Early-life experience, epigenetics, and the developing brain. *Neuropsychopharmacology Reviews, 40*, 141–153.

Kwinto, P., Herman-Sucharska, I., Lesniak, A., Klimek, M., et al. (2015). Relationship between stereoscopic vision, visual perception, and microstructure changes of corpus callosum and occipital white matter in the 4-year-old very low birth weight children. *BioMed Research International*, Article ID 842143, http://dx.doi.org/10.1155/2015/842143.

Lawler, J. M., Hostinar, C. E., Milner, S. B., & Gunnar, M. R. (2014). Disinhibited social engagement in post-institutionalized children: Differentiating normal from atypical behavior. *Developmental Psychopathology, 26*, 451–464.

Leung, B. M. Y., Wiens, K. P., & Kaplan, B. J. (2011). Does prenatal micronutrient supplementation improve children's mental development? A systematic review. *BMC Pregnancy and Childbirth, 11*, 12. Available from http://dx.doi.org/10.1186/1471-2393-11-12.

Lieberman, M. D., Berkman, E. T., & Wagner, T. D. (2012). Correlations in social neuroscience aren't voodoo. *Perspectives in Psychological Science, 33*, 1393–1406.

Lotfipour, S., Ferguson, E., Leonard, G., Perron, M., Pike, B., Richer, L., Seguin, J. R., Toro, R., Veillette, S., Pausova, Z., & Paus, T. (2009). Orbitofrontal cortex and drug use during adolescence: role of prenatal exposure to maternal smoking and BDNF genotype. *Archives of General Psychiatry, 66*, 1244–1252.

Lotfipour, S., Leonard, G., Perron, M., Pike, B., Richer, L., Seguin, J. R., Toro, R., Veillette, S., Pausova, Z., & Paus, T. (2010). Prenatal exposure to maternal cigarette smoking interacts with a polymorphism in the a6 nicotinic acetylcholine receptor gene to influence drug use and striatum volume in adolescence. *Molecular Psychiatry, 15*, 6–8.

McEwen, B. S., & Morrison, J. H. (2013). The brain on stress: Vulnerability and plasticity of the prefrontal cortex over the life course. *Neuron, 79*, 16–29.

McEwen, B. S. (2008). Central effects of stress hormones in health and disease: understanding the protective and damaging effects of stress and stress mediators. *European Journal of Pharmacology, 583*, 174–185.

McGowan, P. O., Suderman, M., Sasaki, A., Huang, T. C., Hallett, M., Meaney, M. J., & Szyf, M. (2011). Broad epigenetic signature of maternal care in the brain of adult rats. *PLoS ONE, 6*, e14739. Available from http://dx.doi.org/10.1371/journal.pone.0014739.

McLaughlin, K. A., Sheridan, M. A., Winter, W., Fox, N. A., Zeanah, C. H., & Nelson, C. A. (2014). Widespread reductions in cortical thickness following early-life deprivation: A neurodevelopmental pathway to attention-deficit/hyperactivity disorder. *Biological Psychiatry, 76*, 629–638.

Meck, W. H., & Williams, C. L. (2003). Metabolic imprinting of choline by its availability during gestation: Implications for memory and attentional processing across the lifespan. *Neuroscience and Biobehavioral Reviews, 27*, 385–399.

Metz, G. A. (2007). Stress as a modulator of motor system function and pathology. *Reviews in Neuroscience, 18*, 209–222.

Mitchell, A. A., Gilboa, S. Z., Werler, M. M., Kelley, K. E., Louik, C., & Hernandez-Diaz, S. (2011). Medication use during pregnancy, with particular focus on prescription drugs: 1976–2008. *American Journal of Obstetrics and Gynecology, 205*, 51.e1–51.e8.

Monfils, H., Driscoll, I., Vavrek, R., Kolb, B., & Fouad, K. (2008). FGF-2 induced functional improvement from neonatal motor cortex injury via corticospinal projections. *Experimental Brain Research, 185*, 453–460.

Monfils, M. H., Driscoll, I., Kamitakahara, H., Wilson, B., Flynn, C., Teskey, G. C., Kleim, J., & Kolb, B. (2006). FGF-2 induced cell proliferation stimulates anatomical neurophysiological and functional recovery from neonatal motor cortex injury. *European Journal of Neuroscience, 24*, 739–749.

Mortera, P., & Herculano-Houzel, S. (2012). Age-related neuronal loss in the rat brain begins at the end of adolescence. *Frontiers in Neuroanatomy,* Available from http://dx.doi.org/10.3389/fnana.2012.00045.

Mudd, A. T., Alexander, L. S., Berding, K., Wawaoruntu, R. V., Berg, B. M., Donovan, S. M., & Dilger, R. N. (2016). Dietary prebiotics, milk fat globule membrane, and lactoferrin affects structural neurodevelopment in the young piglet. *Frontiers in Pediatrics*, Available from http://dx.doi.org/10.3389/fped.2016.00004.

Muhammad, A., & Kolb, B. (2011a). Mild prenatal stress modulated behaviour and neuronal spine density without affecting amphetamine sensitization. *Developmental Neuroscience, 33*, 85–98.

Muhammad, A., Hossain, S., Pellis, S., & Kolb, B. (2011b). Tactile stimulation during development attenuates amphetamine sensitization and structurally reorganizes prefrontal cortex and striatum in a sex-dependent manner. *Behavioral Neuroscience, 125*, 161–174.

Muhammad, A., & Kolb, B. (2011). Maternal separation during development alters dendritic morphology in the nucleus accumbens and prefrontal cortex. *Neuroscience, 216*, 103–109.

Muhammad, A., Mychasiuk, R., Nakahashi, A., Hossain, S., Gibb, R., & Kolb, B. (2012). Prenatal nicotine exposure alters neuroanatomical organization of the developing brain. *Synapse, 66*, 950–954.

Mychasiuk, R., Harker, A., Ilnytskyy, S., & Gibb, R. (2013a). Paternal stress prior to conception alters DNA methylation and behaviour of developing rat offspring. *Neuroscience, 241*, 100–105.

Mychasiuk, R., Gibb, R., & Kolb, B. (2011a). Prenatal bystander stress induces neuroanatomical changes in the prefrontal cortex and hippocampus of developing rat offspring. *Brain Research, 1412*, 55–62.

Mychasiuk, R., Gibb, R., & Kolb, B. (2011b). Prenatal stress produces sexually dimorphic and regionally-specific changes in gene expression in hippocampus and frontal cortex of developing rat offspring. *Developmental Neuroscience, 33*, 531–538.

Mychasiuk, R., Hehar, H., & Esser, M. J. (2015b). A mild traumatic brain injury (mTBI) induces secondary attention-deficit hyperactivity disorder-like symptomology in young rats. *Behavioural Brain Research, 286*, 285–292.

Mychasiuk, R., Hehar, H., Ma, I., Kolb, B., & Esser, M. J. (2015a). The development of lasting impairments: A mild pediatric brain injury alters gene expression, dendritic morphology, and synaptic connectivity in the prefrontal cortex of rats. *Neuroscience, 288*, 145–155.

Mychasiuk, R., Ilnystkyy, S., Kovalchuk, O., Kolb, B., & Gibb, R. (2011c). Intensity matters: Brain, behaviour, and the epigenome of prenatally stressed rats. *Neuroscience, 180*, 105–110.

Mychasiuk, R., Muhammad, A., & Kolb, B. (2014). Environmental enrichment alters structural plasticity of the adolescent brain but does not remediate the effects of prenatal nicotine exposure. *Synapse, 68*, 293–305.

Mychasiuk, R., Muhammad, A., Gibb, R., & Kolb, B. (2013). Long-term alterations to dendritic morphology and synaptic connectivity associated with prenatal exposure to nicotine. *Brain Research, 1499*, 53–60.

Mychasiuk, R., Schmold, N., Ilnystkyy, S., Kovalchuk, O., Kolb, B., & Gibb, R. (2011d). Prenatal bystander stress alters brain, behavior, and the epigenome of developing rat offspring. *Developmental Neuroscience, 33*, 159–169.

Myers, M., Brunelli, S., Squire, J., Shindledecker, R., & Hofer, M. (1989). Maternal behaviour of SHR rats in its relationship to offspring blood pressure. *Developmental Psychobiology, 22*, 29–53.

Nelson, C. A. (2007). A neurobiological perspective on early human deprivation. *Child Development Perspectives, 1*, 13–18.

Nemati, F., & Kolb, B. (2010). Motor cortex injury has different behavioral and anatomical effects in juvenile and adolescent rats. *Behavioral Neuroscience, 24*, 612–622.
Nemati, F., & Kolb, B. (2012). Recovery from medial prefrontal cortex injury during adolescence: Implications for age-dependent plasticity. *Behavioral Brain Research, 229*, 168–175.
Passaro, K. T., Little, R. E., Savitz, D. A., & Noss, J. (1998). Effect of paternal alcohol consumption before conception on infant birth weight. *Teratology, 57*, 294–301.
Patel, S. N., Rose, P. R., & Stewart, M. G. (1988). Training induced dendritic spine density changes are specifically related to memory formation processes in the chick, *Gallus domesticus*. *Brain Research, 463*, 168–173.
Penfield, W., & Boldrey, E. (1937). Somatic motor and sensory representation in the cerebral cortex of man as studied by electrical stimulation. *Brain, 60*, 389–443.
Penn, A. A., & Shatz, C. J. (1999). Brain waves and brain wiring: The role of endogenous and sensory-driven neural activity in development. *Pediatric Research, 45*, 447–458.
Pons, T. P., Garraghty, P. E., Ommaya, K., Kaas, J. H., Taum, E., & Mishkin, M. (1991). Massive cortical reorganization after sensory deafferentation in adult macaques. *Science, 272*, 1857–1860.
Prusky, G. T., Silver, B. D., Tschetter, W. W., Alam, N. M., & Douglas, R. M. (2008). Experience-dependent plasticity from eye opening enables lasting, visual cortex-dependent enhancement of motion vision. *Journal of Neuroscience, 28*, 9817–9827.
Purves, D., & Lichtman, J. W. (1980). Elimination of synapses in the developing nervous system. *Science, 210*, 153–157.
Pusic, K. M., Pusic, A. D., & Kraig, R. P. (2016). Environmental enrichment stimulates immune cell secretion of exosomes that promote CNS myelination and may regulation inflammation. *Cell and Molecular Neurobiology,* Epub ahead of print.
Ramon y Cajal, S. (1928). *Degeneration and regeneration in the nervous system.* London: Oxford University Press.
Rasmussen, T., & Milner, B. (1977). The role of early left-brain injury in determining lateralization of cerebral speech functions. *Annals of the New York Academy of Sciences, 299*, 355–369.
Raza, S., Himmler, B. T., Harker, A., Kolb, B., Pellis, S. M., & Gibb, R. (2015). Effects of prenatal exposure to valproic acid on the development of juvenile-typical social play in rats. *Behavioral Pharmacology, 26*, 707–719.
Rice, F. L., & van der Loos, H. (1977). Development of the barrels and barrel field in the somatosensory cortex of the mouse. *Journal of Comparative Neurology, 171*, 545–560.
Rice, J. P., Suggs, L. E., Lusk, A. V., Parker, M. O., Candelaria-Cook, F. T., Savage, D. D., & Hamilton, D. A. (2012). Effects of exposure to moderate levels of ethanol during prenatal brain development on dendritic length, branching, and spine density in the nucleus accumbens and dorsal striatum of adult rats. *Alcohol, 46*, 577–584.
Robinson, T. E., & Kolb, B. (2004). Structural plasticity associated with drugs of abuse. *Neuropharmacology, 47*(Suppl 1), 33–46.
Rucklidge, J. J., & Kaplan, B. J. (2013). Broad-spectrum micronutrient formulas for the treatment of psychiatric symptoms: A systematic review. *Expert Reviews Neurotherapy, 13*, 49–73.
Rutter, M., & O'Connor, T. G. (2004). Are there biological programming effects for psychological development? Findings from a study of Romanian adoptees. *Developmental Psychology, 40*, 81–94.
Rutter, M., & Pickles, A. (2016). Annual research review: Threats to the validity of child psychiatry and psychology. *Journal of Child Psychology and Psychiatry, 57*, 398–416.
Sackett, G. P. (1984). A nonhuman primate model of risk for deviant development. *American Journal of Mental Deficiency, 88*, 469–476.
Sakamoto, T., Cansev, M., & Wurtman, R. J. (2007). Oral supplementation with docosahexaenoic acid and uridine-5′-monophosphate increases dendritic spine density in adult gerbil hippocampus. *Brain Research, 1182*, 50–59.
Sampaio-Baptista, C., Khrapitchev, A. A., Foxley, S., Schlagheck, T., Scholz, J., Jbabdi, S., DeLuca, G. C., Miller, K. L., Taylor, A., Thomas, N., Kleim, J., Sibson, N. R., Bannerman, D., & Johansen-Berg, H. (2013). Motor skill learning induces changes in white matter microstructure and myelination. *Journal of Neuroscience, 33*, 19499–19503.

Seelke, A. M. H., Perkeybile, A. M., Grunewald, R., Bales, K. L., & Krubitzer, L. A. (2016). Individual differences in cortical connections of somatosensory cortex are associated with parental rearing style in prairie voles (*Microtus ochrogaster*). *Journal of Comparative Neurology, 524*, 564–577.

Segerstrom, S., & Miller, G. (2004). Psychological stress and the human immune system: a meta-analytic study of 30 years of inquiry. *Psychological Bulletin, 1304*, 601–630.

Shatz, C. J. (1992). The developing brain. *Scientific American, 267*, 60–67.

Shatz, C. J. (1996). Emergence of order in visual system development. *Proceedings of the National Academy of Sciences of the United States of America, 93*, 602–608.

Sirevaag, A. M., & Greenough, W. T. (1987). Differential rearing effects on rat visual cortex synapses. III. Neuronal and glial nuclei, boutons, dendrites, and capillaries. *Brain Research, 424*, 320–332.

Sirevaag, A. M., & Greenough, W. T. (1988). A multivariate statistical summary of synaptic plasticity measures in rats exposed to complex, social and individual environments. *Brain Research, 441*, 386–392.

Smythies, L. E., & Smythies, J. R. (2014). Microbiota, the immune system, black moods and the brain–melancholia updated. *Frontiers in Human Neuroscience*, Available from http://dx.doi.org/10.3389/fnhum.2014.00720.

Soubry, A., Hoyo, C., Jirtle, R. L., & Murphy, S. K. (2014). A paternal environmental legacy: Evidence for epigenetic inheritance through the male germ line. *Bioessays, 36*, 359–371.

Spear, L. P. (2000). The adolescent brain and age-related behavioral manifestations. *Neuroscience and Biobehavioral Reviews, 24*, 417–463.

Steinberg, L. (2016). Commentary on special issue on the adolescent brain: Redefining adolescence. *Neuroscience and Biobehavioral Reviews, 70*, 343–346.

Stilling, R. M., Ryan, F. J., Hoban, A. E., Shanahan, F., Clarke, G., Claesson, M. J., Dinan, T. G., & Cryan, J. F. (2015). Microbes and neurodevelopment—Absence of microbiota during early life increases activity-related transcriptional pathways in the amygdala. *Brain, Behavior, and Immunity, 50*, 209–220.

Substance Abuse and Mental Health Services Administration. (2013). Results from the 2013 National Survey on Drug Use and Health: Summary of National Findings, *NSDUH Series H-48, HHS Publication* No. (SMA) 14-4863. Rockville, MD: *Substance Abuse and Mental Health Services Administration*, 2014, p. 26.

Suomi, S. J. (1997). Early determinants of behaviour: evidence from primate studies. *British Medical Bulletin, 53*, 170–184.

Suomi, S. J. (2002). How gene-environment interactions can shape the development of socioemotional regulation in rhesus monkeys (pp. 5–26). In B. S. Zuckerman, A. F. Lieberman, & N. A. Fox (Eds.), *Emotional regulation and developmental health: infancy and early childhood.* New York: Johnson & Johnson Pediatric Institute.

Takesian, A. E., & Hensch, T. K. (2013). Balancing plasticity/stability across brain development. *Progress in Brain Research, 207*, 3–34.

Teskey, G. C. (2016). Revisiting Bliss and Lomo: Long-term potentiation and the synaptic basis of learning and memory. In B. Kolb, & I. Q. Whishaw (Eds.), *Brain and Behaviour: Revisiting the Classic Studies*. London: Sage.

Teskey, G. C., Monfils, M. H., Silasi, G., & Kolb, B. (2006). Neocortical kindling is associated with opposing alterations in dendritic morphology in neocortical layer V and striatum from neocortical layer III. *Synapse, 59*, 1–9.

Thompson, B. L., Levitt, P., & Stanwood, G. D. (2009). Prenatal exposure to drugs: effects on brain development and implications for policy and education. *Nature Reviews Neuroscience, 10*, 303–312.

Thompson, D. K., Chen, J., Beare, R., Adamson, C. L., et al. (2016). Structural connectivity relates to perinatal factors and functional impairment at 7 years in children born very preterm. *NeuroImage, 134*, 328–337.

Tomason, M. E., & Marusak, H. A. (2017). Toward understanding the impact of trauma on the early developing human brain. *Neuroscience, 342*, 55–67.

Tottenham, N., & Galvan, A. (2016). Stress and the adolescent brain: Amygdala-prefrontal cortex circuitry as developmental targets. *Neuroscience and Biobehavioral Reviews, 70*, 217–227.

Van den Bergh, B. R., & Marcoen, A. (2004). High antenatal maternal anxiety is related to ADHD symptoms, externalizing problems, and anxiety in 8- and 9-year-olds. *Child Development, 75*, 1085–1097.
Vargha-Kadem, F., & Watters, G. V. (1985). Development of speech and language following bilateral frontal lesions. *Brain and Language, 25*, 167–183.
Villablanca, J., Hovda, D., Jackson, G., & Infante, C. (1993). Neurological and behavioral effects of a unilateral frontal cortical lesion in fetal kittens, II. Visual system tests and proposing a "critical period" for lesion effects. *Behavioral Brain Research, 57*, 72–92.
Vul, E., & Pashler, H. (2012). Voodoo and circularity errors. *NeuroImage, 62*, 945–948.
Walsh, J. M., Doyle, L. W., Anderson, P. J., & Cheong, J. L. (2014). Moderate and late preterm birth: effect on brain size and maturation at term-equivalent age. *Radiology, 273*, 232–240.
Warren, S. G., Humphreys, A. G., Juraska, J. M., & Greenough, W. T. (1995). LTP varies across the menstrual cycle: Enhanced synaptic plasticity in proestrus rats. *Brain Research, 703*, 26–30.
Weaver, I. C., Meaney, M. J., & Szyf, M. (2006). Maternal care effects on the hippocampal transcriptome and anxiety-mediated behaviors in the offspring that are reversible in adulthood. *Proceedings of the National Academy of Sciences of the United States of America, 103*, 3480–3485.
Weinstock, M. (2008). The long-term behavioural consequences of prenatal stress. *Neuroscience and Biobehavioral Reviews, 32*, 1073–1086.
Whishaw, I. Q., & Miklyaeva, E. I. (1996). A rats's reach should exceed its grasp: analysis of independent limb and digit use in the laboratory rat. In P.-K. Ossenkopp, M. Kavaliers, & P. R. Sanburg (Eds.), *Measuring movement and locomotion: From invertebrates to humans* (pp. 135–169). Austin, TX: Landes.
Wiesel, T. N., & Hubel, D. H. (1963). Single-cell responses in striate cortex of kittens deprived of vision in one eye. *Journal of Neurophysiology, 26*, 1003–1017.
Wiesel, T. N., & Hubel, D. H. (1965). Comparison of the effects of unilateral and bilateral eye closure on cortical unit responses in kittens. *Journal of Neurophysiology, 28*, 1029–1040.
Winnick, M. (1969). Malnutrition and brain development. *Journal of Pediatrics, 74*, 667–679.
Zhang, L. I., Bao, S., & Merzenich, M. M. (2002). Disruption of primary auditory cortex by synchronous auditory inputs during a critical period. *Proceedings of the National Academy of Sciences of the United States of America, 99*, 2309–2314.
Zhang, Y., Inder, T. E., Neil, J. J., Dioerker, D. L., Alexopoulos, D., Anderson, P. J., & van Essen, D. C. (2015). Cortical structural abnormalities in very preterm children at 7 years of age. *NeuroImage, 109*, 469–479.
Zhu, X., Wang, F., Hu, H., Sun, X., Kilgard, M. P., Merzenich, M. M., & Zhou, X. (2014). Environmental acoustic enrichment promotes recovery from developmentally degraded auditory cortical processing. *Journal of Neuroscience, 34*, 5406–5415.

FURTHER READING

Bock, J., Gruss, M., Becker, S., & Braun, K. (2005). Experience-induced changes of dendritic spine densities in the prefrontal and sensory cortex: Correlation with developmental time windows. *Cerebral Cortex, 15*, 802–808.
Bock, J., Poeschel, J., Schindler, J., Borner, F., Shachar-Dadon, A., Ferdman, N., ... Poeggel, G. (2016). Transgenerational sex-specific impact of preconception stress on the development of dendritic spines and dendritic length in the medial prefrontal cortex. *Brain Structure and Function, 221*, 855–863.
Casey, B. J., Galvan, A., & Somerville, L. H. (2015). Beyond simple models of adolescence to an integrated circuit-based account: A commentary. *Developmental Cognitive Neuroscience, 17*, 128–130.
Kolb, B., Cioe, J., & Comeau, W. (2008). Contrasting effects of motor and visual learning tasks on dendritic arborization and spine density in rats. *Neurobiology of Learning and Memory, 90*, 295–300.
Kolb, B., Forgie, M., Gibb, R., Gorny, G., & Rowntree, S. (1998). Age, experience, and the changing brain. *Neuroscience and Biobehavioral Reviews, 22*, 143–159.
Pellis, S. M., & Pellis, V. C. (2009). *The playful brain: Venturing to the limits of neuroscience*. London: Oneworld Publications.

CHAPTER 14

Hormones and Development

Rachel Stark and Robbin Gibb
University of Lethbridge, Lethbridge, AB, Canada

14.1 INTRODUCTION

With the advances in science and technology made in the 20th century, it is remarkable that it was not until 1959 that a group of researchers began the pioneering studies in the field of sexual development. Phoenix and colleagues injected pregnant guinea pigs with testosterone and noted that while the male offspring seemed unaffected, the prenatal exposure of testosterone created females that were phenotypically male in characteristic but genetically female (Phoenix, Goy, Gerall, & Young, 1959). Thus, they published what is known today as the organizational/activational hypothesis of sexual differentiation. Phoenix and colleagues proposed that the androgens act at a specific time or times (critical period) during development to permanently alter the tissue. They suggested a dichotomy between organizational and activational effects. During the prenatal period the androgens acted to organize the tissue, meaning it prepared certain tissues in the body to respond differently to gonadal hormones in adulthood. Then, in adulthood, the gonadal hormones worked to activate the tissues, thus affecting sexual behavior (Arnold, 2009). Their belief that male hormones changed the brain was a new and very controversial idea that caused an explosion in the research of sexual differentiation (Wallen, 2009).

Although this hypothesis has been adapted and changed over time, the main ideas that Phoenix and his colleagues proposed still hold true today. This chapter will focus on sexual differentiation, and how this differentiation affects both brain development and the behaviors that emerge. Factors that impact this development are also discussed. Finally, we delve into our current understanding of sexual differentiation and how this idea has changed over time.

14.2 GENETIC FACTORS INFLUENCING SEXUAL DIFFERENTIATION

14.2.1 Sex differences in gene expression

Gonadal hormones are the defining factor in sexual differentiation of the body and brain, but where do these differences in hormones arise? What causes the gonads to become either testis or ovaries in the first place? This is where genetics plays an

The Neurobiology of Brain and Behavioral Development
DOI: http://dx.doi.org/10.1016/B978-0-12-804036-2.00014-5

important role. Biological sex in humans is determined by the presence or absence of the Y sex chromosome. During gamete production in males, spermatocytes undergo division so that the gametes contain half the parents' genetic material. This results in sperm that either possesses an X or a Y chromosome. The same process happens during female gamete production but since females possess two X chromosomes the gametes that are formed all contain an X chromosome. This gives rise to the typical XX female and XY male sex chromosomes once fertilization has occurred (Cheng & Mruk, 2010).

Gonadal hormones function as secondary factors that act downstream of the primary factors, the X and Y chromosomes (Arnold, 2009). The expression of genes on the sex chromosomes influences the sexual differentiation of the brain, and sequentially behavior prior to the onset of gonadal hormone secretion. This suggests that there is a gene-hormone interaction (Davies & Wilkinson, 2006). The *Ar* gene encodes the androgen receptor and is found on the X chromosome. This receptor is critical for male development. Because it is found on the X chromosome, females carry and express this gene, but circulating progestins, expressed in high levels in females, inhibit the activity of the receptor. Acting as antiandrogen agents, progestins block the male typical mode of development (Raudrant & Rabe, 2003). The *Ar* gene provides a useful example for how genes may encode one mode of development but the interaction with hormones changes it. It is, therefore, important to understand the role sex-linked genes have on development (Fig. 14.1). The three mechanisms by which this may occur include X-linked gene dosage effects, X-linked imprinting, and Y-specific gene expression.

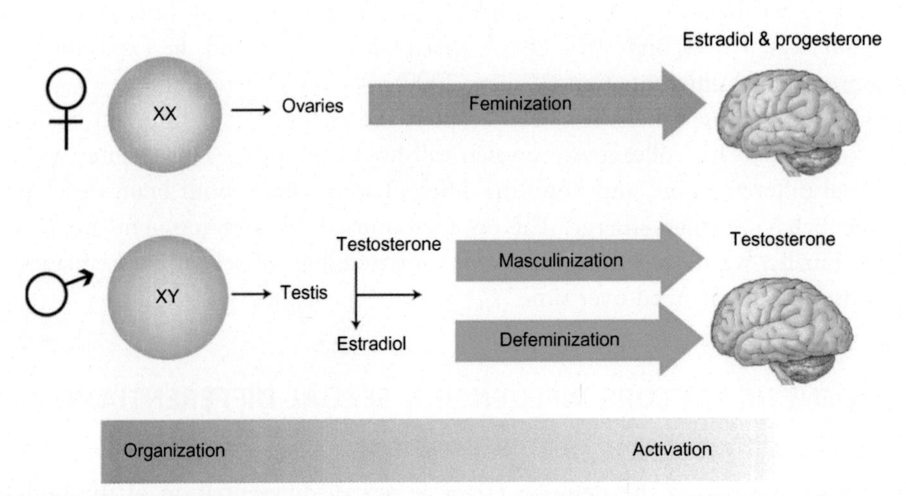

Figure 14.1 The organization/activation affects the sex chromosomes and hormones have on the brain. *Adapted from McCarthy, M.M., & Arnold, A.P. (2011). Reframing sexual differentiation of the brain. Nature Neuroscience, 14(6), 677–683.*

14.2.1.1 X-linked gene dosage

Genetic females, as we have established, possess two X chromosomes. This puts females at risk of potentially fatal levels of gene product from the second X chromosome. A mechanism, called x-inactivation, has evolved to cope with this issue. In this process, one of the X chromosomes is condensed into a Barr body but not all of the genes on the inactivated chromosome are silenced. In fact, approximately 15% escape inactivation (Nieschlag, Werler, Wistuba, & Zitzamm, 2014). The escaped genes will be expressed in higher levels in females than in males, who only have one X chromosome. In other words, the genes that escape X-inactivation are more likely to be phenotypically expressed in females (Davies & Wilkinson, 2006). For example, color vision deficiencies like color blindness occur in about 8% of men (according to the National Eye Institute) while only .5% of the female population is affected. If a female is missing the functional gene on one chromosome, the other is able to compensate for this loss. Furthermore, women are able to express tetrachromacy (although this is fairly rare) as opposed to the normal trichromacy vision due to a double dosing effect (Jornda, Deeb, Bosten, & Mollon, 2010). This suggests that escaped genes may account for some of the sexually dimorphic differences seen between males and females.

14.2.1.2 X-linked imprinting

Genomic imprinting involves the preferential expression of a subset of genes that are marked indicating parental origin. As such, some genes are preferentially expressed from either the paternal or maternal X chromosome (Reik & Walter, 2001). This process serves to sexually differentiate males and females as males inherit their X chromosome solely from their mother and females inherit an X chromosome from both parents. Genetic differences in females arise as the result of expression of the paternal X chromosome that would not occur in males, while the maternal X chromosome would be expressed in both sexes (Davis, Isles, Burgoyne, & Wilkinson, 2006). For example, females with Turner's Syndrome (possessing only one X chromosome; discussed in more detail below) show stark differences depending on whether they received their X chromosome from their mother or father (Iwasa & Pomiankowski, 2001). As such, the parental origin and the preferential expression of genes have been shown to influence fundamental sex differences in the brain and behavior.

14.2.1.3 Y-specific gene expression

The Y chromosome is functionally different from the X chromosome, mainly because it possesses a region that cannot recombine with the X chromosome during division. This nonrecombinant region (NRY) has genes on it that are specific to the Y chromosome and comprises 95% of the chromosome's length. As a result, genes in this region are only expressed in males (Skaletsky et al., 2003). For example, the *Sry* gene is found on the mammalian Y chromosome and is known as the testis determining factor (Prokop et al., 2013). This gene is critical for male typical development of the testis and is another way in which genes account for sexual differentiation.

Originally, it was thought that the *Sry* gene promoted the growth of the male reproductive tracts (Wollfian system) and eliminated the female reproductive tracts (Müllerian system). This in turn caused the production of testosterone, which acts to masculinize and defeminize the tissues of the body, including the brain. Further research into the *Sry* gene has shown that the presence of a Y chromosome may play a larger role in brain development (McCarthy & Arnold, 2011). In one study, researchers created phenotypic female mice. The catch is that they had XY chromosomes. They did this by removing the *Sry* gene from the Y chromosome, and as a result these animals developed in the default female typical pattern (these animals will be denoted XY^-). They furthered this research by mating these genetic XY^- females and found them to be fertile (Lovell-Badge & Robertson, 1990). More recent studies using the XY^- females showed that there may be sex differences in the brain due to the expression of genes found on either the X or Y chromosome. For example, it was demonstrated that when the *Sry* gene was present there were more dopaminergic neurons in the brain and further that the XY^- females had significantly lower numbers of dopaminergic neurons than the XY males. This is thought to be due to the absence of the *Sry* gene in the XY^- females (Carruth, Reisert, & Arnold, 2002). Lastly, research by Dewing et al. (2006) looked at expression of the *Sry* gene and its protein product. They found that there was increased expression of the *Sry* gene in the substantia nigra of the midbrain, the thalamus, and, although not to the same extent, throughout the cortex (Dewing et al., 2006).

14.2.2 Clinical populations

One way in which to understand more about the function of the human body, aside of manipulations to animals (see Chapter 4: The Role of Animal Models in Studying Brain Development), is to study clinical populations. There are several different sex-linked gene abnormalities that have been pivotal in understanding the interactions between genes and hormones, and their effect on sexual differentiation of both the body and brain.

14.2.2.1 Triple X syndrome

Triple X syndrome (XXX) is the addition of an X chromosome, with an incidence rate of 1/1000 females, that is a result of nondisjunction of the sex chromosomes during oogenesis. (Essentially, during gamete development the X chromosomes fail to separate.) As such, the extra chromosome has a maternal origin (May et al., 1990). The number of individuals affected may actually be significantly higher as majority of cases go undiagnosed because some individuals may experience only mild symptoms

(Gustavson, 1999). Phenotypically, individuals with XXX tend to be tall and thin. Behaviorally, they show motor coordination difficulties, auditory processing disorders, psychological and personality disturbances, and an IQ 20 points below the control level. Anatomically, individuals with XXX have lower total brain volumes and larger ventricles with reported asymmetries. Some studies report reductions in amygdala volumes as well (Otter, Schrander-Stumpel, & Crufs, 2009). Most XXX females have normal ovarian and menstrual functions (Stagi et al., 2016) but occasionally cases of birth defects (such as overlapping digits) and some of ovarian and menstrual problems are reported in the literature.

14.2.2.2 Turner syndrome

Turner syndrome (TS) is a sex chromosome disorder resulting from complete or partial loss of the X chromosome, affecting 1 in 2000 females. Nondisjunction in gamete development in either parents results in an individual with only one X chromosome, XO. This causes a reduction in X-linked gene dosing with the individual only expressing male typical levels of gene expression (Uematsu et al., 2002). Physical characteristics include short stature, infantilism, webbed neck, and cubitus valgus (an abnormal carrying position for the arm). Along with these physical signs are cognitive deficiencies including lowered performance IQ, and impairments in visuospatial skills, short-term memory, attention, and social interactions (Nijhuis-van der Sanden, Eling, & Otten, 2003). MRI studies assessing anatomical features of individuals with TS have found smaller bilateral brain volumes in the hippocampus, caudate, lenticular, thalamic nuclei (all smaller in the right hemisphere), and the parieto-occipital regions (Murphy et al., 1993). Moreover, depending on the parental origin of the X chromosome certain symptoms are exaggerated. For example, individuals that inherited their X chromosome from their mother lack social awareness, flexibility, and scored lower on formal tests of social cognitive skills compared to individuals that received their X chromosome from their father (Skuse et al., 1997).

14.2.2.3 Klinefelters syndrome

The most frequent type of congenital chromosomal disorders in males is Klinefelters syndrome (KS), with a prevalence rate of 1 in 426 to 1 in 1000 (Ngun et al., 2014). KS is characterized by an extra X chromosome (XXY) and as in females, the second X chromosome is silenced via X-inactivation and the escaped genes are the cause of the dosing affects that are seen in KS (Nieschlag, Werler, Wistuba, & Zitzamm, 2014). Typical symptoms are due to hypogonadism, mainly infertility and testosterone deficiencies as a result of X-linked dosing. Along with these symptoms are those of a more feminized phenotype including female typical distribution of adipose tissue, and absent or decreased facial hair. In addition, learning difficulties and a below normal verbal IQ have been reported (Bojesen, Juul, & Gravholt, 2003). To further understand the behavioral deficits observed, researchers have conducted MRI studies to evaluate the link between brain and behavior. Studies report enlargement of ventricular volume, and a bilateral reduction of cerebellar hemispheres, as well as a significant reduction in temporal lobe volume (Itti et al., 2006).

14.2.2.4 Congenital adrenal hyperplasia

Congenital adrenal hyperplasia (CAH) occurs when there is a mismatch between gonadal hormones and sexual differentiation. CAH is the result of an autosomal-recessive inherited disorder caused by a mutation in the enzyme, 21-hydroxylase, which is involved in the pathway that converts precursor products into cortisol and aldosterone. In the case of CAH, these precursors are unable to form the respective corticosteroids and are metabolized into androgen. This disorder affects both males and females and occurs in 1 in 15,000 births (White, 2009). Females born with CAH tend to have ambiguous genitalia due to the high levels of androgens in utero, males at birth are harder to diagnose without blood tests and tend to go untreated depending on the severity (Merke & Bornstein, 2005). In the brain, CAH causes a decrease in amygdalar volume (Merke et al., 2003) and individuals affected with CAH have symptoms that range from mild to severe depending on the degree of corticosteroid deficiencies. Behaviorally, there are mixed reviews with some papers reporting increased IQ and some reporting decreased IQ (Nass & Baker, 1991; Wenzel et al., 1978). However, females with CAH are reported to have better spatial abilities, whereas males with CAH are reported to have poorer spatial abilities. This gives some insight into how the overabundance of androgens may act to organize the tissue of the developing fetus. Since the discovery in the 1950s that cortisone was an effective treatment for CAH, the lives of these patients have improved greatly. Screening is done at birth to detect CAH and treatment can begin (Hampson, Rovet, & Altmann, 1998).

14.2.2.5 Androgen insensitivity

Androgen insensitivity (AI) is relatively rare, compared to the other disorders mentioned, with a population prevalence of 1 to 5 in 100,000. Individuals with AI are genetic males, XY. The main pathology of this disorder stems from the lack of functioning androgen receptors. The circulating androgens are at normal levels but the tissues lacking the functioning receptors cannot be activated by androgens and consequentially develop in a more female typical manner (Cohen-Bendahan, van de Beek, & Berenbaum, 2005). There are three categories of AI and they have a varying degree of effects on the body. The first is complete AI syndrome (CAIS), wherein the tissues of the body are insensitive to androgen (i.e., testosterone) and, consequently, the external genitalia differentiate in a female-typical direction. Therefore, CAIS individuals are generally raised as female and are not diagnosed until puberty when they fail to menstruate (Ehrhardt & Meyer-Bahlburg, 1979). The second type is mild AI syndrome (MAIS), this results from the mild impairment of the cells ability to respond to androgens. The male typical genitalia differentiate during fetal development but there is impaired development of secondary sexual characteristics during puberty and infertility (Zuccarello et al., 2008). Partial AI syndrome (PAIS) is the third type. Partial unresponsiveness to androgens results in impairments in the masculinization of the male genitalia during fetal development and impairments to male secondary sexual characteristics. Individuals with PAIS have ambiguous genitalia because not enough testosterone was available during pregnancy to fully complete the development in a typical manner (Hughes & Deeb, 2006). Since there is a spectrum in terms of the etiology of the disorder, there is a varying degree of cognitive sequelae. Impairments in visuospatial tasks and verbal comprehension have been reported (Imperato-McGlnley, Plchardo, Gautier, Voyer, & Bryden, 1991).

14.3 ENVIRONMENTAL FACTORS

The relationship the environment has on development has been well established. Gonadal hormones are an important fetal secondary factor that has the potential to interact with environmental risk factors and experiences that may impact sexual development. Sexual differentiation is vulnerable to pre- and perinatal factors that may have estrogenic-like or antiandrogenic properties. These deviations can interfere with development and cause a masculinization or feminization of the brain and as a result, alter behavior. Here we will consider the effects of prenatal exposures, including maternal exposure to drugs, nutritional status, and environmental contaminants.

14.3.1 Drugs

14.3.1.1 Nicotine

According to the Centers for Disease Control and Prevention (2016), an estimated 15% of US adults report daily smoking. Tobacco use is considered the largest preventable cause of death and disease. It costs an estimated $96 billion dollars in direct medical expenses and causes 443,000 smoking related deaths per year (Centers for Disease Control and Prevention, 2016). Females are particularly susceptible to developing tobacco-related morbidities and mortalities (Allen, Oncken, & Hatsukami, 2014). Resent research has also suggested that women who smoke have a much lower success rate of quitting (Piper et al., 2010). Thus, it has been estimated that 10% of women continue to smoke during pregnancy (Tong et al., 2013).

Alongside a large body of literature indicating the negative impact on overall health, there is a growing evidence of the consequences of nicotine exposure on the developing brain. Women who smoke during pregnancy generally have children born with increased risk for sudden infant death syndrome, low birth weights, attentional and cognitive deficits (including learning and memory impairments), and an increased risk for developing ADHD (Ernst, Moolchan, & Robinson, 2001; Fried, Watkinson, & Gray, 1998). In rodent studies, prenatal exposure of nicotine reduces birth weight, increases locomotor activity, and causes poor performance in maze tasks and permanent changes to dendritic morphology (Ernst et al., 2001; Mychasiuk, Muhammad, Gibb, & Kolb, 2013). These results mimic the deficits observed in the human population and provide insight into the etiology of the effects of nicotine.

In the brain, nicotine from tobacco products acts on nicotinic acetylcholine receptors (nAChR), which are ligand-gated channels that mediate the release of the neurotransmitter, acetylcholine (Ach). These receptors are expressed in the human brain during the first trimester. NAChR expression has also been reported to increase during critical periods of development, leaving the brain extremely susceptible to environmental factors (Dwyer, McQuown, & Leslie, 2009). Sex steroids play a role in the modulation of the nAChR. Specifically, progesterone and estradiol have been shown to increase the expression of the genes encoding the nAChR. While progesterone decreases activity of nAchRs, estradiol increases their activity (Centeno, Henderson, Pau, Bethea, 2006; Gangitano, Salas, Teng, Perez, & De Biasi, 2009; Jin & Steinbach, 2015; Ke & Lukas, 1996). These findings may shed some light on sex differences observed with nicotine use in adulthood such as low success for cessation in women, higher rates of cortisol during withdrawal, and subsequent higher rates of depression and anxiety during periods of nicotine abstinence (Cross, Linker, & Leslie, 2017).

14.3.1.2 Alcohol

Alcohol can readily cross the placenta and as such, it has the potential for affecting fetal development. Prenatal alcohol exposure can result in fetal alcohol spectrum disorder (FASD), with a prevalence rate of 9/1000 births in North America (Thanh & Jonsson, 2010). FASD presents itself in clinical populations with growth retardation, impairments in cognition and self-regulation, and substance-use disorders (O'Connor & Paley, 2009). Prenatal exposure to alcohol has been shown to decrease testosterone surges, which can lead to feminization of the brain and other tissues in males. In females, prenatal alcohol exposure has been shown to cause delays in secondary sexual characteristics later in life, which has been linked to dysregulation of the hypothalamic–pituitary–adrenal (HPA) axis (see Chapter 16: Socioeconomic Status for more information). Furthermore, the symptoms seen with prenatal alcohol exposure may be exacerbated by the changes to the HPA axis in the mother, which acts as an additional perturbation to the developing fetus (Weinberg, Silwowska, Lan, & Hellemans, 2008). Alcohol consumption works to increase the activity of the HPA axis, thereby inducing a stress response in the body. However, chronic alcohol consumption causes the HPA axis to build up a tolerance to alcohol thereby reducing cortisol levels. This tolerance may negatively affect the ability of the HPA axis to respond to future stressors (Spencer & Hutchison, 1999).

14.3.1.3 Marijuana

In 2015, the National Survey on Drug Use and Health found that approximately 20.5 million individuals 18 years of age or older were current users of marijuana, with 117.9 million individuals 12 years of age or older reporting marijuana use at least once in their lifetime (Center for Behavioral Health Statistics and Quality, 2016). Although controversy remains on the effects of chronic use on the brain, marijuana is the most commonly used illegal drug, and its use is on the rise among today's youth (Leatherdale, Hammond, & Ahmed, 2008; Wilson et al., 2000).

Individuals who begin smoking marijuana before the age of 17 have brain changes that include decreases in gray matter and increases white matter. Males who start smoking early have significantly higher global brain volumes. This has been hypothesized to be a result of the rapid growth observed in the brain during this period of adolescence (Wilson et al., 2000).

Delta-9-tetrahydrocannabinol (THC) has been implicated in reducing circulating levels of estradiol and progesterone in females, and testosterone in males. In female rats, administration of THC has been shown to block ovulation. In rhesus monkeys, three weekly injections of THC are enough to disrupt normal menstrual cycling; an effect that lasts for several months. Similar clinical reports have been made in women who chronically use marijuana. In male rats, THC administration drastically lowers circulating levels of testosterone (Murphy, Muñoz, Adrian, & Villanúa, 1998).

THC freely crosses the blood brain barrier, the placenta, and is secreted in breast milk (Kumar et al., 1990). Individuals exposed to marijuana prenatally generally suffer from poor executive function, working memory, poor attention span, and altered acoustic profiles of their cries (Eyler & Behnke, 1999; Fried et al., 1998). In rodents, THC administered to pregnant dams increases reabsorption of pups, and perinatal exposure has been shown to decrease binding capabilities of dopamine receptors. Most striking in its effect is the demasculinization of male rats pre- or perinatally exposed to THC. This could be due to the interaction THC has on GnRH and therefore decreased circulating testosterone during development (Kumar et al., 1990). Imaging studies show functional abnormalities in individuals with THC dependence in the orbitofrontal cortex, insula, basal ganglia, anterior cingulate, with men showing greater activation in left brain regions while women show more activation in right brain regions (Franklin et al., 2002; Li, Kemp, Milivojevic, & Sinha, 2005).

14.3.1.4 Cocaine

According to the National Institute on Drug Abuse, cocaine use has fallen in 2017 compared to 2007, with 1.5 million individuals 12 years of age or older reporting as current users (NIDA, 2015). Cocaine is highly addictive and individuals that have been abstinent show enhanced sensitivity to stress-induced drug/alcohol cue-related responses. Fox et al. (2006) showed that cocaine-dependent women have increased anxiety and negative emotion, along with increases in blood pressure when compared to men. Cocaine- dependent men have greater variability in psychological and physiological responses (Fox et al., 2006). Researchers have shown that during the first 28 days of cocaine abstinence, women have significantly higher levels of cortisol and progesterone indicating possible changes to the HPA axis (Sinha et al., 2007).

Prenatal exposure to cocaine in rodents causes feminization of male genitalia. Vathy, Katay, and Mini (1993) found a significantly shorter ano-genital distance in male rat pups, while females remained unaffected. In adulthood, female rats prenatally exposed to cocaine showed an overall decrease in sexual behavior, while the exposed males had increased mounting and intromission behaviors. Anatomically, there were increases in dopamine and norepinephrine in males in the preoptic area, whereas no differences were found in the females (Vathy et al., 1993).

14.3.2 Diet

Diet is important for overall health and proper development. Maternal and infant diet play a role in hormone levels in the mother, which in turn has an effect on the developing fetus, but formula choices may interfere with infant hormone levels. For example, there has been a major switch in Canada to use soy-based formula, around 20% in 1998. Soy-based formulas contain phytoestrogens; these compounds have estrogen like activity (although weakly so) and may provide doses of 4–11 mg/kg in infants consuming soy-based products. This dose range is much higher than that incurred by traditional Japanese diets (1 mg/kg; "Concerns for the use of soy-based formulas in infant nutrition." 2009). These elevated doses may have effects on the developing male infant. Sharpe et al. (2002) found that when comparing marmoset monkeys hand fed with either soy-based formula or standard cow-based formula, males on the soy diet had significantly decreased testosterone levels. This demonstrates the importance of infant diet on development as it can disturb sex hormone production and/or function, thus leading to changes in the masculinization or feminization of the brain.

The maternal diet plays an important role in the development of the fetus and in the newborn infant through nursing. Low-protein diets during pregnancy in rats result in females that have lower birth weights, and both male and female offspring that become obese in adulthood. Protein restriction after birth slowed the growth of both male and female offspring (Zambrano et al., 2006). Maternal obesity has been linked with lowered sperm counts and decreased sperm quality in male offspring. It has been hypothesized that maternal obesity results in increased fetal exposure to estrogens and this may account for the sperm effects reported in this study (Ramlau-Hansen et al., 2007).

14.3.3 Environmental contaminants

14.3.3.1 Bisphenol A

Bisphenol A (BPA) is used in the manufacturing of plastics and epoxy resins. Estimates on the amount of BPA used each year are upwards of 8 billion pounds, 100 tons of which may be released into the atmosphere (Vandenberg et al., 2010). BPA is a known endocrine disruptor and in the 1930s it was studied for its potential use as a synthetic estrogen. It binds both nuclear and plasma membrane bound estrogen receptors and can be found in detectable levels in urine, blood, amniotic fluid, placenta, and breast milk. The greater the exposure to BPA, the more severe the resulting symptoms (Maffini, Rubin, Sonnenschein, & Soto, 2006). In adolescents, studies have linked exposure of BPA to altered time of puberty, altered estrous cycles, prostate changes, and altered mammary gland development. BPA exposure in adults is linked to diabetes and cardiovascular disease. In women, increased exposure is correlated with recurrent miscarriages, and in men exposure is linked to decreased semen quality and sperm DNA damage (Rubin, 2011). With the substantial perturbations to the heath and development observed in children and adults, exposure during the prenatal period may be more deleterious. BPAs may have permanent organizational effects on the developing fetus. Mean BPA levels reports indicate that children have the highest amount of BPA in their urine as compared to adolescents who have higher levels than adults. This observation is important as it shows that exposure is increasing and that the younger generation is more at risk (Vandenberg et al., 2010).

14.3.3.2 Polychlorinated biphenyls

Although the use of polychlorinated biphenyls (PCBs) has been banned, this class of contaminants has a long half-life and is still found in high levels in the environment. PCBs are fat-soluble and are therefore biomagnified through the food chain, with humans at the top. PCBs are readily transferred to newborns through lactation. The structural similarities of PCBs to thyroid hormones cause a decrease in circulating thyroid hormones in the body through negative feedback loops. This decrease has implications for brain development including consequences on cognitive function, behavioral responses, and decreased brain weights in rodents. PBCs have been correlated with early menarche, abnormal menstrual cycles, increased incidence of endometriosis, spontaneous abortion, fetal death, premature delivery, and low birth weights (Leon-Olea et al., 2014). In men, PCBs have been shown to decrease sperm motility, and to cause sex reversal (male to female) in some animal species with prenatal exposure (Guillette, Crain, Rooney, & Pickford, 1995). More subtly, PCB exposure has been shown to masculinize or defeminize the female hypothalamus whereas in males it has been shown to feminize or demasculinize the hypothalamus (Gore, Martien, Gagnidze, & Pfaff, 2014). Depending on the chemical makeup of the PCB, either an antagonistic or agonistic effect on androgen, progesterone, and estrogen receptors can result therefore confirming the mixed effects observed in both males and females (Hammers et al., 2006).

14.4 SEX DIFFERENCES IN THE BRAIN

As previously discussed, the sex chromosomes have a large impact on sexual dimorphism via the expression of more genes in X-linked gene dosing, as well as the interactions between genes and hormone production. Here, we will discuss sex differences observed in the brain and the mechanisms that may account for these differences. Finally, we will delve into the implications these differences have on sex differences observed in individuals with neurological disorders.

14.4.1 Anatomical differences

A meta-analysis performed by Ruigrok et al. (2014) looked at sex differences in brain structures reported in the literature since 1990. The Ruigrok study found that overall males are reported to have larger brain volumes by between 8% and 13%. With regard to specific brain regions, males appear on average, to have more gray matter in the bilateral amygdalae, hippocampi, putamen, and temporal poles. In addition, the left posterior and anterior cingulate gyri and multiple areas in the cerebellum were larger in males than females. Females, on the other hand, were found to have larger volumes at the right frontal pole, inferior and middle frontal gyri, anterior cingulate gyrus, planum temporale/parietal operculum, insular cortex, and Heschl's gyrus. In addition, the bilateral thalami, precuneus, left parahippocampal gyrus, and lateral occipital cortex are reported to be larger in females (Ruigrok et al., 2014).

14.4.1.1 Sex differences in brain maturation

It is well documented that as humans age, their brain develops and matures. Neurons begin developing in the prenatal period and as a child experiences the world they begin making synapses. Once puberty begins the number of neurons and synapses decrease remarkably. A loss of 100,000 synapses per second (synaptic pruning) is estimated to occur in adolescence. The last phase of brain development is the maturation and myelination of the brain, specifically the cortex. This process continues until at least 30 years of age (Kolb & Whishaw, 2015). Although this process is standard in all humans, sex differences are apparent. Females show more rapid brain growth than their male counterparts and attain their maximum brain volume about 5 years earlier. Myelination of the brain requires more time to complete in males relative to females (Lenroot, Gogtay, & Greenstein, 2007). These results suggest that behavioral development in males should also show delay in some domains and that brain plasticity associated with development persists over a longer time period in males.

There is evidence of sex-specific changes to brain regions during brain maturation. A larger increase in hippocampal and striatal volume in females, and a larger increase in amygdala volume in males, this is thought to result from the activational effects brought on by the changes in hormones levels during puberty. In parallel, it has been shown that in primates that there are more androgen receptors in the amygdala and more estrogen receptors in the hippocampus (Neufang et al., 2009).

Finally, in a diffusion tensor imaging study published in 2013 with more than 400 participants of each sex, it was noted that sex differences arise in brain connectivity with maturation. Adult females show greater interhemispheric connectivity than males whereas males show greater intrahemispheric connectivity than females (Fig. 14.2). These differences were not observed in children, adolescents, or even young adults. As a consequence, males demonstrate enhanced spatial processing, sensorimotor speed and a more modular brain, whereas females show enhanced attention, word and face memory, and a more integrated brain. Overall, these findings suggest that males are likely to complete a task in progress before moving on to another. In contrast, females are multitaskers (Ingalhalikar et al., 2013).

14.4.2 Biological differences

Although sex chromosomes play a huge role in sexual differentiation, they are not the only genetic factor affecting sexual differentiation. In a recent study analyzing over 1100 postmortem brain samples from various brain regions (frontal cortex, occipital cortex, temporal cortex, intralobular white matter, hypothalamus, medulla, cerebral

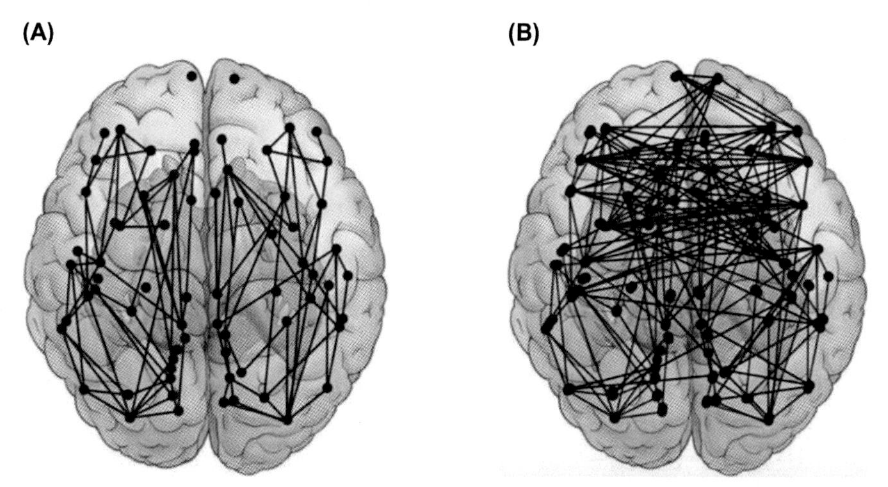

Figure 14.2 Sex differences in brain connectivity: (A) male and (B) female brain. *From Kolb, B., Teskey, G.C., & Whishaw, I.Q. (2016). An introduction to brain and behavior. NY: Worth Publishers.*

cortex, and spinal cord), originating from 137 individuals, it was discovered that 448 genes (2.6%) of the total genes expressed in the human central nervous system are differentially expressed based on sex (Trabzuni et al., 2013). Of these differentially expressed genes, over 85% were detected based on sex-biased splicing. Splicing is a posttranslational modification process where introns are removed from immature mRNA and the exons are then spliced back together to create the mature mRNA (Roy & Gilbert, 2006). This suggests that on top of gene differences, the mechanisms underlying the production of proteins are also different between the sexes. Most notable is the finding from Trabzuni et al. (2013) that 95% of genes with sex-biased splicing and 34% of genes with sex-biased expression were mapped to autosomes. This demonstrates that not all sex differences observed in the brain are due to differences in sex chromosomes alone. It appears that the other chromosomes are also important for observed differences.

14.4.2.1 Sex differences in HPA axis regulation

The HPA axis is an important signaling cascade used to maintain homeostasis in living organisms. The HPA axis is designed to regulate internal and external stressors through the use of various hormones. The response starts in the hypothalamus with the release of corticotropin-releasing hormone (CRH), which acts on the pituitary to release adrenocorticotropic hormone (ACTH). ACTH then stimulates the release of adrenal glucocorticoids into the blood stream, which creates the body's stress response. Finally, glucocorticoids act as negative feedback messengers in the brain to dissipate the stress response (Young, Korszun, Figueiredo, Banks-Solomon, & Herman, 2008).

Animal studies of the development of the HPA axis have provided insight into the importance of the hypothalamic–pituitary–gonadal (HPG) axis and subsequent sex differences. A persistent observation is that female rats have a significantly higher response to stress, and therefore increased corticosterone secretion after stress, compared to males (Panagiotakopoulos & Neigh, 2014). It has been shown that male rats castrated after birth and reared under normal conditions, when supplemented with testosterone in adulthood show female typical patterns in response to stressors. This suggests that testosterone may have an organizational effect on the development of the HPA axis.

In the human literature, sex differences in the HPA axis appear during development with the interaction between the HPA axis and the HPG axis. CRH has an inhibitory impact on the HPG axis directly (acts on specific brain structures) and indirectly (acts on neurotransmission), and has been shown to decrease testosterone levels in men and inhibit ovulation in females. And in a reciprocal fashion, the HPG axis can modulate CRH. For example, estrogen has been shown to increase transcription of the CRH gene whereas androgens have been shown to downregulate CRH gene transcription (Panagiotakopoulos & Neigh, 2014). Dahl et al. (1992) found that prepubescent boys and girls differ in their HPA axis response, with boys reaching higher peak values of cortisol 30 minutes later after CRH infusion than girls. In a study done with adults, response to CRH resulted in a greater increase in cortisol levels in women compared to men, suggestive of "activational" sex differences in the postpuberty period (Born, Ditschuneit, Schreiber, Dodt, & Fehm, 1995). To further the evidence of "organizational" effects on the HPA axis, Heim, Newport, Mletzko, Miller, and Nemeroff (2008) looked at early life stressors (for more information on early life stressors see Chapter 16: Socioeconomic Status) and found that women with childhood stress had an increased ACTH response to stress as adults compared to controls. Men with childhood stress, on the other hand, had increased cortisol levels in response to stress as an adult (Heim et al., 2008).

14.4.3 Differences in immunity

Microglial cells comprise the brain's immune system and play a critical role in the brain during injury, degeneration, and chronic stress (Ajami et al., 2007). More recently, research has uncovered other roles for microglia in the brain including plasticity, neurogenesis, apoptosis, and most notable for this discussion, sexual differentiation of the brain (Nimmerjahn, Kirchhoff, & Helmchen, 2005; Sierra et al., 2010; Lenz, Nugent, Hailyur, & McCarthy, 2013). During development, microglia begin populating the brain and by birth in rodents, there are already significant sex differences in the number and morphology of microglia in various areas of the brain (preoptic area, hippocampus, parietal cortex, and amygdala). For example, in the preoptic area the microglia in the male brain have larger cell body size and a decrease in process length and branching when compared to females (Lenz & McCarthy, 2015). The roles of microglia in the immature brain differ significantly from those attributed to microglia in the mature brain (injury or inflammation response), and these functional differences are thought to support the developing brain (Lenz et al., 2013). Research has shown that microglia regulate processes that are critical for development including synapse elimination, spinogenesis, and spine elimination (Tremblay et al., 2011). Lenz et al. (2013) found that treating 2-day-old female rats with estradiol (the precursor to testosterone) led to complete masculinization of microglial expression. This finding adds to previous findings showing that microglia are critical for sex-specific development.

14.4.4 Sex differences in behavior

It is not far-fetched to believe that with all the sex differences reported in the brain and its development, sex differences exist in behavior. There is compelling evidence for sex differences in at least five cognitive domains; verbal abilities, spatial analysis, motor skills, mathematical aptitude, and perception. In regards to verbal fluency and memory, women are superior. It has long been known that girls begin talking before boys and this may contribute to the observed sex differences (Wallentin, 2009). However, when comparing spatial abilities men outperform women on mental rotation, spatial navigation, and geographical knowledge, while women are better at spatial memory (McBurney, Gaulin, Devineni, & Adams, 1997; Voyer & Voyer, 1995). There seems to be a mix of sex differences when it comes to motor skills; with men being better at throwing and catching and women being better at fine motor skills (Hall & Kimura, 1995; Nicholson & Kimura, 1996). The same is seen with mathematical abilities, where females are better at computation and males better at mathematical reasoning (Hyde, Fennema, & Lamon, 1990). Lastly, when it comes to perception, females seem to have an advantage over males to sensory stimuli (having a lower threshold), sensory speed (faster detection), noticing subtle body and facial cues, and better recognition memory (Kolb & Whishaw, 2015).

14.5 CONCLUSION

The developing brain is remarkable and the processes behind its development are complex. Although the organizational/activational theory originally proposed by Phoenix and colleagues still holds true today, much has been added to expand and understand this phenomenon. As shown, the sex differences seen in humans are a complex interaction between, not only, the sex chromosomes but dimorphic gene expression, sex hormones, and the prenatal and perinatal environment. Environmental contaminants, drugs, and diet can interfere with typical development. In addition, sex differences that arise in response to trauma, addiction, and social influences are well described (see Chapter 16: Socioeconomic Status for more details).

REFERENCES

Ajami, B., Bennett, J. L., Krieger, C., Tetzlaff, W., & Rossi, F. M. (2007). Local self-renewal can sustain CNS microglia maintenance and function throughout adult life. *Nature Neuroscience*, *10*, 1538–1543.

Allen, A. M., Oncken, C., & Hatsukami, D. (2014). Women and smoking: The effect of gender on the epidemiology, health effects, and cessation of smoking. *Current Addiction Reports*, *1*, 53–60.

Arnold, A. P. (2009). The organizational-activational hypothesis as the foundation for a unified theory of sexual differentiation of all mammalian tissues. *Hormones and Behavior*, *55*, 570–578.

Bojesen, A., Juul, S., & Gravholt, C. H. (2003). Prenatal and postnatal prevalence of Linefelter syndrome: A national registry study. *Journal of Clinical Endocrniology and Metabolism*, *88*, 622–626. Available from http://dx.doi.org/10.1210/jc.2002-021491.

Born, J., Ditschuneit, I., Schreiber, M., Dodt, C., & Fehm, H. L. (1995). Effects of age and gender on pituitary–adrenocortical responsiveness in humans. *European Journal of Endocrinology*, *132*(6), 705–711.

Carruth, L. L., Reisert, I., & Arnold, A. P. (2002). Sex chromosome genes directly affect grain sexual differentiation. *Nature Neuroscience*, *5*(10), 933–934. Available from http://dx.doi.org/10.1038/nn922.

Centeno, M. L., Henderson, J. A., Pau, K. Y. F., & Bethea, C. L. (2006). Estradiol increases alpha7 nicotinic receptor in serotonergic dorsal raphe and noradrenergic locus coeruleus neurons of macaques. *Journal of Comparative Neurology, 497*, 489–501.

Center for Behavioral Health Statistics and Quality. (2016). 2015 National survey on drug use and health: Detailed tables. Substance Abuse and Mental Health Services Administration, Rockville, MD.

Centers for Disease Control and Prevention. (2016). Current cigarette smoking among adults—United States, 2005–2015. Morbidity and Mortality Weekly Report.

Cheng, C. Y., & Mruk, D. D. (2010). The biology of spermatogenesis: The past, present and future. *Philosophical Transactions of the Royal Society B—Biological Sciences, 365*(1546), 1459–1463.

Cohen-Bendahan, C. C., van de Beek, C., & Berenbaum, S. A. (2005). Prenatal sex hormone effects on child and adult sex-typed behavior: Methods and findings. *Neuroscience and Biobehavioral Reviews, 29*, 353–384.

Concerns for the use of soy-based formulas in infant nutrition (2009). *Paediatrics & Child Health, 14*(2), 109–113.

Cross, S. J., Linker, K. E., & Leslie, F. M. (2017). Sex-dependent effects of nicotine on the developing brain. *Journal of Neuroscience Research, 95*(1–2), 422–436. Available from http://dx.doi.org/10.1002/jnr.23878.

Dahl, R. E., Siegel, S. F., Williamson, D. E., Lee, P. A., Perel, J., Birmaher, B., & Ryan, N. D. (1992). Corticotropin releasing hormone stimulation test and nocturnal cortisol levels in normal children. *Pediatric Research, 32*(1), 64–68.

Davies, W., Isles, A. R., Burgoyne, P. S., & Wilkinson, L. S. (2006). X-linked imprinting: Effects on brain and behaviour. *BioEssays, 28*, 35–44.

Davies, W., & Wilkinson, L. S. (2006). It is not all hormones: Alternative explanations for sexual differentiation of the brain. *Brain Research, 1126*, 36–45.

Dewing, P., Chiang, C. W. K., Sinchak, K., Sim, H., Fernagut, P.-O., Kelly, S., ... Vilain, E. (2006). Direct regulation of adult brain function by the male-specific factor SRY. *Current Biology, 16*(4), 415–420. <http://dx.doi.org/10.1016/j.cub.2006.01.017>.

Dwyer, J. B., McQuown, S. C., & Leslie, F. M. (2009). The dynamic effects of nicotine on the developing brain. *Pharmacology & Therapeutics, 122*(2), 125–139. <http://dx.doi.org/10.1016/j.pharmthera.2009.02.003>.

Ehrhardt, A. A., & Meyer-Bahlburg, H. F. L. (1979). Prenatal sex hormones and the developing brain: Effects on psychosexual differentiation and cognitive function. *Annual Review of Medicine, 30*, 417–430.

Ernst, M., Moolchan, E. T., & Robinson, M. L. (2001). Behavioral and neural consequences of prenatal exposure to nicotine. *Journal of the American Academy of Child & Adolescent Psychiatry, 40*(6), 630–641. Available from http://dx.doi.org/10.1097/00004583-200106000-00007.

Eyler, F. D., & Behnke, M. (1999). Early development of infants exposed to drugs prenatally. *Clinics in Perinatology, 26*(1), 107–150.

Fox, H. C., Garcia, M., Kemp, K., Milivojevic, V., Kreek, M. J., & Sinha, R. (2006). Gender differences in cardiovascular and corticoadrenal response to stress and drug cues in cocaine dependent individual. *Psychopharmacology, 185*, 348–357.

Franklin, T. R., Acton, P. D., Maldjian, J. A., Gray, J. D., Croft, J. R., Dackis, C. A., & Childress, A. R. (2002). Decreased gray matter concentration in the insular, orbitofrontal, cingulate, and temporal cortices of cocaine patients. *Biological Psychiatry, 51*(2), 134–142. Available from http://dx.doi.org/10.1016/S0006-3223(01)01269-0.

Fried, P. A., Watkinson, B., & Gray, R. (1998). Differential effects on cognitive functioning in 9- to 12-year olds prenatally exposed to cigarettes and marihuana. *Neurotoxicology and Teratology, 20*(3), 293–306. <http://dx.doi.org/10.1016/S0892-0362(97)00091-3>.

Gangitano, D., Salas, R., Teng, Y., Perez, E., & De Biasi, M. (2009). Progesterone modulation of alpha5 nAChR subunits influences anxiety-related behavior during estrus cycle. *Genes, Brain and Behavior, 8*, 398–406.

Gore, A. C., Martien, K. M., Gagnidze, K., & Pfaff, D. (2014). Implications of prenatal steroid perturbations for neurodevelopment, behavior, and autism. *Endocrine Reviews*, *35*(6), 961–991. Available from http://dx.doi.org/10.1210/er.2013-1122.

Guillette, L. J., Crain, D. A., Rooney, A. A., & Pickford, D. B. (1995). Organization versus activation: The role of endocrine-disrupting contaminants (EDCs) during embryonic development in wildlife. *Environmental Health Perspectives*, *103*(Suppl 7), 157–164.

Gustavson, K. H. (1999). Triple X syndrome deviation with mild symptoms. The majority goes undiagnosed. *Lakartidningen*, *96*(50), 5646–5647.

Hall, J. A. Y., & Kimura, D. (1995). Sexual orientation and performance on sexually dimorphic motor tasks. *Archives of Sexual Behavior*, *24*(4), 395–407. Available from http://dx.doi.org/10.1007/bf01541855.

Hamers, T., Kamstra, J. H., Sonneveld, E., Murk, A. J., Kester, M. H. A., Andersson, P. L., … Brouwer, A. (2006). In vitro profiling of the endocrine-disrupting potency of brominated flame retardants. *Toxicological Sciences*, *92*(1), 157–173. Available from http://dx.doi.org/10.1093/toxsci/kfj187.

Hampson, E., Rovet, J. F., & Altmann, D. (1998). Spatial reasoning in children with congenital adrenal hyperplasia due to 21-hydroxylase deficiency. *Developmental Neuropsychology*, *14*(2–3), 299–320.

Heim, C., Newport, D. J., Mletzko, T., Miller, A. H., & Nemeroff, C. B. (2008). The link between childhood trauma and depression: Insights from HPA axis studies in humans. *Psychoneuroendocrinology*, *33*(6), 693–710. <http://dx.doi.org/10.1016/j.psyneuen.2008.03.008>.

Hughes, I. A., & Deeb, A. (2006). Androgen resistance. *Best Practice & Research Clinical Endocrinology & Metabolism*, *20*(4), 577–598. Available from http://dx.doi.org/10.1016/j.beem.2006.11.003.

Hyde, J. S., Fennema, E., & Lamon, S. J. (1990). Gender differences in mathematics performance: A meta-analysis. *Psychological Bulletin*, *107*(2), 139–155.

Imperato-McGlnley, J., Plchardo, M., Gautier, T., Voyer, D., & Bryden, M. P. (1991). Cognitive abilities in androgen-insensitive subjects: Comparison with control males and females from the same kindred. *Clinical Endocrinology*, *34*(5), 341–347. Available from http://dx.doi.org/10.1111/j.1365-2265.1991.tb00303.x.

Itti, E., Gonzalo, I. T. G., Pawlikowska-Haddal, A., Boone, K. B., Mlikotic, A., Itti, L., … Swerdloff, R. S. (2006). The structural brain correlates of cognitive deficits in adults with Klinefelter's syndrome. *The Journal of Clinical Endocrinology & Metabolism*, *91*(4), 1423–1427. Available from http://dx.doi.org/10.1210/jc.2005-1596.

Ingalhalikar, M., Smith, A., Parker, D., Satterthwaite, P. D., Elliot, M. A., Ruparel, K., et al. (2013). Sex differences in the structural connectome of the human brain. *Proceedings of the National Academy of Sciences of the United States of America*, *111*, 823–828.

Iwasa, Y., & Pomiankowski, A. (2001). The evolution of X-linked genomic imprinting. *Genetics*, *158*, 1801–1809.

Jin, X., & Steinbach, J. H. (2015). Potentiation of neuronal nicotinic receptors by 17b-estradiol: Roles of the carboxy-terminal and the amino-terminal extracellular domains. *PLoS ONE*, *10*, e0144631.

Jornda, G., Deeb, S. S., Bosten, J. M., & Mollon, J. D. (2010). The dimensionality of color vision in carriers of anomalous trichromacy. *Journal of Vision*, *10*(8), 1–19.

Ke, L., & Lukas, R. (1996). Effects of steroid exposure on ligand binding and functional activities of diverse nicotinic acetyicholine receptor subtypes. *Journal of Neurochemistry*, *67*, 1100–1112.

Kolb, B., Teskey, G. C., & Whishaw, I. Q. (2016). *An introduction to brain and behavior.* NY: Worth Publishers.

Kolb, B., & Whishaw, I. Q. (2015). Fundamentals of Human Neuropsychology, (7th ed.). New York: Worth Publishers.

Kumar, A. M., Haney, M., Becker, T., Thompson, M. L., Kream, R. M., & Miczek, K. (1990). Effect of early exposure to δ-9-tetrahydrocannabinol on the levels of opioid peptides, gonadotropin-releasing hormone and substance P in the adult male rat brain. *Brain Research*, *525*(1), 78–83. <http://dx.doi.org/10.1016/0006-8993(90)91322-8>.

Leatherdale, S. T., Hammond, D., & Ahmed, R. (2008). Alcohol, marijuana, and tobacco use patterns among youth in Canada. *Cancer Causes & Control*, *19*(4), 361–369. Available from http://dx.doi.org/10.1007/s10552-007-9095-4.

Leon-Olea, M., Martyniuk, C. J., Orlando, E. F., Ottinger, M. A., Rosenfeld, C. S., Wolstenholme, J. T., & Trudeau, V. L. (2014). Current concepts in neuroendocrine disruption. *General and Comparative Endocrinology, 203*, 158–173. Available from http://dx.doi.org/10.1016/j.ygcen.2014.02.005.

Lenroot, R. K., Gogtay, N., Greenstein, D. K., et al. (2007). Sexual dimorphism of brain development trajectories during childhood and adolescence. *NeuroImage, 36*, 1065–1073.

Lenz, K. M., & McCarthy, M. M. (2015). A starring role for microglia in brain sex differences. *Neuroscientist, 21*(3), 306–321. Available from http://dx.doi.org/10.1177/1073858414536468.

Lenz, K. M., Nugent, B. M., Haliyur, R., & McCarthy, M. M. (2013). Microglia are essential to masculinization of brain and behavior. *The Journal of Neuroscience, 33*(7), 2761–2772. Available from http://dx.doi.org/10.1523/JNEUROSCI.1268-12.2013.

Li, C. R., Kemp, K., Milivojevic, V., & Sinha, R. (2005). Neuroimaging study of sex differences in the neuropathology of cocaine abuse. *Gender Medicine, 2*(3), 174–182. <http://dx.doi.org/10.1016/S1550-8579(05)80046-4>.

Lovell-Badge, R., & Robertson, E. (1990). XY female mice resulting from a heritable mutation in the primary testis-determining gene, *Tdy. Development, 109*, 635–646.

Maffini, M. V., Rubin, B. S., Sonnenschein, C., & Soto, A. M. (2006). Endocrine disruptors and reproductive health: The case of bisphenol-A. *Molecular and Cellular Endocrinology, 254–255*, 179–186. http://doi.org/10.1016/j.mce.2006.04.033.

May, K. M., Jacobs, P. A., Lee, M., Ratcliffe, S., Robinson, A., Nielsen, J., & Hassold, T. J. (1990). The parental origin of the extra X chromosome in 47, XXX females. *American Journal of Human Genetics, 46*, 754–761.

McBurney, D. H., Gaulin, S. J. C., Devineni, T., & Adams, C. (1997). Superior spatial memory of women: Stronger evidence for the gathering hypothesis. *Evolution and Human Behavior, 18*(3), 165–174. http://doi.org/10.1016/S1090-5138(97)00001-9.

McCarthy, M. M., & Arnold, A. P. (2011). Reframing sexual differentiation of the brain. *Nature Neuroscience, 14*(6), 677–683. Available from http://dx.doi.org/10.1038/nn.2834.

Merke, D. P., & Bornstein, S. R. (2005). Congenital adrenal hyperplasia. *The Lancet, 365*(9477), 2125–2136.

Merke, D. P., Fields, J. D., Keil, M. F., Vaituzis, A. K., Chrousos, G. P., & Giedd, J. N. (2003). Children with classic congenital adrenal hyperplasia have decreased amygdala volume: Potential prenatal and postnatal hormonal effects. *The Journal of Clinical Endocrinology & Metabolism, 88*(4), 1760–1765. Available from http://dx.doi.org/10.1210/jc.2002-021730.

Murphy, D. G. M., DeCarli, C., Daly, E., Haxby, J. V., Allen, G., McIntosh, A. R., ... Powell, C. M. (1993). X-chromosome effects on female brain: A magnetic resonance imaging study of Turner's syndrome. *The Lancet, 342*(8881), 1197–1200. Available from http://dx.doi.org/10.1016/0140-6736(93)92184-U.

Murphy, L. L., Muñoz, R. M., Adrian, B. A., & Villanúa, M. A. (1998). Function of cannabinoid receptors in the neuroendocrine regulation of hormone secretion. *Neurobiology of Disease, 5*(6), 432–446. Available from http://dx.doi.org/10.1006/nbdi.1998.0224.

Mychasiuk, R., Muhammad, A., Gibb, R., & Kolb, B. (2013). Long-term alterations to dendritic morphology and spine density associated with prenatal exposure to nicotine. *Brain Research, 1499*, 53–60. Available from http://dx.doi.org/10.1016/j.brainres.2012.12.021.

Nass, R., & Baker, S. (1991). Learning disabilities in children with congenital adrenal hyperplasia. *Journal of Child Neurology, 6*(4), 306–312. Available from http://dx.doi.org/10.1177/088307389100600404.

Neufang, S., Specht, K., Hausmann, M., Güntürkün, O., Herpertz-Dahlmann, B., Fink, G. R., & Konrad, K. (2009). Sex differences and the impact of steroid hormones on the developing human brain. *Cerebral Cortex, 19*(2), 464–473. Available from http://dx.doi.org/10.1093/cercor/bhn100.

Ngun, T. C., Ghahramani, N. M., Creek, M. M., Williams-Burris, S. M., Barseghyan, H., Itoh, Y., ... Vilain, E. (2014). Feminized behavior and brain gene expression in a novel mouse model of Klinefelter syndrome. *Archives of Sexual Behavior, 43*(6), 1043–1057.

Nicholson, K. G., & Kimura, D. (1996). Sex differences for speech and manual skill. *Perceptual and Motor Skills, 82*(1), 3–13. Available from http://dx.doi.org/10.2466/pms.1996.82.1.3.

Nieschlag, E., Werler, S., Wistuba, J., & Zitzmann, M. (2014). New approaches to the Klinefelter syndrome. *Annales d'Endocrinologie*, *75*(2), 88–97. Available from http://dx.doi.org/10.1016/j.ando.2014.03.007.

NIDA (2015). Nationwide trends. Retrieved March 24, 2017, from https://www.drugabuse.gov/publications/drugfacts/nationwide-trends.

Nijhuis-van der Sanden, M. W. G., Eling, P. A. T. M., & Otten, B. J. (2003). A review of neuropsychological and motor studies in Turner Syndrome. *Neuroscience & Biobehavioral Reviews*, *27*(4), 329–338.

Nimmerjahn, A., Kirchhoff, F., & Helmchen, F. (2005). Resting microglial cells are highly dynamic surveillants of brain parenchyma in vivo. *Science*, *308*(13), 14–18.

O'Connor, M. J., & Paley, B. (2009). Psychiatric conditions associated with prenatal alcohol exposure. *Developmental Disabilities Research Reviews*, *15*, 225–234.

Otter, M., Schrander-Stumpel, C. T. R. M., & Curfs, L. M. G. (2009). Triple X syndrome: A review of the literature. *European Journal of Human Genetics*, *18*(3), 265–271.

Panagiotakopoulos, L., & Neigh, G. N. (2014). Development of the HPA axis: Where and when do sex differences manifest? *Frontiers in Neuroendocrinology*, *35*(3), 285–302. Available from http://dx.doi.org/10.1016/j.yfrne.2014.03.002.

Phoenix, C. H., Goy, R. W., Gerall, A. A., & Young, W. C. (1959). Organizing action of prenatally administered testosterone propionate on the tissues mediating mating behavior in the female guinea pig. *Endocrinology*, *65*, 369–382.

Piper, M. E., Cook, J. W., Schlam, T. R., Jorenby, D. E., Smith, S. S., Bolt, D. M., & Loh, W. Y. (2010). Gender, race, and education differences in abstinence rates among participants in two randomized smoking cessation trials. *Nicotine and Tobacco Research*, *6*, 647–657.

Prokop, J. W., Underwood, A. C., Turner, M. E., Miller, N., Pietrzak, D., Scott, S., ... Milsted, A. (2013). Analysis of Sry duplications on the *Rattus norvegicus* Y-chromosome. *BMC Genomics*, *14*, 15. Available from http://dx.doi.org/10.1186/1471-2164-14-792.

Ramlau-Hansen, C. H., Nohr, E. A., Thulstrup, A. M., Bonde, J. P., Storgaard, L., & Olsen, J. (2007). Is maternal obesity related to semen quality in the male offspring? A pilot study. *Human Reproduction*, *22*(10), 2758–2762. Available from http://dx.doi.org/10.1093/humrep/dem219.

Raudrant, D., & Rabe, T. (2003). Progestogens with anitandrogenic properties. *Drugs*, *63*(5), 463–492.

Reik, W., & Walter, J. (2001). Genomic imprinting: Parental influence on the genome. *Nature Reviews Genetics*, *2*, 21–32.

Roy, S. W., & Gilbert, W. (2006). The evolution of spliceosomal introns: Patterns, puzzles and progress. *Nature Reviews*, *7*, 211–221. Available from http://dx.doi.org/10.1038/nrg1807.

Rubin, B. S. (2011). Bisphenol A: An endocrine disruptor with widespread exposure and multiple effects. *The Journal of Steroid Biochemistry and Molecular Biology*, *127*(1–2), 27–34. http://doi.org/10.1016/j.jsbmb.2011.05.002.

Ruigrok, A. N. V., Salimi-Khorshidi, G., Lai, M.-C., Baron-Cohen, S., Lombardo, M. V., Tait, R. J., & Suckling, J. (2014). A meta-analysis of sex differences in human brain structure. *Neuroscience & Biobehavioral Reviews*, *39*, 34–50. Available from http://dx.doi.org/10.1016/j.neubiorev.2013.12.004.

Sharpe, R. M., Martin, B., Morris, K., Greig, I., McKinnell, C., McNeilly, A. S., & Walker, M. (2002). Infant feeding with soy formula milk: Effects on the testis and on blood testosterone levels in marmoset monkeys during the period of neonatal testicular activity. *Human Reproduction*, *17*(7), 1692–1703. Available from http://dx.doi.org/10.1093/humrep/17.7.1692.

Sierra, A., Encinas, J. M., Deudero, J. J. P., Chancey, J. H., Enikolopov, G., Overstreet-Wadiche, L. S., & Maletic-Savatic, M. (2010). Microglia shape adult hippocampal neurogenesis through apoptosis-coupled phagocytosis. *Cell Stem Cell*, *7*(4), 483–495. Available from http://dx.doi.org/10.1016/j.stem.2010.08.014.

Sinha, R., Fox, H., Hong, K.-I., Sofuoglu, M., Morgan, P. T., & Bergquist, K. T. (2007). Sex steroid hormones, stress response, and drug craving in cocaine-dependent women: Implications for relapse susceptibility. *Experimental and Clinical Psychopharmacology*, *15*(5), 445–452. Available from http://dx.doi.org/10.1037/1064-1297.15.5.445.

Skaletsky, H., Kuroda-Kawaguchi, T., Minx, P. J., Cordum, H. S., Hillier, L., Brown, L. G., ... Page, D. C. (2003). The male-specific region of the human Y chromosome is a mosaic of discrete sequence classes. *Nature*, *423*, 825.

Skuse, D. H., James, R. S., Bishop, D. V. M., Coppin, B., Dalton, P., Aamodt-Leeper, G., ... Jacobs, P. A. (1997). Evidence form Turner's syndrome of an imprinted X-linked locus affecting cognitive functions. *Nature*, *387*, 705–708.

Spencer, R. L., & Hutchison, K. E. (1999). Alcohol, aging, and the stress response. *Alcohol Research & Health*, *23*(4), 272–283.

Thanh, N. X., & Jonsson, E. (2010). Drinking alcohol during pregnancy: Evidence from Canadian Community Health Survey 2007/2008. *Journal of Population Therapeutics and Clinical Pharmacology*, *17*, 302–307.

Tong, V. T., Dietz, P. M., Morrow, B., D'Angelo, D. V., Farr, S. L., Rockhill, K. M., ... Centers for Disease Control and Prevention (2013). Trends in smoking before, during, and after pregnancy—Pregnancy risk assessment monitoring system, United States, 40 sites, 2000–2010. *MMWR Surveillance Summaries*, *62*, 1–19.

Trabzuni, D., Ramasamy, A., Imran, S., Walker, R., Smith, C., Weale, M. E., & Ryten, M. (2013). Widespread sex differences in gene expression and splicing in the adult human brain. *Nature Communications*, *4*, 1–7. Available from http://dx.doi.org/10.1038/ncomms3771.

Tremblay, M.-È., Stevens, B., Sierra, A., Wake, H., Bessis, A., & Nimmerjahn, A. (2011). The role of microglia in the healthy brain. *The Journal of Neuroscience*, *31*(45), 16064–16069. Available from http://dx.doi.org/10.1523/jneurosci.4158-, 11.2011

Uematsu, A., Yorifuji, T., Muroi, J., Kawai, M., Mamada, M., Kaji, M., & Nakahata, T. (2002). Parental origin of normal X chromosomes in Turner syndrome patients with various karyotypes: Implications for the mechanism leading to generation of a 45,X karyotype. *American Journal of Medical Genetics*, *111*, 134–139.

Vandenberg, L. N., Chahoud, I., Heindel, J. J., Padmanabhan, V., Paumgartten, F. J. R., & Schoenfelder, G. (2010). Urinary, circulating, and tissue biomonitoring studies indicate widespread exposure to bisphenol A. *Environmental Health Perspectives*, *118*(8), 1055–1070.

Vathy, I., Katay, L., & Mini, K. N. (1993). Sexually dimorphic effects of prenatal cocaine on adult sexual-behavior and brain catecholamines in rats. *Developmental Brain Research*, *73*(1), 115–122. http://dx.doi.org/10.1016/0165-3806(93)90053-d.

Voyer, D., & Voyer, S. (1995). Magnitude of sex differences in spatial abilities: A meta-analysis and consideration of critical variables. *Psychological Bulletin*, *117*(2), 250–270.

Wallen, K. (2009). The organizational hypothesis: Reflections on the 50th anniversary of the publication of Phoenix, Goy, Gerall, and Young (1959). *Hormones and Behavior*, *55*(5), 561–565. Available from http://dx.doi.org/10.1016/j.yhbeh.2009.03.009.

Wallentin, M. (2009). Putative sex differences in verbal abilities and language cortex: A critical review. *Brain and Language*, *108*(3), 175–183. http://doi.org/10.1016/j.bandl.2008.07.001.

Weinberg, J., Sliwowska, J. H., Lan, N., & Hellemans, K. G. C. (2008). Prenatal alcohol exposure: Foetal programming, the hypothalamic–pituitary–adrenal axis and sex differences in outcome. *Journal of Neuroendocrinology*, *20*(4), 470–488. Available from http://dx.doi.org/10.1111/j.1365-2826.2008.01669.x.

Wenzel, U., Schneider, M., Zachmann, M., Knorr-Murset, G., Weber, A., & Prader, A. (1978). Intelligence of patients with congenital adrenal hyperplasia due to 21-hydroxylase deficiency, their parents and unaffected siblings. *Helvetica Paediatrica Acta*, *33*(1), 11.

White, P. C. (2009). Neonatal screening for congenital adrenal hyperplasia. *Nature Reviews Endocrinology*, *5*, 490–499.

Wilson, W., Mathew, R., Turkington, T., Hawk, T., Coleman, R. E., & Provenzale, J. (2000). Brain morphological changes and early marijuana use. *Journal of Addictive Diseases*, *19*(1), 1–22. Available from http://dx.doi.org/10.1300/J069v19n01_01.

Young, E. A., Korszun, A., Figueiredo, H. F., Banks-Solomon, M., & Herman, J. P. (2008). Sex differences in HPA axis regulation. In J. B. Becker, K. J. Berkley, N. Geary, E. Hampson, J. P. Herman, & E. A. Young (Eds.), *Sex differences in the brain* (pp. 95–105). New York, NY: Oxford University Press.

Zambrano, E., Bautista, C. J., Deás, M., Martínez-Samayoa, P. M., González-Zamorano, M., Lesesma, H., ... Nathanielsz, P. W. (2006). A low maternal protein diet during pregnancy and lactation has sex- and window of exposure-specific effects on offspring growth and food intake, glucose metabolism and serum leptin in the rat. *The Journal of Physiology*, *571*(1), 221–230. Available from http://dx.doi.org/10.1113/jphysiol.2005.100313.

Zuccarello, D., Ferlin, A., Vinanzi, C., Prana, E., Garolla, A., Callewaert, L., & Foresta, C. (2008). Detailed functional studies on androgen receptor mild mutations demonstrate their association with male infertility. *Clinical Endocrinology*, *68*, 580–588. Available from http://dx.doi.org/10.1111/j.1365-2265.2007.03069.x.

FURTHER READING

Cota, D. (2008). The role of the endocannabiniod system in the regulation of hypothalamic–pituitary–adrenal axis activity. *Journal of Neuroendocrinology*, *20*, 35–38. Available from http://dx.doi.org/10.1111/j.1365-2826.2008.01673.x.

Kirschbaum, C., Schommer, N., Federenko, I., Gaab, J., Neumann, O., Oellers, M., ... Hellhammer, D. H. (1996). Short-term estradiol treatment enhances pituitary–adrenal axis and sympathetic responses to psychosocial stress in healthy young men. *The Journal of Clinical Endocrinology & Metabolism*, *81*, 3639–3643.

Reynolds, R. M., Hii, H. L., Pennell, C. E., McKeague, I. W., Kloet, E. R., Lye, S., ... Foster, J. K. (2013). Analysis of baseline hypothalamic–pituitary– adrenal activity in late adolescence reveals gender specific sensitivity of the stress axis. *Psychoneuroendocrinology*, *38*, 1271–1280.

Uban, K. A., Comeau, W. L., Ellis, L. A., Galea, L. A. M., & Weinberg, J. (2013). Basal regulation of HPA and dopamine systems is altered differentially in males and females by prenatal alcohol exposure and chronic variable stress. *Psychoneuroendocrinology*, *38*(10), 1953–1966. Available from http://dx.doi.org/10.1016/j.psyneuen.2013.02.017.

CHAPTER 15

Injury

Mardee Greenham[1,2,*], Nicholas P. Ryan[1,2,*] and Vicki Anderson[1,2]
[1]Murdoch Childrens Research Institute, Parkville, VIC, Australia
[2]University of Melbourne, Melbourne, VIC, Australia

15.1 INTRODUCTION

Exploration of the consequences of brain insult sustained early in life has a long history, dating back to the 1920s and the seminal works of Lashley (1929), Kennard (1938, 1942), and Hebb (1949), and later comprehensive reviews (e.g., Finger & Stein, 1982; Isaacson, 1975; St James-Roberts, 1975). In this literature, while insults in infancy and early childhood were regarded as qualitatively and quantitatively distinct from those occurring in adulthood, there was little agreement regarding the potential benefits and vulnerabilities of early brain insult (EBI).

Despite lively and continuing debate, recovery from EBI remains imperfectly understood. Insults that would almost certainly result in severe cognitive dysfunction in the adult brain may have quite different consequences for the developing brain (Anderson, Spencer-Smith, & Wood, 2011; Aram, 1988; Bates et al., 2001; Jacobs & Anderson, 2002). For example, early vascular accidents need not preclude normal or higher intellectual and academic achievements (Ballantyne, Spilkin, Hesselink, & Trauner, 2008; Smith & Sugar, 1975). Even after hemispherectomy, when an entire cerebral hemisphere is removed, young children may develop functional cognitive abilities (Dennis & Whittaker, 1976). In contrast, early diffuse insults (e.g., traumatic brain injury) can result in slow recovery and poor outcome compared to similar injuries in adults (Anderson & Moore, 1995; Anderson, Catroppa, Morse, Haritou, & Rosenfeld, 2005, 2004; Yeates et al., 1997). These somewhat unpredictable recovery patterns after EBI are puzzling and require further careful consideration.

The aim of this chapter is to provide a developmental framework for the study of early childhood brain insults, to consider processes of brain development, research findings describing functional outcomes, and finally to explore early work attempting to identify links between these outcomes, and underlying brain correlates that may assist in better understanding outcomes from insult to the developing brain.

* These authors contributed equally to this work.

The Neurobiology of Brain and Behavioral Development
DOI: http://dx.doi.org/10.1016/B978-0-12-804036-2.00015-7

15.2 NORMAL BRAIN DEVELOPMENT

Protracted maturation of the developing brain is mediated by rapid and dynamic increases in neural connectivity that underpins the immature brain's increasing capacity to process and execute a series of highly specialized mental and behavioral functions. As shown in Fig. 15.1, gray matter volume peaks early and then declines, reflecting a process of synaptic pruning of nonmyelinated neurons (Courchesne et al., 2000). These processes coincide with a rapid expansion of white matter volume, a change that reflects increased myelination and a greater number of synaptic connections that drive exponential increases in structural connectivity across the formative decades of life (Giedd et al., 1996). Brain insult at any point across the course of development may irreversibly influence these critical maturational processes, giving rise to a pattern of disorganized, under connected and highly inefficient structural architecture that is likely to exert a cumulative negative influence on the acquisition and refinement of neurobehavioral skills. In line with this hypothesis, evidence

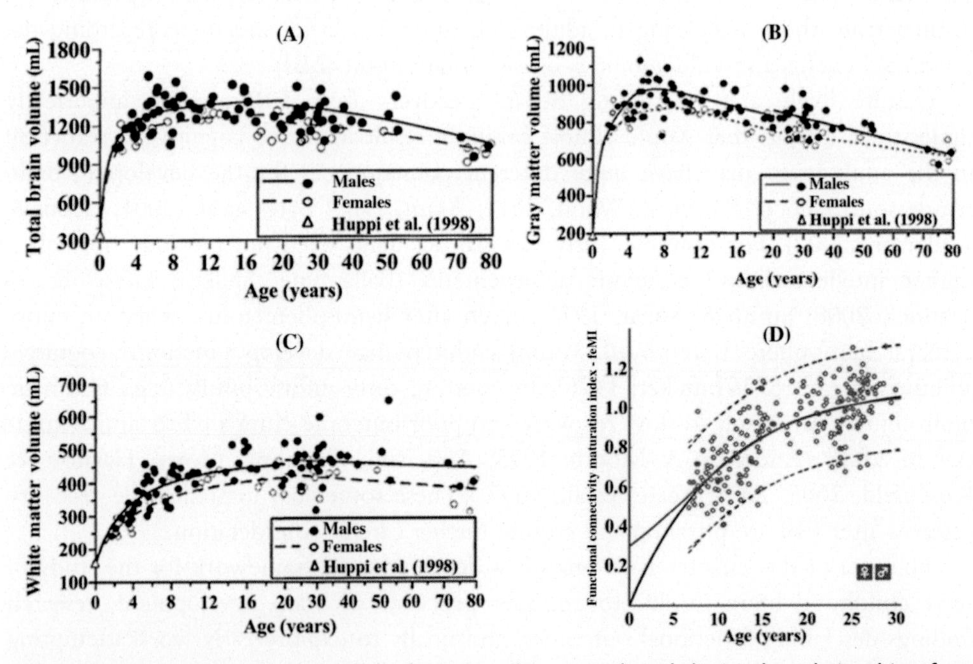

Figure 15.1 Figures (A), (B), and (C) are from Courchesne et al. and depict the relationship of age to total brain development in A which shows an apex in middle childhood but decline thereafter. Part (B) depicts age-dependent changes in gray matter with reductions thought to reflect neuronal pruning and part (C) shows the dynamic increase in white matter through mid-life, thought to reflect myelination and increased connectivity. Part (D) is taken from Vogel et al. and reflects a functional brain maturation curve based on resting state functional connectivity magnetic resonance imaging (rs-fcMRI) studies. Illustrations used with permission.

suggests that the developing brain may be particularly vulnerable to disruption during periods that coincide with critical periods of cognitive and behavioral development (Anderson et al., 2011; Crowe, Catroppa, Babl, Rosenfeld, & Anderson, 2012; Dennis, 1989); however, the pathological substrates of chronic neurobehavioral impairments are less well established.

In contrast to the structural properties of the adult brain which comprises networks of highly connected hub regions that mediate highly specialized high order processes, the lack of specialization within the developing brain confers a unique vulnerability to the effects of diffuse brain insult (Dennis, 1988; Anderson & Moore, 1995; Anderson et al., 2011). The immature status of the developing brain is illustrated by the dynamic nature of synaptic pruning as indexed by the loss of gray matter across the formative years. Fig. 15.2 illustrates the posterior-to-anterior gradient in volume loss of gray matter over the formative years of childhood and adolescence, which inversely correlate with the development of higher cognitive abilities and functioning (Giedd et al., 1996). Since anterior regions of the developing brain are the last to prune and are also among those regions most vulnerable to the focal and diffuse effects of TBI, an early vulnerability model would predict that disruption to these brain networks during critical periods of development is likely to exert cascading, deleterious consequences for subsequent brain and neurobehavioral development (Hebb, 1942; Anderson et al., 2011).

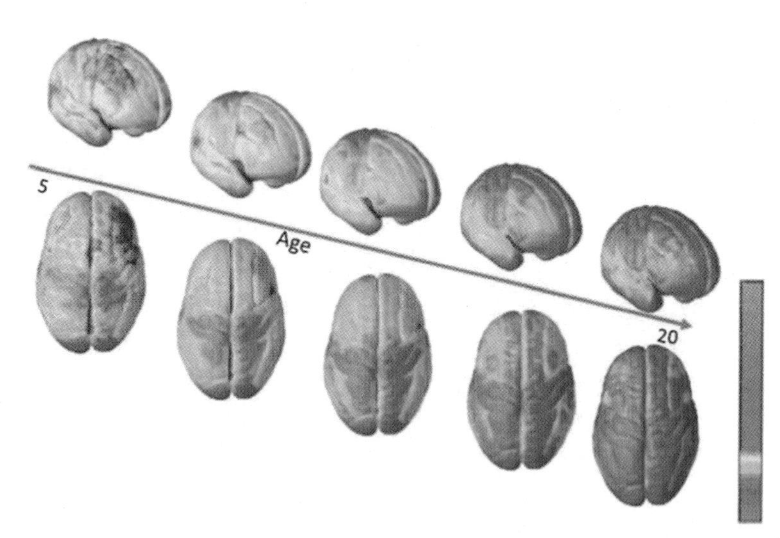

Figure 15.2 Figure from Giedd et al. demonstrating right lateral and top views of the dynamic sequence of gray matter maturation over the cortical surface. The side bar shows a color representation in units of gray matter volume. *Adapted from Gogtay et al. Figure used with permission.*

15.3 PLASTICITY

Plasticity is an intrinsic property of the central nervous system (CNS), reflecting its capacity to respond in a dynamic manner to the environment and experience via modification of neural circuitry. This phenomenon is linked to processes of brain development and function across the lifespan (Duffau, 2006; Kadis et al., 2007; Mosch, Max, & Tranel, 2005; Taupin, 2006). In the context of healthy development, plasticity is considered a beneficial property, facilitating adaptive change in response to environmental stimuli, such as routine learning or specific training and enrichment. In these circumstances, research has documented establishment of new neural connections as well as modifications to the mapping between neural activity and behavior.

In the context of brain injury, and associated disruption to developmental processes, the role of plasticity is less clear, with acknowledgment that the immature brain may not always benefit. While there may be an opportunity to take advantage of the immature brain's lack of functional specificity, via transfer of functions from damaged areas to healthy tissue, the brain's capacity for plasticity might also reflect "vulnerability," with predetermined developmental processes being derailed, neural resources depleted and an absence of a developmental "blueprint" to guide recovery (Anderson et al., 2011; Hebb, 1949; Kolb, 1995; Pascual-Leone, Amedi, Fregni, & Merabet, 2005). Recent progress in the neurosciences provides an exciting opportunity to investigate mechanisms underpinning plasticity and recovery and their behavioral correlates and to advance theoretical paradigms and clinical knowledge.

This debate is best illustrated by the contrasting "plasticity" versus "early vulnerability" approaches, which dispute whether the immature brain has a greater capacity for recovery than the mature or adult brain. This debate is argued at both "*brain*" and "behavior" levels. Plasticity theorists postulate that the young brain is immature, less committed and thus less susceptible to the impact of cerebral damage. Plasticity is thought to be maximal early in development when the CNS is less rigidly specialized (Huttenlocher & Dabholkar, 1997; Kennard, 1938), and synapses and dendritic connections remain unspecified. Such flexibility provides the capacity for transferring or reorganizing functions from damaged brain to healthy tissue. In contrast, early vulnerability proponents postulate that the young brain is uniquely sensitive to insult, and thus EBI is detrimental to development. Hebb (1949) argued that plasticity theories ignored the possibility that brain insult will have different consequences at different times throughout development. He concluded that EBI may be more detrimental than later injury, because cognitive development is critically dependent on the integrity of particular cerebral structures at certain stages of development. Thus, if a cerebral region is damaged at a critical stage of cognitive development it may be that cognitive skills dependent on that region are irreversibly impaired (Kolb, 1995; Luciana, 2003).

15.4 FACTORS CONTRIBUTING TO OUTCOMES FROM EBI

Over recent years, a range of factors have been explored that might reasonably be assumed to influence recovery. However, apart from the established relationship between *insult severity* and outcome, until recently, studies have failed to identify consistent links between outcome and specific insult characteristics (e.g., diffuse versus focal pathology, laterality) (Anderson et al., 2009; Bates et al., 2001; Stiles et al., 2008). With the advent of sophisticated neuroimaging modalities, and the emphasis on functional neural networks rather than only brain structures, findings are emerging that support associations between functional outcomes and disruptions to neural networks (Beauchamp et al., 2011; Ryan et al., 2014). The contribution of other factors has also been explored: presence of residual disability (e.g., hemiparesis, epilepsy) (Ballantyne, Spilkin, & Trauner, 2007), preinsult child and family factors (Anderson et al., 2006; Yeates et al., 1997), and environmental parameters (e.g., sociodemographics, access to interventions, parent/family function) (Anderson et al., 2006; Breslau, 1990; Catroppa & Anderson, 2008; Yeates et al., 1997). While each of these factors appears to contribute incrementally to outcome, we fall short of providing a complete picture of relevant predictors and their interactions.

A further potential piece in the puzzle is the *developmental stage of the child* at time of insult, with major controversy existing regarding the potential impact of this dimension for neurobehavioral and psychological outcome. A review of the literature relevant to these theories provides little clarification. While it is now evident that the young brain has some capacity for neural restitution, via either neural regrowth or anatomical reorganization (Giza & Prins, 2006; Kolb & Gibb, 2001), there is ongoing controversy as to the implications of these processes. Even if neural restitution does occur, full recovery may be limited by either: (1) inappropriate connections being established (Kolb, Pellis, & Robinson, 2004; Stein & Hoffman, 2003) resulting in dysfunctional behavioral recovery or (2) a "crowding effect" (Anderson et al., 2009; Aram & Eisele, 1994; Vargha-Khadem, Isaacs, Papaleloudim, Polkey, & Wilson, 1992), where functions normally subsumed by damaged tissue are crowded into remaining healthy brain areas, with a general depression of all abilities. In support of such concerns, studies of children with prenatal lesions, or those sustaining insults during the first year of life consistently report poorest functional outcomes (Anderson et al., 1997; Duchowny et al., 1996; Ewing-Cobbs et al., 1997; Jacobs, Harvey, & Anderson, 2007; Leventer et al., 1999; Riva & Cassaniga, 1986).

While the debate continues, there is little disagreement that developmental factors play a central role in outcome from EBI. The challenge remains to describe the nature of this relationship. To date, most research has employed single-condition approaches (e.g., dysplasia, stroke, traumatic brain insult), examining age effects within such conditions. Such designs are unable to investigate consequences of insults sustained, across the developmental period, from gestation to adolescence, as these conditions are necessarily age-specific (e.g., dysplasia: prenatal, traumatic brain injury: postnatal). To investigate developmental influences comprehensively, studies need to incorporate conditions occurring throughout gestation and childhood. Further, previous research has often assumed that age effects will be linear, that is, the younger the insult the poorer the outcome. Such an assumption is inconsistent with knowledge of brain maturation, where development is step-wise (Casey, Giedd, & Thompson, 2000; Gogtay et al., 2004), with critical maturational periods for processes such as myelination and synaptogenesis (Gotgay et al., 2004; Klinberg, Vaidya, Gabrieli, Moseley, & Hedehus, 1999), separated by more stable periods. Cognitive theorists describe similar stage-like processes Piaget, 1963). It is likely that disruption during one of these predetermined, neural or cognitive growth periods will cause "flow on" effects, as the establishment of other later emerging skills is thrown off course (Mosch et al., 2005). To date, insufficient evidence is available to pinpoint the timing or scope of these critical periods for humans; however, animal literature provides some insights (Kolb, 1995; Kolb et al., 2005), suggesting that age and recovery are not linearly related, but are associated, via underlying neural processes such as synaptogenesis, dendritic aborization, and myelination.

15.5 OUTCOMES FROM EBI: THE "BEHAVIOR" DIMENSION

15.5.1 Intellectual outcomes

EBI often results in ongoing cognitive impairment with lower intellectual function generally found, regardless of age at insult (Anderson et al., 2009). Within the EBI literature, there are consistent findings that age at insult does impacts long-term cognitive outcomes, yet the relationship appears to be quite complex.

A study from our lab has explored the influence of age at insult in a group of children ($n = 160$) with documented *focal brain lesions* of various etiologies. Results demonstrate that children sustaining EBI were at elevated risk for poorer intellectual function in late childhood (Anderson et al., 2009). Children who were younger than 2 years at insult ("early insult") performed worse than those who 2 years and older ("late insult"), with particular vulnerability in children who sustained an insult prior to and around the time of birth. More than half of the "early insult" group had mildly or significantly impaired full scale IQ scores, compared to around 25% to 30% of the "late insult" group. Similar results have been found in studies exploring specific insult types (e.g., traumatic brain injury, cerebral infection, stroke).

Within the *pediatric traumatic brain injury* (TBI) literature children who sustain an early injury have been consistently found to have poorer intellectual outcomes than children injured at an older age (Anderson & Moore, 1995; Anderson et al., 2005). In particular, these authors reported that children injured younger than 2 years displayed poorer outcomes than older children with similar injury severity (Anderson et al.,

2005). Severity of TBI has also been consistently linked to degree of intellectual impairment, with children sustaining moderate and severe TBI performing significantly below children with mild TBI (Anderson et al., 2004, 2005; Ewing-Cobbs et al., 2006; Ewing-Cobbs, Fletcher, Levin, Iovino, & Miner, 1998).

Following *pediatric stroke*, IQ is consistently skewed towards the lower end of the average range and is significantly lower than controls and standardized test norms (Everts et al., 2008; Gomes, Spencer-Smith, Jacobs, Coleman, & Anderson, 2012; Hajek et al., 2014; Max et al., 2002b; O'Keeffe et al., 2014; Pavlovic et al., 2006; Westmacott, MacGregor, & Askalan, 2009). Specific stroke characteristics also impact intellectual outcome in children poststroke, with larger lesion volume (Everts et al., 2008; Hajek et al., 2014; Hetherington, Tuff, Anderson, & Miles, 2005; Westmacott, Askalan, MacGregor, Anderson, & DeVeber, 2010) and lesions involving both cortical and subcortical regions (Westmacott et al., 2010) associated with poorer intellectual outcomes. Age at insult has also been found to be associated with intellectual outcome in these children, although the findings are somewhat different to those in TBI studies. Earlier age at onset, before 1 year of age, is associated with poorer cognition in some studies (Pavlovic et al., 2006; Westmacott et al., 2010), while others report poorer outcomes in later onset stroke, when brain maturity is more advanced (Allman & Scott, 2013). Cognitive outcomes vary widely between stroke studies due to largely cross-sectional design and differing methodology, such as varying age range of cohorts and wide range in time since diagnosis.

15.5.2 Executive function & attention

Impairments in executive function (EF) have also been reported in EBI. In their study of *focal brain insults*, Anderson et al. (2009) found impairment rates of parent and teacher ratings of everyday executive skills were between 45% and 55% across age at insult groups (Anderson et al., 2009). When age at insult groups were compared (prenatal, perinatal, infancy, preschool, middle childhood, late childhood), outcomes were somewhat different to those for intellectual outcomes and did not adhere to the expected linear pattern of younger age-worse outcome. In keeping with previous research from our group, children who sustained a focal brain insult in the prenatal or perinatal and infancy periods all showed global and significant executive dysfunction. In contrast, children who sustained a brain insult during the preschool period (3–6 years) performed closest to normative expectations. Surprisingly, a focal brain insult in middle childhood (7–9 years) resulted in significant problems and these children performed similarly to the pre- and perinatal insult groups. Finally, those sustaining an insult after age of 10 demonstrated largely intact executive abilities. Of note these findings, suggesting there is a "critical period" for poor outcome in middle childhood, have been replicated by Crowe et al. in a group of children with TBI.

There is also evidence that environmental factors may impact EF following EBI. In another paper from the Anderson et al. (2009) cohort, higher family dysfunction was found to be associated with poorer EF (Anderson et al., 2014a,b). While the reason for this link remains unclear, it is consistent with work exploring the impact of early deprivation (Belsky & de Haan, 2011), where there is evidence of alterations to brain architecture combined with high-level cognitive difficulties in children whose early development has been disrupted by maltreatment, neglect, and impoverished environments (Kempermann, Kuhn, & Gage, 1997).

EF impairments have been consistently reported following *pediatric TBI* (Catroppa & Anderson, 2005; Kaufmann, Fletcher, Levin, Miner, & Ewing-Cobbs, 1993; Kelly & Eyre, 1999; Ornstein et al., 2009), with specific deficits in attentional control (Catroppa, Anderson, Morse, Haritou, & Rosenfeld, 2007; Leblanc et al., 2005), information processing (Tromp & Mulder, 1991), goal setting, and problems solving (Anderson & Catroppa, 2005). Injury severity has been shown to impact EF, with children sustaining a severe TBI often reported to have more impaired skills (Beauchamp et al., 2011a; Muscara, Catroppa, & Anderson, 2008; Nadebaum, Anderson, & Catroppa, 2007). These studies reported that not all EF domains were affected, with some finding impairments in cognitive flexibility, goal setting, and abstract reasoning, not in attention control (Muscara et al., 2008) and others reporting poorer performance in goal setting and processing speed (Beauchamp et al., 2011a).

Few studies have explored executive function in *pediatric stroke*, yet one study has shown impairments in cognitive measures of EF as well as parent and teacher rating of everyday function (Long et al., 2011). This study found 46% of children with arterial ischemic stroke exhibited impaired attentional control, including sustained attention and inhibitory control, and between 40% and 70% showed deficits in cognitive flexibility (working memory, mental flexibility, and divided attention). Impairment rates of 50% or greater were also reported for the information processing domain. Larger lesion size was found to be associated with a number of EF domains, with larger lesion volume related to impaired scores on measures of attentional control, particularly inhibitory control, selective attention, information processing and cognitive flexibility. In contrast, children with a small or medium sized lesion generally performed at age expected levels. Analysis of lesion location in a separate paper found differences on some measures, with poorer cognitive flexibility and regularity aspects of everyday behavior reported in children with frontal, compared to extrafrontal, lesions (Long et al., 2010). However, both groups displayed impairments across a number of domains, suggesting prefrontal brain regions may not play a unique role in EF in the immature brain. Instead damage to nonfrontal regions in childhood also appears to have the capacity to negatively impact functioning, likely due to disruption of diffuse neural networks connecting prefrontal cortex with other key brain regions (Long et al., 2010). There were also few differences found between children with stroke involving cortical or subcortical regions or between left or right-sided lesions.

15.5.3 Psychosocial outcomes

Social and behavioral impairments are common following EBI, and are rated by parents and children as the most debilitating of all sequelae. Dysfunction in psychosocial function will impact many areas of daily life, as well as overall quality of life. While there is some evidence that insult severity may be related to degree of social problems (Andrews, Rose, & Johnson, 1998; Yeates et al., 2004), this dose–response relationship does not appear to be as clear cut for social skills as it is for cognitive function, where it is well established that more severe insult leads to greater deficits (Anderson et al., 2005; Rosema, Crowe, & Anderson, 2012).

One study exploring psychosocial function in children with *focal EBI* reported poorer social and behavioral function in these children than normative expectations, with greater emotional symptoms, hyperactivity, and difficulties with peers (Greenham, Spencer-Smith, Anderson, Coleman, & Anderson, 2010). While the whole group displayed social and behavioral impairment, those who sustained an insult prior to 3 years of age were particularly vulnerable. Interestingly, significant effects were not global, but confined to the domains of peer relations and emotional symptoms. In the exploration of other factors, no significant relationship was found between lesion characteristics and outcome, but the presence of seizures was associated with poorer psychosocial skills. In addition, family functioning was found to impact prosocial behavior and emotional symptoms, with children from more dysfunctional families at higher risk for impairment in these domains (Greenham et al., 2010). These findings indicate that, in contrast to the age-related cognitive deficits documented following EBI (Anderson et al., 2009; Jacobs et al., 2007), psychosocial function is less clearly linked to neurodevelopmental processes. Children sustaining EBI prior to age 3 years do demonstrate a tendency to poorer social and behavioral function; however, outcomes appear less linked to age at insult for later insults. Consistent with these results, emerging research has begun to explore behavioral consequences of EBI and likely risk and resilience factors. Findings suggest that, for domains such as social and behavioral function, environmental influences, child disability, and associated levels of child adjustment may play a greater role (Anderson et al., 2006; Janusz, Kirkwood, Yeates, & Taylor, 2002) than injury-related factors, especially in the long-term postinsult.

A recent systematic review of social outcomes following *pediatric TBI* in children reported that while comparison was difficult due to differing methodology, social function was poorer compared to typically developing children (Rosema et al., 2012). Following TBI children were reported to have poorer self-esteem, fewer friends, and experience more loneliness and social isolation. Deficits in social problem solving were found, including maladaptive, aggressive, and antisocial behavior and poorer emotional and behavioral self-regulation. Reduced social information processing and emotional recognition were also reported. These authors noted that there is a lack of research exploring factors, other than injury severity, that may contribute to these social outcomes.

Similarly, Gomes and colleagues have systematically evaluated research findings for psychosocial function following *pediatric stroke* and report a variety of impairments, including internalizing (anxiety, inattention, psychosomatic complaints) and externalizing (aggression, hyperactivity, emotional lability, impulsivity) problems (Gomes, Rinehart, Greenham, & Anderson, 2014). In a series of studies, Max et al. (2002a,b, 2003) reported 59% of children with stroke developed psychiatric disorders compared to only 14% of controls. The disorders with the highest rate of prevalence in this sample were attention deficit/hyperactivity disorder (ADHD) (49%), anxiety disorder (31%), mood disorders (21%), and personality change (17%). Everts et al. (2008) also report a 50% prevalence of ADHD symptoms (learning difficulties, attention problems, anxiety, and impulsivity) in their child stroke cohort. Externalizing behavior problems (emotional lability, temper tantrums, and aggressive outbursts) have also been reported in around 44% of children (Steinlin, Roellin, & Schroth, 2004). Around 25% have been reported to have emotional problems (Gordon, Ganesan, Towell, & Kirkham, 2002; Neuner et al., 2011). While social function has not been examined in detail, the limited literature has shown poorer peer relationships and reduced social acceptance (De Schryver, Kappelle, Jennekens-Schinkel, & Peters, 2000; Everts et al., 2008; Friefeld, Yeboah, Jones, & deVeber, 2004; Gordon et al., 2002; Neuner et al., 2011; O'Keeffe, Ganesan, King, & Murphy, 2012; Steinlin et al., 2004).

15.6 OUTCOMES FROM EBI: THE "BRAIN" DIMENSION: INSIGHTS FROM ADVANCED STRUCTURAL NEUROIMAGING

15.6.1 Understanding variability in outcome after EBI

Given evidence for substantial variability in outcome and recovery following EBI, there exists a strong clinical imperative to improve prognostic accuracy and identify and target children at increased risk for persisting cognitive, social and behavioral impairment. As shown in Fig. 15.3, the Heuristic Model of Social Competence in Childhood Brain Disorder (HMSC) provides a useful framework for conceptualizing how insult-related, child, and environmental factors may contribute to cognitive, social, and behavioral outcomes (Yeates et al., 2007). More specifically, the model, which is conceived within a child TBI framework, postulates that various injury and noninjury-related risk and resilience factors may independently or interactively contribute to outcome and recovery after child TBI. As previously noted, environmental factors, such as interventions and better family functioning, represent sources of resilience that may buffer against the neurological consequences of TBI, injury-induced structural brain abnormalities and disadvantaged environment are conceptualized as

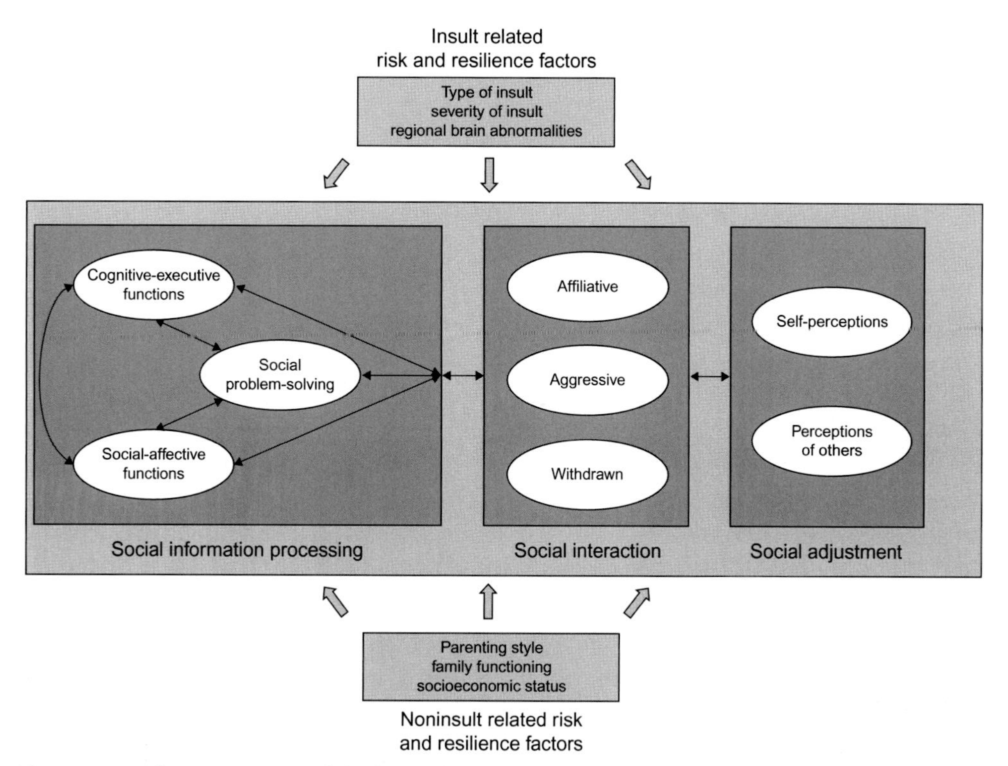

Figure 15.3 The Heuristic Model of Social Competence in Children with Brain Disorder (HMSC) from Yeates et al. (2007). *Reproduced with permission.*

risk factors that increase the likelihood of impairment in social-affective (e.g., theory of mind) and cognitive-executive processes (e.g., working memory, inhibitory), as well as atypical social interaction and poor social adjustment.

The role of environmental factors in moderating outcome receives strong empirical support from both animal and human studies (Giza, Griesbach, & Hovda, 2005; Kolb & Stewart, 1995; Schmidt, Orsten, Hanten, Li, & Levin, 2010; Yeates et al., 1997); however, until the last decade, it has remained difficult to evaluate how injury-induced alterations in large-scale structural brain networks may contribute to substantial variability in outcome following EBI. Dearth of knowledge in this area has stemmed from the inherent weaknesses of computed tomography and conventional MR sequences, which demonstrate poor prognostic utility and that may be unable to detect subtle evidence of traumatic axonal injury that may underlie a child's clinical status (Ashwal, Tong, Ghosh, Bartnik-Olson, & Holshouser, 2014; Ashwal, Tong, Obenaus, & Holshouser, 2010).

Recent advances in neuroimaging may assist to identify alterations in the structural architecture following EBI, in a brain that may appear deceptively "normal" on conventional imaging sequences (Ashwal et al., 2010). In the section that follows, we review the application of these modern neuroimaging techniques to child TBI to elucidate the ways in which structural brain development may deviate from the normative developmental blueprint in children with TBI. In addition, we seek to evaluate the utility of these techniques for prediction of postinjury outcomes, and discuss the potential role of these advanced imaging sequences to identify and target children at increased risk for neurobehavioral deficits.

15.6.2 Application of advanced structural imaging techniques to child TBI

Morphometric MRI permits quantitative characterization of the volume, shape and composition of regions within the immature brain (Caviness, Lange, Makris, Herbert, & Kennedy, 1999). Application of this technique to child TBI has revealed both global and localized reductions in white and gray matter volume in the subacute and chronic phases of injury (Beauchamp et al., 2011c; Bigler et al., 2010; Fearing et al., 2008; Spanos et al., 2007; Wilde et al., 2006, 2007); however, only few studies have examined the specific nature and neuroanatomical distribution of these morphological changes.

In the first of a series of landmark studies in this area, Wilde et al. (2012a,b) employed a longitudinal prospective design to evaluate the extent and neuroanatomical distribution of structural changes to the cortical mantle from 3- to 18-month post-TBI. Results from the 3-month time interval revealed that compared to age-matched orthopedic injury (OI) controls children with moderate-severe TBI demonstrate reduced cortical thickness bilaterally in a distributed network of regions, including superior frontal, dorsolateral frontal, orbital frontal, and anterior cingulate areas, a pattern of changes which were suggested to reflect neuronal loss demonstrated in post mortem brains (Maxwell, MacKinnon, Stewart, & Graham, 2009). Moreover, analysis of longitudinal change in the TBI versus an OI group revealed that, despite evidence for progressive injury-induced atrophic change in large areas of the parietal and occipital lobes, cortical thinning in the anterior brain regions was notably absent following TBI, with some evidence of cortical thickness increases in these areas. This finding suggests that, brain insult during a critical period of brain development may alter the internal signaling processes that initiate programmed cell death at a developmentally typical stage. Furthermore, failure of the medial frontal regions to undergo typical patterns of thinning over time in the TBI group was correlated with poorer emotional control and behavioral regulation (Li & Liu, 2013).

Extending on these earlier findings, an emerging body of recent work highlights the potential value of morphometric MRI to identify children at elevated risk for atypical development of social cognition, referring to the mental processes used to perceive and process, social cues, stimuli and the environment (Beauchamp & Anderson, 2010). These skills emerge relatively early in development and continue to undergo protracted maturation into mid-to-late adolescence, coinciding with the extended functional and structural maturation of areas that contribute to the anatomically distributed social brain network (SBN), which includes the superior temporal sulcus, fusiform gyrus, temporal pole, medial prefrontal cortex, orbitofrontal cortex, amygdala, temporoparietal junction, and inferior parietal cortex (see Fig. 15.4 for schematic representation) (Adolphs, 2009; Beauchamp & Anderson, 2010; Johnson et al., 2005).

Several lines of evidence converge to support the vulnerability of the SBN to disruption from TBI. Deficits in theory of mind (ToM), emotion perception, and moral reasoning are commonly reported in TBI (Dennis et al., 2012; Dennis et al., 2013; Ryan et al., 2014; Tlustos et al., 2011) and have been linked to a range of unfavorable behavioral outcomes including poorer pragmatic communication (Watts & Douglas, 2006), and more frequent externalizing behaviors characterized by rule breaking, aggression, and conduct problems (Ryan et al., 2013). Given the likely consequences of these impairments for social integration and quality of life (Beauchamp & Anderson, 2010), early identification of children at risk for these difficulties remains critical to enable provision of targeted clinical assessments and interventions designed to ameliorate risk for unfavorable chronic social outcomes.

It has been suggested that increased rates of social cognitive impairment among children with TBI may reflect the selective vulnerability of frontal temporo-limbic structures of the SBN to the primary and secondary effects of TBI. Consistent with this hypothesis, using conventional morphometric MRI. Ryan et al. (2016) examined prospective longitudinal relationships between subacute SBN volume, ToM, and

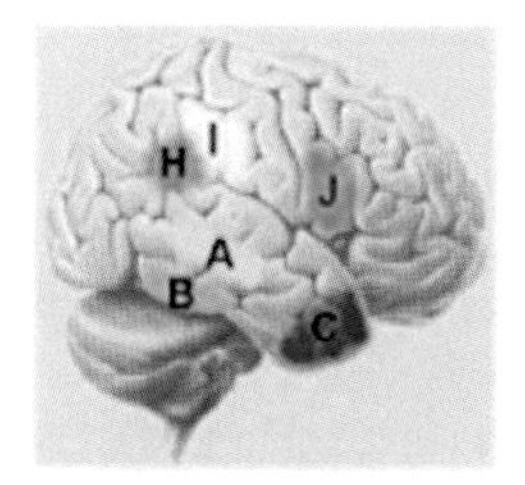

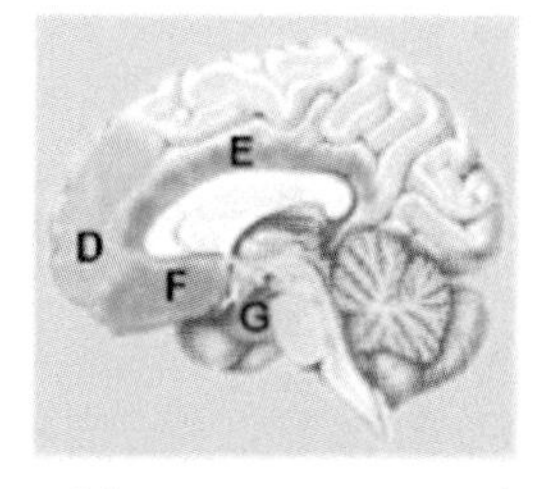

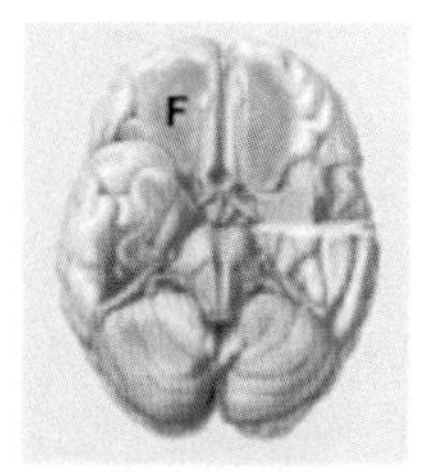

Figure 15.4 Schematic representation of brain regions contributing to the social brain network: (A) superior temporal sulcus; (B) fusiform gyrus; (C) temporal pole; (D) medial prefrontal cortex and frontal pole; (E) cingulate cortex; (F) orbitofrontal cortex; (G) amygdala; (H) temporoparietal junction; (I) inferior parietal cortex; (J) inferior frontal cortex; insula (not shown). *From Beauchamp et al. (2011); Reproduced by permission.*

behavior problems in children and adolescents with mild-severe TBI compared to typically developing children. Relative to typically developing children and those with mild injuries, children with severe TBI showed abnormal SBN morphology, as well local gray matter structural change in the superior temporal sulcus, fusiform gyrus, temporal pole, medial prefrontal cortex, orbitofrontal cortex, temporoparietal cortex, and cingulate. These selective structural alterations were documented alongside relative preservation of global structural network integrity, total, gray, and white matter volume. Abnormalities in SBN morphology were associated with impaired ToM and behavior problems. These findings suggest that SBN pathology may represent a useful imaging biomarker for predicting social cognitive impairment and behavior problems following EBI.

Susceptibility weighted imaging (SWI) is a modified high spatial resolution 3D gradient recalled-echo (GRE) MR technique that accentuates the magnetic properties of blood products, thereby rendering it useful for detecting small amounts of altered blood and blood product in the brain (Haacke, Xu, Cheng, & Reichenbach, 2004; Reichenbach, Venkatesan, Schillinger, Kido, & Haacke, 1997; Sehgal et al., 2005). A number of child TBI studies have recently demonstrated its increased sensitivity to posttraumatic lesions, and its value for prediction of cognitive and intellectual outcomes (Babikian et al., 2010; Beauchamp et al., 2011a; Colbert et al., 2010; Spitz et al., 2013; Tong et al., 2003, 2004).

Early findings suggest that SWI is highly sensitive to traumatic lesions, more so than CT or conventional MR (see e.g., Fig. 15.5) (Beauchamp et al., 2011b). In a recent longitudinal study that examined SWI and its association with social cognition, microhemmorhagic lesions were most commonly identified in frontal temporo-limbic regions commonly associated with the SBN (Fig. 15.6), and more diffuse SWI lesions within this network were prospectively associated with poorer ToM and pragmatic

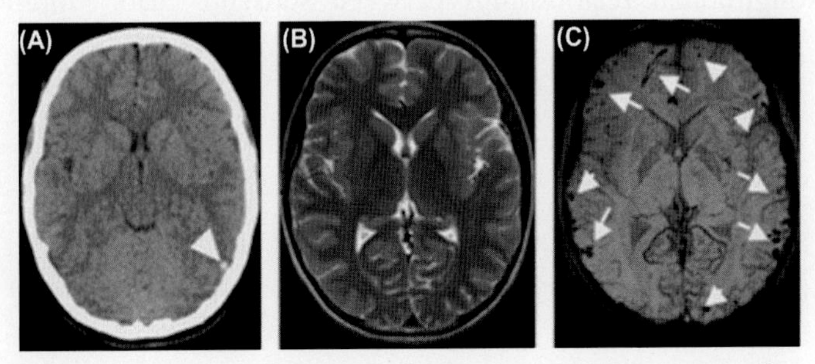

Figure 15.5 Results of (A) CT, (B) T2-weighted MRI, and (C) SWI images for a male patient (10.9 years) with TBI at equivalent anatomical levels. His GCS on presentation to the ED was 13. The CT demonstrates a focus of hemorrhage in the left occipital lobe (arrowhead). The same lesion is poorly seen on the T2-weighted MRI; however, SWI demonstrates multiple foci of hemorrhage (arrows) not identified on the other imaging sequences.

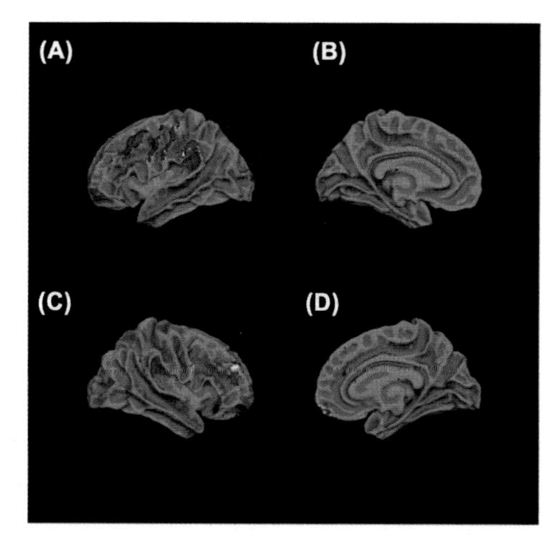

Figure 15.6 Reproduced from Beauchamp et al. (2013) with permission. Probability distribution of brain lesions detected using SWI in the left lateral (A), left medial (B), right lateral (C), and right medial (D) hemispheres. Lesion distributions were created by aligning the individual T1 images to the Montreal Neurological Institute template using the nonlinear normalization procedure in Statistical Parametric Mapping 8 (SPM8). The lesion maps were normalized using the same transformations. The aligned lesion masks were averaged to produce a single image illustrating the distribution of lesions in the study population. Hotter colors indicate the co-location of lesions in multiple subjects. Lesions were most prominent in frontal regions [frontal only = 15 patients, frontal + extrafrontal only = 6, frontal + other regions (CC = 1, deep gray + CC = 1, cerebellum = 1, cerebellum + CC = 1)], followed by extrafrontal regions only ($N = 6$). A small number of patients (4) had lesions in several areas (frontal + extrafrontal + cerebellum = 2, frontal + extrafrontal + deep gray = 1, frontal + extrafrontal + CC = 1). Very few patients had lesions solely in the CC (1), cerebellum (1), or deep gray (0) regions.

language skills at 6- and 24-month postinjury (Ryan et al., 2014). In a related sub study that examined the relation between SWI lesions and higher level aspects of ToM, more diffuse microhemorrhagic lesions contributed to a time-dependent emergence of social cognitive dysfunction in adolescence when brain regions critically involved in these skills are undergoing rapid maturation (Ryan et al., 2015a,b).

Diffusion tensor imaging (DTI) measures the magnitude and directionality of water diffusion in tissue, notably of white matter tracts, and represents a sensitive biomarker of diffuse white matter injury in TBI via detection of traumatic axonal injury (Wilde et al., 2012b). This technique enables evaluation of white matter fiber tracts by taking advantage of the intrinsic directionality of water diffusion in the immature brain (Ashwal et al., 2010). Several commonly derived measures of diffusion anisotropy have proved useful in quantifying white matter injury in child TBI (Alexander, Lee, Lazar, & Field, 2007; Huisman et al., 2004; Wilde et al., 2012b). Fractional anisotropy (FA) is derived from the tendency of water molecules to move preferentially in parallel (rather than perpendicular) to barriers to free diffusion such as axons, fibers, or other support cells. A high degree of anisotropic diffusion is related to homogeneity of fiber orientation, increased fiber density or axonal diameter, and the ratio of intracellular/extracellular space (Wilde et al., 2012b). Conversely, apparent diffusion coefficient (ADC) represents the average diffusivity of free water movement within tissue, which also permits inferences regarding the microstructure in tissue; higher average diffusivity is indicative of decreased fiber density, axonal diameter, decreased myelination, and/or increased extracellular space (Wilde et al., 2012b).

In applying this technique to children and adolescents with moderate-severe injuries there is evidence of decreased FA and/or increased ADC in numerous white matter fiber bundles including the corpus callosum, inferior and superior frontal and supracallosal white matter, internal capsule, superior longitudinal fasciculus, and orbitofrontal white matter, cingulum bundles, and uncinate fasciculus (Oni et al., 2010; Wilde et al., 2012a; Wozniak et al., 2007; Wu et al., 2010; Yuan et al., 2007). It has been hypothesized that reductions in FA and increased ADC reflect axonal disconnection or damage to myelin sheaths (Arfanakis et al., 2002; Wu et al., 2010); however, the implications of these findings remain unclear.

Given there exists substantial variability in outcome across individual children with TBI, it remains critical to establish whether there exists a link between postinjury white matter development and individual variability in subacute and chronic neurobehavioral skills, which continue to undergo extended maturation into mid-to-late adolescence and early adulthood (Blakemore, 2008; Burnett & Blakemore, 2009). A number of studies report robust associations between DTI metrics and outcomes across the domains of memory, executive function, and fine motor and processing speed, suggesting that DTI may be a useful structural biomarker to identify children at elevated risk for a range of unfavorable neurodevelopmental outcomes (Wilde et al., 2006; Babikian et al., 2010; Ewing-Cobbs et al., 2008; Kourtidou et al., 2013; Kurowski et al., 2009; Wozniak & Lim, 2006).

Moreover, preliminary evidence suggests that DTI has potential to identify children at elevated risk for atypical developmental trajectories in the domain of social cognition. As mentioned previously, impairments in these skills are commonly documented after TBI and are shown to relate to a range of adverse outcomes including poorer adaptive behavior, more frequent behavioral problems and impaired pragmatic communication (Robinson et al., 2014; Ryan et al., 2013; Tlustos et al., 2011). On this basis, increased knowledge of neuroanatomical risk factors for impaired social cognition remains a critical area of investigation, and may assist to enable identification of children who may warrant long-term follow-up, assessment and interventions that aim to directly target development of these skills and reduce the likelihood of a range of long-term maladaptive behaviors.

Harvey Levin and colleagues pioneered initial work in this area through a series of studies that demonstrate robust associations between postinjury social cognitive impairment and structural brain abnormalities detected using DTI (Hanten et al., 2008; Levin et al., 2011). Findings revealed strong relations between social problem solving and ADC; such impaired problem solving was associated with increased ADC in the left and right dorsolateral frontal subregions, and right ventromedial frontal region. These robust relationships were observed alongside a robust association between SPS and reduced FA in the right cingulate and dorsolateral prefrontal cortex. Overall, findings from this study suggest that impaired social cognition in child TBI is associated with altered structural connectivity within a prefrontally guided, distributed network including prefrontal, temporal, parietal, and subcortical regions whose development may be selectively vulnerable to the effects of child TBI.

15.6.3 Future directions in neuroimaging of child TBI

Although advanced neuroimaging techniques such DTI, SWI, and morphometric MRI may assist to uncover alterations in structural brain architecture that may underlie aspects of a child's clinical status, there are several limitations to the use of these approaches in pediatric TBI. Since childhood TBI is a highly heterogeneous condition characterized by diffuse alterations in brain structure and function, the development of complex cognitive and behavioral functions is likely to be affected by TBI in ways that are difficult to quantify based on the gross structural integrity regional fiber paths alone (Irimia et al., 2012). More specifically, since neurobehavioral development is likely dependent upon the integration and segregation properties of large-scale networks (Menon, 2011), the challenge remains to quantify the effect of pediatric TBI on these characteristics of structural network topology, and evaluate the prognostic utility of these alterations for real-world cognitive, social, and behavioral impairment (Caeyenberghs et al., 2014).

15.7 CONCLUSIONS

Though preliminary evidence suggests that EBI has potential to disrupt the structural network organization of the developing brain, the specific nature and neuroanatomical locus of these alterations, environmental influences, and child-specific characteristics and their influence on subsequent brain and neurobehavioral development is largely unclear. Further research is needed to explore these domains concurrently and to identify critical interactions and synergies.

ACKNOWLEDGMENTS

The research presented in this chapter was supported by: The Australian Research Council, the Australian National Health and Medical Research Council, the Victorian Neurotrauma Initiative, and the Victorian Government Operating Infrastructure Scheme.

REFERENCES

Adolphs, R. (2009). The social brain: Neural basis of social knowledge. *Annual Review of Psychology, 60*, 693.

Alexander, A. L., Lee, J. E., Lazar, M., & Field, A. S. (2007). Diffusion tensor imaging of the brain. *Neurotherapeutics, 4*(3), 316–329.

Allman, C., & Scott, R. B. (2013). Neuropsychological sequelae following pediatric stroke: A nonlinear model of age at lesion effects. *Child Neuropsychology, 19*(1), 97–107.

Anderson, V., & Catroppa, C. (2005). Recovery of executive skills following paediatric traumatic brain injury (TBI): A 2 year follow-up. *Brain Injury, 19*(6), 459–470.

Anderson, V., Catroppa, C., Dudgeon, P., Morse, S. A., Haritou, F., & Rosenfeld, J. (2006). Understanding predictors of functional recovery and outcome 30 months following early childhood head injury. *Neuropsychology, 20*(1), 42.

Anderson, V., Catroppa, C., Morse, S., Haritou, F., & Rosenfeld, J. (2005). Functional plasticity or vulnerability following early brain injury? *Pediatrics, 116*, 374–382.

Anderson, V., & Moore, C. (1995). Age at injury as a predictor of outcome following pediatric head injury. *Child Neuropsychology, 1*, 187–202.

Anderson, V., Morse, S., Catroppa, C., Haritou, F., & Rosenfeld, J. (2004). Thirty-month outcome from early childhood head injury: A prospective analysis of neurobehavioral recovery. *Brain, 127*, 2608–2620.

Anderson, V., Morse, S., Klug, G., Catroppa, C., Haritou, F., Rosenfeld, J., et al. (1997). Predicting recovery from head injury in young children. *Journal of the International Neuropsychological Society, 3*, 568–580.

Anderson, V., Gomes, A., Greenham, M., Hearps, S., Gordon, A., Rinehart, N., ... Mackay, M. (2014a). Social competence following pediatric stroke: Contributions of brain insult and family environment. *Social Neuroscience, 9*(5), 471–483.

Anderson, V., Spencer-Smith, M., & Wood, A. (2011). Do children really recover better? Neurobehavioural plasticity after early brain insult. *Brain, 134*(8), 2197–2221.

Anderson, V., Spencer-Smith, M., Leventer, R., Coleman, L., Anderson, P., Williams, J., ... Jacobs, R. (2009). Childhood brain insult: Can age at insult help us predict outcome? *Brain, 132*(1), 45–56, Anderson.

Anderson, V., Spencer-Smith, M. M., Coleman, L., Anderson, P. J., Greenham, M., Jacobs, R., et al. (2014b). Predicting neurocognitive and behavioural outcome after early brain insult. *Developmental Medicine & Child Neurology, 56*(4), 329–336.

Andrews, T., Rose, F., & Johnson, D. (1998). Social and behavioural effects of traumatic brain injury in children. *Brain Injury, 12*(2), 133–138.

Aram, D. (1988). Language sequelae of unilateral brain lesions in children. In F. Plum (Ed.), *Language, communication and the brain* (pp. 171–197). New York: Raven Press.

Aram, & Eisele, J. (1994). Intellectual stability in children with unilateral brain lesions. *Neuropsychologia, 32*, 85–95.

Arfanakis, K., Haughton, V. M., Carew, J. D., Rogers, B. P., Dempsey, R. J., & Meyerand, M. E. (2002). Diffusion tensor MR imaging in diffuse axonal injury. *American Journal of Neuroradiology, 23*(5), 794–802.

Ashwal, S., Tong, K., Obenaus, A., & Holshouser, B. (2010). Advanced neuroimaging techniques in children with traumatic brain injury. *New Directions in Pediatric Traumatic Brain Injury: Multidisciplinary and Translational Perspectives*, 68–93.

Ashwal, S., Tong, K. A., Ghosh, N., Bartnik-Olson, B., & Holshouser, B. A. (2014). Application of advanced neuroimaging modalities in pediatric traumatic brain injury. *Journal of Child Neurology*, 0883073814538504.

Babikian, T., Marion, S. D., Copeland, S., Alger, J. R., O'Neill, J., Cazalis, F., et al. (2010). Metabolic levels in the corpus callosum and their structural and behavioral correlates after moderate to severe pediatric TBI. *Journal of Neurotrauma, 27*(3), 473–481.

Ballantyne, A., Spilkin, A., & Trauner, D. (2007). Language outcome after perinatal stroke: Does side matter? *Child Neuropsychology, 13*, 494–509.
Ballantyne, A. O., Spilkin, A. M., Hesselink, J., & Trauner, D. (2008). Plasticity in the developing brain: Intellectual, language and academic functions in children with ischaemic perinatal stroke. *Brain, 131*, 2975–2985.
Bates, E., Reilly, J., Wilfeck, B., Donkers, N., Opie, M., Fenson, J., et al. (2001). Differential effects of unilateral lesions on language production in children and adults. *Brain and Language, 79*, 223–265.
Beauchamp, M. H., & Anderson, V. (2010). SOCIAL: An integrative framework for the development of social skills. *Psychological Bulletin, 136*(1), 39.
Beauchamp, M. H., Ditchfield, M., Maller, J. J., Catroppa, C., Godfrey, C., Rosenfeld, J. V., & Anderson, V. A. (2011). Hippocampus, amygdala and global brain changes 10 years after childhood traumatic brain injury. *International Journal of Developmental Neuroscience, 29*, 137–143.
Beauchamp, M., Catroppa, C., Godfrey, C., Morse, S., Rosenfeld, J. V., & Anderson, V. (2011a). Selective changes in executive functioning ten years after severe childhood traumatic brain injury. *Developmental Neuropsychology, 36*(5), 578–595.
Beauchamp, M. H., Ditchfield, M., Babl, F. E., Kean, M., Catroppa, C., Yeates, K. O., et al. (2011b). Detecting traumatic brain lesions in children: CT versus MRI versus susceptibility weighted imaging (SWI). *Journal of Neurotrauma, 28*(6), 915–927.
Beauchamp, M. H., Ditchfield, M., Catroppa, C., Kean, M., Godfrey, C., Rosenfeld, J. V., et al. (2011c). Focal thinning of the posterior corpus callosum: Normal variant or post-traumatic? *Brain Injury, 25*(10), 950–957.
Beauchamp, M. H., Beare, R., Ditchfield, M., Coleman, L., Babl, F. E., Kean, M., et al. (2013). Susceptibility weighted imaging and its relationship to outcome after pediatric traumatic brain injury. *Cortex, 49*(2), 591–598.
Belsky, J., & de Haan, M. (2011). Annual research review: Parenting and children's brain development: The end of the beginning. *Journal of Child Psychology and Psychiatry, 52*(4), 409–428.
Bigler, E. D., Abildskov, T. J., Wilde, E. A., McCauley, S. R., Li, X., Merkley, T. L., et al. (2010). Diffuse damage in pediatric traumatic brain injury: A comparison of automated versus operator-controlled quantification methods. *NeuroImage, 50*(3), 1017–1026.
Blakemore, S.-J. (2008). The social brain in adolescence. *Nature Reviews Neuroscience, 9*(4), 267–277.
Breslau, N. (1990). Does brain dysfunction increase children's vulnerability to environmental stress? *Archives of General Psychiatry, 47*, 15–20.
Burnett, S., & Blakemore, S. J. (2009). The development of adolescent social cognition. *Annals of the New York Academy of Sciences, 1167*, 51–56.
Caeyenberghs, K., Leemans, A., Leunissen, I., Gooijers, J., Michiels, K., Sunaert, S., et al. (2014). Altered structural networks and executive deficits in traumatic brain injury patients. *Brain Structure and Function, 219*(1), 193–209.
Casey, B., Giedd, J., & Thompson, K. (2000). Structural and functional brain development and its relation to cognitive development. *Biological Psychology, 54*, 241–257.
Catroppa, C., & Anderson, V. (2005). A prospective study of the recovery of attention from acute to 2 years following pediatric traumatic brain injury. *Journal of the International Neuropsychological Society, 11* (01), 84–98.
Catroppa, C., & Anderson, V. (2008). Outcome and predictors of functional recovery five years following pediatric traumatic brain injury. *Journal of Pediatric Psychology, 23*, 1–12.
Catroppa, C., Anderson, V., Morse, S. A., Haritou, F., & Rosenfeld, J. V. (2007). Children's attentional skills 5 years post-TBI. *Journal of Pediatric Psychology, 32*(3), 354–369.
Caviness, V. S., Lange, N. T., Makris, N., Herbert, M. R., & Kennedy, D. N. (1999). MRI-based brain volumetrics: Emergence of a developmental brain science. *Brain and Development, 21*(5), 289–295.
Colbert, C. A., Holshouser, B. A., Aaen, G. S., Sheridan, C., Oyoyo, U., Kido, D., et al. (2010). Value of cerebral microhemorrhages detected with susceptibility-weighted MR imaging for prediction of long-term outcome in children with nonaccidental trauma. *Radiology, 256*(3), 898–905.
Courchesne, E., Chisum, H. J., Townsend, J., Cowles, A., Covington, J., Egaas, B., et al. (2000). Normal brain development and aging: Quantitative analysis at in vivo MR imaging in healthy volunteers 1. *Radiology, 216*(3), 672–682.

Crowe, L., Catroppa, C., Babl, F., Rosenfeld, J., & Anderson, V. (2012). Timing of traumatic brain injury in childhood and intellectual outcome. *Journal of Pediatric Psychology, 37*(7), 745–754.

Dennis, M. (1988). Language and the young damaged brain. In T. Boll, & B. K. Bryant (Eds.), *Clinical neuropsychology and brain function: Research, measurement and practice* (pp. 85–124). Washington: American Psychological Association.

Dennis, M. (1989). Language and the young damaged brain. In T. Boll & B. K. Bryant (Eds.), *Clinical neuropsychology and brain function: Research, measurement and practice* (pp. 85–124). Washington: American Psychological Association, 1989.

Dennis, M., Simic, N., Bigler, E. D., Abildskov, T., Agostino, A., Taylor, H. G., et al. (2013). Cognitive, affective, and conative theory of mind (ToM) in children with traumatic brain injury. *Developmental Cognitive Neuroscience, 5*, 25–39.

Dennis, M., Simic, N., Gerry Taylor, H., Bigler, E. D., Rubin, K., Vannatta, K., et al. (2012). Theory of mind in children with traumatic brain injury. *Journal of the International Neuropsychological Society, 18* (05), 908–916.

Dennis, M., & Whitakker, H. (1976). Language acquisition following hemidecortication: Linguistic superiority of the left over the right hemisphere. *Brain and Language, 3*, 404–433.

De Schryver, E., Kappelle, L. J., Jennekens-Schinkel, A., & Peters, A. C. (2000). Prognosis of ischemic stroke in childhood: A long-term follow-up study. *Developmental Medicine & Child Neurology, 42*(5), 313–318.

Duchowny, M., Jayakar, P., Harvey, A. S., Altman, N., Resnick, T., & Levin, B. (1996). Language cortex representation: Effects of developmental versus acquired pathology. *Annals of Neurology, 40*, 91–98.

Duffau, H. (2006). Brain plasticity: From pathophysiologic mechanisms to therapeutic applications. *Journal of Clinical Neuroscience, 13*, 885–897.

Everts, R., Pavlovic, J., Kaufmann, F., Uhlenberg, B., Seidel, U., Nedeltchev, K., et al. (2008). Cognitive functioning, behavior, and quality of life after stroke in childhood. *Child Neuropsychology, 14*(4), 323–338.

Ewing-Cobbs, L., Fletcher, J., Levin, H., Francis, D., Davidson, K., & Miner, E. (1997). Longitudinal neuropsychological outcome in infants and preschoolers with traumatic brain injury. *Journal of the International Neuropsychological Society, 3*, 581–591.

Ewing-Cobbs, L., Fletcher, J., Levin, H., Iovino, I., & Miner, M. (1998). Academic achievement and academic placement following traumatic brain injury in children and adolescents: A two-year longitudinal study. *Journal of Clinical and Experimental Neuropsychology, 20*(6), 769–781.

Ewing-Cobbs, L., Prasad, M. R., Kramer, L., Cox, C. S., Jr, Baumgartner, J., Fletcher, S., et al. (2006). Late intellectual and academic outcomes following traumatic brain injury sustained during early childhood. *Journal of Neurosurgery, 105*(4 Suppl), 287.

Ewing-Cobbs, L., Prasad, M. R., Swank, P., Kramer, L., Cox, C. S., Jr, Fletcher, J. M., et al. (2008). Arrested development and disrupted callosal microstructure following pediatric traumatic brain injury: Relation to neurobehavioral outcomes. *NeuroImage, 42*(4), 1305–1315.

Fearing, M. A., Bigler, E. D., Wilde, E. A., Johnson, J. L., Hunter, J. V., Li, X., et al. (2008). Morphometric MRI findings in the thalamus and brainstem in children after moderate to severe traumatic brain injury. *Journal of Child Neurology, 23*(7), 729–737.

Finger, S., & Stein, D. G. (1982). *Brain damage and recovery*. New York: Academic Press.

Friefeld, S., Yeboah, O., Jones, J. E., & deVeber, G. (2004). Health-related quality of life and its relationship to neurological outcome in child survivors of stroke. *CNS Spectrums, 9*(06), 465–475.

Giedd, J. N., Rumsey, J. M., Castellanos, F. X., Rajapakse, J. C., Kaysen, D., Catherine Vaituzis, A., et al. (1996). A quantitative MRI study of the corpus callosum in children and adolescents. *Developmental Brain Research, 91*(2), 274–280.

Giza, C., & Prins, M. (2006). Is being plastic really fantastic? Mechanisms of altered plasticity after developmental traumatic brain injury. *Developmental Neuroscience, 28*, 364–379.

Giza, C. C., Griesbach, G. S., & Hovda, D. A. (2005). Experience-dependent behavioral plasticity is disturbed following traumatic injury to the immature brain. *Behavioural Brain Research, 157*(1), 11–22.

Gogtay, N., Giedd, J. N., Lusk, L., et al. (2004). Dynamic mapping of human cortical development during childhood through early adulthood. *Proceedings of the National Academy of Sciences of the United States of America, 101*, 8174–8179.

Gomes, A., Rinehart, N., Greenham, M., & Anderson, V. (2014). A critical review of psychosocial outcomes following childhood stroke (1995–2012). *Developmental Neuropsychology, 39*(1), 9–24.

Gomes, A., Spencer-Smith, M. M., Jacobs, R. K., Coleman, L., & Anderson, V. (2012). Attention and social functioning in children with malformations of cortical development and stroke. *Child Neuropsychology, 18*(4), 392–403.

Gordon, A., Ganesan, V., Towell, A., & Kirkham, F. J. (2002). Functional outcome following stroke in children. *Journal of Child Neurology, 17*(6), 429–434.

Greenham, M., Spencer-Smith, M. M., Anderson, P. J., Coleman, L., & Anderson, V. (2010). Social functioning in children with brain insult. *Frontiers in Human Neuroscience*, 4.

Haacke, E. M., Xu, Y., Cheng, Y.-C. N., & Reichenbach, J. R. (2004). Susceptibility weighted imaging (SWI). *Magnetic Resonance in Medicine, 52*(3), 612–618.

Hajek, C., Yeates, K. O., Anderson, V., Mackay, M., Greenham, M., Gomes, A., et al. (2014). Cognitive outcomes following arterial ischemic stroke in infants and children. *Journal of Child Neurology, 29*(7), 887–894.

Hanten, G., Wilde, E. A., Menefee, D. S., Li, X., Lane, S., Vasquez, C., et al. (2008). Correlates of social problem solving during the first year after traumatic brain injury in children. *Neuropsychology, 22*(3), 357–370.

Hebb, D. (1942). The effect of early and late brain injury upon test scores, and the nature of normal adult intelligence. *Proceedings of the American Philosophical Society, 85*, 275–292.

Hebb, D. O. (1949). *The organization of behaviour.* New York: McGraw-Hill.

Hetherington, R., Tuff, L., Anderson, P., & Miles, B. (2005). Short-term intellectual outcome after arterial ischemic stroke and sinovenous thrombosis in childhood and infancy. *Journal of Child Neurology, 20*(7), 553–559.

Huisman, T. A., Schwamm, L. H., Schaefer, P. W., Koroshetz, W. J., Shetty-Alva, N., Ozsunar, Y., et al. (2004). Diffusion tensor imaging as potential biomarker of white matter injury in diffuse axonal injury. *American Journal of Neuroradiology, 25*(3), 370–376.

Huttenlocher, P., & Dabholkar, A. (1997). Developmental anatomy of the prefrontal cortex. In N. Krasnegor, G. Reid, & P. Goldman-Rakic (Eds.), *Development of the prefrontal cortex: Evolution, neurobiology, and behavior* (pp. 69–83). Baltimore: Paul H Brookes Publishing Company.

Irimia, A., Wang, B., Aylward, S. R., Prastawa, M. W., Pace, D. F., Gerig, G., et al. (2012). Neuroimaging of structural pathology and connectomics in traumatic brain injury: Toward personalized outcome prediction. *NeuroImage: Clinical, 1*(1), 1–17.

Isaacson, R. (1975). The myth of recovery from early brain damage. In N. Ellis (Ed.), *Aberrant development in infancy: Human and animal studies* (pp. 1–25). New Jersey: LEA.

Jacobs, R., & Anderson, V. (2002). Planning and problem solving skills following focal frontal brain lesions in childhood: Analysis using the Tower of London. *Child Neuropsychology, 8*, 93–106.

Jacobs, R., Harvey, A. S., & Anderson, V. (2007). Executive function following focal frontal lobe lesions: Impact of timing of lesion on outcome. *Cortex, 43*(6), 792–805.

Janusz, J. A., Kirkwood, M. W., Yeates, K. O., & Taylor, H. G. (2002). Social problem-solving skills in children with traumatic brain injury: Long-term outcomes and prediction of social competence. *Child Neuropsychology, 8*(3), 179–194.

Johnson, M. H., Griffin, R., Csibra, G., Halit, H., Farroni, T., De Haan, M., et al. (2005). The emergence of the social brain network: Evidence from typical and atypical development. *Development and Psychopathology, 17*(03), 599–619.

Kadis, D., Iida, K., Kerr, E., Logan, W., McAndrews, M., Ochi, A., et al. (2007). Intrahemispheric reorganization of language in children with medically intractable epilepsy of the left hemisphere. *Journal of the International Neuropsychological Society, 13*, 505–516.

Kaufmann, P. M., Fletcher, J. M., Levin, H. S., Miner, M. E., & Ewing-Cobbs, L. (1993). Attentional disturbance after pediatric closed head injury. *Journal of Child Neurology, 8*(4), 348–353.

Kelly, T. P., & Eyre, J. A. (1999). Specific attention and executive function deficits in the long-term outcome of severe closed head injury. *Developmental Neurorehabilitation, 3*(4), 187–192.

Kempermann, G., Kuhn, H. G., & Gage, F. H. (1997). More hippocampal neurons in adult mice living in an enriched environment. *Nature, 386*(6624), 493–495.

Kennard, M. (1938). Reorganization of motor function in the cerebral cortex of monkeys deprived of motor and premotor areas in infancy. *Journal of Neurophysiology, 1*, 477–496.

Kennard, M. (1942). Cortical reorganization of motor function. *Archives of Neurology, 48*, 227–240.

Klinberg, T., Vaidya, C., Gabrieli, J., Moseley, M., & Hedehus, M. (1999). Myelination and organization of the frontal white matter in children: A diffuse tensor MRI study. *NeuroReport, 10*, 2817–2821.

Kolb, B. (1995). *Brain plasticity and behavior.* Mahwah, NJ: LEA.

Kolb, B., & Gibb, R. (2001). Early brain injury, plasticity and behavior. In C. Nelson & M. Luciana (Eds.), *Handbook of developmental cognitive neuroscience.* Cambridge, MA: MIT Press.

Kolb, B., Pellis, S., & Robinson, T. (2004). Plasticity and functions of the orbital frontal cortex. *Brain Cognition, 55*, 104–115.

Kolb, B., Pellis, S., & Robinson, T. E. (2005). Plasticity and functions of the orbital frontal cortex. *Brain and Cognition, 55*, 104–115.

Kolb, B., & Stewart, J. (1995). Changes in the neonatal gonadal hormonal environment prevent behavioral sparing and alter cortical morphogenesis after early frontal cortex lesions in male and female rats. *Behavioral Neuroscience, 109*(2), 285.

Kourtidou, P., McCauley, S. R., Bigler, E. D., Traipe, E., Wu, T. C., Chu, Z. D., et al. (2013). Centrum semiovale and corpus callosum integrity in relation to information processing speed in patients with severe traumatic brain injury. *The Journal of Head Trauma Rehabilitation, 28*(6), 433–441.

Kurowski, B., Wade, S. L., Cecil, K. M., Walz, N. C., Yuan, W., Rajagopal, A., et al. (2009). Correlation of diffusion tensor imaging with executive function measures after early childhood traumatic brain injury. *Journal of Pediatric Rehabilitation Medicine, 2*(4), 273–283.

Lashley, K. (1929). *Brain mechanisms and intelligence.* Chicago: University of Chicago Press.

Leblanc, N., Chen, S., Swank, P. R., Ewing-Cobbs, L., Barnes, M., Dennis, M., et al. (2005). Response inhibition after traumatic brain injury (TBI) in children: Impairment and recovery. *Developmental Neuropsychology, 28*(3), 829–848.

Leventer, R. J., Phelan, E. M., Coleman, L. T., Kean, M. J., Jackson, G. D., & Harvey, A. S. (1999). Clinical and imaging features of cortical malformations in childhood. *Neurology, 53*, 715–722.

Levin, H. S., Wilde, E. A., Hanten, G., Li, X., Chu, Z. D., Vasquez, A. C., et al. (2011). Mental state attributions and diffusion tensor imaging after traumatic brain injury in children. *Developmental Neuropsychology, 36*(3), 273–287.

Li, L., & Liu, J. (2013). The effect of pediatric traumatic brain injury on behavioral outcomes: A systematic review. *Developmental Medicine & Child Neurology, 55*(1), 37–45.

Long, B., Anderson, V., Jacobs, R., Mackay, M., Leventer, R., Barnes, C., et al. (2011). Executive function following child stroke: The impact of lesion size. *Developmental Neuropsychology, 36*(8), 971–987.

Long, B., Spencer-Smith, M. M., Jacobs, R., Mackay, M., Leventer, R., Barnes, C., et al. (2010). Executive function following child stroke: The impact of lesion location. *Journal of Child Neurology*, 0883073810380049.

Luciana, M. (2003). Cognitive development in children born preterm: Implications for theories of brain plasticity following early injury. *Development and Psychopathology, 15*, 1017–1047.

Max, J. E., Fox, P. T., Lancaster, J. L., Kochunov, P., Mathews, K., Manes, F. F., et al. (2002a). Putamen lesions and the development of attention-deficit/hyperactivity symptomatology. *Journal of the American Academy of Child & Adolescent Psychiatry, 41*(5), 563–571.

Max, J. E., Mathews, K., Lansing, A. E., Robertson, B., Fox, P. T., Lancaster, J. L., et al. (2002b). Psychiatric disorders after childhood stroke. *Journal of the American Academy of Child & Adolescent Psychiatry, 41*(5), 555–562.

Max, J. E., Mathews, K., Manes, F. F., Robertson, B., Fox, P. T., Lancaster, J. L., et al. (2003). Attention deficit hyperactivity disorder and neurocognitive correlates after childhood stroke. *Journal of the International Neuropsychological Society, 9*(06), 815–829.

Maxwell, W. L., MacKinnon, M.-A., Stewart, J. E., & Graham, D. I. (2009). Stereology of cerebral cortex after traumatic brain injury matched to the Glasgow outcome score. *Brain*, awp264.

Menon, V. (2011). Large-scale brain networks and psychopathology: A unifying triple network model. *Trends in Cognitive Sciences, 15*(10), 483–506.

Mosch, S., Max, J., & Tranel, D. (2005). A matched lesion analysis of childhood versus adult-onset brain injury due to unilateral stroke. *Cognitive and Behavioral Neurology, 18*, 5–17.

Muscara, F., Catroppa, C., & Anderson, V. (2008). The impact of injury severity on executive function 7–10 years following pediatric traumatic brain injury. *Developmental Neuropsychology, 33*(5), 623–636.

Nadebaum, C., Anderson, V., & Catroppa, C. (2007). Executive function outcomes following traumatic brain injury in young children: A five year follow-up. *Developmental Neuropsychology, 32*(2), 703–728.

Neuner, B., von Mackensen, S., Krümpel, A., Manner, D., Friefeld, S., Nixdorf, S., et al. (2011). Health-related quality of life in children and adolescents with stroke, self-reports, and parent/proxies reports: Cross-sectional investigation. *Annals of Neurology, 70*(1), 70–78.

O'Keeffe, F., Ganesan, V., King, J., & Murphy, T. (2012). Quality-of-life and psychosocial outcome following childhood arterial ischaemic stroke. *Brain Injury, 26*(9), 1072–1083.

O'Keeffe, F., Liégeois, F., Eve, M., Ganesan, V., King, J., & Murphy, T. (2014). Neuropsychological and neurobehavioral outcome following childhood arterial ischemic stroke: Attention deficits, emotional dysregulation, and executive dysfunction. *Child Neuropsychology, 20*(5), 557–582.

Oni, M. B., Wilde, E. A., Bigler, E. D., McCauley, S. R., Wu, T. C., Yallampalli, R., et al. (2010). Diffusion tensor imaging analysis of frontal lobes in pediatric traumatic brain injury. *Journal of Child Neurology, 25*(8), 976–984.

Ornstein, T. J., Levin, H. S., Chen, S., Hanten, G., Ewing-Cobbs, L., Dennis, M., et al. (2009). Performance monitoring in children following traumatic brain injury. *Journal of Child Psychology and Psychiatry, 50*(4), 506–513.

Pascual-Leone, A., Amedi, A., Fregni, F., & Merabet, L. (2005). The plastic human brain cortex. *Annual Review of Neuroscience, 28*, 377–401.

Pavlovic, J., Kaufmann, F., Boltshauser, E., Capone Mori, A., Gubser Mercati, D., Haenggeli, C., et al. (2006). Neuropsychological problems after paediatric stroke: Two year follow-up of Swiss children. *Neuropediatrics, 37*(1), 13–19.

Piaget, J. (1963). *The origins of intelligence in children*. New York: W.W. Norton.

Reichenbach, J. R., Venkatesan, R., Schillinger, D. J., Kido, D. K., & Haacke, E. M. (1997). Small vessels in the human brain: MR venography with deoxyhemoglobin as an intrinsic contrast agent. *Radiology, 204*(1), 272–277.

Riva, D., & Cassaniga, L. (1986). Late effect of unilateral brain lesions before and after the first year of life. *Neuropsychologia, 24*, 423–428.

Robinson, K. E., Fountain-Zaragoza, S., Dennis, M., Taylor, H. G., Bigler, E. D., Rubin, K., et al. (2014). Executive functions and theory of mind as predictors of social adjustment in childhood traumatic brain injury. *Journal of Neurotrauma, 31*(22), 1835–1842.

Rosema, S., Crowe, L., & Anderson, V. (2012). Social function in children and adolescents after traumatic brain injury: A systematic review 1989–2011. *Journal of Neurotrauma, 29*(7), 1277–1291.

Ryan, N. P., Anderson, V., Godfrey, C., Beauchamp, M. H., Coleman, L., Eren, S., et al. (2014). Predictors of very-long-term sociocognitive function after pediatric traumatic brain injury: Evidence for the vulnerability of the immature "social brain". *Journal of Neurotrauma, 31*(7), 649–657.

Ryan, N. P., Anderson, V., Godfrey, C., Eren, S., Rosema, S., Taylor, K., et al. (2013). Social communication mediates the relationship between emotion perception and externalizing behaviors in young adult survivors of pediatric traumatic brain injury (TBI). *International Journal of Developmental Neuroscience, 31*(8), 811–819.

Ryan, N. P., Catroppa, C., Cooper, J. M., Beare, R., Ditchfield, M., Coleman, L., et al. (2015a). Relationships between acute imaging Biomarkers and theory of mind impairment in post-acute pediatric traumatic brain injury: A prospective analysis using susceptibility weighted imaging (SWI). *Neuropsychologia, 66*, 32–38.

Ryan, N. P., Catroppa, C., Cooper, J. M., Beare, R., Ditchfield, M., Coleman, L., et al. (2015b). The emergence of age-dependent social cognitive deficits after generalized insult to the developing brain: A longitudinal prospective analysis using susceptibility-weighted imaging. *Human Brain Mapping, 36*(5), 1677–1691.

Ryan, N.P., Catroppa, C., Beare, R., Silk, T., Crossley, L., Beauchamp, M.H., Yeates, K.O., Anderson, V.A. (2016). Theory of mind mediates the prospective relationship between abnormal social brain network morphology and chronic behavior problems after pediatric traumatic brain injury. *Social Cognitive and Affective Neuroscience*. Advance Access; 14/01/2015. doi: 10.1093/scan/nsw007.

Schmidt, A. T., Orsten, K. D., Hanten, G. R., Li, X., & Levin, H. S. (2010). Family environment influences emotion recognition following paediatric traumatic brain injury. *Brain Injury, 24*(13-14), 1550–1560.

Sehgal, V., Delproposto, Z., Haacke, E. M., Tong, K. A., Wycliffe, N., Kido, D. K., et al. (2005). Clinical applications of neuroimaging with susceptibility-weighted imaging. *Journal of Magnetic Resonance Imaging, 22*(4), 439–450.

Smith, A., & Sugar, C. (1975). Development of above normal language and intelligence 21 years after left hemispherectomy. *Neurology, 25*, 813–818.

Spanos, G., Wilde, E., Bigler, E., Cleavinger, H., Fearing, M., Levin, H., et al. (2007). Cerebellar atrophy after moderate-to-severe pediatric traumatic brain injury. *American Journal of Neuroradiology, 28*(3), 537–542.

Spitz, G., Maller, J. J., Ng, A., O'Sullivan, R., Ferris, N. J., & Ponsford, J. L. (2013). Detecting lesions after traumatic brain injury using susceptibility weighted imaging: A comparison with fluid-attenuated inversion recovery and correlation with clinical outcome. *Journal of Neurotrauma, 30*(24), 2038–2050.

Stein, D., & Hoffman, S. (2003). Concepts of CNS plasticity in the context of brain damage and repair. *The Journal of Head Trauma Rehabilitation, 18*, 317–341.

Steinlin, M., Roellin, K., & Schroth, G. (2004). Long-term follow-up after stroke in childhood. *European Journal of Pediatrics, 163*(4-5), 245–250.

Stiles, J., Stern, C., Appelbaum, M., Nass, R., Trauner, D., & Hesselink, J. (2008). Effects of early focal brain injury on memory for visuospatial patterns: Selective deficits of global-local processing. *Neuropsychology, 22*(1), 61–73.

St James-Roberts, I. (1975). Neurological plasticity, recovery from brain insult and child development. *Advances in Child Development and Behavior, 14*, 253–310.

Taupin, P. (2006). Adult neurogenesis and neuroplasticity. *Restorative Neurology and Neuroscience, 24*, 9–15.

Tlustos, S. J., Chiu, C. Y., Walz, N. C., Taylor, H. G., Yeates, K. O., & Wade, S. L. (2011). Emotion labeling and socio-emotional outcomes 18 months after early childhood traumatic brain injury. *Journal of the International Neuropsychological Society, 17*(6), 1132–1142.

Tong, K. A., Ashwal, S. S., Holshouser, B. A., Nickerson, J. P., Wall, C. J., Shutter, L., et al. (2004). Diffuse axonal injury in children: Clinical correlation with hemorrhagic lesions. *Annals of Neurology, 56*(1), 36–50.

Tong, K. A., Ashwal, S. S., Holshouser, B. A. B., Shutter, L. A. L., Herigault, G. G., Haacke, E. M. E., et al. (2003). Hemorrhagic shearing lesions in children and adolescents with posttraumatic diffuse axonal injury: Improved detection and initial results. *Radiology, 227*(2), 332–339.

Tromp, E., & Mulder, T. (1991). Slowness of information processing after traumatic head injury. *Journal of Clinical and Experimental Neuropsychology, 13*(6), 821–830.

Vargha-Khadem, F., Isaacs, E., Papaleloudim, H., Polkey, C., & Wilson, J. (1992). Development of intelligence and memory in children with hemiplegic cerebral palsy. *Brain, 115*, 315–329.

Watts, A. J., & Douglas, J. M. (2006). Interpreting facial expression and communication competence following severe traumatic brain injury. *Aphasiology, 20*(8), 707–722.

Westmacott, R., Askalan, R., MacGregor, D., Anderson, P., & DeVeber, G. (2010). Cognitive outcome following unilateral arterial ischaemic stroke in childhood: Effects of age at stroke and lesion location. *Developmental Medicine & Child Neurology, 52*(4), 386–393.

Westmacott, R., MacGregor, D., & Askalan, R. (2009). Late emergence of cognitive deficits after unilateral neonatal stroke. *Stroke, 40*(6), 2012–2019.

Wilde, E. A., Ayoub, K. W., Bigler, E. D., Chu, Z. D., Hunter, J. V., Wu, T. C., et al. (2012a). Diffusion tensor imaging in moderate-to-severe pediatric traumatic brain injury: Changes within an 18 month post-injury interval. *Brain Imaging and Behavior, 6*(3), 404–416.

Wilde, E. A., Bigler, E. D., Hunter, J. V., Fearing, M. A., Scheibel, R. S., Newsome, M. R., et al. (2007). Hippocampus, amygdala, and basal ganglia morphometrics in children after moderate-to-severe traumatic brain injury. *Developmental Medicine & Child Neurology, 49*(4), 294–299.

Wilde, E. A., Chu, Z., Bigler, E. D., Hunter, J. V., Fearing, M. A., Hanten, G., et al. (2006). Diffusion tensor imaging in the corpus callosum in children after moderate to severe traumatic brain injury. *Journal of Neurotrauma, 23*(10), 1412–1426.

Wilde, E. A., Merkley, T. L., Bigler, E. D., Max, J. E., Schmidt, A. T., Ayoub, K. W., et al. (2012b). Longitudinal changes in cortical thickness in children after traumatic brain injury and their relation to behavioral regulation and emotional control. *International Journal of Developmental Neuroscience, 30*(3), 267–276.

Wozniak, J. R., Krach, L., Ward, E., Mueller, B. A., Muetzel, R., Schnoebelen, S., et al. (2007). Neurocognitive and neuroimaging correlates of pediatric traumatic brain injury: A diffusion tensor imaging (DTI) study. *Archives of Clinical Neuropsychology, 22*(5), 555–568.

Wozniak, J. R., & Lim, K. O. (2006). Advances in white matter imaging: A review of in vivo magnetic resonance methodologies and their applicability to the study of development and aging. *Neuroscience & Biobehavioral Reviews, 30*(6), 762–774.

Wu, T. C., Wilde, E. A., Bigler, E. D., Li, X., Merkley, T. L., Yallampalli, R., et al. (2010). Longitudinal changes in the corpus callosum following pediatric traumatic brain injury. *Developmental Neuroscience, 32*(5-6), 361–373.

Yeates, K., Taylor, H. G., Drotar, D., Wade, S., Stancin, T., & Klein, S. (1997). Pre-injury environment as a determinant of recovery from traumatic brain injuries in school-aged children. *Journal of the International Neuropsychological Society, 3*, 617–630.

Yeates, K. O., Bigler, E. D., Dennis, M., Gerhardt, C. A., Rubin, K. H., Stancin, T., et al. (2007). Social outcomes in childhood brain disorder: A heuristic integration of social neuroscience and developmental psychology. *Psychological Bulletin, 133*(3), 535–556.

Yeates, K. O., Swift, E., Taylor, H. G., Wade, S. L., Drotar, D., Stancin, T., et al. (2004). Short- and long-term social outcomes following pediatric traumatic brain injury. *Journal of the International Neuropsychological Society, 10*(3), 412–426.

Yuan, W., Holland, S., Schmithorst, V., Walz, N., Cecil, K., Jones, B., et al. (2007). Diffusion tensor MR imaging reveals persistent white matter alteration after traumatic brain injury experienced during early childhood. *American Journal of Neuroradiology, 28*(10), 1919–1925.

FURTHER READING

Anderson, V., & Catroppa, C. (2006). Advances in post-acute rehabilitation after childhood-acquired brain injury – A focus on cognitive, behavioural, and social domains. *American Journal of Physical Medicine & Rehabilitation, 85*, 767–778.

Crowe, L., Catroppa, C., Babl, F., & Anderson, V. (2012). Intellectual, behavioral, and social outcomes of accidental traumatic brain injury in early childhood. *Pediatrics, 129*(2), e262–e268.

Flavell, J. H. (1999). Cognitive development: Children's knowledge about the mind. *Annual Review of Psychology, 50*, 21–45.

Hertz-Pannier, L., Chiron, C., Jambaque, I., Renaux-Kieffer, V., Van de Moortele, P., Delalande, O., et al. (2002). Late plasticity for language in a child's non-dominant hemisphere: A pre- and post-surgery fMRI study. *Brain, 125*, 361–372.

CHAPTER 16

Social Dysfunction: *The Effects of Early Trauma and Adversity on Socialization and Brain Development*

Allonna Harker
University of Lethbridge, Lethbridge, AB, Canada

16.1 INTRODUCTION

Research over the past 60 years has shown that early stress predisposes individuals to a range of maladaptive behaviors and psychopathologies including drug addiction, depression, and schizophrenia (SCZ) (e.g., Charil, Laplante, Vaillancourt, & King, 2010; McEwen & Morrison, 2013). Although there is likely a genetic predisposition for individuals to respond poorly to early stressors, genome-wide association scans have failed to identify major gene(s) linked to psychiatric disease (Abdolmaledy, 2014). Rather, as described in detail in Chapter 7, Epigenetics and Genetics of Brain Development, the emerging field of epigenetics has provided a mechanism to account for the effect of stress, including early aversive experiences, on brain and behavioral development. In this chapter, we will review the behavioral effects of different forms of stress in development, in addition to adverse early experiences during childhood and adolescence. We then consider the consequences of adverse experiences on developing brain structure and neurocognition. In addition, we consider the role of the family and socioeconomic status (SES) on brain and behavioral development. Finally, we examine the complex interplay between genetics, epigenetics, and experience in relation to resiliency.

16.2 STRESS

Stress in an unavoidable condition of life. The human body is designed to react and adapt to an environment to survive and thrive. A delicate balance of opposing forces, or equilibrium must be maintained. Maintenance of equilibrium is referred to as homeostasis. This physiological state of balance is vulnerable to diverse perturbations by internal or external events or perceptions. The term "stress" is used to describe an organism's method of reacting to an intrinsic or extrinsic challenge to homeostasis, whether actual or perceived. Once homeostasis has been threatened, a complex sequence of behavioral and physiological processes responds to reestablish desired

The Neurobiology of Brain and Behavioral Development
DOI: http://dx.doi.org/10.1016/B978-0-12-804036-2.00016-9

equilibrium, known as an "adaptive stress response" or simply "stress response" (Kirschbaum & Hellhammer, 1994).

At the heart of the stress response is the hypothalamic–pituitary–adrenal axis (HPA) connecting the central nervous system and the endocrine system. The hypothalamus produces and secretes the hormone corticotropin-releasing factor in response to stress. The more stressed the organism is, the more corticotropin-releasing hormone (CRH) will be released. The pituitary gland is stimulated to release adrenocorticotropic hormone (ACTH) that travels through the blood to the kidneys stimulating the adrenal glands to increase glucocorticoid production. Glucocorticoids, including cortisol, are required for regulation of metabolic rate, inflammation, and immune response. Cortisol activates a physical response to stress and prepares the body for fight-or-flight. Central to the stress response is the understanding that the HPA axis operates on feedback loops, designed to calm the system down after a stress response. In a synergistic and harmonious response, cortisol receptors message the hypothalamus to inhibit production of CRH and the pituitary to inhibit ACTH, thereby ending the stress response through a negative feedback loop. However, not all stress responses are created equally. HPA axis dysregulation occurs when stressful events become persistent or overwhelming and cortisol and norepinephrine are chronically overproduced, as occurs with traumatic and/or chronic (toxic) stress. According to the National Scientific Council on the Developing Child, stress is divided into the following three categories: positive, tolerable, and toxic.

16.2.1 Positive stress

Positive stress or eustress is adaptive and necessary for healthy development. This type of stress causes a moderate physiological response (increased heart rate and blood pressure, mild changes in stress hormone levels), is short-lived, and occurs frequently. According to the Council on the Developing Child, positive stress occurs within relationships that are viewed to be stable and supportive, permitting and even inducing stress hormones to return to the normal range of function. This context inspires a sense of mastery and heightened self-regulation. Examples of positive stress include everyday experiences such as a math test or job interview (Franke, 2014).

16.2.2 Tolerable stress

Stressors that are more challenging, longer in duration, and may alter the everyday routine are believed to induce tolerable stress. If unrestrained, tolerable stress can impact brain development, however, this can be mitigated by the presence of caring and supportive relationships. As with positive stress, there can be a heightened sense of mastery and self-control within this protective environment. Common examples of tolerable stress include divorce or natural disaster. If tolerable stress remains unchecked, it can become toxic.

16.2.3 Toxic stress

Adversity that is most often severe, frequent, and/or prolonged is considered to produce toxic stress. This type of stress often occurs in a context devoid of adequate nurturing and support. Chronic neglect, physical, emotional, or sexual abuse, exposure to violence are some examples of stressors that can be toxic to the brain. Extensive research has shown that sustained activation of the stress response system caused by toxic stress can significantly impair proper brain development, and increase the risk for stress-related diseases and psychopathologies later in life. The programming effects of early exposure (preconception/prenatal) to toxic stress on the HPA axis are believed to impact the regulation of corticosteroid receptors in the hippocampus, playing an important role memory function (Huang et al., 2010). Beyond this imprinted influence on HPA reactivity, toxic stress can detrimentally alter the developing fetus in several ways different areas (see Chapters 3 and 13 for more discussion on the stress response).

16.3 ADVERSE EXPERIENCES DURING CHILDHOOD

Children who experience early childhood maltreatment (toxic stress) are at risk of developing a variety of adverse health conditions throughout their lifespan. The American Academy of Pediatrics Policy Report on the effects of child abuse propose that children exposed to early adversity or trauma in the form of abuse and/or neglect may express behavioral patterns later in life that include: depression, emotional instability, and a propensity for aggressive or violent behavior toward others (Stirling & Amaya-Jackson, 2008). Further, these behaviors may continue long after the adverse environment has changed or the child has been removed from this environment and placed in foster care. In addition to changes in observable behavior, early adversity during childhood can lead to less visible yet permanent changes to brain architecture and function, epigenetic modifications and alterations in gene function. A cascade of events resulting from these changes lead to multiple risk factors and adverse health conditions during adulthood (Franke, 2014; Harker, Raza, Williamson, Kolb, & Gibb, 2015; Kolb, Harker, & Gibb, 2017; Mychasiuk, Harker, Ilnytskyy, & Gibb, 2013). (See Chapter 13, Brain Plasticity and Experience for a longer discussion of brain plasticity.)

16.3.1 Definitions of child maltreatment (Centers for Disease Control & Prevention, CDC)

16.3.1.1 Child abuse and neglect

Child abuse and neglect is any act or series of acts of commission or omission by a parent or other caregiver (e.g., clergy, coach, and teacher) that results in harm, potential for harm, or threat of harm to a child (Table 16.1).

Table 16.1 Acts of commission/omission (CDC)

Acts of Commission *(Child Abuse)*
Acts of commission are deliberate and intentional: harm to a child might not be the intended consequence.
Intention only applies to caregiver acts, not the consequences of those acts. For example, a caregiver might intend to hit a child as punishment (i.e., hitting the child is not
The following types of maltreatment involve acts of commission: · Physical abuse · Sexual abuse · Psychological abuse

Acts of Omission *(Child Neglect)*
Acts of omission are the failure to provide for a child's basic physical, emotional, or educational needs or to protect a child from harm or potential harm.
Like acts of commission, harm to a child might not be the intended consequence.
The following types of maltreatment involve acts of omission: · Physical neglect · Emotional neglect · Medical and dental neglect · Educational neglect · Inadequate supervision · Exposure to violent environments

16.3.2 Prevalence of child maltreatment

Unfortunately, the incidence of childhood maltreatment is alarmingly frequent. A report by the Canadian Medical Association advises that the percentage of those who have suffered child abuse is as high as 34%. According to United States Child Protective Services (CPS) there were 702,000 victims of child abuse and neglect in 2014. The true occurrence of abuse and neglect may be underestimated by CPS, as a non-CPS study estimated that 25% of children experience some form of child maltreatment during their lifetimes. Approximately 1580 children died from child maltreatment in 2014 in the United States, with a total lifetime cost of $124 billion per year resulting from child abuse and/or neglect [Centers for Disease Control & Prevention (CDC)].

16.3.3 What predicts individual risk of child maltreatment?

It is apparent that certain populations of children are at higher risk for victimization than others. This risk is influenced by a combination of individual, family, and environmental factors that either heighten or lessen risk over time within a given situation. The following are predictors of child maltreatment risk for the individual:

- Children younger than 4 years of age.
- Special needs that may increase caregiver burden (e.g., developmental and intellectual disabilities, mental health issues, and chronic physical illnesses).
- Parents' lack of understanding of children's needs, child development and parenting skills.
- Parents' history of child maltreatment in family of origin.
- Substance abuse and/or mental health issues including depression in the family.
- Parental characteristics such as young age, low education, single parenthood, large number of dependent children, and low income.
- Nonbiological, transient caregivers in the home (e.g., mother's male partner).
- Parental thoughts and emotions that tend to support or justify maltreatment behaviors.

Although these risk factors provide valuable information regarding those most at risk for victimization, they are correlational, and cannot predict those who will become a victim or a perpetrator (CDC). Vincent Felitti has stated that "adult medicine begins in pediatrics" (Van Niel et al., 2014). Felitti et al. (1998) performed a landmark study examining the relationship of childhood abuse and household dysfunction to many of the leading causes of death in adults. This study has become known as the Adverse Childhood Experiences (ACE) Study, and was one of the first research projects to examine the impact of multiple childhood stressors and how they contribute to the development of adverse health consequences later in life.

16.3.4 Adverse childhood experiences

The ACE study was first conducted in 1998 to determine: retrospectively and prospectively, the relationship between the breadth of exposure to traumatic childhood experiences, and the propensity for risky behavior and disease in adulthood. Of the 13,494 adults in the first wave of participants who received the ACE questionnaire, 9508 adults responded (70.5%). A second wave of participants brought the overall participants to 17,337 respondents that were involved in the study. Researchers divided the ACE questionnaire into the following classifications of abuse, household challenges, and neglect (wave 2 only) for analysis.

16.3.5 ACE score significance

Researchers found that there was a strong graded correlation between the extent of childhood maltreatment or household dysfunction (number of ACEs 0–7), and the risk factors associated with adult disease and mortality. At least one ACE was reported by almost two thirds of the participants, two or more ACEs were reported by one-fourth of the participants, and more than one in five participants reported three or more ACEs. Persons who had experienced 0 categories of ACEs were compared to those who experience four or more categories of ACEs, with the later having a 4- to 12-fold increase in health risks for drug abuse, alcoholism, depression, and suicide attempt.

The breadth of exposure to abuse or household dysfunction (the number of categories of ACEs) revealed a graded relationship to the occurrence of disease during adulthood. The ACE score, a total sum of the different categories of ACEs reported by participants, is used to assess cumulative childhood stress. Study findings repeatedly reveal a graded dose-response relationship between ACEs and negative health and well-being outcomes across the life course. These findings suggest a significant and cumulative effect on adult health outcomes due to persistent/chronic stressful experiences (toxic stress) during childhood (CDC) (Tables 16.2, 16.3).

As the number of ACEs increase so does the risk for the following:

- Chronic obstructive pulmonary disease
- Depression
- Fetal death
- Health-related quality of life
- Illicit drug use
- Ischemic heart disease
- Liver disease
- Poor work performance
- Financial stress
- Risk for intimate partner violence

16.3.6 The link between ACEs and health risk behaviors and adult disease

According to Felitti et al. (1998), the mechanisms responsible for linking ACEs to risky behaviors and adult diseases center on certain conscious or unconscious behaviors such as smoking, alcohol or drug abuse, overeating, and sexual behaviors. These behaviors are used as coping mechanisms and are believed to induce prompt psychological or pharmacological impact when faced with the stress of household dysfunction, including all forms of abuse and domestic violence. For example, we understand that nicotine has specific psychoactive effects that are known to be beneficial in regulating affect, and that those who suffer with depression are more likely to smoke (Glassman, Covey, Stetner, & Rivelli, 2001; Murphy et al., 2003). Felitti and colleagues found that exposure to a higher number of ACE categories increased the probability of smoking by age 14, chronic smoking during adulthood, and finally, the presence of adult diseases related to smoking.

Table 16.2 ACE questionnaire

All ACE questions refer to the respondent's first 18 years of life.
Abuse
o ***Emotional abuse:*** A parent, stepparent, or adult living in your home swore at you, insulted you, put you down, or acted in a way that made you afraid that you might be physically hurt.
o ***Physical abuse:*** A parent, stepparent, or adult living in your home pushed, grabbed, slapped, threw something at you, or hit you so hard that you had marks or were injured.
o ***Sexual abuse:*** An adult, relative, family friend, or stranger who was at least 5 years older than you ever touched or fondled your body in a sexual way, made you touch his/her body in a sexual way, attempted to have any type of sexual intercourse with you.
Household Challenges
o ***Mother treated violently:*** Your mother or stepmother was pushed, grabbed, slapped, had something thrown at her, kicked, bitten, hit with a fist, hit with something hard, repeatedly hit for over at least a few minutes, or ever threatened or hurt by a knife or gun by your father (or stepfather) or mother's boyfriend.
o ***Household substance abuse:*** A household member was a problem drinker or alcoholic or a household member used street drugs.
o ***Mental illness in household:*** A household member was depressed or mentally ill or a household member attempted suicide.
o ***Parental separation or divorce:*** Your parents were ever separated or divorced.
o ***Criminal household member:*** A household member went to prison.
Neglect
o ***Emotional neglect:*** No person in your family helped you feel important or special, you did not feel loved, people in your family did not look out for each other and did not feel close to each other, and your family was not a source of strength and support.
o ***Physical neglect:*** There was no person to take care of you, protect you, and take you to the doctor if you needed it, you didn't have enough to eat, your parents were too drunk or too high to take care of you, and you had to wear dirty clothes.

An increasing body of research documents the potent and enduring relationship between stressful experiences in early childhood and array of adverse complications, both physical and psychological, that develop throughout childhood and later into adulthood. The ACE studies have demonstrated that ACE such as child abuse and/or neglect or household dysfunction result in persistent toxic stress. The residual effects of toxic stress can have aberrant consequences on adult health including: depression, suicide, hypertension, diabetes, cigarette smoking, substance abuse, and bone fractures. These conditions enhance the probability of acquiring one of the leading causes of

Table 16.3 ACE prevalence rates

ABUSE	Women Percent (N = 9,367)	Men Percent (N = 7,970)	Total Percent (N = 17,337)
ABUSE			
Emotional Abuse	13.10%	7.60%	10.60%
Physical Abuse	27%	29.90%	28.30%
Sexual Abuse	24.70%	16%	20.70%
HOUSEHOLD CHALLENGES			
Mother Treated Violently	13.70%	11.50%	12.70%
Household Substance Abuse	29.50%	23.80%	26.90%
Household Mental Illness	23.30%	14.80%	19.40%
Parental Separation or Divorce	24.50%	21.80%	23.30%
Incarcerated Household Member	5.20%	4.10%	4.70%
NEGLECT			
Emotional Neglect[a]	16.70%	12.40%	14.80%
Physical Neglect[a]	9.20%	10.70%	9.90%

The prevalence estimates reported below are from the entire ACE Study sample ($n = 17{,}337$). Prevalence of ACEs by Category for CDC-Kaiser ACE Study Participants by Sex, Waves 1 and 2.
Note: Collected during Wave 2 only ($N = 8{,}629$). Research papers that use Wave 1 and/or Wave 2 data may contain slightly different prevalence estimates.

death in adulthood such as stroke, cancer, and heart disease. In follow-up studies throughout the past two decades, researchers have concluded that ACEs are correlated with an increased risk of premature death, although the increased risk was only partially explained by confirmable ACE-related health and social problems. The implication is that there are likely other mechanisms by which ACEs may impact the final consequence of premature death (Brown et al., 2009).

16.4 ADVERSE EXPERIENCES DURING ADOLESCENCE

The adolescent can be impacted by both child abuse and/or neglect occurring during childhood, as well as the continuance of maltreatment into adolescence, or the commencement of maltreatment during adolescence. Thornberry, Henry, Ireland, and Smith (2010) conducted the Rochester Youth Development Study with 907 participants (72% male) to examine whether developmental time periods during which maltreatment occurred, were *causally* related to behavioral outcomes later in adulthood. Based on results from their study they arrived at two important findings. First, they concluded that maltreatment was not only a risk factor for future behavioral and developmental outcomes, but also a *causal* agent. Second, that the developmental stage at the time of maltreatment conditioned the impact of that maltreatment. Researchers found that maltreatment limited to childhood significantly affected drug use and abuse, suicidal ideation, and symptoms of depression; more inwardly directed reactions to stress. In contrast, adolescent-limited maltreatment impacted early adult development in a more pervasive manner, affecting 10 of the 11 outcomes. These outcomes included participation in criminal behavior, violent crime and official arrests/incarcerations, alcohol and drug use and abuse, risky sexual behaviors and diagnosis of STD = Sexually transmitted disease IQ = Intelligence Quotient, and suicidal ideation.

16.4.1 Definition of youth violence

Interpersonal violence is defined as "the intentional use of physical force or power, threatened or actual, against another person or against a group or community that results in or has a high likelihood of resulting in injury, death, psychological harm, maldevelopment, or deprivation." This definition associates intent with committing the act regardless of the outcome (CDC). Youth violence can emerge during childhood and continue into adulthood. Violent behaviors can cause physical harm, emotional harm, or a combination of both. Behaviors such as bullying, slapping, or hitting can have a more emotional impact on the victim, while acts such as robbery and assault can cause the victim serious bodily harm or even death. The most important objective when it comes to youth violence is to *stop it before it starts* (Table 16.4).

16.4.2 What predicts individual risk of youth violence?

Research examining youth violence generally include persons between the ages of 10 and 24, although patterns of youth violence can appear in early childhood (CDC). Research suggests that exposure to family violence and physical abuse are identified as the most salient predictors of youth violence. During adolescence, children previously exposed to abuse and neglect resulting in pain and suffering are at a significantly increased risk of perpetrating pain on others through aggression and violent behavior (Gilbert et al., 2009; Haapasalo & Pokela, 1999; Mass, Herrenkohl, & Sousa, 2008; Trickett, Negriff, Ji, & Peckins, 2011). A study by the National Institute of Justice in the United States suggests maltreated children were 11 times more likely to be arrested for criminal activity during adolescence (English, Widom, & Brandford, 2004).

Continued research on youth violence has expanded our awareness of the factors that make certain populations more susceptible to victimization and perpetration. According to the CDC, the following risk factors may increase the tendency that an

Table 16.4 Prevalence of youth violence

Prevalence of Youth Violence
Violence or threat: In a 2015 nationally-representative sample of youth in grades 9-12
• 22.6% reported being in a physical fight in the 12 months preceding the survey; the prevalence was higher among males (28.4%) than females (16.5%).
• 16.2% reported carrying a weapon (gun, knife or club) on one or more days in the 30 days preceding the survey; the prevalence was higher among males (24.3%) than females (7.5%).
• 5.3% reported carrying a gun on one or more days in the 30 days preceding the survey; the prevalence was higher among males (8.7%) than females (1.6%).
Bullying: In a 2015 nationally-representative sample of youth in grades 9-12
• 20.2% reported being bullied on school property in the 12 months preceding the survey; the prevalence was higher among females (24.8%) than males (15.8%).
• 15.5% reported being bullied electronically (email, chat room, instant messaging, website, texting) in the 12 months preceding the survey.
• The prevalence of electronic bullying was higher among females (21.7%) than males (9.7%).
Youth Violence In 2014, people ages 10 to 24
• 4,300 were victims of homicide—an average of 12 each day.
• Homicide is the 3rd leading cause of death for young people of this age.
• Among homicide victims, 86% (3,703) were male and 14% (597) were female.
• Among homicide victims, 86% were killed with a firearm.
• Youth homicides and assault-related injuries result in an estimated $18.2 billion in combined medical and work loss costs.

adolescent will become violent, but they are not considered the *cause* of youth violence, they are viewed as contributing factors:

- History of violent victimization
- Attention deficits, hyperactivity or learning disorders
- History of early aggressive behavior
- Involvement with drugs, alcohol, or tobacco
- Low IQ
- Poor behavioral control
- Deficits in social cognitive or information-processing abilities
- High emotional distress
- History of treatment for emotional problems

- Antisocial beliefs and attitudes
- Exposure to violence and conflict in the family

Several longitudinal studies that have focused on the occurrence of maltreatment prior to adulthood have concluded significant risk factors for many behavioral outcomes including alcohol and drug abuse, risky sexual behaviors, early pregnancy, depression, suicidal ideation, crime, and violence. According to Thornberry et al. (2010) there is an increased impact on adjustment during adulthood *caused* by adolescent maltreatment. Whereas childhood-limited maltreatment tends to center on internalizing problems, adolescent maltreatment focuses on problems that are both internally and externally expressed, creating maladjustment that pervades many areas of development and functioning later in life. Further, behavioral problems are magnified as the adolescent faces peer rejection due to an inability to form positive friendships, a result commonly observed in maltreated children and adolescents (Trickett et al., 2011).

In addition to the longitudinal studies that have been performed by individual research teams, there have been national studies that have looked at maltreatment during childhood and adolescence: Longitudinal Study of Child Abuse and Neglect; National Study of Child and Adolescent Well-Being; Study of Adolescent Health. Adolescent evaluations are just beginning to emerge from this research as participants reach the age of adolescence. As these studies and others continue to examine the consequences of adolescent maltreatment, we stand to gain a richer understanding of the impact on brain structure and function, and behavioral outcomes due to maltreatment experienced during this critical developmental period (Trickett et al., 2011).

16.5 CONSEQUENCES OF ADVERSE EXPERIENCES ON DEVELOPING BRAIN STRUCTURE

Neurobiological research has shown that early adverse events can result in a toxic amount of stress that can elevate levels of hormones and neurotransmitters and alter HPA reactivity. Chronic increases of these fundamental substances may be a contributing factor to abnormal pruning or apoptosis, interruption in myelination, inhibition of cell growth, or reduction in the expression of brain-derived neurotrophic factor. This neurobiological response to childhood trauma increases the risk of structural and functional brain abnormalities, leading to adverse behavioral outcome later in life (De Bellis et al., 2002; Kavanaugh, Dupont-Frechette, Jerskey, & Holler, 2017). The developing brain is extremely malleable or "plastic" and is uniquely sensitive to environmental perturbations during periods of rapid growth, observed during both childhood and again in puberty.

Both animal and human studies demonstrate that developing brain architecture can be disrupted and impaired by the persistent elevation of stress hormones, setting up a cascade of events leading to debilitating health conditions and psychopathologies later in life (Bick and Nelson, 2016; Harker et al., 2015; Harker et al., 2017; Kolb et al., 2017; Mychasiuk et al., 2013; Shonkoff & Garner, 2011). De Bellis (2001) suggests that the psychobiological sequelae resulting from maltreatment during childhood and adolescence could, and should, be considered an *environmentally induced complex neurodevelopmental disorder*. Considerable research has been conducted examining the consequences of childhood maltreatment on brain architecture and function. Some researchers have discovered global changes in brain structures, while others reveal differences in discrete brain areas, or changes in connectivity, or chemical transmission. Postmaltreatment differences in data across studies may be related to differences in various research parameters including; the developmental time-period (age) of the participant, sex of participant, the type and duration of toxic stress, the support system, and tools available to participant. Analysis of disparate findings across studies is beyond the scope of this chapter, the important point is that there *are* findings, child maltreatment changes the brain. A brief overview examining the impact of early maltreatment on brain development follow.

16.5.1 Chemical mediators of maltreatment

Dysregulation of HPA axis reactivity has been observed in maltreated children and adolescents. This neuroendocrine system produces the necessary physiological response to stress through the secretion of cortisol, a glucocorticoid used to maintain the body's homeostasis. Cortisol acts in multiple ways to maintain homeostatic balance and is purported to be the "final common pathway" of the stress response (Levine, 1993). Current literature contains many inconsistencies regarding the regulation of the HPA axis following childhood abuse and/or neglect. Bruce, Fisher, Pears, and Levine (2009) found differing basal (nonstress) cortisol levels in relation to the type of abuse. Severe emotionally maltreated foster children presented with high morning cortisol levels, whereas foster children who experienced more severe physical neglect presented with low morning cortisol levels.

A prospective longitudinal study conducted by Trickett et al. (2011) examined the normative developmental course of nonstress cortisol levels at six time points from childhood through early adulthood. They found that, on average, normative nonstress cortisol levels steadily increased from middle childhood into early adulthood, followed by a leveling off period. Researchers examined the nonstress cortisol levels of 84 maltreated women against this normative timeline and found that victims of maltreatment experience hypersecretion of cortisol subsequent the initial incident, followed by an attenuation of cortisol in adolescence and hypersecretion of cortisol by early adulthood. This research provides support for the attenuation hypothesis posited by Susman (2006), whereby a heightened stress response occurs subsequent to severe maltreatment early in life, followed by suppression of the stress response over time. Understanding the known deleterious consequences of chronic exposure to glucocorticoids on developing brain structure, suppression of the stress sequelae may be viewed as an adaptive response (Trickett et al., 2011).

White et al. (2017) performed the first comprehensive investigation examining cortisol concentrations subsequent to maltreatment using hair cortisol concentrations (HCC) to quantify levels of cortisol. This new cortisol assessment technique allows for reliable evaluation of the cumulative cortisol secretions over an extended period, instead of previously researched moment-to-moment, short-term fluctuations in cortisol secretion. Assessing cortisol in scalp hair may prove a breakthrough by permitting accurate and potent retrospective analysis of cumulative concentrations of cortisol over several months (Stalder & Kirschbaum, 2012). This method has been used to confirm altered HCC levels in severely traumatized or chronically depressed individuals, or in adult psychiatric populations. Results of this study indicate that from middle childhood onwards, there is a gradual dose-dependent reduction in HCC secretion that corresponds with maltreatment. Researchers suggest that the impact of this pattern of HCC secretion may bias toward externalizing symptoms. These findings further support the attenuation hypothesis. Susman (2006) suggests that observance of attenuated cortisol levels following early maltreatment is a risk marker for persistent psychopathologies later in life.

16.5.2 Frontal regions

Using structural magnetic resonance imaging (MRI), researchers have found evidence of altered brain development and functional impairment in children diagnosed with maltreatment-related posttraumatic stress disorder (PTSD). PTSD is a psychiatric disorder often observed in abused and/or neglected children (Famularo, Fenton, & Kinscherff, 1993, 1996). Compared with control participants, the brains of participants diagnosed with maltreatment-related PTSD presented with reduced intracranial, cerebral, and prefrontal cortex volumes. Reductions were also found in prefrontal cortical white matter and right temporal lobe volumes, as well as areas within the corpus callosum. The lateral ventricles were larger in maltreated subjects, reflecting reduced gray and white matter volumes (Bick & Nelson, 2016; De Bellis et al., 2002).

In a study conducted by Poletti et al. (2016), regional gray matter volumes were examined and compared to subjects diagnosed with bipolar disorder (BD), SCZ, and ACEs. Research revealed that both BD and SCZ showed regional volumetric decreases in orbitofrontal cortex, insula, and thalamic gray matter volume when exposed to high ACEs, but not when exposed to low ACEs, compared to healthy controls. These results support the long-standing belief that ACEs change brain architecture and play a pivotal role in psychopathologies later in life.

16.5.3 Limbic circuitry

16.5.3.1 Hippocampus

The limbic system is primarily responsible for our emotional life and participates in various higher order mental functions such as learning, motivation, and memory formation. The hippocampus belongs to the limbic system and plays an important role in the formation, consolidation, and retention of memories, spatial navigation, and the regulation of stress. Traumatic stress is associated with aberrations in hippocampal volumes.

In a longitudinal brain-imaging study, Whittle et al. (2016) investigated the impact of childhood maltreatment on the volume of hippocampal subregions during adolescence, an important period of plasticity. Using structural MRI, 166 (85 male) adolescent participants took part in three assessments during adolescence (12, 16, and 19 years). Participants were given a self-report of childhood maltreatment and assessed using the Diagnostic and Statistical Manual of Mental Disorders Axis I psychopathology index. Researchers observed that childhood maltreatment was significantly correlated with the development of psychopathologies later in life. Further, both child maltreatment and early onset psychiatric disorder impact the development of different hippocampal subregions. Although researchers observed a significant correlation between early maltreatment and psychopathologies, they could not confirm that hippocampal volume alterations mediated this correlation.

Rinne-Albers et al. (2013) conducted a review of imaging studies looking at traumatized juveniles and young adults. They found that structural MRI studies of hippocampal volume observed in children are inconsistent and do not correlate with the decrease in volume that is detected in adults with childhood trauma-induced PTSD. Researchers hypothesized that hippocampal volume decreases observed in adulthood appeared over time and therefore could not be detected during childhood. Furthermore, Rinne-Albers et al. postulate that a consequence of childhood-trauma induced PTSD results in hippocampal atrophy, or alternatively, a smaller hippocampus may make one more sensitive to early perturbations thereby increasing vulnerability for development of PTSD.

Childhood maltreatment is correlated with significantly increased rates of major depressive disorder (MDD) and is a risk factor for substance abuse and psychopathologies later in life (Kaufman & Charney, 2001; Teicher, Anderson, & Polcari, 2012). Likely the most frequently reported neuroimaging discovery correlated with MDD is the structural finding of reduced hippocampal volumes. Although subjects impacted by childhood maltreatment frequently present with reduced hippocampal volumetric measures, it is unclear whether the observed reduction is a consequence of MDD or a possible risk factor.

Opel et al. (2014) sought to differentiate the diagnostic influence of MDD from the impact of childhood maltreatment on hippocampal morphology. Using a structural

MRI and the Childhood Trauma Questionnaire, 85 depressed patients and 85 sex and age matched healthy controls were examined. Results indicate that there was a robust correlation between the level of previous maltreatment and the degree of hippocampal atrophy in depressed patients. Researchers suggest that rather than a diagnosis of MDD, hippocampal loss in MDD patients could be a result of early adverse experiences. Data from this study suggest that early maltreatment may generate hippocampal atrophy that may create a trait-like risk factor, a "limbic scar" for acquiring MDD later in life. Correlations between hippocampal atrophy, childhood maltreatment, and MDD have been observed and reported in several other imaging studies (Teicher et al., 2012; Vythilingam et al., 2002). Despite differences in research parameters and results, it is evident that early adverse life events alter hippocampal volumetric measures, and increase the risk of psychopathologies later in life.

16.5.3.2 Amygdala

The amygdala is involved in the formation and storage of memories, as well as decision making. Further, the amygdala plays a key role in the perception of emotions, threat appraisal, and fear conditioning. In a literature review examining the impact of early adverse experiences on brain development, Bick and Nelson (2016) found that studies investigating maltreated children, with or without maltreatment-induced psychiatric disorders, generally do not show differences in amygdala volumes compared to nonmaltreated controls. However, a few recent studies of children with a history of abuse and/or neglect have revealed a reduction in amygdala volume associated with emotional neglect. In one study, retrospective reports of emotional neglect provided by adolescent subjects were inversely affiliated with amygdala volume, the more pervasive the neglect, the smaller the amygdala volume. Another study investigating maltreated preadolescent children revealed an association between smaller amygdala volumes and physical abuse. Observed reductions suggest possible risk of behavioral problems. A study by Whittle et al. (2013) examined retrospectively reported maltreatment by youth exhibiting psychiatric difficulties and found that the left amygdala had faster growth from early to mid-adolescence, compared to maltreated youth without psychiatric symptoms. According to Bick and Nelson, these results suggest that there may be significant differences in the neurodevelopment of children who meet psychiatric disorder criteria and those who do not. In a prospective longitudinal study of infants exposed to maternal depression, a positive association between amygdala size and the severity of maternal depression was revealed using MRI scans obtained during late childhood.

16.5.4 Corpus callosum

The corpus callosum is a large bundle of more than 200 million myelinated nerve fibers that connect the two brain hemispheres, permitting communication between the right and left sides of the brain. Abnormalities within the corpus callosum have been identified in maltreated children. Teicher et al. (2004) examined the corpus callosum of children with a history of maltreatment. Researchers discovered that maltreated subjects had a 17% reduction in total corpus callosum area compared to controls, and an 11% reduction compared to nonmaltreated psychiatric patients. Observed decreased corpus callosum size was associated with traumatic experiences early in life, rather than diagnosed psychiatric illness. Neglect was found to have a greater impact on corpus callosum size in boys. Alternatively, girls exhibited the most significant reduction in corpus callosum size in relation to sexual abuse. Researchers suggest that a possible explanation is that female subjects may be less dependent on securing adequate early stimulation/attention than male subjects, leading to an early period of vulnerability for boys. Whereas vulnerability for girls may occur later in development, as sexual abuse tends to occur at a later developmental period than neglect.

16.6 CONSEQUENCES OF ADVERSE EXPERIENCES ON DEVELOPING BRAIN FUNCTION (NEUROCOGNITION)

A literature review investigating neurocognitive deficits in children and adolescents' subsequent early adverse events identified several important implications related to brain development (see review by Kavanaugh et al., 2017). This comprehensive review examined 23 studies that assessed the following neuropsychological measures observed in children and adolescents following early childhood abuse and/or neglect.

16.6.1 Intelligence

Researchers found that although intellectual functioning of maltreated children generally revealed a lowering of IQ when compared to healthy controls, mean group scores tended to fall between the *low average* to *average* range. Variations between studies were observed, likely a result of inconsistent demographic variables. However, researchers did find correlations between intelligence and the following: severity of abuse, developmental period at time of maltreatment, type and duration of maltreatment, as well as the presence and severity of PTSD. As a result, researchers suggest that the impact of maltreatment on IQ would best be conceptualized on a continuum rather than as a categorical model.

16.6.2 Attention and executive functions

As routinely studied aspects of neurocognition, attention/executive functions (EF) were reported to produce the most frequent and severe impairments resulting from maltreatment during childhood. With the exception of one study, executive function performance in maltreated children and adolescents was found to be significantly lower than performance in nonmaltreated controls. Researchers observed that certain types of maltreatment were differentially correlated with attention and executive function. For example, cognitive flexibility and problem solving are associated with physical and sexual abuse, whereas, there is an association with attention and working memory and emotional abuse. Additional weaknesses in the areas of problem solving, abstraction and planning/problem solving are observed in children with a history of both neglect and physical abuse, compared to those with a history of neglect alone. Further, researchers found deficits in inhibitory control and working memory resulting from maltreatment during infancy. However, there were no observed EF weaknesses resulting if the child experienced maltreatment during only one developmental time-period, when compared to healthy controls. As with intelligence, the risk for weaknesses in attention and executive function are related to specific factors such as the type and amount of maltreatment, the presence of PTSD, the developmental time-period, and the duration and frequency of maltreatment.

16.6.3 Visual-spatial

Children and adolescents with a history of maltreatment display significant weaknesses is visual-motor, visual-perceptual, and visual-constructional skills when compared to nonmaltreated controls. Further evidence reveals that the observed weaknesses can be linked to the severity and the duration of PTSD.

16.6.4 Language

Numerous research studies have failed to find deficits in language skills in children and adolescents following maltreatment when compared to control groups. Alternatively, research examining group-based differences found deficits in aspects of speeded naming, receptive vocabulary, language comprehension, and confrontational naming. These types of language deficits have been associated with the existence of PTSD, anxious/depressive symptoms, and sexual abuse. In a longitudinal study by Noll et al. (2010), researchers established that while maltreated children started with comparable receptive language skills compared to nonmaltreated children, the rate at which they acquired receptive language skills was significantly lower than their nonabused peers. Further, compared to healthy controls, the receptive skills of maltreated children peaked at an earlier age. Overall, research findings in this area are inconsistent in relation to the language abilities of maltreated children. Researchers suggest that a possible explanation for the varied findings is that the rate of acquisition in maltreated children is delayed, and further impacted by an early peak in language development.

16.6.5 Memory

Although literature examining the impact on memory of adult stress and trauma is plentiful, far less is known about the effect of childhood trauma and maltreatment on memory. While some studies fail to show any impact on memory function resulting from childhood maltreatment, others have identified memory deficits in aspects of verbal/visual immediate and delayed recall. In one study, sexual abuse was associated with overall memory, whereas another study found that PTSD was related to delayed visual recall. The length of child protection involvement was correlated with learning retention. Weaknesses in verbal and visual memory may be observed in maltreated children and adolescents in relation to the presence of sexual abuse and the quantity of child protection involvement.

16.6.6 Motor/psychomotor

Once again, results in this area are inconsistent. As some studies show slower performance on a fine motor speed task in children with a history of maltreatment, others found no differences between the maltreated group and nonmaltreated group. Research findings in this area are lacking, and future studies are necessary.

16.6.7 Summary

While weaknesses resulting from childhood or adolescent maltreatment were observed across nearly all neurocognitive domains, the most frequent and severe impairments were identified within the EF domain. Researchers found that neurocognitive deficits were associated with factors such as the type, duration, and severity of maltreatment, the developmental time-period during which maltreatment took place, as well as the presence and severity of PTSD and anxious/depressive symptomatology. As with any risk factor, neurocognitive weaknesses are not destined to occur following maltreatment, rather, researchers suggest that neurocognitive risk would be better interpreted within a continuum.

16.7 THE "FAMILY"

Defined as *"a basic social unit consisting of parents and their children, considered as a group, whether dwelling together or not: the traditional family"* or perhaps one of the many other definitions of family that are available, the family is required by the offspring to provide the necessary protection and guidance needed for survival during development. Beyond purely survival, the parent–child relationship is paramount to future offspring health and development. Early caregiving experiences can set in motion a cascade of psychological and developmental events that impact developing brain architecture and behavioral outcomes. Early positive caregiving bestowed by a stable caregiver provides a foundation for normative brain growth and adaptive cognitive, social, and emotional development (Bick & Nelson, 2015; Tottenham, 2014).

16.7.1 What predicts risk to the family unit? (CDC)

- Family disorganization
- Parenting stress
- Intimate partner violence
- Poor parent–child relationships
- Community violence
- Concentrated neighborhood disadvantage
- High poverty and residential instability, high unemployment rates

16.7.2 Socioeconomic status

The measure of SES is most often a combination of income, education and occupation. It is conceptualized as the *class* or social standing of a particular individual or group of people. SES impacts individuals, families, and society at all levels, the most vulnerable of which are the very young and the very old. The American Psychological Association (APA) suggests that a family's well-being including parenting practices, and family stability, is impacted by SES. The following studies reported by the APA reveal a scattering of the issues associated with families living in lower SES circumstances.

- Poverty reliably predicts child abuse and neglect (Ondersma, 2002).
- Higher psychological stress and poor health outcomes for families have been linked to lower SES, due to domestic crowding (Melki, Beydoun, Khogali, Tamim, & Unis, 2004).
- All members of a family are more likely to be victims of violence if they are living in poverty (Pearlman, Zierler, Gjelsvik, & Verhok-Oftedahl, 2003).

SES impacts our physical and mental health, daily activities, choice of neighborhood to live in, and ability to access programs and resources. The consequences of these "choices" can be observed across the lifespan, starting with the most vulnerable, children.

16.7.2.1 SES and children

Extensive research chronicles the negative influence of poverty on a child's social, emotional, and cognitive development. Low SES during childhood is associated with a reduction in adequate nutrition, lower scores on cognitive tests, higher rates of academic failure or grade retention, and higher incidences of school dropout. Hart and Risely (1995) followed a group of children from 42 families of varying demographics for 2.5 years (ages 7–9 months to 36 months) and found a large discrepancy in the size of the vocabulary related to SES. Children from high-SES families have nearly twice the vocabulary as those from low-SES families. When measured again at age 9 to 10, the SES-related differences had actually increased. Hanson et al. (2013)

repeatedly examined the brains of children (newborn to 3 year old) from disparate demographics using noninvasive imaging techniques. Infants had similar levels of gray matter throughout all demographic levels however, by age 4, children from low-SES families had lower gray matter volumes in both frontal and parietal cortex than those from a higher SES standing. Differences in volume appeared as early as 12 months of age. Behavioral problems, such as hyperactivity, rule breaking, and excessive aggression, were correlated with lower levels of total gray matter by age 4 years.

16.7.2.2 SES and youth

Escarce (2003) reviews the literature examining SES and income, health care, and health status of adolescents in the United States. Researchers found that when compared to the good or excellent health of affluent peers, low-income adolescents were more inclined to find themselves in fair or poor health. Further, this discrepancy in health status was found to be a limiting factor related to the decrease in activity of low-income adolescents. An increase in behavioral and emotional problems were also observed in the adolescents from low-SES families. Concerns about the health of low-SES adolescents were reinforced when researchers discovered that compared to affluent peers, these adolescents tended to be uninsured, deprived of medical and dental care, had no dependent source of care, and faced financial and nonfinancial barriers during the previous year. Numerous studies support these findings (Ford, Bearman, & Moody, 1999; Molnar, Cerda, Roberts, & Buka, 2008; Montgomery, Kiely, & Pappas, 1996; Newacheck, Hung, Park, Brindis, & Irwin, 2003). According to the National Research Council (1995), it is likely that the single most important factor that determines quality of families, neighborhoods, and schools is family income.

16.7.2.3 SES and education

Educational opportunities, according to Escarce (2003), are profoundly influenced by family finances, thereby limiting long-term choices available to low-SES adolescents. A result of residential stratification and segregation, schools that low-income students routinely attend have lower funding levels, as well as reduced access to textbooks, laboratory and instructional materials, books and educational resources. These schools also tend to employ less-qualified teachers and administrators, thereby generating lower level curricula. Serious consequences evolve from the combination of the aforementioned school characteristics and limited parental involvement including disciplinary problems, chaotic environments, lower grades, decreased scores on standardized tests, and higher drop-out rates. As adulthood arrives, low-income adolescents are in worse health, engage in more dangerous and risky behaviors, attain lower education, and have less opportunity than their more affluent counterparts.

Although research studies examining the impact of SES on the developing brains of children and adolescents appears bleak, there are strategies that can provide necessary interventions for children and teens living in poverty. Early interventions have been shown to alter the developmental trajectory of at-risk children.

16.7.2.4 SES and early intervention

Using early intensive intervention, two longitudinal studies highlight the importance of investing in disadvantaged families to promote optimal development of at-risk children. The Perry program enrolled 64 low-SES children (age 3 and 4) whose IQs tested between 70 and 85 points. Children were provided 2.5 hours of intensive daily educational intervention in a full-day childcare setting. In addition, one weekly 90-minute home visit was implemented to build parental involvement, lasting for 30 weeks each year, for 4 years. The Abecedarian Program followed 111 disadvantaged children starting at an average age of 4.4 months until children were 8 years of age. This program was more intensive than the Perry program and involved full-day instruction from 7:30 am to 5:30 pm, 5 days/week, for 50 weeks/year. The longitudinal follow-up data from both studies revealed that children enrolled in either program performed much better than comparable nonenrolled children on several important measures. Participants acquired higher scores on tests of reading and mathematics skills, increased number of years of education, and higher attendance at a 4-year college or university. Further, participants were less likely to become teen parents and more likely to own their own homes.

Although brain-imaging techniques were not available at the time of these two studies, Hair and colleagues provided documentation of atypical structural development in several important brain areas of children from low-SES households. Researchers conducted a longitudinal imagining study analyzing 823 MRI scans of 389 economically diverse, typically developing children and adolescents (aged 4—22 years). Participants were followed up and rescanned at 24-month intervals across three periods. SES levels were correlated to structural differences in brain areas related to school readiness skills (attention, planning, cognitive flexibility). The largest impact was observed among children from the poorest households. Children living below the federal poverty level exhibited regional gray matter volumes in the frontal lobe, temporal lobe, and hippocampus 8—10 percentage points below developmental norms for their gender and age. This maturational lag in gray matter volume was correlated with poor performance on achievement tests.

Early intervention provides support and reinforcement for children from all demographics and would help to level the playing field for disadvantaged children. Learning begins early and can dramatically impact the developmental trajectory and brain function. SES impacts both brain development and plasticity.

16.8 "THE COMMUNITY"

16.8.1 Definition of community

Community is defined as *"a group of people living in the same place or having a particular characteristic in common"* as well as *"a feeling of fellowship with others, as a result of sharing common attitudes, interests, and goals."*

16.8.2 What predicts risk of trauma for communities?

- Living in a disadvantaged neighborhood
- Disorganization and neglect
- Availability of drugs
- High population turn-over
- Lack of neighborhood attachment

Community resources can have a positive influence on children and adolescents. For example, youth are more likely to be exposed to good adult role models other than their parents when communities have informal sources of adult supervision, when there is a strong sense of community, when neighborhoods are perceived to be safe, and when neighborhood and city services are functioning properly.

16.9 RESILIENCE

16.9.1 What is resilience?

Resilience is the ability to bounce back or adjust to a stressful situation, challenge, tragedy, trauma, or adversity. Resiliency is not a trait that one either possesses or does not possess, it is not rare, nor is it an exclusive product of the genetically gifted. All children are capable of exceptional accomplishments. Resilient children are confident, inquisitive, flexible, and can competently adapt to experiences that impact their world.

16.9.2 Are infants born resilient?

Babies are born with the capacity to be resilient. While it is true that not one specific gene or combination of genes are responsible for resilience, genetic factors do play an important role. As discussed at the beginning of this chapter, the stress response that an infant inherits can impact adaptive responses and influence resiliency. Resiliency is a product of the complex interaction between genetic, experiential, and environmental factors. The "Resilience Scale" is a metaphor that can be used to help visualize the factors involved in this complex interaction (Fig. 16.1).

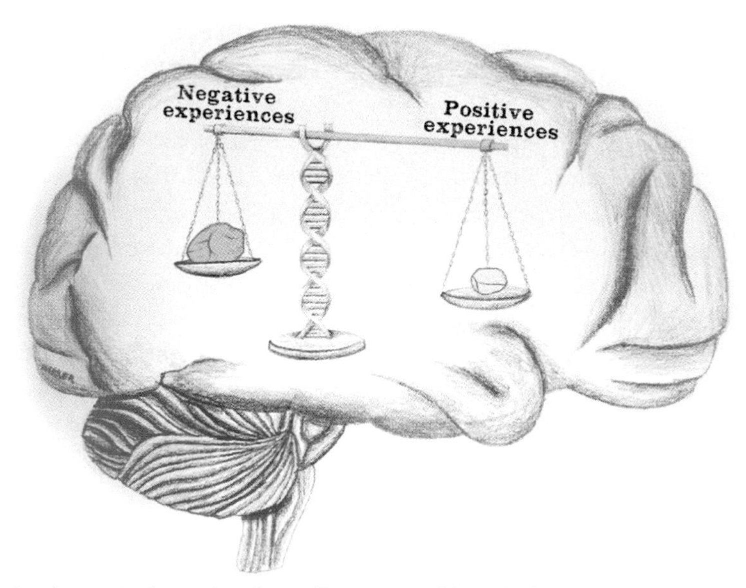

Figure 16.1 Resiliency Scale: In the above illustration, although the negative experiences (risk factors) outweigh the positive experiences (protective factors), the scale continues to favor the positive experiences. This is possible due to the *position* of the fulcrum (DNA strand). Through a combination of genetic inheritance and epigenetic influences the fulcrum has shifted in a more resilient direction. This position is protective and permits compensation of a heavier load of negative experiences. Inside forces (DNA and epigenetics) and outside forces (experiences) work together to create **resiliency**. This is the reason some people appear to be remarkably resilient despite a heavy load of risk factors. The fulcrum can also move in the opposite direction using the same forces, thereby setting the stage for possible psychopathologies later in life.

16.9.3 Resiliency scale, a metaphor (adapted from the Alberta Family Wellness Initiative)

We start with the loading of the scale. Every child will experience both positive and adverse experiences/factors that impact resiliency. During development, experiences pile up on both sides of the scale, positive and negative. Negative/adverse experiences such as stress, neglect, abuse, and parental addiction can be viewed as risk factors that cause toxic stress. Supportive relationships, community resources, and proper role models can be viewed as positive experiences and protective factors. As was discussed earlier in the chapter, the developmental time-period is an important factor to consider in respect to resiliency. Early adverse experiences can have a significant effect on developing brain architecture and the possibility of psychopathologies later in life.

However, experience is only part of the complex interaction that contributes to resiliency. For example, it is not uncommon to observe people who have experienced significant childhood trauma but have thrived, despite early struggles. Alternatively, some people struggle even though their scales appear to have been positively loaded. This is where genetic inheritance comes into play. Some people are genetically hypersensitive to the effects of toxic stress, whereas others have the ability to endure copious amounts of toxic stress without any deleterious effects to brain architecture. The fulcrum on the scale, illustrated here as a DNA strand, represents genetic inheritance. Epigenetic research has determined that genes can be modified during critical developmental time periods through experience. In our illustration, this means that the fulcrum is not fixed, and can be moved (epigenetic influences) toward either end (positive or adverse), altering the *way* that the scale responds to experiences, and determining *how easily* these factors can tip the scale in either direction. The fulcrum will start in a different position for each child, determined by genetic inheritance, but can be moved one way or the other epigenetically.

In the above illustration, although the negative experiences (risk factors) outweigh the positive experiences (protective factors), the scale continues to favor the positive experiences. This is possible owing to the *position* of the fulcrum (DNA strand). Through a combination of genetic inheritance and epigenetic influences the fulcrum has shifted in a more resilient direction. This position is protective and permits compensation of a heavier load of negative experiences. Inside forces (DNA and epigenetics) and outside forces (experiences) work together to create **resiliency**. This is the reason why some people appear to be remarkably resilient despite a heavy load of risk factors. The fulcrum can also move in the opposite direction using the same forces, thereby setting the stage for possible psychopathologies later in life.

16.9.4 Can you build resilience?

YES! It is possible to build and strengthen resilience at any age. The APA maintains that it is not only possible to build resilience, it is a fundamental strategy that can help children and adolescents manage situations that cause stress and anxiety in their lives. Resilience develops as children are allowed to experience stressful situations and are guided by supportive caregivers to resolve the incident in a competent manner. It is not about protecting children and adolescents from all stress and pain, it's about letting them resolve or manage the situation for themselves with skills and strategies they have developed over time. Each time a child or adolescent successfully manages a stressful situation, resilience develops. An early foundation of resilience will prepare the child or adolescent with skills that will help them tackle challenges that they will most certainly encounter throughout their lives. The APA provides the following 10 strategies for building resilience in children and teens (Table 16.5).

It is important to realize that the road to resilience is most often paved with emotional pain and distress!

Table 16.5 Strategies for building resilience in children and teens (APA)

Strategies For Building Resilience in Children and Teens (APA)	
1	**Make connections**
	Teach your child how to make friends and how to be a friend
	Build a strong family network to support the child
	Connect with community for social support
2	**Help your child by having him or her help others**
	Volunteer work can empower a child and create a sense of mastery
3	**Maintain a daily routine**
	Routine can be comforting and children crave structure
4	**Take a break**
	Focus on non-stressful activities
	Have unstructured time to allow the child time to be creative
5	**Teach your child self-care**
	Be a good example, eat properly, exercise and get lots of rest
	Stay balanced - allow your child "down time" to relax and have fun
6	**Move toward goals**
	Teach your child to set reasonable goals and work towards them in steps
	Break large school assignments down into smaller achievable goals
	Every goal achieved, no matter how small, creates resilience
7	**Nurture a positive self-view**
	Help your child learn to trust themselves to solve problems and make decisions
	Teach your child to see humor in life and to have the ability to laugh at themselves
8	**Keep things in perspective and maintain a hopeful outlook**
	Help your child to see a positive future beyond the current situation
	An optimistic and positive outlook allows your child to see good things in life
9	**Look for opportunities for self-discovery**
	With a parents help, tough times can help children learn the most about themselves
	Help your child see what he/she is "made of"
10	**Accept that change is part of living**
	Help your child see that change is a part of life and new goals can replace old goals
	Point out how other students at school change as they move up grade levels

16.10 SUMMARY

A heavy cost is paid by maltreated children and adolescents. In addition, this debt is bestowed on families, communities, and society, to be paid in full by generations to come. The research in this chapter has underscored the significant toll that early victimization has on brain development, function, and behavioral outcomes. Early maltreatment propagates a host of deleterious consequences specific to the child and the type and severity of maltreatment, that can impact physical and mental health throughout the lifespan. The risks to families and communities were investigated, including the role that SES plays in relation to child and adolescent well-being and ability to thrive. There is no doubt that the burden of early trauma, adversity, and status is extensive and complex, but there are some positive aspects to acknowledge. First, not all traumatized children or adolescents experience the developmental sequelae associated with maltreatment. Second, we have a plethora of current research examining the impact of trauma and maltreatment that can aid researchers and clinicians in the development of prevention and intervention procedures and strategies. Finally, we have *resilience*. Resilience can be developed and encouraged in every child, adolescent and adult, regardless of past history. This is a chapter of hope!

REFERENCES

Abdolmaleky, H. M. (2014). Horizons of psychiatric genetics and epigenetics: Where are we and where are we heading? *Iranian Journal of Psychiatry and Behavioral Sciences*, *8*, 1–10.

American Psycological Association (APA). http://www.apa.org/.

Bick, J., & Nelson, C. A. (2016). Early adverse experiences and the developing brain. *Neuropsychopharmacology Reviews*, *41*, 177–196. Available from http://dx.doi.org/10.1038/npp.2015.252.

Brown, D. W., Anda, R. F., Tiemeier, H., Felitti, V. J., Edwards, V. J., Croft, J. B., & Giles, W. H. (2009). Adverse childhood experiences and the risk of premature mortality. *American Journal of Preventative Medicine*, *37*, 389–396. Available from http://dx.doi.org/10.1016/j.amepre.2009.06.021.

Bruce, J., Fisher, P., Pears, K. C., & Levine, S. (2009). Morning cortisol levels in preschool-aged foster children: Differential effects of maltreatment type. *Development and Psychopathology*, *51*, 14–23. Available from http://dx.doi.org/10.1002/dev.20333.

Centers for Disease Control and Prevention (CDC). https://www.cdc.gov.

Charil, A., Laplante, D. P., Vaillancourt, C., & King, S. (2010). Prenatal stress and brain development. *Brain Research Reviews*, *65*, 56–79.

De Bellis, M. D. (2001). Developmental traumatology: The psychobiological development of maltreated children and its implications for research, treatment, and policy. *Development and Psychopathology*, *13*(3), 539–564 . Available from http://dx.doi.org/10.1017/, s0954579401003078.

De Bellis, M. D., Keshavan, M. S., Shifflett, H., Iyengar, S., Beers, S. R., Hall, J., & Moritz, G. (2002). Brain structures in pediatric maltreatment-related posttraumatic stress disorder: A sociodemographically matched study. *Biological Psychiatry*, *52*, 1066–1078.

English, D. J., Widom, C. S., & Brandford, C. (2004). Another look at the effects of child abuse. *NIJ Journal*, *251*, 23–24, National Institute of Justice, 251, 23-24.

Escarce, J. J. (2003). Socioeconomic status and the fates of adolescents. *Health Services Research*, *35*, 1229–1233.

Famularo, R., Fenton, T., Augustyn, M., & Zuckerman, B. (1996). Persistence of pediatric post- traumatic stress disorder after 2 years. *Child Abuse & Neglect*, *20*, 1245–1248. Available from http://dx.doi.org/10.1016/S0145-2134(96)00119-6.

Famularo, R., Fenton, T., & Kinscherff, R. (1993). Child maltreatment and the development of post-traumatic stress disorder. *The American Journal of Diseases of Children*, *147*, 755–760. Available from http://dx.doi.org/10.1001/archpedi.1993.02160310057018.

Felitti, V. J., Anda, R. F., Nordenberg, D., Williamson, D. F., Spitz, A. M., Edwards, V., . . . Marks, J. S. (1998). Relationship of childhood abuse and household dysfunction to many of the leading causes of death in adults. *American Journal of Preventive Medicine*, *14*, 245–258. Available from http://dx.doi.org:10.1016/S0749-3797(98)00017-8.

Ford, C. A., Bearman, P. S., & Moody, J. (1999). Forgone health care among adolescents. *Journal of the American Medical Association, 282*, 2227–2234.

Franke, H. A. (2014). Toxic stress: Effects, prevention and treatment. *Children, 1*, 390–402. Available from http://dx.doi.org/10.3390/children1030390.

Gilbert, R., Widom, C. S., Browne, K., Fergusson, D., Webb, E., & Janson, S. (2009). Burden and consequences of child maltreatment in high-income countries. *The Lancet, 373*, 68–81.

Glassman, A. H., Covey, L. S., Stetner, F., & Rivelli, S. (2001). Smoking cessation and the course of major depression: a follow-up study. *The Lancet, 357*, 1929–1932. Available from http://dx.doi.org/10.1016/S0140-6736(00)05064-9.

Haapasalo, J., & Pokela, E. (1999).). Child-rearing and child abuse antecedents of criminality. *Aggression and Violent Behavior, 4*, 107–127.

Hanson, J. L., Hair, N., Shen, D. G., Shi, F., Gilmore, J. H., & Wolfe, B. L. (2013). Family poverty affects the rate of human infant brain growth. *PLoS ONE, 8*(12), 1–8. e80954. Available from http://dx.doi.org/10.1371/journal.pone.0080954. pmid:24349025.

Harker, A., Raza, S., Williamson, K., Kolb, B., & Gibb, R. (2015). Preconception stress in Long-Evans rats alters dendritic morphology and connectivity in the brain of developing offspring. *Neuroscience, 303*, 200–210. Available from http://dx.doi.org/10.1016/j.neuroscience.2015.06.058.

Hart, B., & Risley, T. R. (1995). *Meaningful differences in the everyday experience of young American children*. Paul H Brookes Publishing, Baltimore.

Huang, Y., Shi, Xuechuan, Xu, H., Yang, H., Chen, T., Chen, S., & Chen, X. (2010). Chronic unpredictable stress before pregnancy reduce the expression of brain-derived neurotrophic factor and N-methyl-D-aspartate reception in hippocampus of offspring rats associated with impairment of memory. *Neurochemical Research, 35*, 1038–1049. Available from http://dx.doi.org/10.1007/s11064-010-0152-0.

Kaufman, J., & Charney, D. (2001). Effects of early stress on brain structure and function: Implications for understanding the relationship between child maltreatment and depression. *Development and Psychopathology, 13*, 451–471.

Kavanaugh, B. C., Dupont-Frechette, J. A., Jerskey, B. A., & Holler, K. A. (2017). Neurocognitive deficits in children and adolescents following maltreatment: Neurodevelopmental consequences and neuropsychological implications of traumatic stress. *Applied Neuropsychology: Child, 6*(1), 64–78. Available from http://dx.doi.org/10.1080/21622965.2015.1079712.

Kirschbaum, C., & Hellhammer, D. H. (1994). Salivary cortisol in psychoneuroendocrine research: Recent developments and applications. *Psychoneuroendocrinology, 19*, 313–333. Available from http://dx.doi.org/10.1016/0306-4530(94)90013-2.

Kolb, B., Harker, A., & Gibb, R. (2017). Principles of plasticity in the developing brain. *Developmental Medicine & Child Neurology*, in press.

Levine, S. (1993). The psychoendocrinology of stress. *Annals of the New York Academy of Sciences, 697*, 61–69. Available from http://dx.doi.org/10.1111/j.1749-6632.1993.tb49923.x.

Mass, C., Herrenkohl, T. I., & Sousa, C. (2008). Review of research on child maltreatment and violence in youth. *Trauma, Violence & Abuse, 9*, 56–67. Available from http://dx.doi.org/10.1177/1524838007311105.

McEwen, B. S., & Morrison, J. H. (2013). The brain on stress: Vulnerability and plasticity of the prefrontal cortex over the life course. *Neuron, 79*, 16–29.

Melki, I. S., Beydoun, H. A., Khogali, M., Tamim, H., & Unis, K. A. (2004). Household crowding index: A correlate of socioeconomic status and inter-pregnancy pacing in an urban setting. *Journal of Epidemiology & Community Health, 58*, 476–480. Available from http://dx.doi.org/10.1136/jech.2003.012690.

Molnar, B. E., Cerda, M., Roberts, A. L., & Buka, S. L. (2008). Effects of neighborhood resources on aggressive and delinquent behaviors among urban youths. *American Journal of Public Health, 98*, 1086–1093.

Montgomery, L. E., Kiely, J. L., & Pappas, G. (1996). The effects of poverty, race, and family structure on US children's health. *American Journal of Public Health, 86*, 1401–1405.

Murphy, J. M., Horton, N. J., Monson, R. R., Laird, N. M., Sobol, A. M., & Leighton, A. H. (2003). Cigarette smoking in relation to depression: Historical trends form the Stirling County Study. *The American Journal of Psychiatry, 160*, 1663–1669. Available from http://dx.doi.org/10.1176/appi.ajp.160.9.1663.

Mychasiuk, R., Harker, A., Ilnytskyy, S., & Gibb, R. (2013). Paternal stress prior to conception alters DNA methylation and behaviour of developing rat offspring. *Neuroscience, 241*, 100–105. Available from http://dx.doi.org/10.1016/j.neuroscience.2013.03.025.

National Research Council, Commission on Behaivoral and Social Sciences and Education, 1995. Losing generations: adolescents in high-risk settings. Washington, DC: U.S. Government Printing Office.

Newacheck, P. W., Hung, U. Y., Park, M. J., Brindis, C. D., & Irwin, C. E. (2003). Disparities in adolescent health and health care: Does socioeconomic status matter? *Health Services Research, 38*, 1235–1252.

Ondersma, S. J. (2002). Predictors of neglect within low-SES families: The importance of substance abuse. *American Journal of Orthopsychiatry, 72*, 383–391. Available from http://dx.doi.org/10.1037/0002-9432.72.2.383.

Opel, N., Redlich, R., Zwanzger, P., Grotegerd, D., Arolt, V., Heindel, W., ... Dannlowski, U. (2014). Hippocampal atrophy in major depression: A function of childhood maltreatment rather than diagnosis? *Neuropsychopharmacology, 39*, 2723–2731. Available from http://dx.doi.org/10.1038/npp.2014.145.

Pearlman, D. M., Zierler, S., Gjelsvik, A., & Verhok-Oftedahl, W. (2003). Neighborhood environment, racial position and risk of police reported domestic violence: A contextual analysis. *Public Health Report, 118*, 44–58. Available from http://dx.doi.org/10.1093.phr/118.1.44.

Poletti, S., Vai, B., Smeraldi, E., Cavallaro, R., Colombo, C., & Beneditti, F. (2016). Adverse childhood experiences influence the detrimental effect of bipolar disorder and schizophrenia on cortico-limbic grey matter volumes. *Journal of Affective Disorders, 189*, 290–297. Available from http://dx.doi.org/10.1016/j.jad.2015.09.049.

Shonkoff, J. P., & Garner, A. S. (2011). The lifelong effects of early childhood adversity and toxic stress. *Pediatrics, 129*, 232–246. Available from http://dx.doi.org/10.1542/peds.2011-2663.

Stalder, T., & Kirschbaum, C. (2012). Analysis of cortisol in hair – State of the art and future directions. *Brain, Behavior, and Immunity, 12*, 1019–1029. Available from http://dx.doi.org/10.1016/j.bbi.2012.02.002.

Stirling, J., & Amaya-Jackson, L. (2008). Understanding the behavioral and emotional consequences of child abuse. *Pediatrics, 122*, 667–673. Available from http://dx.doi.org/10.1542/peds.2008-1885.

Susman, E. J. (2006). Psychobiology of persistent antisocial behavior: Stress, early vulnerabilities and the attenuation hypothesis. *Neuroscience and Biobehavioral Reviews, 30*, 376–389. Available from http://dx.doi.org/10.1016/j.neubiorev.2005.08.002.

Teicher, M. H., Anderson, C. M., & Polcari, A. (2012). Childhood maltreatment is associated with reduced volume in the hippocampal subfields CA3, dentate gyrus, and subiculum. *PNAS, 109*, E563–E572. Available from http://dx.doi.org/10.1073/pnas.1115396109.

Teicher, M. H., Dumont, N. L., Ito, Y., Vaituzis, C., Giedd, J. N., & Andersen, S. L. (2004). Childhood neglect is associated with reduced corpus callosum area. *Biological Psychiatry, 56*, 80–85. Available from http://dx.doi.org/10.1016/j.biopsych.2004.03.016.

Thornberry, T. P., Henry, K. L., Ireland, T. O., & Smith, S. A. (2010). The causal impact of childhood-limited maltreatment and adolescent maltreatment on early adult adjustment. *Journal of Adolescent Health, 14*, 359–365. Available from http://dx.doi.org/10.1016/j.jadohealth.2009.09.011.

Tottenham, N. (2014). *The importance of early experiences for neuro-affective development.* The Neurobiology of Childhood (pp. 109–129). Springer, Berlin Heidelberg.

Trickett, P. K., Negriff, S., Ji, J., & Peckins, M. (2011). Child maltreatment and adolescent development. *Journal of Research on Adolescence, 21*, 3–20. Available from http://dx.doi.org/10.1111/j.1532-7795.2010.00711.x.

Vythilingam, M., Heim, C. M., Newport, J., Miller, A. H., Anderson, E., & Bronen, R. (2002). Childhood trauma associated with smaller hippocampal volume in women with major depression. *American Journal of Psychiatry, 159*, 2072–2080. Available from http://dx.doi.org/10.1176/appi.ajp.159.12.2072.

White, L. O., Ising, M., von Klitzing, K., Sierau, S., Michel, A., Klein, A. M., ... Stalder, T. (2017). Reduced hair cortisol after maltreatment mediates externalizing symptoms in middle childhood and adolescence. *Journal of Child Psychology and Psychiatry*, 1–10. Available from http://dx.doi.org/10.1111/jcpp.12700.

Whittle, S., Simmons, J. G., Hendriksma, S., Vijayakumar, N., Byrne, M. L., Dennison, M., & Allen, N. B. (2016). Childhood maltreatment, psychopathology, and the development of hippocampal subregions during adolescence. *Brain and Behavior*, 1–9. Available from http://dx.doi.org/10.1002/brb3.607.

FURTHER READING

Daskalakis, N. P., Bagot, R. C., Parker, K. J., Vinkers, C. H., Kloet, & de Kloet, E. R. (2013). The three-hit concept of vulnerability and resilience: Toward understanding adaptation to early-life adversity outcome. *Psychoneuroendocrinology*, 1–16. Available from http://dx.doi.org/10.1016/j.psyneuen.2013.06.008.

De Bellis, M. D., & Keshavan, M. S. (2003). Sex differences in brain maturation in maltreatment-related pediatric posttraumatic stress disorder. *Neuroscience and Biobehavioral Reviews*, *27*, 103–117. Available from http://dx.doi.org/10.1016/0145-2134(91)90067-N.

De Bellis, M. D., Lefter, L., Trickett, P. K., & Putnam, F. W. (1994). Urinary catecholamine excretion in sexually abused girls. *Journal of the American Academy of Child & Adolescent Psychiatry*, *33*, 320–327.

Gapp, K., Soldado-Magraner, S., Alvarez-Sanchez, M., Bohacek, J., Vernaz, G., Shu, H., ... Mansuy, I. M. (2014). Early life stress in fathers improves behavioral flexibility in their offspring. *Nature Communications*, *5*(5466), 1–8. Available from http://dx.doi.org/10.1038/ncomms6466.

Gibb, R., Gonzalez, C., & Kolb, B. (2014). Prenatal enrichment and recovery from perinatal cortical damage: Effects of maternal complex housing. *Frontiers in Behavioral Neuroscience*, *8*, 223. Available from http://dx.doi.org/10.3389/fnbeh.2014.00223.

Hair, N. L., Hanson, J. L., Wolfe, B. L., & Pollak, S. D. (2015). Association of child poverty, brain development, and academic achievement. *The Journal of the American Medical Association Pediatrics*, *169*, 822–829. Available from http://dx.doi.org/10.1001/jamapediatrics.2015.1475.

Kuhn, M., Scharfenort, R., shumann, D., Schiele, M. A., Munsterkotter, A. L., Deckert, J., ... Lonsdorf, T. B. (2016). Mismatch or allostatic load? Timing of life adversity differentially shapes gray matter volume and anxious temperament. *Social Cognitive and Affective Neuroscience*, 537–547. Available from http://dx.doi.org/10.1093/scan/nsv137.

Putnam, F.W., Trickett, P.K., Helmers, K., Dorn, L., Everett, B. (1991). *Cortisol Abnormalities in Sexually Abused Girls*. Paper presented at the 144th Annual Meeting, Washington, DC.

Queiroz, E. A., Lombardi, A. B., Santos Furtado, C. R. H., Peixoto, C. C. D., Soares, T. A., Fabre, Z. L., ... Lippi, J. R. S. (1991). Biochemical correlate of depression in children. *Arquivos de Neuro-Psiquiatra*, *49*, 418–425.

Rando, O. J., & Verstrepen, K. J. (2007). Timescales of genetic and epigenetic inheritance. *Cell*, *128*, 655–668.

Sullivan, R., Wilson, D. A., Feldon, J., Yee, B. K., Meyer, U., Richter-Levin, G., ... Braun, K. (2006). The international society for developmental psychobiology annual meeting symposium: Impact of early life experiences on brain and behavioral development. *Developmental Psychobiology*, *48*, 583–602. Available from http://dx.doi.org/10.1002/dev.20170.

Teicher, M. H., Samson, J. A., Polcari, A., & McGreenery, C. E. (2006). Sticks, stones and hurtful words: Relative effects of various forms of childhood maltreatment. *The American Journal of Psychiatry*, *163*, 993–1000. Available from http://dx.doi.org/10.1176/ajp.2006.163.6.993.

Trickett, P. K., Noll, J. G., Susman, E. J., Shenk, C. E., & Putnam, F. W. (2010). Attenuation of cortisol across development for victims of sexual abuse. *Developmental Psychopathology*, *22*, 165–175. Available from http://dx.doi.org/10.1017/S0954579409990332.

Turecki, G., & Meaney, M. J. (2014). Effects of the social environment and stress on glucocorticoid receptor gene methylation: A systematic review. *Biological Psychiatry*, in press (Epub ahead of print). http://dx.doi.org/10.1016/j.biopsych.2014.11.022.

Vissing, Y. M., Straus, M. A., Gelles, R. J., & Harrop, J. W. (1991). Verbal Aggression by parents and psychosocial problems of children. *Child Abuse and Neglect*, *15*, 223–238. Available from http://dx.doi.org/10.1016/0145-2134(91)90067-N.

INDEX

Note: Page numbers followed by "*f*", "*t*", and "*b*" refer to figures, tables, and boxes, respectively.

A

B

D

E

K

L

M

N

O

Printed in the United States
By Bookmasters